NF-κB

A Network Hub Controlling Immunity, Inflammation, and Cancer

A subject collection from *Cold Spring Harbor Perspectives in Biology*

EDITED BY

Michael Karin
University of California, San Diego

Louis M. Staudt
National Cancer Institute

COLD SPRING HARBOR LABORATORY PRESS
Cold Spring Harbor, New York • www.cshlpress.com

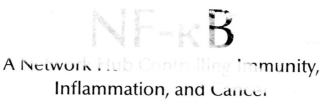

NF-κB

A Network Hub Controlling Immunity, Inflammation, and Cancer

A subject collection from *Cold Spring Harbor Perspectives in Biology*

ALSO FROM COLD SPRING HARBOR LABORATORY PRESS

OTHER SUBJECT COLLECTIONS FROM COLD SPRING HARBOR
PERSPECTIVES IN BIOLOGY

Cell–Cell Junctions
Generation and Interpretation of Morphogen Gradients
Symmetry Breaking in Biology

RELATED LABORATORY MANUALS

The TGF-β Family
Single-Molecule Techniques: A Laboratory Manual
Transcriptional Regulation in Eukaryotes: Concepts, Strategies, and Techniques, Second Edition

OTHER RELATED TITLES

Epigenetics

NF-κB: A Network Hub Controlling Immunity, Inflammation, and Cancer

A Subject Collection from *Cold Spring Harbor Perspectives in Biology*
Articles online at www.cshperspectives.org

Publisher	John Inglis	**Production Editor**	Kaaren Kockenmeister
Acquisition Editor	Richard Sever	**Production Manager**	Denise Weiss
Director of Development, Marketing & Sales	Jan Argentine	**Book Marketing Manager**	Ingrid Benirschke
		Sales Account Managers	Jane Carter and
Project Coordinator	Mary Cozza		Elizabeth Powers
Permissions Coordinator	Carol Brown	**Cover Designer**	Mike Albano

Front cover artwork: Schematic illustration of canonical and non-canonical pathways for activation of the transcription factor NF-κB. Canonical signals activate NEMO-containing IKK complexes (green), which degrade the canonical IκB proteins (IκBα, β, ε) and IκBγ complexes with associated NF-κB dimers. Released Rel-containing dimers move to the nucleus to activate gene expression programs and further expression of IκBα, IκBε, p105, p100, and RelB proteins. Non-canonical signals activate IKKα complexes (blue), which degrade IκBλ complexes associated with NF-κB dimers. The resulting increase in synthesis of p100 and RelB, concomitant with IKKα complex activity, causes increased p100 processing to p52 and dimerization with RelB to generate active RelB:p52 dimers in the nucleus. Stress signals can activate the eIF2α kinases, causing phosphorylation of eIF2α and resulting in inhibited translation. A block in IκB synthesis, in combination with constitutive IKK activity, results in the loss of IκB proteins and subsequent NF-κB dimer activation. See chapter by Ellen O'Dea and Alexander Hoffmann in this volume.

Library of Congress Cataloging-in-Publication Data

NF-[kappa] B : a network hub controlling immunity, inflammation, and cancer
/ edited by Louis M. Staudt, Michael Karin.
       p.; cm.
  Includes bibliographical references and index.
  ISBN 978-0-87969-902-4 (hardcover : alk. paper)--ISBN 978-1-936113-55-2 (pbk. : alk. paper)
  1. NF-kappa B (DNA-binding protein) I. Staudt, Louis. II. Karin, Michael.
III. Title.
  [DNLM: 1. NF-kappa B–genetics–Collected Works. 2. NF-kappa
B–immunology–Collected Works. 3. Signal Transduction–Collected Works.
QU 475 N575 2010]

QP552.N46N4 2010
572.8′6–dc22

                                                    2009048884

10 9 8 7 6 5 4 3 2 1

All World Wide Web addresses are accurate to the best of our knowledge at the time of printing.

All Cold Spring Harbor Laboratory Press publications may be ordered directly from Cold Spring Harbor Laboratory Press, 500 Sunnyside Blvd., Woodbury, New York 11797–2924. Phone: 1-800-843-4388 in Continental U.S. and Canada. All other locations: (516) 422-4100. FAX: (516) 422-4097. E-mail: cshpress@cshl.edu. For a complete catalog of all Cold Spring Harbor Laboratory Press publications, visit our website at http://www.cshpress.com/.

Contents

Contents

Preface

THE NF-κB PATHWAY BROUGHT MAMMALIAN SIGNAL TRANSDUCTION into the modern era, allowing the analysis of cell signaling to become inclusive and holistic. Prior to its definition, biologists had focused on receptor-proximal events, largely through study of receptor tyrosine kinases and G-protein-coupled receptors or the study of isolated protein kinases. The gaping hole in our understanding at that point was the connection to the nucleus. It was clear that many of the physiological effects of receptor signaling were achieved by altering gene expression, but how signals were transduced from the plasma membrane to the nucleus was by-and-large a mystery. The elucidation of NF-κB signaling in all of its molecular glory provided a new paradigm for understanding how receptor signaling can elicit transcriptional responses in mammalian cells. Even more importantly, NF-κB helped transform the study of cell signaling from dealing with cultured cells to consideration of whole animal systems.

As David Baltimore points out in his introduction, NF-κB was born rather innocently in a search for transcriptional regulators of the immunoglobulin locus. At the time, this effort was directed towards understanding the orchestrated changes in gene expression that take place during lymphocyte development. However, by following their noses, scientists in the Baltimore lab soon realized that the nuclear accumulation of the NF-κB DNA binding proteins could be induced by a number of extracellular stimuli and set out to understand how this molecular connection was made. The discovery that NF-κB exists in a latent form in the cytoplasm and translocates to the nucleus in response to various stimuli changed the way mammalian signal transduction was conceptualized.

What followed was a breathtaking explosion of research that uncovered pivotal roles for NF-κB in a host of biological processes critical for normal physiology of animals and their ability to counter stress, disease, and infection. As Jules Hoffman discusses, NF-κB emerged early in animal evolution as an important regulator of development and as a defense mechanism against invading pathogens, themes that have been embellished in vertebrate evolution. Although mammals no longer count on NF-κB in the control of general development and morphogenesis, they use it to signal downstream of a host of receptors, many of which are members of extended gene families to control responses to stress, infection, and injury.

This collection includes reviews of the molecular mechanisms by which NF-κB transcription factors are activated and exert their function in the nucleus, as well as reviews that summarize certain realms of biology that are particularly influenced by NF-κB signaling. Ingrid Wertz and Vishva Dixit focus on the mechanisms whereby receptors that detect antigens, inflammatory cytokines, and foreign organisms utilize protein–protein interactions and the ubiquitin system to engage IkB kinase (IKK) to initiate NF-κB signaling. The structure of IKK complex components and the elaborate regulation of its protein kinase activity are described by Alain Israël in his contributon. Andrea Oeckinghaus and Sankar Ghosh summarize the process by which IkBs sequester NF-κB in a latent state in the cytoplasm and how IKK action relieves this inhibition. Yinon Ben-Neriah and colleagues discuss the role of ubiquitination in the regulated degradation of the IkBs. The Rel-homology DNA binding domains of NF-κB transcription factors are discussed in structural detail by Tom Huxford and Goury Ghosh. The various NF-κB heterodimers have distinct target genes, as discussed by Ranjan Sen and Steve Smale, and can be altered in their transcription regulatory activities by post-translational modifications and association with co-factors, as discussed by

Fengyi Wan and Mike Lenardo. Chromatin structure further shapes the transcriptional output of NF-κB dimers, as reviewed by Gioacchino Natoli. The overall biological output of NF-κB signaling must be viewed from a systems biology perspective, as argued by Ellen O'Dea and Alex Hoffmann, whereas Steve Gerondakis and Uli Siebenlist recount the manifold ways in which NF-κB signaling controls lymphocyte differentiation, activation, and function in vivo. Equally important is the regulation of innate immunity and inflammatory responses, as presented by both Toby Lawrence and Michael Karin. Inflammation can promote cancer development, and Michael Karin outlines the compelling evidence for the critical pathogenic role played by NF-κB signaling in this process. Not surprisingly, cancers of many varieties accumulate genetic lesions that subvert NF-κB signaling to protect against cell death and promote proliferation, offering many possible avenues for therapy, as discussed by Lou Staudt. Barbara and Christian Kaltschmidt remind us that without NF-κB in our neurons, we would suffer learning disabilities and memory loss and would gain little from reading collections like this!

Twenty-four years after the discovery of mammalian NF-κB and more than 25,000 publications later, there remain mysteries and challenges in the NF-κB field. The contribution by Tao Lu and George Stark demonstrates that unbiased genetic screens continue to yield new regulators of NF-κB, so it will be years before we have a complete parts list for this system and a full understanding of its working. Given the dysregulation of NF-κB in inflammatory and autoimmune diseases, as well as in cancer, it is imperative that precise methods to manipulate NF-κB are developed. This will be a challenge given the baroque regulation of the NF-κB signaling system and its diverse biological functions, but one that can be met by building on the strong edifice of knowledge presented in this volume.

Finally, we wish to thank the Cold Spring Harbor Laboratory Press project manager, Mary Cozza, for her excellent support in pulling this book together, as well as Alex Hoffmann, whose NF-κB wiring diagram formed the basis of the front cover illustration.

MICHAEL KARIN
University of California, San Diego

LOUIS M. STAUDT
National Cancer Institute

Discovering NF-κB

David Baltimore

California Institute of Technology, Pasadena, California 91125

Correspondence: baltimo@caltech.edu

NF-κB is a protein transcription factor that can orchestrate complex biological processes, such as the inflammatory response. It was discovered in a very different and very limited context, and only over time has its protean nature become evident.

It was actually a very logical process that led us to NF-κB. Although the discovery was very exciting, NF-κB was not the protein we were seeking at the time. To explain this, I must go back to when my laboratory first became interested in immunology.

My first independent position was at the Salk Institute, where I arrived in the spring of 1965. For the previous 4 years that I had been in research, my interests had revolved around the biochemistry of viruses. At Salk, although my work continued to be on viruses, I was exposed to the fascinating questions of immunology. The main issue was how the enormous diversity of antibodies is generated from a limited amount of genetic information. Like so many others, I thought about the question, but it took the experimental attention of Susumu Tonegawa, in 1976, to crack the problem and show that the solution involved DNA rearrangement.

In 1974, the methods of recombinant DNA technology were first developed and it was clear that previously intractable complex systems, like the immune system, could be examined with these methods. In 1976, knowing that the methods were available and that the paradigm of DNA rearrangement had been established, some postdoctoral students in my laboratory and I decided to plunge into this field. I wanted to apply our biochemical skills to this suddenly tractable system. We were already working on one enzyme that was involved in immunoglobulin gene specification, terminal transferase, and had a useful viral transformation system in the laboratory that affected lymphoid cells, the Abelson mouse leukemia virus. So, immunology was not totally new to us.

We had to develop our skills with recombinant DNA methods, become familiar with the awful lingo of immunology, and define some questions for ourselves, but all of that came to pass. In time, I began to see the question of how immune cells develop as the key one for my laboratory. It seemed likely that the problem would come down to understanding the control of transcription factors. So, we focused on transcription of immune cell genes as our primary interest. We had produced evidence that in the development of B lymphocytes, the heavy-chain locus is first to rearrange its DNA, followed by the light-chain locus (Siden et al. 1981). Cary Queen joined the laboratory and studied the transcription of the κ light-chain gene and demonstrated that it contains an intragenic transcriptional enhancer (Queen and Baltimore 1983). These developments led us to ask whether it might be possible to understand the transition of a cell from heavy-chain only to heavy-plus-light chain by understanding the transcription factors that bind to the κ light-chain enhancer. Understanding the proteins that bind to the regulatory sites in both the heavy- and the light-chain genes became the project of a new postdoctorate, Ranjan Sen. He worked closely with people in Phil Sharp's laboratory, who had similar interests.

Ranjan and Harinder Singh, from the Sharp laboratory, worked out how to use mobility shift assays to find transcription factors, and first published on the existence of the Oct factors (Singh et al. 1986). Then Ranjan applied the methods to enhancers and found multiple factors binding to both the heavy- and κ light-chain enhancers (Sen and Baltimore 1986a). Among the factors he discovered was one that bound only to the κ light-chain enhancer—it covered the sequence GGGACTTTCC. We called it NF-κB because it was a nuclear factor that bound selectively to the κ enhancer and was found in extracts of B-cell tumors but not other cell lines (Sen and Baltimore 1986a).

The next step was supposed to be the killer experiment. 70Z/3 cells were known to have a rearranged κ light chain but not to express it and did not have detectable NF-κB. We knew also that treatment of the cells with lipopolysaccharide (LPS) induced transcription of the κ gene. The killer result would be that LPS induced NF-κB. Sure enough, it did (Sen and Baltimore 1986b). Furthermore, it did so without the need for new protein synthesis. Thus, we concluded that NF-κB is a factor that pre-exists in an apparently inhibited state and is released from that inhibition by LPS treatment. It looked like we had found a factor that might cause cells to go from making only heavy chain to making heavy and light chains, which could thus explain a step in differentiation. However, history has treated this optimistic conclusion with total disrespect, as is shown below.

I will not describe all that we have done on the NF-κB system but will only take this story one step further. That step was taken by Patrick Baeuerle, who joined my laboratory as a postdoctoral fellow. He found that the inactive form of NF-κB is in the cytoplasm of 70Z/3 cells and can be liberated from its inhibited form by treatment of cytoplasmic extracts with a detergent (Baeuerle and Baltimore 1988a). This discovery allowed us to purify the inhibitor, which we named IκB (Baeuerle and Baltimore 1988b). That set the stage for a detailed biochemical study of the activation process, an effort that has involved many investigators and is not complete to this day.

The seeds of doubt about the role of NF-κB as a regulator of B-cell development were sown in these early papers. We showed that the inhibited NF-κB is not specific to B-lineage cells: it was evident in T cells and even HeLa cells (Sen and Baltimore 1988b)—we know now that virtually all cells have it. Another paper showed this even more directly (Baeuerle and Baltimore 1988a). Thus, it was evident that NF-κB could be active in a wide range of cells and further work has borne this out.

A later postdoctorate, Yang Xu, provided the *coup de grace* for the notion that NF-κB is critical to κ-chain transcription. He knocked out the intronic κ enhancer—containing the NF-κB binding site—in mice and showed that, in those cells that rearrange κ, the gene is transcribed at a normal rate (Xu et al. 1996). There is a second enhancer, lacking an NF-κB binding site, that can control κ gene transcription. Each enhancer plays a quantitative role in κ-gene rearrangement, but not a qualitative one (Inlay et al. 2002).

Meanwhile, over the 24 years since its discovery, NF-κB has been implicated in a wide range of normal and disease processes. No transcription factor has attracted more experimental attention. Its role in inflammatory processes is especially important. Yet, its role in the transcription of the κ light chain, for which it was named, remains uncertain.

REFERENCES

Baeuerle PA, Baltimore D. 1988a. Activation of DNA-binding activity in an apparently cytoplasmic precursor of the NF-κB transcription factor. *Cell* **53**: 211–217.

Baeuerle PA, Baltimore D. 1988b. IκB: a specific inhibitor of the NF-κB transcription factor. *Science* **242**: 540–546.

Inlay M, Alt FW, Baltimore D, Xu Y. 2002. Essential roles of the κ light chain intronic enhancer and 3' enhancer in κ rearrangement and demethylation. *Nature Immunology* **3:** 463–468.

Queen C, Baltimore D. 1983. Immunoglobulin gene transcription is activated by downstream sequence elements. *Cell* **33:** 741–748.

Sen R, Baltimore D. 1986a. Multiple nuclear factors interact with the immunoglobulin enhancer sequences. *Cell* **46:** 705–716.

Sen R, Baltimore D. 1986b. Inducibility of κ immunoglobulin enhancer-binding protein NF-κB by a posttranslational mechanism. *Cell* **47:** 921–928.

Siden E, Alt FW, Shinefeld L, Sato V, Baltimore D. 1981. Synthesis of immunoglobulin μ chain gene products precedes synthesis of light chains during B-lymphocyte development. *Proc Natl Acad Sci* **78:** 1823–1827.

Singh H, Sen R, Baltimore D, Sharp PA. 1986. A nuclear factor that binds to a conserved sequence motif in transcriptional control elements of immunoglobulin genes. *Nature* **319:** 154–158.

Xu Y, Davidson L, Alt FW, Baltimore D. 1996. Deletion of the Igκ light chain intronic enhancer/matrix attachment region impairs but does not abolish VκJκ rearrangement. *Immunity* **4:** 377–385.

The NF-κB Family of Transcription Factors and Its Regulation

Andrea Oeckinghaus[1,2] and Sankar Ghosh[1,2]

[1]Department of Immunobiology and Department of Molecular Biophysics and Biochemistry, Yale University School of Medicine, New Haven, Connecticut 06520

[2]Department of Microbiology and Immunology, Columbia University, College of Physicians and Surgeons, New York 10032

Correspondence: sg2715@columbia.edu

Nuclear factor-κB (NF-κB) consists of a family of transcription factors that play critical roles in inflammation, immunity, cell proliferation, differentiation, and survival. Inducible NF-κB activation depends on phosphorylation-induced proteosomal degradation of the inhibitor of NF-κB proteins (IκBs), which retain inactive NF-κB dimers in the cytosol in unstimulated cells. The majority of the diverse signaling pathways that lead to NF-κB activation converge on the IκB kinase (IKK) complex, which is responsible for IκB phosphorylation and is essential for signal transduction to NF-κB. Additional regulation of NF-κB activity is achieved through various post-translational modifications of the core components of the NF-κB signaling pathways. In addition to cytosolic modifications of IKK and IκB proteins, as well as other pathway-specific mediators, the transcription factors are themselves extensively modified. Tremendous progress has been made over the last two decades in unraveling the elaborate regulatory networks that control the NF-κB response. This has made the NF-κB pathway a paradigm for understanding general principles of signal transduction and gene regulation.

Following the identification of NF-κB (nuclear factor-κB) as a regulator of κB light chain expression in mature B and plasma cells by Sen and Baltimore, inducibility of its activity in response to exogenous stimuli was demonstrated in various cell types (Sen et al. 1986b; Sen et al. 1986a). Years of intense research that followed demonstrated that NF-κB is expressed in almost all cell types and tissues, and specific NF-κB binding sites are present in the promoters/enhancers of a large number of genes. It is now well-established that NF-κB plays a critical role in mediating responses to a remarkable diversity of external stimuli, and thus is a pivotal element in multiple physiological and pathological processes. The progress made in the past two decades in understanding how different stimuli culminate in NF-κB activation and how NF-κB activation is translated into a cell-type- and situation-specific response has made the NF-κB pathway a paradigm for understanding signaling mechanisms and gene regulation. Coupled with the large number of diseases in which dysregulation of NF-κB has been implicated, the continuing interest into the regulatory mechanisms that govern the activity of this important transcription factor can be easily explained.

Because of its ability to influence expression of numerous genes, the activity of NF-κB is tightly regulated at multiple levels. The primary mechanism for regulating NF-κB is through inhibitory IκB proteins (IκB, inhibitor of NF-κB), and the kinase that phosphorylates IκBs, namely, the IκB kinase (IKK) complex. A number of post-translational modifications also modulate the activity of the IκB and IKK proteins as well as NF-κB molecules themselves. In this article, we introduce the major players in the NF-κB signaling cascade and describe the key regulatory steps that control NF-κB activity. We highlight the basic principles that underlie NF-κB regulation, and many of these topics are discussed in depth in other articles on the subject.

NF-κB STIMULI AND κB-DEPENDENT TARGET GENES

NF-κB transcription factors are crucial players in an elaborate system that allows cells to adapt and respond to environmental changes,

a process pivotal for survival. A large number of diverse external stimuli lead to activation of NF-κB and the genes whose expression is regulated by NF-κB play important and conserved roles in immune and stress responses, and impact processes such as apoptosis, proliferation, differentiation, and development.

Bacterial and viral infections (e.g., through recognition of microbial products by receptors such as the Toll-like receptors), inflammatory cytokines, and antigen receptor engagement, can all lead to activation of NF-κB, confirming its crucial role in innate and adaptive immune responses. In addition, NF-κB activation can be induced upon physical (UV- or γ-irradiation), physiological (ischemia and hyperosmotic shock), or oxidative stresses (Fig. 1) (Baeuerle and Henkel 1994; Hayden et al. 2006).

Consistent with the large number of signals that activate NF-κB, the list of target genes controlled by NF-κB is also extensive (Pahl 1999). Importantly, regulators of NF-κB such as IκBα, p105, or A20 are themselves NF-κB-dependent, thereby generating auto-regulatory feedback loops in the NF-κB response. Other

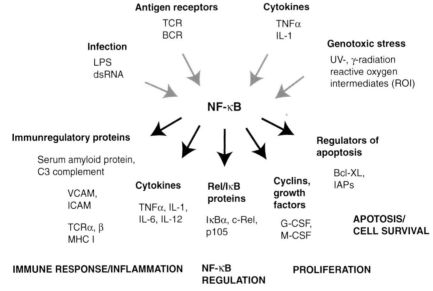

Figure 1. NF-κB stimuli and target genes. NF-κB acts as a central mediator of immune and inflammatory responses, and is involved in stress responses and regulation of cell proliferation and apoptosis. The respective NF-κB target genes allow the organism to respond effectively to these environmental changes. Representative examples are given for NF-κB inducers and κB-dependent target genes.

NF-κB target genes are central components of the immune response, e.g., immune receptor subunits or MHC molecules. Inflammatory processes are controlled through NF-κB-dependent transcription of cytokines, chemokines, cell adhesion molecules, factors of the complement cascade, and acute phase proteins. The list of κB-dependent target genes can be further extended to regulators of apoptosis (antiapoptotic Bcl family members and inhibitor of apoptosis proteins/IAPs) and proliferation (cyclins and growth factors), thereby substantiating its role in cell growth, proliferation, and survival (Chen and Manning 1995; Kopp and Ghosh 1995; Wissink et al. 1997). Furthermore, crucial functions of NF-κB in embryonic development and physiology of the bone, skin, and central nervous system add to the importance of this pleiotropic transcription factor (Hayden and Ghosh 2004).

Because so many key cellular processes such as cell survival, proliferation, and immunity are regulated through NF-κB-dependent transcription, it is not surprising that dysregulation of NF-κB pathways results in severe diseases such as arthritis, immunodeficiency, autoimmunity, and cancer (Courtois and Gilmore 2006).

THE NF-κB TRANSCRIPTION FACTOR FAMILY

The NF-κB transcription factor family in mammals consists of five proteins, p65 (RelA), RelB, c-Rel, p105/p50 (NF-κB1), and p100/52 (NF-κB2) that associate with each other to form distinct transcriptionally active homo- and heterodimeric complexes (Fig. 2). They all share a conserved 300 amino acid long amino-terminal Rel homology domain (RHD) (Baldwin 1996; Ghosh et al. 1998), and sequences within the RHD are required for dimerization, DNA binding, interaction with IκBs, as well as nuclear translocation. Crystal structures of p50 homo- and p50/p65 heterodimers bound to DNA revealed that the amino-terminal part of the RHD mediates specific DNA binding to the NF-κB consensus sequence present in regulatory elements of NF-κB target genes (5′ GGGPuNNPyPyCC-3′), whereas the

carboxy-terminal part of the RHD is mainly responsible for dimerization and interaction with IκBs (Ghosh et al. 1995; Muller et al. 1995; Chen et al. 1998).

In most unstimulated cells, NF-κB dimers are retained in an inactive form in the cytosol through their interaction with IκB proteins. Degradation of these inhibitors upon their phosphorylation by the IκB kinase (IKK) complex leads to nuclear translocation of NF-κB and induction of transcription of target genes. Although NF-κB activity is inducible in most cells, NF-κB can also be detected as a constitutively active, nuclear protein in certain cell types, such as mature B cells, macrophages, neurons, and vascular smooth muscle cells, as well as a large number of tumor cells.

Through combinatorial associations, the Rel protein family members can form up to 15 different dimers. However, the physiological existence and relevance for all possible dimeric complexes has not yet been demonstrated. The p50/65 heterodimer clearly represents the most abundant of Rel dimers, being found in almost all cell types. In addition, dimeric complexes of p65/p65, p65/c-Rel, p65/p52, c-Rel/c-Rel, p52/c-Rel, p50/c-Rel, p50/p50, RelB/p50, and RelB/p52 have been described, some of them only in limited subsets of cells (Hayden and Ghosh 2004). RelB seems to be unique in this regard as it is only found in p50- or p52-containing complexes (Ryseck et al. 1992; Dobrzanski et al. 1994). p50/c-Rel dimers are the primary component of the constitutively active NF-κB observed in mature B cells (Grumont et al. 1994b; Miyamoto et al. 1994). The NF-κB family of proteins can be further divided into two groups based on their transactivation potential because only p65, RelB, and c-Rel contain carboxy-terminal transactivation domains (TAD) (Fig. 2). RelB is unique in that it requires an amino-terminal leucine zipper region in addition to its TAD to be fully active (Dobrzanski et al. 1993). p50 and p52 are generated by processing of the precursor molecules p105 and p100, respectively. The amino-termini of these precursors contain the RHDs of p50 or p52, followed by a glycine rich region (GRR) and multiple

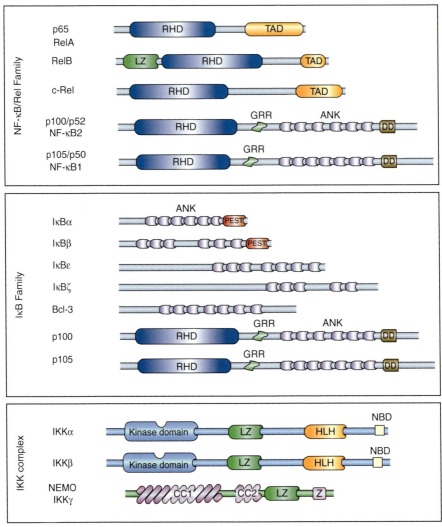

Figure 2. Members of the NF-κB, IκB, and IKK protein families. The Rel homology domain (RHD) is characteristic for the NF-κB proteins, whereas IκB proteins contain ankyrin repeats (ANK) typical for this protein family. The precursor proteins p100 and p105 can therefore be assigned to and fulfill the functions of both the NF-κB and IκB protein families. The domains that typify each protein are indicated schematically. CC, coiled-coil; DD, death domain; GRR, glycine-rich region; HLH, helix-loop-helix; IKK, IκB kinase; LZ, leucine-zipper; NBD, NEMO binding domain; PEST, proline-, glutamic acid-, serine-, and threonine-rich region; TAD, transactivation domain; ZF, zinc finger.

copies of ankyrin repeats that are characteristic for the IκB protein family. Therefore, not all combinations of Rel dimers are transcriptionally active: DNA-bound p50 and p52 homo- and heterodimers have been found to repress κB-dependent transcription, most probably by preventing trancriptionally active NF-κB dimers from binding to κB sites, or through

recruitment of deacetylases to promoter regions (Zhong et al. 2002). An intriguing feature of p50 and p52 is their ability to associate with atypical IκB proteins such as IκBζ and Bcl-3, which most likely provide transcriptional activation properties. Because of their carboxy-terminal ankyrin repeats, the precursors p105 and p100 can also inhibit nuclear localization and transcriptional activity of NF-κB dimers that they are associated with, and hence can be classified as IκB proteins.

The specific physiological role of individual NF-κB dimers remains to be fully understood. Crystallographic structural analysis has helped delineate some of the principles that govern selection of particular κB-sites by NF-κB dimers, and biochemical experiments revealed preferential affinities of individual dimers for particular sites in vitro. However, knockout studies demonstrated that biochemical affinity data do not fully explain the specificity of NF-κB dimers in vivo and it has become obvious that dimer selection cannot be reduced to the κB sequence alone (Udalova et al. 2002; Hoffmann et al. 2003; Hoffmann et al. 2006; Schreiber et al. 2006; Britanova et al. 2008). The complex structures of target promoters together with the potential of particular NF-κB dimers to trigger specific protein–protein interactions at the promoter is likely to be more important as transcriptional control involves the concerted action of the transcription factor, coactivators, and corepressors in the context of nucleosomal chromatin. The κB sequence may also affect the conformation of the bound NF-κB dimers and thus influence their ability to recruit coactivators. Furthermore, the kinetics of NF-κB dimer binding and release from DNA is based on the presence of sufficient IκBs, as well as the nature of post-translational modifications of NF-κB subunits. The combinatorial diversity of the NF-κB dimers certainly adds to their ability to regulate distinct but overlapping sets of genes by influencing all these aspects of κB site selection.

IκB PROTEINS

In most resting cells, the association of NF-κB dimers with one of the prototypical IκB proteins IκBα, IκBβ, and IκBε, or the precursor Rel proteins p100 and p105, determines their cytosolic localization, which makes the IκB proteins essential for signal responsiveness. In addition, expression of two atypical members of this protein family, namely Bcl-3 and

IκBζ, can be induced upon stimulation and regulate the activity of NF-κB dimers in the nucleus. Finally, an alternative transcript of the p105 gene, apparently only expressed in some murine lymphoid cells, has been identified and termed IκBγ, but its physiological function remains unclear (Inoue et al. 1992; Gerondakis et al. 1993; Grumont and Gerondakis 1994a).

All IκB proteins are characterized by the presence of five to seven ankyrin repeat motifs, which mediate their interaction with the RHD of NF-κB proteins (Fig. 2). In general, individual IκB proteins are thought to preferentially associate with a particular subset of NF-κB dimers, however surprisingly little experimental evidence concerning this important subject exists. IκBα and IκBβ are primarily believed to inhibit c-Rel and p65-containing complexes, with IκBα having a higher affinity for p65:p50 than for p65:p65 complexes (Malek et al. 2003). RelB exclusively binds p100, whereas Bcl-3 and IκBζ preferentially associate with homodimers of p50 and p52 (Hoffmann et al. 2006).

The Canonical IκBs

The amino-terminal signal responsive regions (SRR) of the typical IκB proteins IκBα, IκBβ, and IκBε contain conserved serine residues (e.g., Ser32 and Ser36 in IκBα) that are targeted by the IKKβ subunit of the IKK complex in the course of canonical signaling to NF-κB (see also paragraph: The IKK complex). Phosphorylated IκB proteins are then modified with K48-linked ubiquitin chains by ubiquitin ligases of the SCF or SCRF (Skp1-Culin-Roc1/Rbx1/Hrt-1-F-box) family, upon recognition of the DS^PGXXS^P motif of phosphorylated IκB proteins through their receptor subunit β-TRCP (β-tranducin repeat containing protein) (Yaron et al. 1998; Fuchs et al. 1999; Hatakeyama et al. 1999; Kroll et al. 1999; Wu and Ghosh 1999). Ubiquitination of IκBs ultimately results in their degradation by the proteasome.

IκBα is the best-studied member of the IκB family of proteins and displays all defining characteristics of an NF-κB inhibitor. The canonical p65/p50 heterodimer is primarily associated with IκBα and rapid, stimulus-induced degradation of IκBα is critical for nuclear translocation and DNA binding of NF-κB p65/p50. The traditional model of IκB function posits that IκB proteins sequester NF-κB in the cytosol; however, the actual situation is more complex. Crystallographic structural analysis of an IκBα:p65:p50 complex revealed that the IκBα molecule masks only the nuclear localization sequence (NLS) of p65, whereas the NLS of p50 remains accessible (Huxford et al. 1998; Jacobs and Harrison 1998). This exposed NLS together with the nuclear export sequence (NES) present in IκBα leads to constant shuttling of IκBα: NF-κB complexes between the nucleus and cytosol, despite a steady-state localization that appears to be exclusively cytosolic. Degradation of IκBα shifts this dynamic equilibrium because it eliminates contribution of the NES in IκBα and exposes the previously masked NLS of p65. As active NF-κB promotes IκBα expression, an important negative feedback regulatory mechanism is generated that critically influences the duration of the NF-κB response. Consistent with such a role, termination of the NF-κB response upon TNFα stimulation is notably delayed in the case of IκBα deficiency (Klement et al. 1996).

The roles of the other IκB family members are less well established. Although IκBβ shows NF-κB binding specificity similar to IκBα, it also possesses a number of unique properties. In contrast to IκBα, IκBβ:NF-κB complexes do not undergo nuclear-cytoplasmic shuttling (Tam and Sen 2001) and IκBβ can associate with NF-κB complexes bound to DNA, suggesting a possible regulatory function in the nucleus (Thompson et al. 1995; Suyang et al. 1996). Even though IκBβ is degraded and resynthesized in a stimulus-dependent manner—although with considerably delayed kinetics compared to IκBα—IκBβ deletion does not affect the kinetics of NF-κB responses significantly (Hoffmann et al. 2002). Surprisingly, knocking in of IκBβ to replace IκBα, thereby placing IκBβ expression under the regulation of the

IκBα promoter, leads to normal kinetics of NF-κB activation and termination despite the fact that IκBβ normally does not function in the early dampening of the NF-κB response (Cheng et al. 1998). This suggests that the functional characteristics of the different canonical IκBs are at least in part caused by temporal distinctions in their resynthesis based on the specific transcriptional regulation of their promoters (Hoffmann et al. 2002).

Similar to IκBβ, IκBε is also degraded and resynthesized in response to NF-κB signaling with significantly delayed kinetics (Kearns et al. 2006). Besides, IκBε expression and function seem to be limited to cells of the hematopoetic lineage, reinforcing the hypothesis that distinct IκBs exert unique roles in NF-κB responses in different cellular scenarios.

The Precursor IκBs

The precursor proteins p100 and p105 are exceptional in their ability to function as Rel protein inhibitors, because they can dimerize with other NF-κB molecules via their RHDs, whereas their carboxy-terminal ankyrin repeats serve the function of IκB proteins (Rice et al. 1992; Naumann et al. 1993; Solan et al. 2002). Although their overall sequence and organization is similar, generation of p50 and p52 from their respective precursors occurs through distinct processing mechanisms. p105 is constitutively processed, resulting in a preset ratio of p105 and p50 containing Rel dimers in each cell. This constitutive processing has been suggested to occur co- or post-translationally (Lin et al. 1998; Moorthy et al. 2006) and depends on a GRR located carboxy-terminal to the RHD that serves as termination signal for the proteosomal proteolysis (Lin and Ghosh 1996). However, p105 can, like IκBα, also undergo complete degradation upon phosphorylation of carboxy-terminal serine residues by IKKβ, a process that is favored when p105 is bound to other NF-κB subunits (Harhaj et al. 1996; Heissmeyer et al. 1999; Cohen et al. 2001; Cohen et al. 2004; Perkins 2006). Interestingly, this event has been suggested to be independent of ubiquitination (Moorthy et al. 2006).

 Cite this article as *Cold Spring Harb Perspect Biol* 2009;1:a000034

p100 processing represents a predominantly stimulus-dependent event that defines the non-canonical NF-κB activation pathway (see also paragraph: The IKK complex). It is regulated through phosphorylation of p100 by the IKKα subunit of the IKK complex, which results in ubiquitination and subsequent partial degradation of p100 by the proteasome, releasing transcriptionally active p50/RelB complexes (Senftleben et al. 2001). However, p100 can also regulate p65-containing complexes downstream of IKKα and limit p65-dependent responses in T cells in a negative feedback loop similar to IκBα, thereby acting as an inhibitor in canonical signaling pathways (Ishimaru et al. 2006; Basak et al. 2007). Constitutive processing of p100 has only been described to occur at a low level in certain cell types (Heusch et al. 1999). The ability of p100 to regulate RelB is of particular importance, as RelB-containing dimers exclusively associate with p100 and require p100 binding for stabilization (Solan et al. 2002). Therefore, p100 is able to affect NF-κB transcriptional responses at a level that extends beyond its role as an IκB protein.

The Atypical IκBs

The role of the atypical IκB protein Bcl3 in NF-κB signaling is not fully understood and its mechanism of action is still unclear. Bcl3 contains a TAD and is found primarily in the nucleus, where it associates with p50 and p52 homo- or heterodimers. Bcl3 has been suggested to provide transcriptional activation function on otherwise repressive p50 or p52 homodimers (Bours et al. 1993). Furthermore, Bcl3 may remove repressive dimers from promoters, thereby allowing transcriptionally active p65/50 to access and facilitate transcription in an indirect manner through release of repression (Wulczyn et al. 1992; Hayden and Ghosh 2004; Perkins 2006). Conversely, induction of Bcl3 expression has been shown to inhibit subsequent NF-κB activation. In such a model, Bcl3 may stabilize repressing dimers at κB sites, thereby preventing access of TAD containing NF-κB dimers and consequent transcription (Wessells et al. 2004; Carmody

et al. 2007). The potential of Bcl3 to either promote or impede transcription has been shown to depend to some extent on different post-translational modifications (Nolan et al. 1993; Bundy and McKeithan 1997; Massoumi et al. 2006).

Within the IκB family, IκBζ is most similar to Bcl3. It is not constitutively expressed, but induced upon engagement of IL-1 and TLR4 receptors and is localized to the nucleus (Yamazaki et al. 2005). For the most part, IκBζ associates with p50 homodimers. Expression of a subset of NF-κB target genes, including IL-6, is not induced in IκBζ-deficient cells upon IL-1 or LPS stimulation. IκBζ is hypothesized to act as a coactivator of p50 homodimers even though IκBζ does not contain an apparent TAD (Yamamoto et al. 2004). IκBζ has also been suggested to negatively regulate p65-containing complexes and may thus also be able to selectively activate or impede specific NF-κB activity (Yamamoto et al. 2004; Motoyama et al. 2005).

It is obvious that our perception of IκB proteins has changed significantly over the years from relatively simple inhibitory molecules to complex regulators of NF-κB-driven gene expression. As described previously, IκBs can influence the makeup of cellular NF-κB through specific stabilization of unstable dimers, stabilize DNA-bound dimers, thereby repressing or prolonging transcriptional responses, or even act as transcriptional coactivators. Therefore, it is probably more appropriate to consider IκBs as multifaceted NF-κB regulators.

THE IKK COMPLEX: CANONICAL AND NON-CANONICAL SIGNALING TO NF-κB

The first biochemical experiments assigned the cytosolic IκB kinase activity to a large protein complex of 700–900 kDa capable of specifically phosphorylating IκBα on serines 32 and 36. Purification of this complex revealed the presence of two catalytically active kinases, IKKα and IKKβ, and the regulatory subunit IKKγ (NEMO) (Chen et al. 1996; Mercurio et al. 1997; Woronicz et al. 1997; Zandi et al.

1997). IKKα and IKKβ are ubiquitously expressed and contain an amino-terminal kinase domain, a leucine zipper, and a carboxy-terminal helix-loop-helix domain (HLH) (Fig. 2). The leucine zipper is responsible for dimerization of the kinases and mutations in this region render the kinases inactive, whereas the HLH is dispensable for dimerization, but essential for optimal kinase activity (Zandi et al. 1998). The carboxy-terminal portions of IKKα and -β are critical for their interaction with the regulatory subunit IKKγ, which is mediated by a hexapeptide sequence (LDWSWL) on IKKs termed the NEMO binding domain (NBD) (May et al. 2000; May et al. 2002). Activation of the IKK complex is dependent on the phosphorylation of two serines in the sequence motif SLCTS of the T-loop regions in at least one of the IκB kinases (Mercurio et al. 1997; Ling et al. 1998). Whether these phosphorylation events occur through trans-autophosphorylation or through phosphorylation by an upstream kinase continues to be debated (Hayden and Ghosh 2004; Scheidereit 2006).

The physiological significance of IKKα and IKKβ has been analyzed extensively in knockout mice. These studies have revealed two general types of signal propagation pathways to NF-κB activation, which are distinct in respect to the inducing stimuli, the IKK subunits involved, and the NF-κB/IκB substrates targeted (Fig. 3). In the canonical NF-κB pathway, which is induced by inflammatory cytokines, pathogen-associated molecules, and antigen receptors, IKKβ is both necessary and sufficient to phosphorylate IκBα or IκBβ in an IKKγ-dependent manner. Thus, IKKβ-deficient mice exhibit embryonic lethality caused by severe liver apoptosis because of defective TNF signaling to NF-κB in the developing liver (Beg et al. 1995; Li et al. 1999a; Li et al. 1999b). Congruently, IKKβ-deficient cells were shown to be unable to activate NF-κB upon stimulation with proinflammatory cytokines such as TNFα or interleukin-1 (IL-1) (Li et al. 1999a; Li et al. 1999b). The role of IKKα in canonical NF-κB signaling remains unclear; however, more recent studies

have established a role for IKKα in regulating gene expression by modifying the phosphorylation status of histones and p65 (Anest et al. 2003; Yamamoto et al. 2003; Chen and Greene 2004).

In contrast to the canonical pathway, the non-canonical pathway that is induced by specific members of the TNF cytokine family, such as BAFF, lymphotoxin-β, or CD40 ligand relies on IKKα, but not IKKβ or IKKγ. IKKα is believed to selectively phosphorylate p100 associated with RelB (Senftleben et al. 2001; Scheidereit 2006). Although the upstream signal propagation events are poorly understood, IKKα clearly requires activation of NIK (NF-κB-inducing kinase), which seems to function as both an IKKα-activating kinase as well as a scaffold linking IKKα and p100 (Xiao et al. 2004). Phosphorylation of p100 entails recruitment of SCF^{βTrCP} and subsequent polyubiquitination of Lys855, which is situated in a peptide sequence homologous to that targeted in IκBα (Fong and Sun 2002; Amir et al. 2004). Similar to p105 processing, the GRR of p100 is required for partial processing to generate p52. In addition to non-canonical signaling to NF-κB, IKKα regulates developmental processes that are both dependent, as well as independent, of NF-κB-induced gene transcription (Sil et al. 2004).

POST-TRANSLATIONAL MODIFICATIONS OF NF-κB

Given the wide range of biological processes that are affected by NF-κB, and the disastrous consequences of dysregulated NF-κB signaling, intricate and highly regulated mechanisms exist for controlling NF-κB activity. These added layers of regulation involve a variety of posttranslational modifications of various core components of this signaling pathway, thereby influencing NF-κB activity at multiple levels. The critical phosphorylation and ubiquitination events of IκB proteins and the IKK complex have been described previously. In addition, the NF-κB subunits themselves are subject to a wide range of regulatory

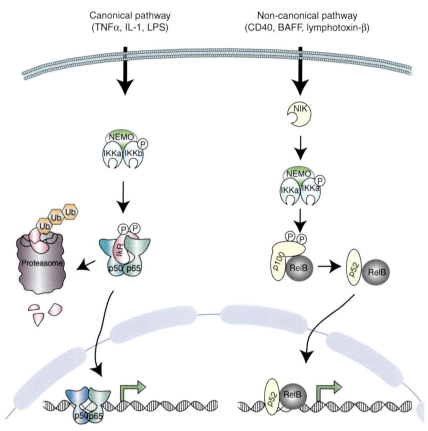

Figure 3. Canonical and non-canonical signaling to NF-κB. The canonical pathway is, e.g., induced by TNFα, IL-1, or LPS and uses a large variety of signaling adaptors to engage IKK activity. Phosphorylation of serine residues in the signal responsive region (SRR) of classical IκBs by IKKβ leads to IκB ubiquitination and subsequent proteosomal degradation. This results in release of the NF-κB dimer, which can then translocate to the nucleus and induce transcription of target genes. The non-canonical pathway depends on NIK (NF-κB-inducing kinase) induced activation of IKKα. IKKα phosphorylates the p100 NF-κB subunit, which leads to proteosomal processing of p100 to p52. This results in the activation of p52-RelB dimers, which target specific κB elements.

protein modifications, including phosphorylation, ubiquitination, or acetylation (Perkins 2006). Some of these modifications also represent important means for cross talk between different signaling pathways (Perkins 2007).

Importantly, degradation of IκBs and nuclear translocation of the NF-κB dimers alone are not sufficient in eliciting a maximal NF-κB response, which has been shown to also critically depend on protein modifications of NF-κB subunits. Post-translational modifications of p65 have been most thoroughly investigated in this context. Phosphorylation by protein kinase A (PKA) was the first

modification of p65 to be recognized and has since been demonstrated to be critical for p65-driven gene expression of many but not all target genes (Naumann and Scheidereit 1994; Neumann et al. 1995; Chen and Greene 2004). The catalytic subunit of PKA (PKAc) exists in complex with cytosolic IκB:NF-κB complexes and is activated upon IκB degradation to phosphorylate Ser276 of p65. p50 and c-Rel are also modified by PKAc at equivalent sites to RelA (Perkins 2006). In p65, Ser 276 phosphorylation is thought to break an intramolecular interaction between the carboxy-terminal part of the protein and the RHD, thereby facilitating

DNA binding and enabling interaction with the transcriptional coactivators p300 and CBP (CREB-binding protein) (Zhong et al. 1998; Zhong et al. 2002). In addition to PKA, MSK1 and MSK2 (mitogen- and stress-activated protein kinase) have been shown to target Ser276, and in doing so possibly mediating cross talk with the ERK and p38 MAPK pathways (Vermeulen et al. 2003; Perkins 2006). Several additional phosphorylation events have been described but the physiological consequences in most cases remain less clear. Ser536 within the TAD of p65 seems to be a phosphorylation site that is targeted by several kinases, including IKKβ. Numerous effects have been ascribed to this modification, including regulation of p65 nuclear localization and CBP/p300 interaction (Chen et al. 2005; Perkins 2006). CK2 has been shown to inducibly modify Ser529; however, it remains unclear whether this phosphorylation affects p65 transcriptional activity (Wang et al. 2000). Ser311 can be phosphorylated by the atypical protein kinase PKCζ in response to TNF stimulation, which has also been demonstrated to positively influence p65:CBP interaction (Duran et al. 2003). Interestingly, of the four phosphorylation sites mentioned above, two are found in the RHD (Ser276 and Ser311), whereas the other two are located within the TAD (Ser529 and Ser 536).

The diversity of the p65 phosphorylation sites detected so far suggests that possibly even more modification sites will be discovered in the future, in particular in other NF-κB subunits. It is likely that these modifications would serve to fine-tune NF-κB transcriptional activity, e.g., in determining target-gene specificity and timing of gene expression, rather than acting as simple on–off switches.

p65 can also be inducibly acetylated at different sites (Lys122, 123, 218, 221, and 310) with variable effects on its activity, probably by the action of CBP/p300 and associated histone acetyltransferases (HAT). While acetylation of lysines 218, 221, and 310 has been shown to promote p65 transactivation, acetylated lysines 122 and 123 act in an inhibitory manner. Acetylation at Lys310, promoted by phosphorylation of Ser276, has been shown to augment transcriptional activity without affecting DNA or IκB binding, presumably through generating a binding site for a coactivator protein. In contrast, acetylated Lys221 impedes association with IκBα, thereby enhancing DNA binding (Chen et al. 2002; Chen and Greene 2004). Different histone deacetylases (HDAC) such as HDAC-3 and SIRT1 have been implicated in removing p65 acetylations (Chen and Greene 2004; Yeung et al. 2004).

TERMINATION OF THE NF-κB RESPONSE

Our knowledge about the mechanisms leading to NF-κB activation far exceeds what we know about the regulatory mechanisms that determine its inactivation. The best studied and well accepted mechanism for termination of the NF-κB response involves the resynthesis of IκB proteins induced by activated NF-κB (Pahl 1999). Newly synthesized IκBα can enter the nucleus, remove NF-κB from the DNA, and relocalize it to the cytosol (Hayden and Ghosh 2004). Also, the precursors p105 and p100, as well as c-Rel and RelB, are NF-κB inducible genes, and hence the composition of NF-κB complexes in a cell can vary over time. Significant progress has also been made in understanding how negative regulators influence signaling upstream of the IKK complex (Hayden and Ghosh 2004; Krappmann and Scheidereit 2005; Hacker and Karin 2006). In the recent years, additional inhibitory mechanisms have been identified that function later in the pathway and directly affect the active, DNA-bound NF-κB. These include post-translational modifications of NF-κB subunits and have been shown to participate in shutting off the NF-κB response by altering cofactor binding or mediating displacement and degradation of NF-κB dimers. Because acetylation of p65 has been demonstrated to negatively impact its DNA-binding affinity (Kiernan et al. 2003), regulation of p65 association with HATs and HDACs influences termination of the NF-κB response (Ghosh and Hayden 2008). An alternative scenario involves ubiquitin-dependent proteosomal degradation of promoter-bound p65, facilitated by the ubiquitin ligase SOCS-1 (Ryo et al. 2003; Saccani et al. 2004). In parallel, the nuclear ubiquitin ligase PDLIM2 has been shown to target p65 for ubiquitination and proteasomal degradation, as well as to transport p65 to promyelocytic leukemia (PML) nuclear bodies, thereby supporting transcriptional silencing (Tanaka et al. 2007). In addition, IKKα has been demonstrated to promote p65 and c-Rel turnover and removal from proinflammatory gene promoters in macrophages (Lawrence et al. 2005). Finally, NF-κB subunits have been shown to be inhibited by oxidation and alkylation of a redox-sensitive cysteine located in the DNA-binding loop (Perkins 2006). This list is far from being complete and it seems that we have just begun to understand the complex network of modifications regulating NF-κB transcriptional activity in the nucleus. As a detailed knowledge of these mechanisms is crucial for our understanding of the function of NF-κB, this area of investigation is likely to continue to attract significant attention in the future.

CONCLUSIONS

The transcription factors of the NF-κB family control the expression of a large number of target genes in response to changes in the environment, thereby helping to orchestrate inflammatory and immune responses. As aberrant NF-κB activation underlies various disease states, precise activation and termination of NF-κB is ensured by multiple regulatory processes. It is fair to say that the IκB proteins represent one of the primary means of NF-κB regulation, although their versatile effects on NF-κB activity make them more complex than simple inhibitory proteins. A diverse array of post-translational modifications of IKK, IκB, and NF-κB proteins provides the means to fine-tune the NF-κB response at multiple levels. Tremendous progress has been made in understanding the regulatory mechanisms shaping the NF-κB response, yet it is striking how much still remains to be discovered. Because of the well-characterized links between NF-κB and diseases like cancer or arthritis, unraveling

the complexity of NF-κB regulation remains a major goal to help act on specific steps of the NF-κB pathway, thereby avoiding the risk of harmful side effects that could result from general inhibition of NF-κB.

REFERENCES

Amir RE, Haecker H, Karin M, Ciechanover A. 2004. Mechanism of processing of the NF-κ B2 p100 precursor: Identification of the specific polyubiquitin chain-anchoring lysine residue and analysis of the role of NEDD8-modification on the SCF(β-TrCP) ubiquitin ligase. *Oncogene* **23:** 2540–2547.

Anest V, Hanson JL, Cogswell PC, Steinbrecher KA, Strahl BD, Baldwin AS. 2003. A nucleosomal function for IκB kinase-α in NF-κB-dependent gene expression. *Nature* **423:** 659–663.

Baeuerle PA, Henkel T. 1994. Function and activation of NF-κ B in the immune system. *Annu Rev Immunol* **12:** 141–179.

Baldwin AS Jr. 1996. The NF-κ B and I κ B proteins: New discoveries and insights. *Annu Rev Immunol* **14:** 649–683.

Basak S, Kim H, Kearns JD, Tergaonkar V, O'Dea E, Werner SL, Benedict CA, Ware CF, Ghosh G, Verma IM, et al. 2007. A fourth IκB protein within the NF-κB signaling module. *Cell* **128:** 369–381.

Beg AA, Sha WC, Bronson RT, Ghosh S, Baltimore D. 1995. Embryonic lethality and liver degeneration in mice lacking the RelA component of NF-κ B. *Nature* **376:** 167–170.

Bours V, Franzoso G, Azarenko V, Park S, Kanno T, Brown K, Siebenlist U. 1993. The oncoprotein Bcl-3 directly transactivates through κ B motifs via association with DNA-binding p50B homodimers. *Cell* **72:** 729–739.

Britanova LV, Makeev VJ, Kuprash DV. 2008. In vitro selection of optimal RelB/p52 DNA-binding motifs. *Biochem Biophys Res Commun* **365:** 583–588.

Bundy DL, McKeithan TW. 1997. Diverse effects of BCL3 phosphorylation on its modulation of NF-κB p52 homodimer binding to DNA. *J Biol Chem* **272:** 33132–33139.

Carmody RJ, Ruan Q, Palmer S, Hilliard B, Chen YH. 2007. Negative regulation of toll-like receptor signaling by NF-κB p50 ubiquitination blockade. *Science* **317:** 675–678.

Chen CC, Manning AM. 1995. Transcriptional regulation of endothelial cell adhesion molecules: A dominant role for NF-κ B. *Agents Actions Suppl* **47:** 135–141.

Chen LF, Greene WC. 2004. Shaping the nuclear action of NF-κB. *Nat Rev Mol Cell Biol* **5:** 392–401.

Chen LF, Mu Y, Greene WC. 2002. Acetylation of RelA at discrete sites regulates distinct nuclear functions of NF-κB. *EMBO J* **21:** 6539–6548.

Chen ZJ, Parent L, Maniatis T. 1996. Site-specific phosphorylation of IκB by a novel ubiquitination-dependent protein kinase activity. *Cell* **84:** 853–862.

Chen FE, Huang DB, Chen YQ, Ghosh G. 1998. Crystal structure of p50/p65 heterodimer of transcription factor NF-κB bound to DNA. *Nature* **391:** 410–413.

Chen LF, Williams SA, Mu Y, Nakano H, Duerr JM, Buckbinder L, Greene WC. 2005. NF-κB RelA phosphorylation regulates RelA acetylation. *Mol Cell Biol* **25:** 7966–7975.

Cheng JD, Ryseck RP, Attar RM, Dambach D, Bravo R. 1998. Functional redundancy of the nuclear factor κ B inhibitors I κ B and I κ B β. *J Exp Med* **188:** 1055–1062.

Cohen S, Orian A, Ciechanover A. 2001. Processing of p105 is inhibited by docking of p50 active subunits to the ankyrin repeat domain, and inhibition is alleviated by signaling via the carboxyl-terminal phosphorylation/ ubiquitin-ligase binding domain. *J Biol Chem* **276:** 26769–26776.

Cohen S, Achbert-Weiner H, Ciechanover A. 2004. Dual effects of IκB kinase β-mediated phosphorylation on p105 Fate: SCF(β-TrCP)-dependent degradation and SCF(β-TrCP)-independent processing. *Mol Cell Biol* **24:** 475–486.

Courtois G, Gilmore TD. 2006. Mutations in the NF-κB signaling pathway: Implications for human disease. *Oncogene* **25:** 6831–6843.

Dobrzanski P, Ryseck RP, Bravo R. 1993. Both N- and C-terminal domains of RelB are required for full transactivation: Role of the N-terminal leucine zipper-like motif. *Mol Cell Biol* **13:** 1572–1582.

Dobrzanski P, Ryseck RP, Bravo R. 1994. Differential interactions of Rel-NF-κ B complexes with I κ B determine pools of constitutive and inducible NF-κ B activity. *EMBO J* **13:** 4608–4616.

Duran A, Diaz-Meco MT, Moscat J. 2003. Essential role of RelA Ser311 phosphorylation by zetaPKC in NF-κB transcriptional activation. *EMBO J* **22:** 3910–3918.

Fong A, Sun SC. 2002. Genetic evidence for the essential role of β-transducin repeat-containing protein in the inducible processing of NF-κ B2/p100. *J Biol Chem* **277:** 22111–22114.

Fuchs SY, Chen A, Xiong Y, Pan ZQ, Ronai Z. 1999. HOS, a human homolog of Slimb, forms an SCF complex with Skp1 and Cullin1 and targets the phosphorylation-dependent degradation of IκB and β-catenin. *Oncogene* **18:** 2039–2046.

Gerondakis S, Morrice N, Richardson IB, Wettenhall R, Fecondo J, Grumont RJ. 1993. The activity of a 70 kilodalton I κ B molecule identical to the carboxyl terminus of the p105 NF-κ B precursor is modulated by protein kinase A. *Cell Growth Differ* **4:** 617–627.

Ghosh S, Hayden MS. 2008. New regulators of NF-κB in inflammation. *Nat Rev Immunol* **8:** 837–848.

Ghosh S, May MJ, Kopp EB. 1998. NF-κ B and Rel proteins: Evolutionarily conserved mediators of immune responses. *Annu Rev Immunol* **16:** 225–260.

Ghosh G, van Duyne G, Ghosh S, Sigler PB. 1995. Structure of NF-κ B p50 homodimer bound to a κ B site. *Nature* **373:** 303–310.

Grumont RJ, Gerondakis S. 1994a. Alternative splicing of RNA transcripts encoded by the murine p105 NF-κ B gene generates I κ B γ isoforms with different inhibitory activities. *Proc Natl Acad Sci* **91:** 4367–4371.

Grumont RJ, Gerondakis S. 1994b. The subunit composition of NF-κ B complexes changes during B-cell development. *Cell Growth Differ* **5:** 1321–1331.

Hacker H, Karin M. 2006. Regulation and function of IKK and IKK-related kinases. *Sci STKE* **2006:** re13.

Harhaj EW, Maggirwar SB, Sun SC. 1996. Inhibition of p105 processing by NF-κB proteins in transiently transfected cells. *Oncogene* **12:** 2385–2392.

Hatakeyama S, Kitagawa M, Nakayama K, Shirane M, Matsumoto M, Hattori K, Higashi H, Nakano H, Okumura K, Onoe K, et al. 1999. Ubiquitin-dependent degradation of IκB is mediated by a ubiquitin ligase Skp1/Cul 1/F-box protein FWD1. *Proc Natl Acad Sci* **96:** 3859–3863.

Hayden MS, Ghosh S. 2004. Signaling to NF-κB. *Genes Dev* **18:** 2195–2224.

Hayden MS, West AP, Ghosh S. 2006. NF-κB and the immune response. *Oncogene* **25:** 6758–6780.

Heissmeyer V, Krappmann D, Wulczyn FG, Scheidereit C. 1999. NF-κB p105 is a target of IκB kinases and controls signal induction of Bcl-3-p50 complexes. *EMBO J* **18:** 4766–4778.

Heusch M, Lin L, Geleziunas R, Greene WC. 1999. The generation of nfkb2 p52: Mechanism and efficiency. *Oncogene* **18:** 6201–6208.

Hoffmann A, Leung TH, Baltimore D. 2003. Genetic analysis of NF-κB/Rel transcription factors defines functional specificities. *EMBO J* **22:** 5530–5539.

Hoffmann A, Natoli G, Ghosh G. 2006. Transcriptional regulation via the NF-κB signaling module. *Oncogene* **25:** 6706–6716.

Hoffmann A, Levchenko A, Scott ML, Baltimore D. 2002. The IκB-NF-κB signaling module: Temporal control and selective gene activation. *Science* **298:** 1241–1245.

Huxford T, Huang DB, Malek S, Ghosh G. 1998. The crystal structure of the IκB/NF-κB complex reveals mechanisms of NF-κB inactivation. *Cell* **95:** 759–770.

Inoue J, Kerr LD, Kakizuka A, Verma IM. 1992. I κ B γ, a 70 kd protein identical to the C-terminal half of p110 NF-κ B: A new member of the I κ B family. *Cell* **68:** 1109–1120.

Ishimaru N, Kishimoto H, Hayashi Y, Sprent J. 2006. Regulation of naive T cell function by the NF-κB2 pathway. *Nat Immunol* **7:** 763–772.

Jacobs MD, Harrison SC. 1998. Structure of an IκB/NF-κB complex. *Cell* **95:** 749–758.

Kearns JD, Basak S, Werner SL, Huang CS, Hoffmann A. 2006. IκBepsilon provides negative feedback to control NF-κB oscillations, signaling dynamics, and inflammatory gene expression. *J Cell Biol* **173:** 659–664.

Kiernan R, Bres V, Ng RW, Coudart MP, El Messaoudi S, Sardet C, Jin DY, Emiliani S, Benkirane M. 2003. Post-activation turn-off of NF-κ B-dependent transcription is regulated by acetylation of p65. *J Biol Chem* **278:** 2758–2766.

Klement JF, Rice NR, Car BD, Abbondanzo SJ, Powers GD, Bhatt PH, Chen CH, Rosen CA, Stewart CL. 1996. IκB deficiency results in a sustained NF-κB response and severe widespread dermatitis in mice. *Mol Cell Biol* **16:** 2341–2349.

Kopp EB, Ghosh S. 1995. NF-κ B and rel proteins in innate immunity. *Adv Immunol* **58**: 1–27.

Krappmann D, Scheidereit C. 2005. A pervasive role of ubiquitin conjugation in activation and termination of IκB kinase pathways. *EMBO Rep* **6**: 321–326.

Kroll M, Margottin F, Kohl A, Renard P, Durand H, Concordet JP, Bachelerie F, Arenzana-Seisdedos F, Benarous R. 1999. Inducible degradation of IκB by the proteasome requires interaction with the F-box protein h-βTrCP. *J Biol Chem* **274**: 7941–7945.

Lawrence T, Bebien M, Liu GY, Nizet V, Karin M. 2005. IKK limits macrophage NF-κB activation and contributes to the resolution of inflammation. *Nature* **434**: 1138–1143.

Li Q, Van Antwerp D, Mercurio F, Lee KF, Verma IM. 1999a. Severe liver degeneration in mice lacking the IκB kinase 2 gene. *Science* **284**: 321–325.

Li ZW, Chu W, Hu Y, Delhase M, Deerinck T, Ellisman M, Johnson R, Karin M. 1999b. The IKKβ subunit of IκB kinase (IKK) is essential for nuclear factor κB activation and prevention of apoptosis. *J Exp Med* **189**: 1839–1845.

Lin L, Ghosh S. 1996. A glycine-rich region in NF-κB p105 functions as a processing signal for the generation of the p50 subunit. *Mol Cell Biol* **16**: 2248–2254.

Lin L, DeMartino GN, Greene WC. 1998. Cotranslational biogenesis of NF-κB p50 by the 26S proteasome. *Cell* **92**: 819–828.

Ling L, Cao Z, Goeddel DV. 1998. NF-κB-inducing kinase activates IKK- by phosphorylation of Ser-176. *Proc Natl Acad Sci* **95**: 3792–3797.

Malek S, Huang DB, Huxford T, Ghosh S, Ghosh G. 2003. X-ray crystal structure of an IκBβ × NF-κB p65 homodimer complex. *J Biol Chem* **278**: 23094–23100.

Massoumi R, Chmielarska K, Hennecke K, Pfeifer A, Fassler R. 2006. Cyld inhibits tumor cell proliferation by blocking Bcl-3-dependent NF-κB signaling. *Cell* **125**: 665–677.

May MJ, Marienfeld RB, Ghosh S. 2002. Characterization of the Iκ B-kinase NEMO binding domain. *J Biol Chem* **277**: 45992–46000.

May MJ, D'Acquisto F, Madge LA, Glockner J, Pober JS, Ghosh S. 2000. Selective inhibition of NF-κB activation by a peptide that blocks the interaction of NEMO with the IκB kinase complex. *Science* **289**: 1550–1554.

Mercurio F, Zhu H, Murray BW, Shevchenko A, Bennett BL, Li J, Young DB, Barbosa M, Mann M, Manning A, et al. 1997. IKK-1 and IKK-2: Cytokine-activated IκB kinases essential for NF-κB activation. *Science* **278**: 860–866.

Miyamoto S, Schmitt MJ, Verma IM. 1994. Qualitative changes in the subunit composition of κ B-binding complexes during murine B-cell differentiation. *Proc Natl Acad Sci* **91**: 5056–5060.

Moorthy AK, Savinova OV, Ho JQ, Wang VY, Vu D, Ghosh G. 2006. The 20S proteasome processes NF-κB1 p105 into p50 in a translation-independent manner. *EMBO J* **25**: 1945–1956.

Motoyama M, Yamazaki S, Eto-Kimura A, Takeshige K, Muta T. 2005. Positive and negative regulation of nuclear factor-κB-mediated transcription by IκB-zeta, an inducible nuclear protein. *J Biol Chem* **280**: 7444–7451.

Muller CW, Rey FA, Sodeoka M, Verdine GL, Harrison SC. 1995. Structure of the NF-κ B p50 homodimer bound to DNA. *Nature* **373**: 311–317.

Naumann M, Scheidereit C. 1994. Activation of NF-κ B in vivo is regulated by multiple phosphorylations. *Embo J* **13**: 4597–4607.

Naumann M, Nieters A, Hatada EN, Scheidereit C. 1993. NF-κ B precursor p100 inhibits nuclear translocation and DNA binding of NF-κ B/rel-factors. *Oncogene* **8**: 2275–2281.

Neumann M, Grieshammer T, Chuvpilo S, Kneitz B, Lohoff M, Schimpl A, Franza BR Jr, Serfling E. 1995. RelA/p65 is a molecular target for the immunosuppressive action of protein kinase A. *Embo J* **14**: 1991–2004.

Nolan GP, Fujita T, Bhatia K, Huppi C, Liou HC, Scott ML, Baltimore D. 1993. The bcl-3 proto-oncogene encodes a nuclear I κ B-like molecule that preferentially interacts with NF-κ B p50 and p52 in a phosphorylation-dependent manner. *Mol Cell Biol* **13**: 3557–3566.

Pahl HL. 1999. Activators and target genes of Rel/NF-κB transcription factors. *Oncogene* **18**: 6853–6866.

Perkins ND. 2006. Post-translational modifications regulating the activity and function of the nuclear factor κ B pathway. *Oncogene* **25**: 6717–6730.

Perkins ND. 2007. Integrating cell-signalling pathways with NF-κB and IKK function. *Nat Rev Mol Cell Biol* **8**: 49–62.

Rice NR, MacKichan ML, Israel A. 1992. The precursor of NF-κ B p50 has I κ B-like functions. *Cell* **71**: 243–253.

Ryo A, Suizu F, Yoshida Y, Perrem K, Liou YC, Wulf G, Rottapel R, Yamaoka S, Lu KP. 2003. Regulation of NF-κB signaling by Pin1-dependent prolyl isomerization and ubiquitin-mediated proteolysis of p65/RelA. *Mol Cell* **12**: 1413–1426.

Ryseck RP, Bull P, Takamiya M, Bours V, Siebenlist U, Dobrzanski P, Bravo R. 1992. RelB, a new Rel family transcription activator that can interact with p50-NF-κ B. *Mol Cell Biol* **12**: 674–684.

Saccani S, Marazzi I, Beg AA, Natoli G. 2004. Degradation of promoter-bound p65/RelA is essential for the prompt termination of the nuclear factor κB response. *J Exp Med* **200**: 107–113.

Scheidereit C. 2006. IκaB kinase complexes: Gateways to NF-κB activation and transcription. *Oncogene* **25**: 6685–6705.

Schreiber J, Jenner RG, Murray HL, Gerber GK, Gifford DK, Young RA. 2006. Coordinated binding of NF-κB family members in the response of human cells to lipopolysaccharide. *Proc Natl Acad Sci* **103**: 5899–5904.

Sen R, Baltimore D. 1986a. Inducibility of κ immunoglobulin enhancer-binding protein NF-κ B by a posttranslational mechanism. *Cell* **47**: 921–928.

Sen R, Baltimore D. 1986b. Multiple nuclear factors interact with the immunoglobulin enhancer sequences. *Cell* **46**: 705–716.

Senftleben U, Cao Y, Xiao G, Greten FR, Krahn G, Bonizzi G, Chen Y, Hu Y, Fong A, Sun SC, et al. 2001. Activation by IKK of a second, evolutionary conserved, NF-κ B signaling pathway. *Science* **293**: 1495–1499.

Sil AK, Maeda S, Sano Y, Roop DR, Karin M. 2004. IκB kinase- acts in the epidermis to control skeletal and craniofacial morphogenesis. *Nature* **428**: 660–664.

Solan NJ, Miyoshi H, Carmona EM, Bren GD, Paya CV. 2002. RelB cellular regulation and transcriptional activity are regulated by p100. *J Biol Chem* **277:** 1405–1418.

Suyang H, Phillips R, Douglas I, Ghosh S. 1996. Role of unphosphorylated, newly synthesized I κ B β in persistent activation of NF-κ B. *Mol Cell Biol* **16:** 5444–5449.

Tam WF, Sen R. 2001. IκB family members function by different mechanisms. *J Biol Chem* **276:** 7701–7704.

Tanaka T, Grusby MJ, Kaisho T. 2007. PDLIM2-mediated termination of transcription factor NF-κB activation by intranuclear sequestration and degradation of the p65 subunit. *Nat Immunol* **8:** 584–591.

Thompson JE, Phillips RJ, Erdjument-Bromage H, Tempst P, Ghosh S. 1995. I κ B-β regulates the persistent response in a biphasic activation of NF-κ B. *Cell* **80:** 573–582.

Udalova IA, Mott R, Field D, Kwiatkowski D. 2002. Quantitative prediction of NF-κ B DNA-protein interactions. *Proc Natl Acad Sci* **99:** 8167–8172.

Vermeulen L, De Wilde G, Van Damme P, Vanden Berghe W, Haegeman G. 2003. Transcriptional activation of the NF-κB p65 subunit by mitogen- and stress-activated protein kinase-1 (MSK1). *EMBO J* **22:** 1313–1324.

Wang D, Westerheide SD, Hanson JL, Baldwin AS Jr. 2000. Tumor necrosis factor-induced phosphorylation of RelA/p65 on Ser529 is controlled by casein kinase II. *J Biol Chem* **275:** 32592–32597.

Wessells J, Baer M, Young HA, Claudio E, Brown K, Siebenlist U, Johnson PF. 2004. BCL-3 and NF-κB p50 attenuate lipopolysaccharide-induced inflammatory responses in macrophages. *J Biol Chem* **279:** 49995–50003.

Wissink S, van de Stolpe A, Caldenhoven E, Koenderman L, van der Saag PT. 1997. NF-κ B/Rel family members regulating the ICAM-1 promoter in monocytic THP-1 cells. *Immunobiology* **198:** 50–64.

Woronicz JD, Gao X, Cao Z, Rothe M, Goeddel DV. 1997. IκB kinase-β: NF-κB activation and complex formation with IκB kinase- and NIK. *Science* **278:** 866–869.

Wu C, Ghosh S. 1999. β-TrCP mediates the signal-induced ubiquitination of IκBβ. *J Biol Chem* **274:** 29591–29594.

Wulczyn FG, Naumann M, Scheidereit C. 1992. Candidate proto-oncogene bcl-3 encodes a subunit-specific inhibitor of transcription factor NF-κ B. *Nature* **358:** 597–599.

Xiao G, Fong A, Sun SC. 2004. Induction of p100 processing by NF-κB-inducing kinase involves docking IκB kinase (IKK) to p100 and IKK-mediated phosphorylation. *J Biol Chem* **279:** 30099–30105.

Yamamoto M, Yamazaki S, Uematsu S, Sato S, Hemmi H, Hoshino K, Kaisho T, Kuwata H, Takeuchi O, Takeshige K, et al. 2004. Regulation of Toll/IL-1-receptor-mediated gene expression by the inducible nuclear protein IκBzeta. *Nature* **430:** 218–222.

Yamamoto Y, Verma UN, Prajapati S, Kwak YT, Gaynor RB. 2003. Histone H3 phosphorylation by IKK- is critical for cytokine-induced gene expression. *Nature* **423:** 655–659.

Yamazaki S, Muta T, Matsuo S, Takeshige K. 2005. Stimulus-specific induction of a novel nuclear factor-κB regulator, IκB-zeta, via Toll/Interleukin-1 receptor is mediated by mRNA stabilization. *J Biol Chem* **280:** 1678–1687.

Yaron A, Hatzubai A, Davis M, Lavon I, Amit S, Manning AM, Andersen JS, Mann M, Mercurio F, Ben-Neriah Y. 1998. Identification of the receptor component of the IκBα-ubiquitin ligase. *Nature* **396:** 590–594.

Yeung F, Hoberg JE, Ramsey CS, Keller MD, Jones DR, Frye RA, Mayo MW. 2004. Modulation of NF-κB-dependent transcription and cell survival by the SIRT1 deacetylase. *EMBO J* **23:** 2369–2380.

Zandi E, Chen Y, Karin M. 1998. Direct phosphorylation of IκB by IKK and IKKβ: Discrimination between free and NF-κB-bound substrate. *Science* **281:** 1360–1363.

Zandi E, Rothwarf DM, Delhase M, Hayakawa M, Karin M. 1997. The IκB kinase complex (IKK) contains two kinase subunits, IKK and IKKβ, necessary for IκB phosphorylation and NF-κB activation. *Cell* **91:** 243–252.

Zhong H, Voll RE, Ghosh S. 1998. Phosphorylation of NF-κ B p65 by PKA stimulates transcriptional activity by promoting a novel bivalent interaction with the coactivator CBP/p300. *Mol Cell* **1:** 661–671.

Zhong H, May MJ, Jimi E, Ghosh S. 2002. The phosphorylation status of nuclear NF-κ B determines its association with CBP/p300 or HDAC-1. *Mol Cell* **9:** 625–636.

Specification of DNA Binding Activity of NF-κB Proteins

Fengyi Wan and Michael J. Lenardo

Laboratory of Immunology, National Institute of Allergy and Infectious Diseases, National Institutes of Health, Bethesda, Maryland 20892

Correspondence: lenardo@nih.gov

Nuclear factor-κB (NF-κB) is a pleiotropic mediator of inducible and specific gene regulation involving diverse biological activities including immune response, inflammation, cell proliferation, and death. The fine-tuning of the NF-κB DNA binding activity is essential for its fundamental function as a transcription factor. An increasing body of literature illustrates that this process can be elegantly and specifically controlled at multiple levels by different protein subsets. In particular, the recent identification of a non-Rel subunit of NF-κB itself provides a new way to understand the selective high-affinity DNA binding specificity of NF-κB conferred by a synergistic interaction within the whole complex. Here, we review the mechanism of the specification of DNA binding activity of NF-κB complexes, one of the most important aspects of NF-κB transcriptional control.

Nuclear factor-κB (NF-κB), a collective term for a family of transcription factors, was originally detected as a transcription-enhancing, DNA-binding complex governing the immunoglobulin (Ig) light chain gene intronic enhancer (Sen and Baltimore 1986; Lenardo et al. 1987). NF-κB is evolutionarily and structurally conserved and has representative members in a wide range of species. In essentially all unstimulated nucleated cells, NF-κB complexes are retained in latent cytoplasmic form through binding to a member of the inhibitor of NF-κB (IκB) proteins (Lenardo and Baltimore 1989; Lenardo et al. 1989; Hayden and Ghosh 2004; Hayden and Ghosh 2008). NF-κB induction typically occurs following the activation of the IκB kinase (IKK) signalosome, resulting in the phosphorylation and subsequent dispatch of the inhibitory IκBs to the proteasome for protein degradation (Hacker and Karin 2006). This cytoplasmic "switch" liberates NF-κB complexes for subsequent nuclear translocation and target gene transcription (Scheidereit 2006). It provides a pre-established genetic switch that is independent of new protein synthesis and triggered by a biochemical change in the cell. This adaptability and versatility no doubt underlies its broad use. A diverse spectrum of modulating stimuli can activate this pleiotropic transcription factor; furthermore, the fundamental use of NF-κB has been highlighted with an ever-increasing array of genetic targets, responsible for diverse biological activities including immune response, inflammation, cell proliferation, and death (Grilli et al. 1993) (also see http://www.nf-kb.org).

The best known subunits of mammalian NF-κB consist of five proteins in the Rel family: RelA (p65), RelB, c-Rel, p50, and p52, which are capable forming homo- and hetero-dimeric complexes in almost any combination (Hayden and Ghosh 2004). Each of these subunits harbors a prototypical amino-terminal sequence of roughly 300 amino acids, termed the Rel homology domain (RHD), that mediates dimerization, DNA-binding, nuclear localization, and cytoplasmic retention by IκBs (Rothwarf and Karin 1999; Chen and Greene 2004). In contrast, the transcription activation domain (TAD) necessary for the target gene expression is present only in the carboxyl terminus of p65, c-Rel, and RelB subunits. NF-κB complexes have long been thought to function dimerically; but functional and biochemical information belied this simple conceptualization. The native complex of NF-κB from nuclear extracts is more than 200 kDa, significantly higher than that reconstituted from purified p50 and p65 proteins (115 kDa) (Urban et al. 1991). Moreover, native NF-κB complexes have a >100-fold higher affinity for Ig κB motif DNA than reconstituted p65–p50 heterodimers (Phelps et al. 2000). A new study shows that another essential subunit of NF-κB complex, ribosomal protein S3 (RPS3), cooperates with Rel dimers to achieve full binding and transcriptional activity (Wan et al. 2007). As an integral component, RPS3 plays a critical role in determining the DNA binding affinity and specificity of NF-κB, which will be discussed in more detail in the following discussion (Wan et al. 2007). Therefore, the molecular machine known as NF-κB consists of both Rel and non-Rel subunits that actually comprise multiple protein complexes with different gene activation specificities, masquerading as a single NF-κB complex in the nucleus.

NF-κB exerts its fundamental role as transcription factor by binding to variations of the consensus DNA sequence of 5′-GGGRNYYYCC-3′ (in which R is a purine, Y is a pyrimidine, and N is any nucleotide) known as κB sites (Chen et al. 1998). How NF-κB selectively recognizes a small subset of relevant κB sites from the large excess of potential binding sites (about 1.4×10^4 estimated in human genome) is a critical step for stimulus-specific gene transcription. Increasing evidence suggests that specific chromatin modifications and configurations are required for NF-κB proteins to access the chromosomally embedded cognate κB motifs (Natoli et al. 2005; Natoli 2006). The presence of κB sites, however, appears to be a minimal requirement for NF-κB regulation but not sufficient for gene induction (Wan et al. 2007). We will attempt to decipher the elegant but recondite control of DNA binding activity of NF-κB proteins at multiple levels, which is one of the most important, yet complex, aspects of NF-κB function. This process, more abstruse than initially considered, involves IκBs, Rel subunits, Rel-associating proteins, and non-Rel subunits in both the cytoplasm and the nucleus, as well as complicated associations in the nuclear chromatin. This synopsis will highlight our current knowledge of the DNA binding activity of NF-κB complex, focusing on its liberation from cytoplasmic sequestration complexes to recruitment to cognate κB regulatory sites in the genome.

IκB FAMILY PROTEINS

The evidence that induction of NF-κB activity does not require new protein synthesis, and that the detergent deoxycholate liberated active κB-site DNA binding activity in cytosolic extracts from unstimulated cells, revealed a key property of this gene regulatory system (Baeuerle and Baltimore 1988). This established a new regulatory paradigm involving the specific and reversible DNA binding of NF-κB proteins governed by NF-κB inhibitor protein(s), known as IκB. The IκB protein family currently consists of eight members: IκBα, IκBβ, IκBγ, IκBε, IκBζ, Bcl-3, and the Rel protein precursors p105 and p100, all of which possess a characteristic structural feature of ankyrin repeats. The centerpiece of both classical and alternative NF-κB pathways is the IκBs, functioning at the primary level to regulate the DNA binding activity of NF-κB proteins.

Cite this article as *Cold Spring Harb Perspect Biol* 2009;1:a000067

Specific modification and subsequent degradation of IκBs is critical in regulating NF-κB DNA binding activity, particularly in the classic or canonical pathway (Table 1). In unstimulated cells, NF-κB complexes are sequestered with various binding preferences by the IκBs in the cytoplasm because IκBs physically mask the nuclear localization signal (NLS) of NF-κB. On stimulation, whether intra- or extracellular, multiple intracellular signaling pathways converge on a tripartite IκB kinase (IKK) complex consisting of two functionally nonredundant kinases IKKα and IKKβ, as well as a regulatory subunit IKKγ (Fig. 1A). These mediate the phosphorylation of IκBs at specific amino acid residues (for instance, Ser-32 and Ser-36 in IκBα) predominantly through the action of the IKKβ subunit in the activated IKK complex. These phosphorylation events are a prerequisite for their successive

Table 1. Post-translational modification of IκB family proteins

IκB family proteins	Target residues	Enzymes	References
Phosphorylation			
IκBα	S32 and S36	IKKβ	(Hayden and Ghosh 2004)
IκBα	S283, S289, T291, S293, and T299	CKII	(Lin et al. 1996; McElhinny et al. 1996; Schwarz et al. 1996)
IκBα	Y42	p56-lck, Syk, and c-Src	(Koong et al. 1994)
	Y42	Unknown	(Schoonbroodt et al. 2000)
	Y42 and Y305	Unknown	(Waris et al. 2003)
IκBα	Unknown	PI3K/Akt	(Sizemore et al. 1999)
IκBβ	S19 and S23	IKKβ	(Hayden and Ghosh 2004)
IκBβ	S313 and S315	CKII	(Chu et al. 1996)
IκBε	S18 and S22	IKKβ	(Hayden and Ghosh 2004)
Bcl-3	S394 and S398	GSK3β	(Viatour et al. 2004)
p100	S99, S108, S115, S123, S866, S870, and S872	IKKα	(Senftleben et al. 2001; Xiao et al. 2001; Xiao et al. 2004)
p105	S927 and S932	IKKβ	(Lang et al. 2003)
p105	S903 and S907	GSK3β	(Demarchi et al. 2003)
p105	S337	PKAc	(Hou et al. 2003)
Ubiquitination			
IκBα	K21 and K22	βTrCP	(Hayden and Ghosh 2004)
IκBβ	K6	βTrCP	(Hayden and Ghosh 2004)
IκBε	K6	βTrCP	(Hayden and Ghosh 2004)
p100	K856	βTrCP	(Amir et al. 2004)
p105	Multiple Ks	βTrCP	(Cohen et al. 2004)
Sumoylation			
IκBα	K21	Unknown	(Desterro et al. 1998)
p100	K90, K298, K689, and K863	Ubc9	(Vatsyayan et al. 2008)
Acetylation			
p100	Unknown	P300	(Hu and Colburn 2005; Deng et al. 2006)
p105	K431, K440, and K441	P300	(Furia et al. 2002; Deng and Wu 2003)
S-nitrosylation			
p105	C62		(Matthews et al. 1996)

IKKβ, IκB kinase β; CKII, Casein kinase II; p56-lck, lymphocyte-specific protein tyrosine kinase; Syk, spleen tyrosine kinase; c-Src, normal cellular Src kinase; PI3K, phosphoinositide 3-kinases; GSK3β, glycogen synthase kinase 3 β; IKKα, IκB kinase α; PKAc, catalytic subunit of protein kinase A; βTrCP, β-transducin repeat-containing protein; Ubc9, E2 small ubiquitin-like modifier (SUMO)-conjugating enzyme Ubc9; p300, E1A binding protein p300.

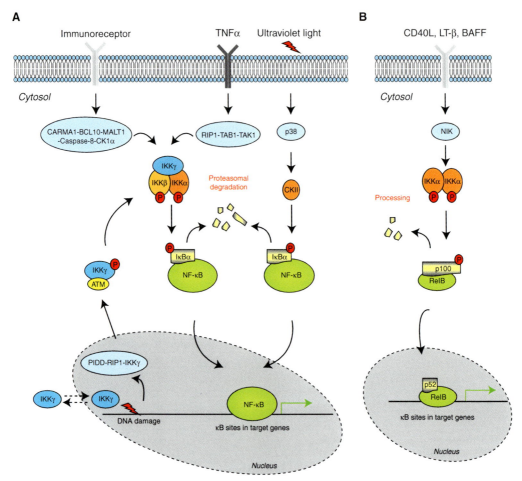

Figure 1. A schematic representation of how NF-κB DNA binding is regulated by IκB degradation and processing following activating stimuli. (*A*) IκB degradation mechanism that depends on site-specific phosphorylation of IκBs mediated by activated IKKβ in the IKK complex as well as CKII. Unique intracellular signaling complexes (which contains more components as shown, e.g., CARMA1-BCL10-MALT1-Caspase-8-CK1α complex downstream of T cell receptor [TCR] ligation [Bidere et al. 2009]) phosphorylate and activate the IKK complex depending on the initial stimulus (e.g., TNF, immunoreceptors, and DNA damage). IKKβ phosphorylates serine residues located in the amino-terminus of IκBs following TNF and TCR stimulation. In contrast, NF-κB induction by ultraviolet light and select chemotherapeutic agents requires the MAP kinase p38 and CKII. CKII phosphorylates serine residues located in the carboxyl terminus of IκBs. Phosphorylation of IκBs leads to successive ubiquitinylation and proteasomal degradation. The removal of IκBs liberates NF-κB complexes for nuclear translocation and binding to cognate regions in target genes. (*B*) The alternative pathway, which is induced by a few stimuli including CD40L, LT-β, and BAFF. NIK and IKKα are critical kinases for the phosphorylation of p100 at regulatory serine residues within its carboxyl terminus. This causes proteasomal degradation of the carboxyl terminus of the p100 molecule, generating a DNA binding NF-κB complex containing RelB and processed p52.

Cite this article as *Cold Spring Harb Perspect Biol* 2009;1:a000067

ubiquitinylation by β-transducin repeat-containing protein (βTrCP) and degradation in the 26S proteasome. The removal of IκBs liberates NF-κB for nuclear translocation and binding to cognate sites in target genes.

Additionally, various kinases other than the IKK complex may also mediate site-specific modification of IκBs, contributing to the diversity of signals regulating the latent DNA binding activity of NF-κB. In ultraviolet (UV) light-induced NF-κB activation, the phosphorylation-dependent IκBα degradation process does not rely on the IKK complex but the alternative serine/threonine kinase, casein kinase II (CKII) (Fig. 1A). Several studies have shown that CKII phosphorylates IκBα at a cluster of serine and threonine residues in the carboxyl terminus through a p38-MAPK kinase-dependent pathway, which is critical for degradation of IκBα and induced NF-κB DNA binding activity in cells exposed to UV light (Lin et al. 1996; McElhinny et al. 1996; Schwarz et al. 1996). Moreover, tyrosine kinases p56-lck, Syk, and c-Src and the serine/threonine kinase GSK3β have also been reported to phosphorylate IκBs and control the DNA binding activity of NF-κB (Table 1) (Koong et al. 1994; Viatour et al. 2004). Along with phosphorylation, modifications such as sumoylation, acetylation, and *s*-nitrosylation of the IκBs at specific residues can also trigger the degradation process (Table 1). However, these modifications lead to the final common pathway of ubiquitination of IκBs at specific lysine residues as the essential requirement for IκB degradation and release of NF-κB for nuclear translocation (Table 1).

p100 and p105, the precursor proteins for NF-κB subunits p52 and p50 respectively, contain multiple ankyrin repeats (similar to those of the IκB family) in their carboxy-terminal domains. In some circumstances, these keep NF-κB complexes cytoplasmically anchored. The regulation of NF-κB DNA binding activity by p105 mirrors IκBs of the classical pathway by undergoing inducible, complete degradation, particularly when it is bound to NF-κB complexes (Heissmeyer et al. 2001). By contrast, partial processing, but not full degradation, of p100 represents another mechanism by which IκB regulates the DNA binding activity of NF-κB proteins. This was discovered recently in an alternative pathway of NF-κB regulation involving RelB and p52 (Fig. 1B) (Hayden and Ghosh 2004). A few select stimuli including CD40 ligand (CD40L), lymphotoxin β (LT-β), and B-cell activating factor (BAFF) phosphorylate p100 on carboxy-terminal serine residues, triggering successive polyubiquitination and proteasomal degradation of the carboxyl terminus of the p100 molecule. The resulting complex, containing RelB and processed p52, functions as the main NF-κB species with selected DNA binding activity in the alternative pathway. Of interest, the kinase signalosome mediating p100 phosphorylation contains the NF-κB-inducing kinase (NIK) and IKKα, and functions completely independently of IKKβ and IKKγ. Moreover, recent evidence suggests that IKKα phosphorylates p100 at several serine residues within the RHD of p52, and these modifications can augment p52 DNA binding, in addition to inducing the efficient processing of p105 (Xiao et al. 2004).

Besides their hallmark sequestration function in cytoplasm, IκBs can also regulate the DNA binding properties of NF-κB proteins in the nucleus. IκBζ localizes to the nucleus following its inducible expression (Yamamoto et al. 2004) and Bcl-3 is also found in the nucleus associated with p50- and p52-containing NF-κB complexes (Cogswell et al. 2000). IκBs such as IκBα, IκBβ, and IκBε are transcriptional targets of NF-κB itself, creating a negative feedback loop. This facilitates restoration of latent NF-κB complexes in the cytoplasm and maintains the cell's responsiveness to subsequent stimuli. With variant kinetics of degradation and resynthesis, IκBs compete NF-κB complexes off their chromosomal locations and export them back into cytoplasm. By affecting the cytoplasmic–nuclear shuttling of NF-κB proteins, these titrate the occupancy of gene regulatory elements in nuclear DNA. Strikingly, IκBβ may also function to directly regulate NF-κB DNA binding activity on relevant promoters or enhancers, as evident by its stable nuclear association with NF-κB complexes that are already bound to κB sites in chromatin (Thompson et al. 1995; Suyang et al. 1996). Moreover, both IκBζ and Bcl-3 primarily interact with specific NF-κB complexes and subvert their DNA binding activity. IκBζ was reported to associate preferentially with p50 homodimers and also negatively regulate p65-containing NF-κB complexes (Kang et al. 1992; Motoyama et al. 2005; Hayden and Ghosh 2008), which suggests it may possess the capability to selectively inhibit or activate specific NF-κB species. However, the role of Bcl-3 in regulating DNA binding seems controversial. It has been proposed to remove the repressive p50 homodimers from κB DNA, allowing transcriptionally active NF-κB species access to those cognate elements (Lenardo and Siebenlist 1994; Hayden and Ghosh 2008). However, recent evidence indicates that Bcl-3 might also stabilize p50 homodimers and inhibit NF-κB activation by preventing transcriptionally active NF-κB complexes from binding to κB DNA (Carmody et al. 2007). Regardless of the inhibited or enhanced NF-κB transactivation consequences, IκBs regulate the DNA binding activity of NF-κB proteins by controlling the high-affinity interactions of NF-κB species to cognate DNA regulatory sites.

POST-TRANSLATIONAL MODIFICATIONS OF REL PROTEINS

Like IκB proteins, Rel subunits are subject to numerous post-translational modifications, which represents another mechanism of regulating NF-κB activation that has been extensively studied recently. These modifications, within conserved RHD, TAD or linker sequences, convey various physiological functions of NF-κB (Viatour et al. 2005; Perkins 2006; Neumann and Naumann 2007; O'Shea and Perkins 2008). This includes control of NF-κB DNA binding activity through modifications, particularly phosphorylation and acetylation, of Rel subunits or adjusting the association of NF-κB complexes to either IκBs or κB DNA (Table 2). This could be caused by steric alternations of the Rel proteins

themselves or to controlling their associations with other proteins in transcriptional regulatory complexes.

The phosphorylation of p65 under basal and activated conditions is by far the best-characterized modification among Rel proteins, and p65 is the transcriptionally active component of the NF-κB species that is most abundant and has the broadest function. So far, nine phosphorylation sites have been identified in p65 located in both the RHD and TAD, with several sites remarkably critical for DNA binding capability. Ser-536 phosphorylation by various kinases leads to a substantially weaker interaction of p65 with newly synthesized IκBα, leading to a dramatic decrease of nuclear p65 export, which enhances the duration of κB DNA access (Table 2). Furthermore, Ser-536-phosphorylated p65 does not interact with cytosolic IκBα at all, which could accelerate p65 nuclear localization and attachment to DNA (Sasaki et al. 2005). Phosphorylation of p65 at Ser-276 and Ser-536 together with successive acetylation at Lys-310 alters its affinity to IκB proteins, resulting in different kinetics of p65 cytoplasmic-nuclear shuttling and subsequent DNA binding activity (Chen et al. 2005). Phosphorylation of Thr-254 by the nuclear peptidylprolyl isomerase Pin1, followed by isomerization of specific amino acid residues, also strongly increases p65 nuclear translocation and DNA binding activity. Pin1-induced conformational changes increase the structural stability of p65 with dramatic decrease in IκBα affinity (Ryo et al. 2003). Recently, phosphorylation of Ser-276 by the mitogen- and stress-activated protein kinase-1 (MSK1) was shown to be essential for p65 binding to the κB intronic enhancer site of the mast cell growth factor *SCF* (stem cell factor) gene in inflammation (Reber et al. 2009). Phosphorylation of p65 can also repress the transcriptional activity of other Rel subunits. For instance, Ser-536-phosphorylated p65 was reported to specifically interact with RelB in the nucleus, forming a complex that cannot bind to κB DNA (Jacque et al. 2005).

Post-translational modification of c-Rel and its effect on DNA binding activity is similar

Table 2. Post-translational modifications of Rel subunits that regulate NF-κB DNA binding activity

Rel family proteins	Target residues	Enzymes	Functional effect on DNA binding activity	References
Phosphorylation				
p65	T254	Unknown	Enhancement	(Ryo et al. 2003)
	S529	CKII	Unknown	(Bird et al. 1997; Wang et al. 2000; O'Mahony et al. 2004)
	Unknown	PI3K/Akt	Enhancement	(Sizemore et al. 1999)
	S276	MSK1	Enhancement	(Reber et al. 2009)
	S536	IKKα	Enhancement	(Jiang et al. 2003; O'Mahony et al. 2004)
		IKKβ	Enhancement	(Sakurai et al. 1999)
		IKKε	Enhancement	(Buss et al. 2004; Adli and Baldwin 2006; Mattioli et al. 2006)
		TBK1	Enhancement	(Fujita et al. 2003; Buss et al. 2004)
		RSK1	Enhancement	(Bohuslav et al. 2004)
c-Rel	Unknown	TBK1	Enhancement	(Harris et al. 2006)
	Unknown	IKKε	Enhancement	(Sanchez-Valdepenas et al. 2006)
p50	S337	PKAc	Enhancement	(Guan et al. 2005)
Acetylation				
p65	K218, K221, and K310	p300	Enhancement	(Chen et al. 1998)
	K122 and K123	PCAF and p300	Reduce	(Kiernan et al. 2003)
p50	K431, K440, and K441	p300	Enhancement	(Furia et al. 2002)

CKII, Casein kinase II; MSK1, Mitogen- and stress-activated protein kinase-1; PI3K, Phosphoinositide 3-kinases; IKKα, IκB kinase α; IKKβ, IκB kinase β; IKKε, IκB kinase epsilon; TBK1, TANK-binding kinase 1; RSK1, Ribosomal protein S6 kinase; 90 kDa, polypeptide 1; PKAc, catalytic subunit of protein kinase A; p300, E1A binding protein p300; PCAF, p300/CREB binding protein-associated factor.

to p65, but with fewer characterized kinase(s) or target residue(s). c-Rel can be specifically phosphorylated by the serine/threonine kinases NIK, TBK1, and IKKε on its carboxyl terminus, which dissociates the IκBα-c-Rel complex and enhances binding to cognate DNA (Harris et al. 2006; Sanchez-Valdepenas et al. 2006). Furthermore, tyrosine phosphorylation of c-Rel within its carboxyl terminus also increased its ability to bind κB sites in human neutrophils stimulated with granulocyte-colony-stimulating factor (Druker et al. 1994). More importantly, dysregulated Ser-525 phosphorylation located within c-Rel TAD, caused by a serine to proline point mutation discovered in two patients with follicular and mediastinal B-cell lymphoma, might slightly enhance c-Rel DNA binding activity, at least

for κB motifs (Starczynowski et al. 2007). These data show that the carboxy-terminal modifications, especially in the TAD, are vital in regulating c-Rel DNA binding activity.

The RelB monomer is easily degraded and preferentially binds to p52 and p50, making it an unusual NF-κB subunit (Hayden and Ghosh 2004). Although the kinases have not been identified yet, phosphorylation of RelB at multiple residues including Thr-84, Ser-254, Ser-368, and Ser-552, have been reported as essential for its dimerization and degradation (Viatour et al. 2005; Neumann and Naumann 2007). It certainly deserves further investigation whether these or other unknown modifications of RelB regulate DNA binding, given the crucial role of RelB in the alternative pathway of NF-κB activation. Although inducible phosphorylation of p50, similar to its precursor p105, has been noticed in various cell types (Neumann and Naumann 2007), site-specific modifications of transactivation repressive p50 and p52 have been rarely reported. One recent study showed the catalytic subunit of PKA (PKAc) specifically phosphorylated mature p50 at Ser-337 located within RHD and augmented DNA binding activity, which may be important for maintaining stable negative regulation of NF-κB gene expression in unstimulated cells (Guan et al. 2005).

Acetylation is another key modification of Rel subunits that regulates their association with IκBs and direct binding to cognate DNA (Table 2). Five residues, Lys-122, Lys-123, Lys-218, Lys-221, and Lys-310, can be specifically acetylated in p65 in response to TNFα stimulation. Acetylation at Lys-218 and Lys-221 markedly diminishes the binding of p65 to IκBα thereby increasing its DNA binding (Chen et al. 2002). In particular, Lys-221 of p65 directly interacts with Met-279 of IκBα in the crystal structure of p65-p50-IκBα complex (Huxford et al. 1998); therefore, acetylation of Lys-221 may result in a conformational change in p65 and lessen its interaction with IκBα (Chen et al. 2002). In contrast, acetylation of p65 at Lys-122 and Lys-123 accelerates its dissociation from DNA and successive nuclear export by IκBα, attenuating NF-κB transactivation (Kiernan et al. 2003). In addition, acetylation of Rel subunits may modulate their binding with DNA directly. Acetylation of p65 at Lys-221 apparently causes a conformation change favoring κB DNA binding. This is supported by the crystal structure of p65-p50-κB DNA complex illustrating that Lys-221 directly contacts the DNA backbone (Chen et al. 1998). Various mutational analyses strengthen the concept that this acetylation influences DNA-binding properties. Mutation of Lys-221 produces a sharp decline in the DNA binding affinity of p65 homodimers and a substantial decrease in that of p65-p50 heterodimers (Chen et al. 2002). Strikingly, p50 can be acetylated in vitro by p300/CBP at Lys-431, Lys-440, and Lys-441, and this modification augments its DNA binding properties, as evidenced by pull-down assays using κB oligonucleotides (Furia et al. 2002). Collectively, site-specific acetylation of Rel proteins, at least p65 and p50, augments their intrinsic DNA binding activity, as shown for other transcription factors (Boyes et al. 1998).

REL SUBUNIT-ASSOCIATING PROTEINS

For full NF-κB transactivation to occur at a given target DNA locus, a successive enhanceosome complex needs to be assembled by multiple coactivators, corepressors, other transcription factors, and basal transcription machinery proteins through protein–protein interactions with Rel subunits (Hayden and Ghosh 2004; Hayden and Ghosh 2008). For instance, over 100 proteins can modulate NF-κB association with chromatin or the assembly of an NF-κB enhanceosome via their interaction with full length, RHD, TAD, or the central region of p65 (O'Shea and Perkins 2008). Until a short time ago, Rel subunits were widely regarded as the only executers to recognize and bind κB DNA. However, several Rel-associating proteins, albeit with no κB DNA binding capability themselves, have recently been illustrated to be essential in regulating NF-κB binding to cognate DNA (Table 3). These Rel-associating proteins feature restricted tissue distribution, specific expression profiles, or unique pathophysiological circumstances and inhibit, as currently reported, the DNA binding activity of NF-κB proteins. This establishes Rel-associating proteins as a group of tissue- or cell context-specific negative regulators of NF-κB beyond the IκB family, adding another level of complexity to the exquisite spatiotemporal control of DNA binding activity of NF-κB proteins.

Some Rel-associating proteins attenuate NF-κB DNA binding in cotransfection experiments. RelA-associated inhibitor (RAI), isolated in a yeast two-hybrid screen utilizing the central region of p65 as bait, interacts with p65 in vitro and in vivo, and specifically inhibits p65 binding to the cognate DNA when cotransfected in human embryonic kidney 293 cells (Yang et al. 1999). A similar function was ascribed to the aryl hydrocarbon receptor

Cite this article as *Cold Spring Harb Perspect Biol* 2009;1:a000067

Table 3. Rel-associating proteins that regulate NF-κB DNA binding activity

NF-κB subunits	Associating/ interacting proteins	Domain/portion of Rel proteins sufficient for binding	Evidence for integration into DNA binding complexes	Functional effect on DNA binding activity	References
p65					
	RAI	RHD, aa 176–405	No	Inhibition	(Yang et al. 1999)
	AhR	Not mapped	No	Inhibition	(Tian et al. 1999; Kim et al. 2000; Ruby et al. 2002)
	β-Catenin	Not mapped	No	Inhibition	(Deng et al. 2002)
	Cdk9	Not mapped	No	Inhibition	(Amini et al. 2002)
	PIAS1	TAD, aa 299–551	No	Inhibition	(Liu et al. 2005)
	VP1686	Not mapped	No	Inhibition	(Bhattacharjee et al. 2006)
	Myocardin	RHD, aa 1–276	No	Inhibition	(Tang et al. 2008)
	RPS3[a]	RHD, aa 21–186	Yes	Enhancement	(Wan et al. 2007)
p50					
	β3-Endonexin	Not mapped	No	Inhibition	(Besta et al. 2002)
	Cdk9	Not mapped	No	Inhibition	(Amini et al. 2002)
	β-Catenin	Not mapped	No	Inhibition	(Deng et al. 2002)

RAI, RelA associated inhibitor; AhR, Aryl hydrocarbon receptor; Cdk9, Cyclin-dependent kinase 9; PIAS1, Protein inhibitor of activated STAT1; VP1686, *Vibrio parahaemolyticus* type III secretion protein; RPS3, Ribosomal protein S3; β3-Endonexin, Integrin β 3 binding protein.

[a]RPS3, an integral subunit of certain NF-κB DNA binding complex that also physically interacts with p65, is listed here to compare with Rel-associating proteins in their DNA-binding regulatory functions.

(AhR), a ligand-activated transcription factor. When cotransfected in 293 cells, AhR specifically inhibited the DNA binding activity of p65, but not the p50–p50 complex (Tian et al. 1999; Ruby et al. 2002), suggesting that synergic interaction between AhR and NF-κB, mainly the p65 subunit, is critical for suppression of immune responses and xenobiotic metabolism. Some pathogen-encoded proteins also possess the ability to regulate host NF-κB binding to cognate DNA, hampering immune responses directed against the microbes that express them. For instance, *Vibrio parahaemolyticus* is a causative agent of human gastrointestinal diseases and significantly suppresses the induction of the DNA binding activity of NF-κB via the physical interaction between its effector protein VP1686 and p65. Such attenuation of NF-κB DNA binding activity by VP1686 is sufficient to sensitize infected-macrophages for death, because of the diminished expression of many antiapoptosis-related NF-κB target genes (Bhattacharjee et al. 2006).

The regulatory effects of Rel-associating proteins have been studied not only in the cellular context of NF-κB signaling, but also in in vitro biochemical investigations with purified proteins. One of the first studies described a surprising cross-regulation of NF-κB by β-catenin (Deng et al. 2002). Both p65 and p50 could complex with β-catenin independently of IκBα, but seemed to require additional cellular factors. β-catenin markedly attenuated the DNA binding of both p50–p65 and p50–p50 complexes as shown by electrophoretic mobility shift assays (EMSA), causing reduced NF-κB target gene expression. Of note, β-catenin is not directly integrated into the NF-κB-DNA complex in spite of its strong association with NF-κB subunits, suggesting it may interact with NF-κB proteins to disrupt their DNA binding ability (Deng et al. 2002).

Inhibition of NF-κB DNA binding activity by β-catenin could be important in oncogenesis, as hinted at by an inverse correlation between the β-catenin expression levels and levels of the NF-κB target gene *Fas* in colon and breast tumor tissues (Deng et al. 2002). Furthermore, the inhibitory effects on NF-κB DNA binding by Cyclin-dependent kinase 9 (Cdk9), the PIAS1 (protein inhibitor of activated STAT [signal transducers and activator of transcription] 1) protein, and myocardin were also shown using EMSA, in which these proteins significantly reduced the DNA binding activity of NF-κB (Amini et al. 2002; Liu et al. 2005; Tang et al. 2008). Cdk9 markedly suppressed the association of the p50–p50, p65–p65, and p50–p65 complexes to κB DNA, and the ability of NF-κB to modulate HIV-1 gene transcription was controlled by this inhibitory function (Amini et al. 2002). Both PIAS1 and myocardin specifically inhibited the DNA binding activity of p65-containing complexes, serving as negative regulators of NF-κB in certain conditions. PIAS1 was originally identified in the Jak/STAT signaling pathway, whereas myocardin is expressed specifically in cardiac and smooth muscle cells (Liu et al. 2005; Tang et al. 2008). Of interest, their effects on NF-κB DNA binding appear essential for tuning cytokine-induced NF-κB target gene expression and cardiomyocyte proliferation and differentiation, respectively (Liu et al. 2005; Tang et al. 2008).

As described above, β-catenin and Cdk9 both complex with p50 and dramatically inhibit its DNA binding activity (Amini et al. 2002; Deng et al. 2002). β3-endonexin is another p50-interacting molecule that inhibits p50–p65 complex binding to κB DNA. Moreover, binding of β3-endonexin to p50 was inhibited in the presence of wild-type but not mutated κB oligonucleotides, suggesting a steric competition between β3-endonexin and κB DNA for the p50–p65 complex. Despite the association with the transcription repressive p50 subunit, β3-endonexin negatively regulates expression of the urokinase-type plasminogen activator receptor that is essential for endothelial migration (Besta et al. 2002).

INTEGRAL NON-REL SUBUNITS IN NF-κB COMPLEXES

As for the unresolved and important question of how regulatory specificity of NF-κB is achieved, it has long been regarded that the variability of κB sequences may govern the usage of certain Rel dimers at specific promoters. Principally, each κB site variant could preferentially recruit one type of Rel dimer over other species (Natoli et al. 2005; Natoli 2006). Selective NF-κB gene expression, however, cannot be completely explained by a simple correlation between the sequence of κB sites in target genes and the requirement for a specific Rel dimer (Hoffmann et al. 2003). However, κB sites may still impart a specific configuration for NF-κB binding, because a single nucleotide change in an NF-κB binding site affected the formation of productive interactions between Rel dimer and coactivators (Leung et al. 2004). Therefore, other protein components beyond Rel subunits could form integral parts of the NF-κB binding complex, thus controlling its recognition and action on target genes. Support for this hypothesis comes from observations made in seminal previous studies. First, the contradiction in the size of native NF-κB from nuclear extracts (> 200 kDa) and that reconstituted from purified p50 and p65 proteins (115 kDa) implies the presence of other proteins in the native complex (Urban et al. 1991). Second, reconstituted p65–p50 heterodimers from purified proteins have a >100-fold lower affinity for DNA than native NF-κB complexes, at least for binding to the Ig κB motif (Phelps et al. 2000). Third, distinct variants of the κB motifs display different responses to various NF-κB inducers in different cell types implying selective regulation. Finally, a large number of NF-κB binding motifs beyond either canonical or variant κB site sequences were revealed throughout the human genome by mapping with chromatin immunoprecipitation-coupled microarray or sequencing (Martone et al. 2003; Schreiber et al. 2006; Lim et al. 2007). Collectively, these findings strongly suggested that other non-Rel proteins could not only regulate NF-κB DNA binding activity, but also participate in DNA binding as an essential component.

This hypothesis was confirmed by a recent study showing that RPS3, an integral non-Rel subunit in certain NF-κB DNA binding complexes, is essential for the recruitment of NF-κB p65 to selected κB sites (Wan et al. 2007). RPS3 prominently features a heterogeneous nuclear protein K (hnRNP K) homology (KH) domain, a structural motif that binds single-stranded RNA and DNA with some sequence specificity (Siomi et al. 1993). Indeed, the KH domain within RPS3 is essential for association with p65 (Wan et al. 2007). This study underscored the inherent complexity of NF-κB binding to κB sites by demonstrating that DNA binding capability is not conferred strictly by Rel subunits, as has been long assumed. Rather, the integral non-Rel subunit RPS3 represents a newly recognized subunit that potently contributes to DNA binding

activity. These observations suggest a new regulatory paradigm in which DNA binding activity could be regulated within NF-κB complexes through synergistic interactions between Rel and non-Rel subunits (Fig. 2).

RPS3 was found to physically interact with p65 in a proteomic screen, and shown to be critical in NF-κB transactivation. Under conditions of reduced RPS3, the DNA binding activity of NF-κB complexes was significantly attenuated. Knockdown of RPS3 resulted in failed recruitment of p65 to selected endogenous gene regulatory sites and abortive induction, despite normal p65 nuclear translocation. Therefore, RPS3 facilitates p65 binding to cognate DNA, which is essential for normal expression of specific NF-κB target genes involved in key physiological processes. This was dramatically shown for the induction of immunoglobulin κ light chain gene expression in B cells and cell proliferation and cytokine secretion in T cells that were

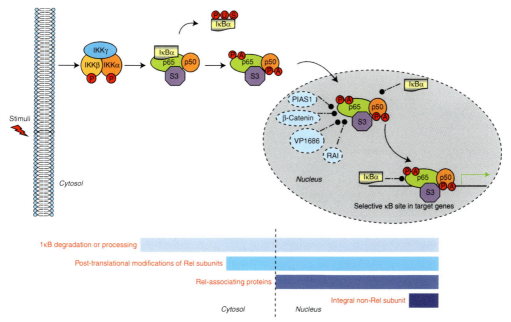

Figure 2. Complex regulation of NF-κB DNA binding activity. As illustrated in one of the most abundant NF-κB complexes, the DNA binding activity of NF-κB is controlled in both the cytoplasm and nucleus at multiple levels: involving degradation or processing of inhibitory IκBs; the post-translational modification of Rel subunits; Rel-associating proteins that modulate NF-κB DNA binding potential; and an integral non-Rel subunit RPS3 required for selective NF-κB target gene transcription based on enhanced DNA binding affinity. The bars represent the intracellular locations where the DNA binding activity of NF-κB proteins is regulated by indicated proteins.

markedly impaired by RPS3 knockdown (Wan et al. 2007). Strikingly, purified RPS3 protein exerted a dramatic synergistic effect on the DNA binding activity of both p65–p65 and p50–p65, but not p50–p50 complexes in EMSA (Wan et al. 2007). As a DNA-binding facilitator, RPS3 dramatically stabilizes NF-κB association with certain cognate sites. This could explain the extremely high affinity of semipurified NF-κB complexes for DNA, which is not manifested by complexes formed solely of purified p50 and p65 subunits. By contrast, none of the aforementioned Rel-associating proteins has been shown to integrate into NF-κB DNA binding complexes, and all of them inhibit the p65 DNA binding activity, although several were assessed in vitro using recombinant proteins in EMSA (Amini et al. 2002; Deng et al. 2002; Liu et al. 2005; Tang et al. 2008).

It is important to recognize that RPS3 is not an NF-κB-associated transcriptional coactivator, which by definition reorganizes chromatin templates and recruits the basal transcriptional machinery to the promoter region. RPS3 possesses little, if any, intrinsic transcriptional activating ability in a standard coactivation assay (Wan et al. 2007). Furthermore, the ability of RPS3-specific antibodies to dramatically supershift or diminish p65-containing DNA complexes in EMSAs strongly suggests that RPS3 is an integral part of NF-κB DNA binding complexes (Wan et al. 2007). By contrast, administration of a specific antibody against p300, one of well-characterized transcriptional coactivators that complex with NF-κB, did not alter NF-κB-DNA complexes, suggesting that p300 is not incorporated into the DNA binding complex (Deng et al. 2002). Further lines of evidence support the notion that RPS3 is an integral subunit of NF-κB: RPS3 physically associates with p65, p50, and IκBα in resting cells (guided through its interaction with p65); RPS3 can specifically translocate to the nucleus in response to T cell receptor (TCR) and TNFα stimulation; RPS3 is recruited to κB sites in a large number of NF-κB-driven genes in vivo upon stimulation; and RPS3 and p65 are significantly correlated in transcribing

a subset of TCR ligation-induced NF-κB genes (Wan et al. 2007). Whether the RPS3 subunit is essential in other stimuli-induced NF-κB signal pathways and whether it targets certain NF-κB complexes to specific κB sites in different cell types certainly deserves further investigation. Because RPS3 is only required for selected particular genomic κB sites to be activated under certain conditions and preferentially directs binding to κB sites with some sequence specificity, we call it a "specifier" subunit of NF-κB. This also lends credence to the idea that there are actually multiple molecular complexes containing RPS3-like "specifier" subunits with different gene activation specificities that all masquerade as single NF-κB complex in the nucleus. Indeed, another KH domain protein, Src-associated in mitosis, 68 kDa (Sam68), was found to be essential for some NF-κB gene transcription where RPS3 is not required (F. Wan and M. Lenardo, unpubl.). Sequence specificity preferred by various KH domains could confer different gene activation patterns via diverse NF-κB complexes. These findings may unveil a novel regulatory paradigm in which KH domain proteins serve as essential functional components in regulating the DNA binding activity of not only NF-κB, but also other transcription factors, because several other KH domain proteins have been shown to bind to DNA recognition motifs and to promote transcription (Tomonaga and Levens 1996; Ostrowski et al. 2003; Moumen et al. 2005).

CONCLUDING REMARKS

Since it was originally identified as a regulator of κ light chain expression in B cells over 20 years ago, NF-κB has served as a paradigm for signaling associated with inflammation, autoimmunity, and cancer. Recruitment of NF-κB proteins to regulatory DNA sites within the chromatin is fundamental and crucial for their target gene transcription. The complexity inherent in the DNA binding activity of NF-κB proteins is essential for achieving a fine-tuned regulatory specificity. An increasing body of literature illustrates that the DNA binding activity of NF-κB proteins can be elegantly and specifically controlled at multiple levels by different protein subsets, including IκBs, Rel subunits, Rel-associating proteins, and integral non-Rel subunits (Fig. 2). In particular, the expanding list of Rel-associating proteins constitutes a unique category of tissue- or cell type-specific negative regulators of NF-κB beyond IκBs that control DNA binding activity under certain circumstances. Furthermore, the recent identification of a non-Rel subunit of NF-κB itself provides a new way to understand the selective high-affinity DNA binding specificity of NF-κB conferred by a synergistic interaction within the whole complex.

Despite extensive studies on the control of NF-κB DNA binding activity, numerous issues are still unresolved and warrant further investigation. These include the identity of regulatory kinases and site-specific modulating target residues in IκBs that mediate their degradation/processing; additional post-translational modifications of Rel subunits beyond p65 and their affect on DNA binding; and specific associating and regulatory proteins for RelB and c-Rel. Because the inclusion of RPS3 does not fully explain the size of native NF-κB or its high affinity to cognate κB sites, the identification of other unknown non-Rel subunits of NF-κB is well worth further investigation. Undoubtedly, a more complete understanding of the complex control of the DNA binding activity of NF-κB proteins will not only revise or add to our fundamental knowledge of gene regulation, but also elucidate novel target molecules for pharmacological interventions.

ACKNOWLEDGMENTS

We thank Andrew Snow and Amanda Weaver for critical reading of the manuscript. The work in the authors' laboratory is supported by the Intramural Research Program of the National Institutes of Health (NIH), NIAID. F.W. is a recipient of NIH grant K99CA137171. We apologize to those who made also important contributions to the issues discussed and who could not be cited because of space limits.

REFERENCES

Adli M, Baldwin AS. 2006. IKK-i/IKKepsilon controls constitutive, cancer cell-associated NF-κB activity via regulation of Ser-536 p65/RelA phosphorylation. *J Biol Chem* **281:** 26976–26984.

Amini S, Clavo A, Nadraga Y, Giordano A, Khalili K, Sawaya BE. 2002. Interplay between cdk9 and NF-κB factors determines the level of HIV-1 gene transcription in astrocytic cells. *Oncogene* **21:** 5797–5803.

Amir RE, Haecker H, Karin M, Ciechanover A. 2004. Mechanism of processing of the NF-κB2 p100 precursor: Identification of the specific polyubiquitin chain-anchoring lysine residue and analysis of the role of NEDD8-modification on the SCF(β-TrCP) ubiquitin ligase. *Oncogene* **23:** 2540–2547.

Baeuerle PA, Baltimore D. 1988. I κB: A specific inhibitor of the NF-κB transcription factor. *Science* **242:** 540–546.

Besta F, Massberg S, Brand K, Muller E, Page S, Gruner S, Lorenz M, Sadoul K, Kolanus W, Lengyel E, et al. 2002. Role of β(3)-endonexin in the regulation of NF-κB-dependent expression of urokinase-type plasminogen activator receptor. *J Cell Sci* **115:** 3879–3888.

Bhattacharjee RN, Park KS, Kumagai Y, Okada K, Yamamoto M, Uematsu S, Matsui K, Kumar H, Kawai T, Iida T, et al. 2006. VP1686, a *Vibrio* type III secretion protein, induces toll-like receptor-independent apoptosis in macrophage through NF-κB inhibition. *J Biol Chem* **281:** 36897–36904.

Bidere N, Ngo VN, Lee J, Collins C, Zheng L, Wan F, Davis RE, Lenz G, Anderson DE, Arnoult D, et al. 2009. Casein kinase 1α governs antigen-receptor-induced NF-κB activation and human lymphoma cell survival. *Nature* **458:** 92–96.

Bird TA, Schooley K, Dower SK, Hagen H, Virca GD. 1997. Activation of nuclear transcription factor NF-κB by interleukin-1 is accompanied by casein kinase II-mediated phosphorylation of the p65 subunit. *J Biol Chem* **272:** 32606–32612.

Bohuslav J, Chen LF, Kwon H, Mu Y, Greene WC. 2004. p53 induces NF-κB activation by an IκB kinase-independent mechanism involving phosphorylation of p65 by ribosomal S6 kinase 1. *J Biol Chem* **279:** 26115–26125.

Boyes J, Byfield P, Nakatani Y, Ogryzko V. 1998. Regulation of activity of the transcription factor GATA-1 by acetylation. *Nature* **396:** 594–598.

Buss H, Dorrie A, Schmitz ML, Hoffmann E, Resch K, Kracht M. 2004. Constitutive and interleukin-1-inducible phosphorylation of p65 NF-κB at serine 536 is mediated by multiple protein kinases including IκB kinase (IKK)-α, IKKβ, IKKepsilon, TRAF family member-associated (TANK)-binding kinase 1 (TBK1), and an unknown kinase and couples p65 to TATA-binding protein-associated factor II31-mediated interleukin-8 transcription. *J Biol Chem* **279:** 55633–55643.

Carmody RJ, Ruan Q, Palmer S, Hilliard B, Chen YH. 2007. Negative regulation of toll-like receptor signaling by NF-κB p50 ubiquitination blockade. *Science* **317:** 675–678.

Chen LF, Greene WC. 2004. Shaping the nuclear action of NF-κB. *Nat Rev Mol Cell Biol* **5:** 392–401.

Chen FE, Huang DB, Chen YQ, Ghosh G. 1998. Crystal structure of p50/p65 heterodimer of transcription factor NF-κB bound to DNA. *Nature* **391:** 410–413.

Chen LF, Mu Y, Greene WC. 2002. Acetylation of RelA at discrete sites regulates distinct nuclear functions of NF-κB. *Embo J* **21:** 6539–6548.

Chen LF, Williams SA, Mu Y, Nakano H, Duerr JM, Buckbinder L, Greene WC. 2005. NF-κB RelA phosphorylation regulates RelA acetylation. *Mol Cell Biol* **25:** 7966–7975.

Chu ZL, McKinsey TA, Liu L, Qi X, Ballard DW. 1996. Basal phosphorylation of the PEST domain in the IκBβ regulates its functional interaction with the c-rel proto-oncogene product. *Mol Cell Biol* **16:** 5974–5984.

Cogswell PC, Guttridge DC, Funkhouser WK, Baldwin AS Jr. 2000. Selective activation of NF-κB subunits in human breast cancer: Potential roles for NF-κB2/p52 and for Bcl-3. *Oncogene* **19:** 1123–1131.

Cohen S, Achbert-Weiner H, Ciechanover A. 2004. Dual effects of IκB kinase β-mediated phosphorylation on p105 Fate: SCF(β-TrCP)-dependent degradation and SCF(β-TrCP)-independent processing. *Mol Cell Biol* **24:** 475–486.

Demarchi F, Bertoli C, Sandy P, Schneider C. 2003. Glycogen synthase kinase-3 β regulates NF-κB1/p105 stability. *J Biol Chem* **278:** 39583–39590.

Deng WG, Wu KK. 2003. Regulation of inducible nitric oxide synthase expression by p300 and p50 acetylation. *J Immunol* **171:** 6581–6588.

Deng J, Miller SA, Wang HY, Xia W, Wen Y, Zhou BP, Li Y, Lin SY, Hung MC. 2002. β-catenin interacts with and inhibits NF-κB in human colon and breast cancer. *Cancer Cell* **2:** 323–334.

Deng WG, Tang ST, Tseng HP, Wu KK. 2006. Melatonin suppresses macrophage cyclooxygenase-2 and inducible nitric oxide synthase expression by inhibiting p52 acetylation and binding. *Blood* **108:** 518–524.

Desterro JM, Rodriguez MS, Hay RT. 1998. SUMO-1 modification of IκBα inhibits NF-κB activation. *Mol Cell* **2:** 233–239.

Druker BJ, Neumann M, Okuda K, Franza BR Jr, Griffin JD. 1994. rel Is rapidly tyrosine-phosphorylated following granulocyte-colony stimulating factor treatment of human neutrophils. *J Biol Chem* **269:** 5387–5390.

Fujita F, Taniguchi Y, Kato T, Narita Y, Furuya A, Ogawa T, Sakurai H, Joh T, Itoh M, Delhase M, et al. 2003. Identification of NAP1, a regulatory subunit of IκB kinase-related kinases that potentiates NF-κB signaling. *Mol Cell Biol* **23:** 7780–7793.

Furia B, Deng L, Wu K, Baylor S, Kehn K, Li H, Donnelly R, Coleman T, Kashanchi F. 2002. Enhancement of nuclear factor-κ B acetylation by coactivator p300 and HIV-1 Tat proteins. *J Biol Chem* **277:** 4973–4980.

Grilli M, Chiu JJ, Lenardo MJ. 1993. NF-κB and Rel: Participants in a multiform transcriptional regulatory system. *Int Rev Cytol* **143:** 1–62.

Guan H, Hou S, Ricciardi RP. 2005. DNA binding of repressor nuclear factor-κB p50/p50 depends on phosphorylation of Ser337 by the protein kinase A catalytic subunit. *J Biol Chem* **280:** 9957–9962.

Hacker H, Karin M. 2006. Regulation and function of IKK and IKK-related kinases. *Sci STKE* **2006:** re13.

Harris J, Oliere S, Sharma S, Sun Q, Lin R, Hiscott J, Grandvaux N. 2006. Nuclear accumulation of cRel following C-terminal phosphorylation by TBK1/IKK epsilon. *J Immunol* **177:** 2527–2535.

Hayden MS, Ghosh S. 2004. Signaling to NF-κB. *Genes Dev* **18:** 2195–2224.

Hayden MS, Ghosh S. 2008. Shared principles in NF-κB signaling. *Cell* **132:** 344–362.

Heissmeyer V, Krappmann D, Hatada EN, Scheidereit C. 2001. Shared pathways of IκB kinase-induced SCF(βTrCP)-mediated ubiquitination and degradation for the NF-κB precursor p105 and IκBα. *Mol Cell Biol* **21:** 1024–1035.

Hoffmann A, Leung TH, Baltimore D. 2003. Genetic analysis of NF-κB/Rel transcription factors defines functional specificities. *Embo J* **22:** 5530–5539.

Hou S, Guan H, Ricciardi RP. 2003. Phosphorylation of serine 337 of NF-κB p50 is critical for DNA binding. *J Biol Chem* **278:** 45994–45998.

Hu J, Colburn NH. 2005. Histone deacetylase inhibition down-regulates cyclin D1 transcription by inhibiting nuclear factor-κB/p65 DNA binding. *Mol Cancer Res* **3:** 100–109.

Huxford T, Huang DB, Malek S, Ghosh G. 1998. The crystal structure of the IκBα/NF-κB complex reveals mechanisms of NF-κB inactivation. *Cell* **95:** 759–770.

Jacque E, Tchenio T, Piton G, Romeo PH, Baud V. 2005. RelA repression of RelB activity induces selective gene activation downstream of TNF receptors. *Proc Natl Acad Sci U S A* **102:** 14635–14640.

Jiang X, Takahashi N, Matsui N, Tetsuka T, Okamoto T. 2003. The NF-κB activation in lymphotoxin β receptor signaling depends on the phosphorylation of p65 at serine 536. *J Biol Chem* **278:** 919–926.

Kang SM, Tran AC, Grilli M, Lenardo MJ. 1992. NF-κB subunit regulation in nontransformed CD4+ T lymphocytes. *Science* **256:** 1452–1456.

Kiernan R, Bres V, Ng RW, Coudart MP, El Messaoudi S, Sardet C, Jin DY, Emiliani S, Benkirane M. 2003. Post-activation turn-off of NF-κB-dependent transcription is regulated by acetylation of p65. *J Biol Chem* **278:** 2758–2766.

Kim DW, Gazourian L, Quadri SA, Romieu-Mourez R, Sherr DH, Sonenshein GE. 2000. The RelA NF-κB subunit and the aryl hydrocarbon receptor (AhR) cooperate to transactivate the c-myc promoter in mammary cells. *Oncogene* **19:** 5498–5506.

Koong AC, Chen EY, Giaccia AJ. 1994. Hypoxia causes the activation of nuclear factor κB through the phosphorylation of I κ B α on tyrosine residues. *Cancer Res* **54:** 1425–1430.

Lang V, Janzen J, Fischer GZ, Soneji Y, Beinke S, Salmeron A, Allen H, Hay RT, Ben-Neriah Y, Ley SC. 2003. βTrCP-mediated proteolysis of NF-κB1 p105 requires phosphorylation of p105 serines 927 and 932. *Mol Cell Biol* **23**: 402–413.

Lenardo MJ, Baltimore D. 1989. NF-κB: A pleiotropic mediator of inducible and tissue-specific gene control. *Cell* **58**: 227–229.

Lenardo M, Siebenlist U. 1994. Bcl-3-mediated nuclear regulation of the NF-κB trans-activating factor. *Immunol Today* **15**: 145–147.

Lenardo M, Pierce JW, Baltimore D. 1987. Protein-binding sites in Ig gene enhancers determine transcriptional activity and inducibility. *Science* **236**: 1573–1577.

Lenardo MJ, Fan CM, Maniatis T, Baltimore D. 1989. The involvement of NF-κB in β-interferon gene regulation reveals its role as widely inducible mediator of signal transduction. *Cell* **57**: 287–294.

Leung TH, Hoffmann A, Baltimore D. 2004. One nucleotide in a κB site can determine cofactor specificity for NF-κB dimers. *Cell* **118**: 453–464.

Lim CA, Yao F, Wong JJ, George J, Xu H, Chiu KP, Sung WK, Lipovich L, Vega VB, Chen J, et al. 2007. Genome-wide mapping of RELA(p65) binding identifies E2F1 as a transcriptional activator recruited by NF-κB upon TLR4 activation. *Mol Cell* **27**: 622–635.

Lin R, Beauparlant P, Makris C, Meloche S, Hiscott J. 1996. Phosphorylation of IκBα in the C-terminal PEST domain by casein kinase II affects intrinsic protein stability. *Mol Cell Biol* **16**: 1401–1409.

Liu B, Yang R, Wong KA, Getman C, Stein N, Teitell MA, Cheng G, Wu H, Shuai K. 2005. Negative regulation of NF-κB signaling by PIAS1. *Mol Cell Biol* **25**: 1113–1123.

Martone R, Euskirchen G, Bertone P, Hartman S, Royce TE, Luscombe NM, Rinn JL, Nelson FK, Miller P, Gerstein M, et al. 2003. Distribution of NF-κB-binding sites across human chromosome 22. *Proc Natl Acad Sci U S A* **100**: 12247–12252.

Matthews JR, Botting CH, Panico M, Morris HR, Hay RT. 1996. Inhibition of NF-κB DNA binding by nitric oxide. *Nucleic Acids Res* **24**: 2236–2242.

Mattioli I, Geng H, Sebald A, Hodel M, Bucher C, Kracht M, Schmitz ML. 2006. Inducible phosphorylation of NF-κB p65 at serine 468 by T cell costimulation is mediated by IKKepsilon. *J Biol Chem* **281**: 6175–6183.

McElhinny JA, Trushin SA, Bren GD, Chester N, Paya CV. 1996. Casein kinase II phosphorylates IκBα at S-283, S-289, S-293, and T-291 and is required for its degradation. *Mol Cell Biol* **16**: 899–906.

Motoyama M, Yamazaki S, Eto-Kimura A, Takeshige K, Muta T. 2005. Positive and negative regulation of nuclear factor-κB-mediated transcription by IκB-zeta, an inducible nuclear protein. *J Biol Chem* **280**: 7444–7451.

Moumen A, Masterson P, O'Connor MJ, Jackson SP. 2005. hnRNP K: an HDM2 target and transcriptional coactivator of p53 in response to DNA damage. *Cell* **123**: 1065–1078.

Natoli G. 2006. Tuning up inflammation: How DNA sequence and chromatin organization control the induction of inflammatory genes by NF-κB. *FEBS Lett* **580**: 2843–2849.

Natoli G, Saccani S, Bosisio D, Marazzi I. 2005. Interactions of NF-κB with chromatin: The art of being at the right place at the right time. *Nat Immunol* **6**: 439–445.

Neumann M, Naumann M. 2007. Beyond IκBs: Alternative regulation of NF-κB activity. *Faseb J* **21**: 2642–2654.

O'Mahony AM, Montano M, Van Beneden K, Chen LF, Greene WC. 2004. Human T-cell lymphotropic virus type 1 tax induction of biologically Active NF-κB requires IκB kinase-1-mediated phosphorylation of RelA/p65. *J Biol Chem* **279**: 18137–18145.

O'Shea JM, Perkins ND. 2008. Regulation of the RelA (p65) transactivation domain. *Biochem Soc Trans* **36**: 603–608.

Ostrowski J, Kawata Y, Schullery DS, Denisenko ON, Bomsztyk K. 2003. Transient recruitment of the hnRNP K protein to inducibly transcribed gene loci. *Nucleic Acids Res* **31**: 3954–3962.

Perkins ND. 2006. Post-translational modifications regulating the activity and function of the nuclear factor κB pathway. *Oncogene* **25**: 6717–6730.

Phelps CB, Sengchanthalangsy LL, Malek S, Ghosh G. 2000. Mechanism of κ B DNA binding by Rel/NF-κ B dimers. *J Biol Chem* **275**: 24392–24399.

Reber L, Vermeulen L, Haegeman G, Frossard N. 2009. Ser276 phosphorylation of NF-κB p65 by MSK1 controls SCF expression in inflammation. *PLoS ONE* **4**: e4393.

Rothwarf DM, Karin M. 1999. The NF-κB activation pathway: A paradigm in information transfer from membrane to nucleus. *Sci STKE* **1999**: RE1.

Ruby CE, Leid M, Kerkvliet NI. 2002. 2,3,7,8-Tetrachlorodibenzo-p-dioxin suppresses tumor necrosis factor-α and anti-CD40-induced activation of NF-κB/Rel in dendritic cells: p50 homodimer activation is not affected. *Mol Pharmacol* **62**: 722–728.

Ryo A, Suizu F, Yoshida Y, Perrem K, Liou YC, Wulf G, Rottapel R, Yamaoka S, Lu KP. 2003. Regulation of NF-κB signaling by Pin1-dependent prolyl isomerization and ubiquitin-mediated proteolysis of p65/RelA. *Mol Cell* **12**: 1413–1426.

Sakurai H, Chiba H, Miyoshi H, Sugita T, Toriumi W. 1999. IκB kinases phosphorylate NF-κB p65 subunit on serine 536 in the transactivation domain. *J Biol Chem* **274**: 30353–30356.

Sanchez-Valdepenas C, Martin AG, Ramakrishnan P, Wallach D, Fresno M. 2006. NF-κB-inducing kinase is involved in the activation of the CD28 responsive element through phosphorylation of c-Rel and regulation of its transactivating activity. *J Immunol* **176**: 4666–4674.

Sasaki CY, Barberi TJ, Ghosh P, Longo DL. 2005. Phosphorylation of RelA/p65 on serine 536 defines an IκBα-independent NF-κB pathway. *J Biol Chem* **280**: 34538–34547.

Scheidereit C. 2006. IκB kinase complexes: Gateways to NF-κB activation and transcription. *Oncogene* **25**: 6685–6705.

Schoonbroodt S, Ferreira V, Best-Belpomme M, Boelaert JR, Legrand-Poels S, Korner M, Piette J. 2000. Crucial role of the amino-terminal tyrosine residue 42 and the carboxyl-terminal PEST domain of IκBα in NF-κB activation by an oxidative stress. *J Immunol* **164**: 4292–4300.

Schreiber J, Jenner RG, Murray HL, Gerber GK, Gifford DK, Young RA. 2006. Coordinated binding of NF-κB family members in the response of human cells to lipopolysaccharide. *Proc Natl Acad Sci U S A* **103**: 5899–5904.

Schwarz EM, Van Antwerp D, Verma IM. 1996. Constitutive phosphorylation of IκBα by casein kinase II occurs preferentially at serine 293: Requirement for degradation of free IκBα. *Mol Cell Biol* **16**: 3554–3559.

Sen R, Baltimore D. 1986. Inducibility of κ immunoglobulin enhancer-binding protein NF-κB by a posttranslational mechanism. *Cell* **47**: 921–928.

Senftleben U, Cao Y, Xiao G, Greten FR, Krahn G, Bonizzi G, Chen Y, Hu Y, Fong A, Sun SC, Karin M. 2001. Activation by IKKα of a second, evolutionary conserved, NF-κB signaling pathway. *Science* **293**: 1495–1499.

Siomi H, Matunis MJ, Michael WM, Dreyfuss G. 1993. The pre-mRNA binding K protein contains a novel evolutionarily conserved motif. *Nucleic Acids Res* **21**: 1193–1198.

Sizemore N, Leung S, Stark GR. 1999. Activation of phosphatidylinositol 3-kinase in response to interleukin-1 leads to phosphorylation and activation of the NF-κB p65/RelA subunit. *Mol Cell Biol* **19**: 4798–4805.

Starczynowski DT, Trautmann H, Pott C, Harder L, Arnold N, Africa JA, Leeman JR, Siebert R, Gilmore TD. 2007. Mutation of an IKK phosphorylation site within the transactivation domain of REL in two patients with B-cell lymphoma enhances REL's in vitro transforming activity. *Oncogene* **26**: 2685–2694.

Suyang H, Phillips R, Douglas I, Ghosh S. 1996. Role of unphosphorylated, newly synthesized IκBβ in persistent activation of NF-κB. *Mol Cell Biol* **16**: 5444–5449.

Tang RH, Zheng XL, Callis TE, Stansfield WE, He J, Baldwin AS, Wang DZ, Selzman CH. 2008. Myocardin inhibits cellular proliferation by inhibiting NF-κB(p65)-dependent cell cycle progression. *Proc Natl Acad Sci U S A* **105**: 3362–3367.

Thompson JE, Phillips RJ, Erdjument-Bromage H, Tempst P, Ghosh S. 1995. IκB-β regulates the persistent response in a biphasic activation of NF-κB. *Cell* **80**: 573–582.

Tian Y, Ke S, Denison MS, Rabson AB, Gallo MA. 1999. Ah receptor and NF-κB interactions, a potential mechanism for dioxin toxicity. *J Biol Chem* **274**: 510–515.

Tomonaga T, Levens D. 1996. Activating transcription from single stranded DNA. *Proc Natl Acad Sci U S A* **93**: 5830–5835.

Urban MB, Schreck R, Baeuerle PA. 1991. NF-κB contacts DNA by a heterodimer of the p50 and p65 subunit. *Embo J* **10**: 1817–1825.

Vatsyayan J, Qing G, Xiao G, Hu J. 2008. SUMO1 modification of NF-κB2/p100 is essential for stimuli-induced p100 phosphorylation and processing. *EMBO Rep* **9**: 885–890.

Viatour P, Dejardin E, Warnier M, Lair F, Claudio E, Bureau F, Marine JC, Merville MP, Maurer U, Green D, et al. 2004. GSK3-mediated BCL-3 phosphorylation modulates its degradation and its oncogenicity. *Mol Cell* **16**: 35–45.

Viatour P, Merville MP, Bours V, Chariot A. 2005. Phosphorylation of NF-κB and IκB proteins: Implications in cancer and inflammation. *Trends Biochem Sci* **30**: 43–52.

Wan F, Anderson DE, Barnitz RA, Snow A, Bidere N, Zheng L, Hegde V, Lam LT, Staudt LM, Levens D, et al. 2007. Ribosomal protein S3: A KH domain subunit in NF-κB complexes that mediates selective gene regulation. *Cell* **131**: 927–939.

Wang D, Westerheide SD, Hanson JL, Baldwin AS Jr. 2000. Tumor necrosis factor α-induced phosphorylation of RelA/p65 on Ser529 is controlled by casein kinase II. *J Biol Chem* **275**: 32592–32597.

Waris G, Livolsi A, Imbert V, Peyron JF, Siddiqui A. 2003. Hepatitis C virus NS5A and subgenomic replicon activate NF-κB via tyrosine phosphorylation of IκBα and its degradation by calpain protease. *J Biol Chem* **278**: 40778–40787.

Xiao G, Harhaj EW, Sun SC. 2001. NF-κB-inducing kinase regulates the processing of NF-κB2 p100. *Mol Cell* **7**: 401–409.

Xiao G, Fong A, Sun SC. 2004. Induction of p100 processing by NF-κB-inducing kinase involves docking IκB kinase α (IKKα) to p100 and IKKα-mediated phosphorylation. *J Biol Chem* **279**: 30099–30105.

Yamamoto M, Yamazaki S, Uematsu S, Sato S, Hemmi H, Hoshino K, Kaisho T, Kuwata H, Takeuchi O, Takeshige K, et al. 2004. Regulation of Toll/IL-1-receptor-mediated gene expression by the inducible nuclear protein IκBzeta. *Nature* **430**: 218–222.

Yang JP, Hori M, Sanda T, Okamoto T. 1999. Identification of a novel inhibitor of nuclear factor-κB, RelA-associated inhibitor. *J Biol Chem* **274**: 15662–15670.

A Structural Guide to Proteins of the NF-κB Signaling Module

Tom Huxford[1] and Gourisankar Ghosh[2]

[1]Department of Chemistry and Biochemistry, San Diego State University, San Diego, California 92182-1030

[2]Department of Chemistry and Biochemistry, University of California, San Diego, La Jolla, California 92093-0375

Correspondence: gghosh@ucsd.edu

The prosurvival transcription factor NF-κB specifically binds promoter DNA to activate target gene expression. NF-κB is regulated through interactions with IκB inhibitor proteins. Active proteolysis of these IκB proteins is, in turn, under the control of the IκB kinase complex (IKK). Together, these three molecules form the NF-κB signaling module. Studies aimed at characterizing the molecular mechanisms of NF-κB, IκB, and IKK in terms of their three-dimensional structures have lead to a greater understanding of this vital transcription factor system.

NF-κB is a master transcription factor that responds to diverse cell stimuli by activating the expression of stress response genes. Multiple signals, including cytokines, growth factors, engagement of the T-cell receptor, and bacterial and viral products, induce NF-κB transcriptional activity (Hayden and Ghosh 2008). A point of convergence for the myriad of NF-κB inducing signals is the IκB kinase complex (IKK). Active IKK in turn controls transcription factor NF-κB by regulating proteolysis of the IκB inhibitor protein (Fig. 1). This nexus of three factors, IKK, IκB, and NF-κB, forms the NF-κB signaling module—a molecular relay switch mechanism that is conserved across diverse species (Ghosh et al. 1998; Hoffmann et al. 2006). In this article, we introduce the human NF-κB, IκB, and IKK proteins, and discuss how they function from the perspective of their three-dimensional structures.

NF-κB

Introduction to NF-κB

NF-κB was discovered in the laboratory of David Baltimore as a nuclear activity with binding specificity toward a ten-base-pair DNA sequence 5'-GGGACTTTCC-3' present within the enhancer of the immunoglobin κ light chain gene in mature antibody-producing B cells (Sen and Baltimore 1986). The biochemically purified activity was found to be composed of 50 and 65 kilodalton (kDa) subunits. Cloning of the p50 subunit revealed significant amino-acid sequence homology between its amino-terminal 300 amino acids and the oncogene

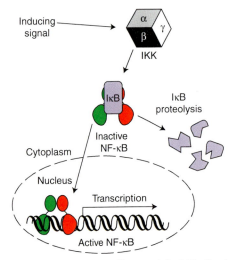

Figure 1. The NF-κB signaling module. NF-κB exists in the cytoplasm of resting cells by virtue of its noncovalent association with an IκB inhibitor protein. The IκB kinase (IKK) responds to diverse stimuli by catalyzing the phosphorylation-dependent 26 S proteasome-mediated degradation of complex-associated IκB. Active NF-κB accumulates in the nucleus where it binds with DNA sequence specificity in the promoter regions of target genes and activates their transcription.

from the reticuloendotheliosis virus of turkeys and shared by v-Rel and its cellular proto-oncogene c-Rel. This portion of conserved amino acid sequence was termed the Rel homology region (RHR). As is discussed later, the mRNA responsible for producing the p50 subunit was found to encode a longer precursor protein of 105 kDa in size that possesses the entire p50 amino-acid sequence at its amino-terminal end and its own inhibitor within its carboxy-terminal region. Once the cDNA encoding the p65 subunit (also known as RelA) was sequenced, it was also found to contain an amino-terminal RHR. Two additional NF-κB family subunits, RelB and p52 (the processed product of a longer 100-kDa precursor), were also discovered to harbor the conserved RHR within their amino-terminal regions. These five polypeptides, p50, p65/RelA, c-Rel, p52, and RelB, constitute the entire family of NF-κB subunits encoded by the human genome (Fig. 2A).

Rel Homology Region Structure

The first glimpse at the structure of the RHR was afforded by the successful determination of two x-ray crystal structures of the NF-κB p50:p50 homodimer in complex with related κB DNA (Ghosh et al. 1995; Müller et al. 1995). The structures uncovered a symmetrical protein: DNA complex structure reminiscent of a butterfly with double-stranded DNA comprising the "body" and two-protein subunit "wings" (Fig. 2B,C). These structures revealed a completely novel DNA binding motif in which one entire 300 amino acid RHR from each p50 subunit in the dimer are involved in contacting one whole turn along the major groove of double-stranded DNA (Baltimore and Beg 1995; Müller et al. 1996).

As revealed by the NF-κB p50:DNA complex structures, the RHR consists of two folded domains. The amino-terminal domain (or Rel-N) is approximately 160–210 amino acids in length and exhibits a variant of the immunoglobulin fold. The carboxy-terminal dimerization domain (referred to as Rel-C) spans roughly 100 amino acids and also adopts an immunoglobulin-like fold. The five RHR-containing NF-κB subunits assemble to form various homo- and heterodimer combinations to form active NF-κB transcription factors. The two domains are joined by a short flexible linker approximately 10 amino acids in length. A carboxy-terminal flexible region in which is embedded the nuclear localization signal terminates the conserved RHR portion of the NF-κB subunits. Besides dimerization, this RHR is responsible for sequence-specific DNA binding, nuclear localization, and interaction with IκB proteins.

Outside of the conserved RHR, three NF-κB subunits, p65/RelA, c-Rel, and RelB, contain a transcription activation domain (TAD) at their extreme carboxy-terminal ends. This region, which is poorly understood in protein structural terms, is responsible for the increase in target gene expression that results from induction of NF-κB and, consequently, NF-κB dimers that possess at least one of these subunits function as activators of transcription.

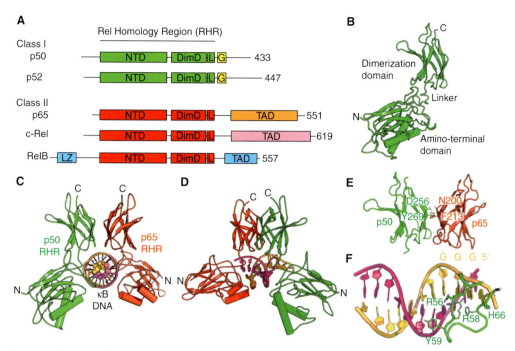

Figure 2. The NF-κB family. (*A*) The human genome encodes five polypeptides that assemble in various dimer combinations to form active NF-κB transcription factors. Each of the subunits contains the Rel homology region (RHR) near its amino terminus. The RHR consists of two folded domains, the amino-terminal domain (NTD) and the dimerization domain (DimD), that are joined by a short flexible linker and a carboxy-terminal flexible region that contains the nuclear localization signal (L). Three of the subunits, p65, c-Rel, and RelB, also contain a transcription activation domain (TAD) at their carboxy-terminal ends. RelB contains a predicted leucine zipper motif (LZ) amino-terminal to its RHR. The NF-κB subunits p50 and p52 lack transactivation domains and have glycine-rich regions (G). (*B*) A ribbon diagram representation of the RHR from p50 in its DNA-bound conformation. (*C*) The NF-κB p50:p65/RelA heterodimer bound to κB DNA. (*D*) Another view of the complex. (*E*) The NF-κB p50:p65/RelA heterodimer dimerization domains with key amino acid side chains labeled. (*F*) κB DNA from the NF-κB:DNA complex with key base-contacting amino acid residues labeled.

RelB also contains a predicted leucine zipper motif amino-terminal to its RHR. The NF-κB subunits p50 and p52 lack transactivation domains. Rather, their extreme carboxy-terminal ends are rich in glycine. Consequently, NF-κB dimers consisting exclusively of p50 and p52 subunits are capable of nuclear localization and DNA binding but fail to activate target gene expression and, in fact, function both in vitro and in vivo as repressors of transcription (Franzoso et al. 1992).

NF-κB Dimerization

Assembly of individual NF-κB subunits into dimers capable of sequence-specific DNA binding and activating target gene expression is mediated entirely by the dimerization domain. In theory, a total of 15 unique homo- and heterodimers are possible from combinatorial dimerization of the five NF-κB subunits (Hoffmann et al. 2006). Of these, 12 have been identified in vivo. Three that are not known to exist are the RelB:RelB, RelB:c-Rel, and p52:c-Rel. However, of these three, only RelB homodimer fails to exist in vivo, and the status of the other two dimers is uncertain; they could exist under specialized conditions such as the p65/RelA:RelB heterodimer (Marienfeld et al. 2003).

The homo- and heterotropic interactions between NF-κB family subunits follow the central dogma of all protein–protein complexes: A set of amino acid residues from each polypeptide participate in direct contacts with one another, forming the interface, whereas a different set of residues present outside of the interface indirectly affects the stability of the interface by modulating the local environment. One of the most stable NF-κB dimer interfaces is created by the p50:p65/RelA heterodimer (Fig. 2D). The p50:p65/RelA heterodimer assembles with a significantly greater stability than either of the respective p50:p50 or p65/RelA:p65/RelA homodimers (Huang et al. 1997). Differences in the amino acid sequences of p50 and p65/RelA at two positions partially explain the variation in dimerization stability observed in the three dimers. The positions of Tyr-269 and Asp-256 in p50 are occupied by Phe-213 and Asn-200, respectively, in p65. Within the heterodimer, the Asp and Asn approach one another symmetrically at the interface and stabilize the heterodimer through formation of a highly stable hydrogen bond. In contrast, in the homodimers, the juxtaposition of Asp-Asp and Asn-Asn at the interface are detrimental to dimer stability. Similarly, the hydroxyl group on Tyr-269 of p50 hydrogen bonds at the dimer interface, contributing to stabilization of both the p50:p50 homodimer as well as the p50:p65/RelA heterodimer. The substitution of Phe at this position serves to weaken the p65/RelA:p65/RelA homodimer relative to the other two. Taken together, these observations help to explain why the p65/RelA:p65/RelA homodimer forms with lower stability than does the p50:p50 homodimer and why the p50:p65/RelA heterodimer is more stable than both homodimers.

However, direct contact between complementary amino acids at the interface fails to completely explain the observed trends of NF-κB subunit dimerization. Mutation of noninterfacial residues contained within the dimerization domain has been shown to modulate dimerization. For example, changing p65/RelA Cys-216 to Ala affects homodimer formation (Ganchi et al. 1993). The role of noninterfacial amino acid residues in dimerization is most strikingly illustrated in the case of RelB. All

interfacial residues in RelB are either identical or homologous to those of other NF-κB subunits. And yet, RelB assembles into a completely unique domain-swapped homodimer (Huang et al. 2005). Domain swapping occurs as a consequence of the destabilization of the folded RelB dimerization domain, suggesting that domain stability is an important determinant of protein–protein interaction. In cells, decreased folding stability in both the amino-terminal and dimerization domains contributes to its degradation by the proteasome, which explains why the RelB homodimer does not exist in vivo (Marienfeld et al. 2001).

Post-translational modification may also play a role in modulating dimerization propensity. One study has shown that RelB forms a dimer with a p65/RelA that is phosphorylated at position Ser-276 (Jacque et al. 2005). This serine is located in a loop within the dimerization domain, projected away from the dimer interface. It is unclear as to how phosphorylation of this serine positively impacts p65/RelA:RelB heterodimer formation. One explanation could be that the increased negative charge may indirectly modulate dimer-forming residues of both p65/RelA and RelB.

NF-κB Recognition of κB DNA

X-ray structures of NF-κB in complex with κB DNA revealed a new mode of DNA recognition wherein a dimer composed of the RHR of two NF-κB subunits intimately contacts double-stranded DNA within the major groove through one complete turn (Ghosh et al. 1995; Müller et al. 1995). NF-κB employs both its amino-terminal and dimerization domains to encircle its target DNA. DNA contacts are mediated by amino acids emanating from loops that connect β-strand elements of secondary protein structure. The p50:p50 homodimer structures revealed that this NF-κB subunit employs its amino acids His-66, Arg-58, and Arg-56 to contact three guanine nucleotide bases at the extreme 5′ ends of its consensus κB DNA. However, X-ray analyses of NF-κB p65/RelA:p65/RelA homodimers bound to κB DNA revealed that, when bound to a canonical

10-base-pair κB DNA, one p65/RelA subunit contacts DNA in an analogous manner as observed in the p50 homodimer structures, whereas the second p65/RelA subunit significantly repositions its entire amino-terminal domain to mediate interactions with the DNA backbone (Chen et al. 1998b). Such a binding mode, which is afforded by flexibility in the short linker region that connects the two structured domains of the RHR, preserves binding affinity at the cost of fewer contacts to DNA bases. A close inspection of the homodimer:DNA complex structures leads to the suggestion that a homodimer of p50 might optimally bind to an 11-base-pair sequence composed of two 5′-GGGPuN half sites bracketing a central A:T base pair. In contrast, the NF-κB p65/RelA homodimer optimally recognizes a nine-base-pair target sequence containing two 5′-GGPuN half sites and a central A:T base pair. Determination of a second X-ray structure of NF-κB p65/RelA homodimer in complex with the κB DNA sequence from the promoter of IL-8 (5′-GGAA T TTCC-3′) confirmed this hypothesis (Chen et al. 2000).

The lessons learned from structural analyses of p50 and p65/RelA homodimers bound to diverse κB DNA sequences suggested that the canonical NF-κB p50:p65/RelA heterodimer might recognize its 10-base-pair target sequence with the p50 subunit binding specifically to a 5′-GGGPyN half site, whereas the p65/RelA subunit binds to a 5′-GGPyN half site separated from the p50 site by one A:T base pair. X-ray structure determination of a p50:p65/RelA RHR heterodimer bound to κB DNA from the original κ light chain gene promoter, which is identical to a κB sequence that is present in the promoter of HIV genome (5′-GGGAC T TTCC-3′), served to confirm this speculation (Chen et al. 1998a). Additional crystal structures of NF-κB p50: p65/RelA heterodimer bound to different κB DNA further support the basic rules of DNA half-site recognition developed from the homodimer:DNA structures (Berkowitz et al. 2002; Escalante et al. 2002).

X-ray crystal structures of several additional NF-κB:DNA complexes have now been determined (Cramer et al. 1997; Cramer et al. 1999; Moorthy et al. 2007; Panne et al. 2007; Fusco et al. 2009). Taken together, these structures suggest a model for how NF-κB dimers recognize κB DNA that contain significant deviations from the consensus sequence. Changes of κB DNA sequence can be of two types. In the first, alteration occurs within the G:C base pairs that occupy positions on the outside of the κB DNA and that are directly contacted by the amino-terminal domain. An example is the altered κB sequence 5′-GGGAC T TTTC-3′ (change from immunoglobin κB sequence is underlined). Because of the loss of an important C:G base pair, the NF-κB p65/RelA subunit conformation when bound to TTCC-3′ is drastically different than to TTTC-3′. The flexible linker region and modular domain architecture within the RHR allows the amino-terminal domain to reposition itself and bind the DNA backbone to accommodate such variations in κB DNA sequence. A second type of κB DNA sequence alteration involves changes within the central five base pairs. Of these, the central A:T is not directly contacted by the protein and the others are recognized nonspecifically through van der Waals contacts by p50 Tyr-59 (Tyr-36 in p65/RelA). The third base pair from the 5′-end or its symmetric pair are less sensitive to change, as illustrated by a comparison of immunoglobin κB and IFN-β κB sequences (5′-GGGAC T TTCC-3′ and 5′-GGGAA T TTCC-3′, respectively). The C:G to A:T base-pair change does not affect DNA recognition by the protein but may influence stability of binding because of the more rigid DNA structure of IFN-β κB sites. In this second case, overall conformations of the protein:DNA complexes are similar, but binding affinity could differ significantly.

Molecular dynamics simulations have revealed intriguing structural transitions of a 20-base-pair DNA containing the κB site of IL-2 promoter (AGAA A TTCC). The central A:T base pair (underlined) undergoes cross-stand stacking and the central A:T base pair flips out of the DNA axis (Mura and McCammon 2008). This dynamic behavior of the DNA suggests a highly complex mechanism of DNA recognition by the NF-κB dimers, in which DNA sequence variations may play a significant role in the recognition process. In general, one can confidently say that the sequences at the center of a κB sequence may profoundly affect the binding affinity and specificity.

IκB

Introduction to IκB

Almost immediately on detecting NF-κB in immune cells, researchers discovered that a latent κB DNA binding activity was present in the cytoplasm of all resting cells and that this pool of NF-κB could be activated by treatment of cell lysates with the weak detergent deoxycholate (Baeuerle and Baltimore 1988a). This suggested that noncovalent interaction with an inhibitor protein was responsible for maintaining NF-κB in an inactive state. Purification of the inhibitor activity led to the cloning of the inhibitory proteins IκBα and IκBβ (Baeuerle and Baltimore 1988b; Thompson et al. 1995). A third IκB gene was later identified by sequence homology in an EST database and was called IκBε (Li and Nabel 1997; Simeonidis et al.

1997; Whiteside et al. 1997). Subsequent experiments demonstrated that it also exhibits NF-κB inhibitory activity.

Classical IκB Sequence and Structure

Both IκBα and IκBβ, as well as the more recently discovered IκBε, contain a central ankyrin repeat domain (ARD) that contains six ankyrin repeats (Fig. 3A). The ankyrin repeat (ANK) is a roughly 33-amino-acid consensus amino acid sequence that appears in multiple copies in numerous proteins (Fig. 3B) (Sedgwick and Smerdon 1999). Ankyrin repeats are part of a greater superfamily of helical repeat motifs, which include HEAT repeats, armadillo repeats, and leucine-rich repeats, and are common to proteins involved in protein–protein interactions (Groves and Barford 1999). At their amino-terminal ends, classical IκB proteins contain a sequence of amino acids that do not adopt a folded structure in solution (Jaffray et al. 1995). Contained within this signal response region are the conserved serine sites of phosphorylation by IKK. Roughly 10 amino acids amino-terminal to this pair of serines

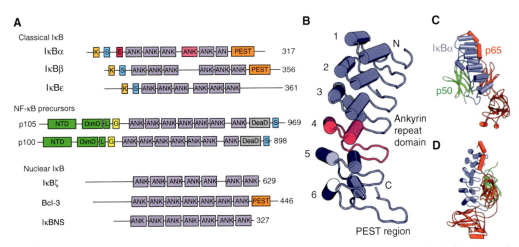

Figure 3. The family of human IκB proteins. (*A*) IκB proteins are classified as in text. Classical IκB proteins possess ankyrin repeats (ANK) flanked by an amino-terminal signal response region and carboxy terminal PEST region. The signal response regions contain sites of phosphorylation by IKK (S), ubiquitination (K), and nuclear export (E). The NF-κB precursors serve as IκB proteins as well as the source of the mature p50 and p52 NF-κB subunits. (*B*) Ribbon diagram of the IκBα structure from the NF-κB:IκBα complex crystal structure. Individual ankyrin repeats are numbered, ANK 4 is colored magenta, and the PEST region is labeled. (*C*) Ribbon diagram of the NF-κB:IκBα complex. (*D*) Another view of the complex.

Cite this article as *Cold Spring Harb Perspect Biol* 2009;1:a000075

reside conserved lysine amino acid sites of poly-ubiquitination. This amino-terminal region of IκBα also contains a functional nuclear export sequence (Johnson et al. 1999; Huang et al. 2000). It is not masked on binding to NF-κB and contributes to the observed cytoplasmic localization of NF-κB:IκBα complexes. Neither IκBβ nor IκBε possess this inherent nuclear export potential. Consequently, stable NF-κB:IκBβ complexes reside stably either in the cytoplasm or the nucleus, whereas IκBε appears to function as a negative feedback regulator of cytoplasmic NF-κB (Malek et al. 2001; Tam and Sen 2001; Kearns et al. 2006). At their carboxy-terminal ends, the three classical IκB proteins contain a short sequence rich in the amino acids proline, glutamic acid, serine, and threonine. This so-called PEST region is common to many proteins that, like IκB, display rapid turnover in cells (Rogers et al. 1986; Pando and Verma 2000). The PEST region of IκBα, however, is also required for its ability to disrupt preformed NF-κB:DNA complexes (Ernst et al. 1995).

IκB Interactions with NF-κB

The X-ray structure of IκBα in complex with the NF-κB p50:p65/RelA heterodimer was determined independently by two separate laboratories in 1998 (Huxford et al. 1998; Jacobs and Harrison 1998). Both groups relied on a similar strategy of removing the signal response region of IκBα and the amino-terminal domain of the p50 subunit to stabilize the conformationally dynamic complex for cocrystallization (Huxford et al. 2000). The structure reveals how IκBα uses its entire ankyrin repeat-containing domain as well as its carboxy-terminal PEST sequence to mediate an extensive protein–protein interface of roughly 4300 Å^2 (Fig. 3C,D). The carboxy-terminal 30 amino acids from the NF-κB p65/RelA subunit RHR, which were disordered in NF-κB:DNA complex structures, adopt an ordered helical structure that contacts the first two ankyrin repeats and forms significant hydrophobic contacts with the amino-terminal face of the IκBα ankyrin repeat stack. This interaction

masks the p65/RelA nuclear localization signal. Ankyrin repeats three through five participate in multiple van der Waals contacts with one surface of the p50:p65/RelA heterodimer dimerization domains. The sixth ankyrin repeat and PEST region of IκBα present a vast acidic patch, which opposes the largely positively charged DNA binding surfaces of the p65/RelA amino-terminal domain. As a consequence of this electrostatic interaction, the p65/RelA amino-terminal domain occupies a position relative to the dimerization domain that is rotated roughly 180° and translated 40 Å when compared with its DNA bound structures. The transition of p65/RelA to the conformation observed in the NF-κB:IκB complex does not disrupt the amino-terminal domain structure and is afforded entirely by the flexible linker region that connects the amino-terminal and dimerization domains. The structure of a similar construct of IκBβ bound to the dimerization domain from the NF-κB p65/RelA:p65/RelA homodimer suggests that IκBβ uses a similar strategy in binding to NF-κB, although it relies less on interactions with the p65/RelA amino-terminal domain for complex stability (Malek et al. 2003).

IκBα Dynamics

Protein dynamics, or the rates with which a protein exchanges between quasi-stable folded states, is an extremely important aspect of protein structure and function. Several independent lines of investigation, including thermal and chemical denaturation, computer simulations, NMR spectroscopy, and failed crystallization attempts, have led to the conclusion that the IκBα protein exhibits a high degree of structural dynamics in solution (Huxford et al. 2000; Pando and Verma 2000; Croy et al. 2004; Bergqvist et al. 2006). This runs counter to the data that have emerged from protein engineering studies that clearly show that ankyrin repeat proteins designed after consensus sequences or those that appear in nature are extremely stably folded (Binz et al. 2003; Binz et al. 2004). IκBα has, therefore, evolved as an inherently unstable ankyrin repeat-containing protein. The consequences of this are twofold. First of all, free IκBα is easily degraded in cells. This signal-independent degradation involves the 20S proteasome and the carboxy-terminal PEST of IκBα (Mathes et al. 2008). Moreover, recent computational modeling of the NF-κB pathway through a systems biology approach has confirmed that regulation of NF-κB activation can be controlled by small changes in the rate of degradation of a constitutively expressed free cytoplasmic IκBα (O'Dea et al. 2007). On binding to NF-κB, the dynamic IκBα ankyrin repeat fold becomes stable and the PEST region protein turnover signal sequence is adopted as a DNA-inhibitory functional element (Sue et al. 2008). Degradation of IκBα is shifted from the steady state to a signal-dependent pathway that requires phosphorylation within the flexible amino-terminal signal response region. It is likely that the inherent folding instability of IκBα contributes to it being targeted by the proteasome, whereas NF-κB, which is composed of two stably folded domains and with which IκBα is associated in a complex at subnanomolar dissociation binding constant, remains intact.

Nonclassical IκB Proteins

Through the efforts of investigators attempting to understand regulation of NF-κB, a more diverse family of IκB proteins has emerged. Proteins of the IκB family are all linked by the fact that they contain ankyrin repeats and interact with NF-κB subunits to affect gene expression. The classical IκB proteins, IκBα, IκBβ, and IκBε, have been described. A second class of IκB proteins is represented by the NF-κB precursor proteins p105 and p100 (Basak et al. 2007). These two proteins act both as precursors of NF-κB p50 and p52 subunits, respectively, and as inhibitors of NF-κB (Fig. 3A). The NF-κB precursor proteins are responsible for inhibiting nearly half of the NF-κB in resting cells. However, unlike classical IκB proteins that inhibit NF-κB by forming 1:1 complexes, these NF-κB precursors participate in large multiprotein assemblies, wherein more than one NF-κB dimer can bind to multiple

p100 and/or p105 subunits. The assembly of p100 and p105 into larger complexes is mediated by an oligomerization domain located immediately amino-terminal to their ankyrin-repeat domains. This assembly is heterogeneous, i.e., several different NF-κB subunits can be inhibited in a single inhibitory complex (Savinova et al. 2009). Therefore, stimulus-specific degradation of p100 or p105 can in principle release different NF-κB dimers. Together, these dimers exhibit a much broader spectrum of gene regulatory activities (Savinova et al. 2009, Shih et al. 2009). The large heterogeneous NF-κB inhibitory complex assemblies have been dubbed NF-κBsomes to distinguish them from the smaller NF-κB inhibitory complexes formed by IκBα, -β, and -ε.

Together with IκBα and IκBε, the NF-κB precursors are targets of NF-κB-driven transcription. Newly synthesized IκB serve to block NF-κB activity postinduction. This negative feedback regulation of NF-κB by IκB proteins is critical for control of inflammation and other diseases (Hoffmann et al. 2002).

Nuclear IκB Proteins

A third, entirely different class of IκB is represented by the proteins Bcl-3, IκBζ/MAIL, and IκBNS (Fig. 3A). Bcl-3 was cloned as a consequence to its proximity to a breakpoint mutation in some leukemias (Ohno et al. 1990). IκBζ was discovered in a screen of genes that displayed increased expression after induction of immune cells with bacterial lipopolysaccharide (LPS) or the inflammatory cytokine interleukin-1 (IL-1) (Kitamura et al. 2000; Haruta et al. 2001; Yamazaki et al. 2001). IκBNS was identified as a gene that is expressed in T cells during negative selection (Fiorini et al. 2002). Unlike classical IκB, these proteins do not contain amino-terminal signal-dependent phosphorylation sites or carboxy-terminal PEST regions. Their classification as "nuclear IκB" derives from the fact that they contain ankyrin repeats, bind NF-κB subunits, and concentrate within the nucleus when expressed in cells (Michel et al. 2001).

Cite this article as *Cold Spring Harb Perspect Biol* 2009;1:a000075

Each of the three nuclear IκB proteins are themselves the products of NF-κB-dependent genes (Eto et al. 2003; Ge et al. 2003; Hirotani et al. 2005). They bind to NF-κB, but, whereas the classical IκB proteins prefer dimers that possess at least one p65/RelA or c-Rel subunit, or nonclassical p105 and p100 binds all NF-κB subunits, nuclear IκBs bind specifically to homodimers of p50 (Hatada et al. 1992; Yamazaki et al. 2001; Trinh et al. 2008). Finally, association of nuclear IκB proteins with nuclear NF-κB can have diverse but important consequences on gene expression (Franzoso et al. 1992; Muta et al. 2003; Hirotani et al. 2005; Motoyama et al. 2005; Riemann et al. 2005). In peritoneal macrophages derived from mice lacking the gene encoding IκBζ, for example, a complete inability to produce the NF-κB-dependent cytokine interleukin-6 (IL-6) in response to LPS treatment was observed (Yamamoto et al. 2004). As IL-6 is a gene that is expressed in a later phase of NF-κB induction, it is apparent that the early induction and nuclear accumulation of IκBζ plays a vital role in the LPS-dependent expression of this pluripotent cytokine.

IKK

Introduction to IKK

In a thrilling conclusion to a search to identify an enzymatic activity that was capable of phosphorylating the two serine amino acids of the signal response region of IκBα, researchers from three labs reported in 1997 the IκB Kinase complex (IKK) (DiDonato et al. 1997; Mercurio et al. 1997; Regnier et al. 1997). This IKK was purified from cytokine-induced HeLa cells and exhibited an apparent molecular mass of 700–900 kDa. Microsequencing revealed two related kinase domain-containing subunits, referred to as IKKα and IKKβ (or alternatively as IKK1 and IKK2, respectively) (DiDonato et al. 1997; Zandi et al. 1997). These exhibit molecular masses of 85 and 87 kDa, respectively, and display 50% identity at the amino-acid level. The IKKα subunit was recognized to be the

same protein that was originally cloned in 1995 as CHUK, a kinase with homology to the helix-loop-helix transcription factors, and IKKβ was subsequently identified as a hit in a yeast two-hybrid screen with NIK, a kinase suspected to function upstream of IKK, as bait (Connelly and Marcu 1995; Regnier et al. 1997; Woronicz et al. 1997). A third subunit, known as IKKγ (also known as NEMO), was also identified as a 49 kDa member of the IKK complex (Rothwarf et al. 1998; Yamaoka et al. 1998; Li et al. 1999b; Mercurio et al. 1999).

The function of IKK1/IKKα in cells is somewhat exotic and continues to be elucidated (Hu et al. 2001; Senftleben et al. 2001; Anest et al. 2003; Yamamoto et al. 2003; Sil et al. 2004; Lawrence et al. 2005). However, multiple studies including mouse knockouts have revealed that IKK2/IKKβ is responsible for phosphorylating IκB in response to NF-κB-inducing signals (Li et al. 1999a; Li et al. 1999c; Tanaka et al. 1999). Because of its role as the primary inducer of the NF-κB transcription factor, IKK2/IKKβ plays the critical role in promoting inflammation and cell survival in response to proinflammatory stimuli. The NEMO/IKKγ subunit does not possess any kinase domain or enzymatic activity and instead it acts as an adapter subunit that links the catalytic subunits to receptor proximal signaling molecules. Mouse knockout studies clearly reveal that proper NF-κB signaling through IKK is not possible without this subunit (Makris et al. 2000; Rudolph et al. 2000; Schmidt-Supprian et al. 2000).

IKK Catalytic Subunit Domain Organization

The IKK1/IKKα and IKK2/IKKβ subunits exhibit an uncommon domain organization (Fig. 4A). Their first roughly 300 amino acids contain a clearly recognizable kinase domain. This is followed by a short region that exhibits distant homology to ubiquitin (Ikeda et al. 2007). A central region contains a leucine zipper motif followed by a region with slight homology to the helix-loop-helix transcription factors. This is followed by a serine-rich region. Finally, the carboxy-terminal element of IKKβ

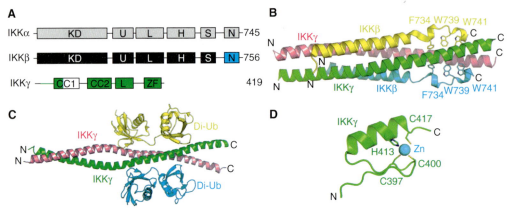

Figure 4. Subunits of the human IKK complex. (*A*) Domain organization of IKK subunits. Catalytic subunits contain a kinase domain (KD), ubiquitin-like domain (U), leucine zipper (L), helix-loop-helix (H), serine-rich (S), and NEMO-binding motif (N). The NEMO/IKKγ subunit contains two predicted coiled-coil motifs (CC1 and -2), a leucine zipper (L), and a carboxy-terminal zinc-finger (ZF). (*B*) Ribbon diagram of the IKK2/IKKβ:NEMO/IKKγ complex. Individual polypeptides are labeled as well as some of the conserved hydrophobic amino acid side chains from IKK2/IKKβ. (*C*) Ribbon diagram of the NEMO/IKKγ:di-ubiquitin complex. (*D*) The NEMO/IKKγ carboxy-terminal zinc-finger motif structure.

has been shown to interact directly with NEMO/IKKγ (May et al. 2000; May et al. 2002). A genome-wide analysis has identified a small clade of proteins with domain organization reminiscent of the catalytic IKK1/IKKα and IKK2/IKKβ subunits (Manning et al. 2002). This kinase subgroup includes the proteins IKKε and TBK1/NAK. Both have been characterized as upstream modulators of IKK activity and are, therefore, both structurally and functionally related to the catalytic IKK subunits (Tojima et al. 2000; Peters and Maniatis 2001).

With the exception of the extreme carboxy-terminal NEMO/IKKγ-interacting motif, the structures of IKK1/IKKα and IKK2/IKKβ are unknown. Although it is clear that the kinase domain of IKK2/IKKβ is necessary for catalyzing phospho-transfer, regions of the protein outside of this domain are necessary for directing specificity to the amino-terminal serines of IκBα. Deletion of the leucine zipper and helix-loop-helix has been shown to yield a mutant enzyme that, although it is capable of catalyzing phospho-transfer to IκBα, fails to recognize the amino-terminal serines required for NF-κB activation in response to inflammatory signaling (Shaul et al. 2008).

NEMO/IKKγ Domain Organization

Structural interest in NEMO/IKKγ arises from its lack of a kinase domain or, for that matter, homology to any other protein of known structure. Secondary-structure prediction methods suggest that NEMO/IKKγ is a mostly helical protein with two signature coiled coil (CC) elements and a leucine zipper motif in the middle are flanked by a helical dimerization domain near the amino-terminal end and a Zn-finger motif at the carboxyl terminus. Three-dimensional structures of several fragments of the NEMO/IKKγ subunit either as free polypeptides or bound to ligands have recently been elucidated. These NEMO/IKKγ structures allow one to envision that with the exception of the very ends, NEMO/IKKγ consists of long helices that are punctuated by short unstructured regions. These helical segments wrap around each other forming a long coiled-coil dimer with fraying of the monomers at each end. The carboxy-terminal end contains a Zn-finger motif and the 40-residue long amino terminus is likely to be flexible and unstructured.

The NEMO/IKKγ fragment structures suggest how IKKγ might be involved in cellular signaling. NEMO/IKKγ was previously thought to form a multimer with different segments shown to form dimers, trimers, or tetramers (Agou et al. 2002; Tegethoff et al. 2003; Marienfeld et al. 2006; Drew et al. 2007; Herscovitch et al. 2008). With new structural information, it now appears that previous conclusions might not be accurate. It is not surprising as flexible coiled-coil motifs migrate through the gel filtration beads differently than the stably folded globular proteins of identical mass.

Two Distinct IKK Activation Pathways

The IKK1/IKKα and IKK2/IKKβ subunits are activated through distinct signaling pathways (Senftleben et al. 2001). These pathways are activated by two distinct sets of stimuli. The canonical pathway is triggered by LPS or inflammatory cytokines TNF-α or IL-1 and signals through IKK2/IKKβ to activate p65/RelA and c-Rel dimers through the degradation of IκBα, IκBβ, IκBε, and p105. The noncanonical pathway results from BAFF, LT-β, and CD40 signaling through IKKα, leading to activation of the p52:RelB heterodimer through processing of p100 into p52 (Ghosh and Karin 2002). Activation of both pathways requires interactions of signaling molecules through specific poly-ubiquitin moieties that are covalently linked to some of these molecules. In addition, upstream protein kinases are also essential for activation of both kinases. The major distinction between these two pathways is the involvement of a single upstream kinase, known as NF-κB inducing kinase (NIK), to activate IKKα, whereas multiple kinases can activate IKKβ. IKKα and IKKβ are both present in the same particle in vivo. One puzzling question, however, is whether the different pathways target these two distinct subunits in a single complex or if there is a separate pool of IKKα present in cells that is activated by the noncanonical pathway. For the canonical pathway, however, the IKKβ subunit of the heterodimer must be activated. The role for IKKα in canonical signaling is unclear. The presence of the IκB kinases in multiple signaling pathways suggests that they are capable of participating in diverse signaling complexes throughout the cell.

IKK Complex Oligomerization

Although several combinations of the IKK1/IKKα, IKK2/IKKβ, and NEMO/IKKγ subunits have been proposed to account for the original 700–900 kDa complex, no successful reconstitution from purified components has resulted in an active complex of this size. Therefore, 10 years after isolation of the complex, even the oligomerization state of the IKK complex remains unclear. Given the presence of multiple conserved elements that can potentially mediate homo- and heteromeric interactions between subunits, there exist many possibilities. Early attempts to determine the arrangement of subunits in the complex led to the model that, independent of its NEMO/IKKγ scaffolding protein, IKK1/IKKα and IKK2/IKKβ are capable of forming stable homo- or heterodimers (Zandi et al. 1998). More recent work with human IKK2/IKKβ purified in milligram quantities from recombinant baculovirus-infected sf9 insect cells has shown that the full length subunit purifies as a tetramer. Removal of the carboxy-terminal 100 amino acids containing the serine-rich and NEMO/IKKγ-binding regions results in a dimeric enzyme, whereas removal of the entire carboxy-terminal portion beginning at the leucine zipper renders the kinase monomeric. Interestingly, it was observed that although this monomeric IKK2/IKKβ kinase domain remains catalytically active, it fails to specifically phosphorylate the signal response region of IκBα and instead catalyzes phosphorylation of the PEST (Shaul et al. 2008). The presence of multiple domains linked to flexibility in all three subunits shows why it is difficult to structurally characterize functional IKK complexes.

The Emerging Structure of NEMO/IKKγ

A working structural model of the NEMO/IKKγ subunit has begun to take shape with the recent successful determination of several X-ray and NMR structures of discrete functional portions. The structure of an IKK2/IKKβ polypeptide bound the amino-terminal helical region (residues 44 to 111) of NEMO/IKKγ revealed the complex forms a parallel four-helix bundle where two IKK2/IKKβ peptides associate with the NEMO/IKKγ dimer (Rushe et al. 2008). The IKK2/IKKβ peptides appear to fold on binding to the NEMO/IKKγ-dimer scaffold, making several contacts (Fig. 4B). The affinity of interaction between NEMO/IKKγ and IKK2/IKKβ subcomplex is high (K_D is in low nanomolar range) and requires a large array of contacts. However, few of the hydrophobic residues in IKK2/IKKβ that are observed to line the hydrophobic pocket formed by the NEMO/IKKγ dimer are required for complex formation. Although no clear experimental data is available, the IKK:IKKγ complexes might exist as a 2:2 complex, where a dimeric NEMO/IKKγ binds to an IKK dimer. The amino-terminal helical dimerization domain of NEMO/IKKγ interacts with both the IKK1/IKKα and IKK2/IKKβ carboxy-terminal peptides, forming a 1:1:2 (IKK1/IKKα:IKK2/IKKβ:NEMO/IKKγ) IKK complex. This is possibly the basal state of the IKK complex in cells.

Several reports demonstrated an interaction between poly-ubiquitin chains and the CC2-LZ region of NEMO/IKKγ (Lo et al. 2009). Although its was previously thought that the poly-ubiquitin chain is linked through lysine 63 and glycine 76 of ubiquitin, experiments have now shown that linear ubiquitin chains bind to NEMO/IKKγ with 100-fold tighter binding affinity than do K63-linked chains. The X-ray structure of the complex between the CC2-LZ of NEMO/IKKγ and a linear di-ubiquitin motif has been recently elucidated (Rahighi et al. 2009). Two di-ubiquitin motifs are bound to two chains of the LZ motif (Fig. 4C). It is striking that this central region of NEMO/IKKγ also exhibits the elongated coiled-coil motif that was observed in the previously determined IKK2/IKKβ:NEMO/IKKγ complex as well as in the X-ray structure of a complex between the carboxy-terminal helical region of NEMO/IKKγ and the viral protein vFLIP from Kaposi's sarcoma virus (Bagnéris et al. 2008).

The extreme carboxy-terminal region of NEMO/IKKγ adopts a CCHC-type zinc-finger motif. The solution structure of this region has been determined by multidimensional NMR spectroscopy (Cordier et al. 2008). Its structure adopts the familiar fold of a zinc-finger motif (Fig. 4d). Furthermore, this motif has been characterized as a ubiquitin-binding motif required for NF-κB signaling in response to TNF-α (Cordier et al. 2009).

When taken together, the NEMO/IKKγ substructures suggest that this subunit adopts an elongated helical structure. Two long helices bind one another through extended coiled-coil interactions to form the docking site for a pair of kinase subunits at the amino terminus, binding sites for di-ubiquitin and other proteins throughout the central region, and a pair of zinc-finger ubiquitin binding sites at the carboxyl terminus. This modular arrangement of docking sites for diverse proteins seems appropriate for a subunit that is thought to function principally as a scaffold for inducible signaling. Although significant structural work on IKK complexes remains to be carried out, this oddly elongated NEMO/IKKγ model serves to explain some of the complications associated with early studies that relied primarily on size exclusion chromatography for characterization of complex size and subunit stoichiometry.

ACKNOWLEDGMENTS

T.H. is supported by American Cancer Society grant RSG-08-287-01-GMC. G.G. is supported by National Institutes of Health (NIH) grants (GM085490, AI064326, and NCI141722).

REFERENCES

Agou F, Ye F, Goffinont S, Courtois G, Yamaoka S, Israël A, Véron M. 2002. NEMO trimerizes through its coiled-coiled C-terminal domain. *J Biol Chem* **277:** 17464–17475.

Anest V, Hanson JL, Cogswell PC, Steinbrecher KA, Strahl BD, Baldwin AS. 2003. A nucleosomal function for IκB kinase-α in NF-κB-dependent gene expression. *Nature* **423:** 659–663.

Cite this article as *Cold Spring Harb Perspect Biol* 2009;1:a000075

Baeuerle PA, Baltimore D. 1988a. Activation of DNA-binding activity in an apparently cytoplasmic precursor of the NF-κB transcription factor. *Cell* **53**: 211–217.

Baeuerle PA, Baltimore D. 1988b. IκB: A specific inhibitor of the NF-κB transcription factor. *Science* **242**: 540–546.

Bagnéris C, Ageichik AV, Cronin N, Wallace B, Collins M, Boshoff C, Waksman G, Barrett T. 2008. Crystal structure of a vFlip-IKKγ complex: Insights into viral activation of the IKK signalosome. *Mol Cell* **30**: 620–631.

Baltimore D, Beg AA. 1995. DNA-binding proteins. A butterfly flutters by. *Nature* **373**: 287–288.

Basak S, Kim H, Kearns JD, Tergaonkar V, O'Dea E, Werner SL, Benedict CA, Ware CF, Ghosh G, Verma IM, Hoffmann A. 2007. A fourth IκB protein within the NF-κB signaling module. *Cell* **128**: 369–381.

Bergqvist S, Croy CH, Kjaergaard M, Huxford T, Ghosh G, Komives EA. 2006. Thermodynamics Reveal that Helix Four in the NLS of NF-κB p65 Anchors IκBα, Forming a Very Stable Complex. *J Mol Biol* **360**: 421–434.

Berkowitz B, Huang DB, Chen-Park FE, Sigler PB, Ghosh G. 2002. The x-ray crystal structure of the NF-κB p50:p65 heterodimer bound to the interferon-β κB site. *J Biol Chem* **277**: 24694–24700.

Binz HK, Amstutz P, Kohl A, Stumpp MT, Briand C, Forrer P, Grütter MG, Plückthun A. 2004. High-affinity binders selected from designed ankyrin repeat protein libraries. *Nat Biotechnol* **22**: 575–582.

Binz HK, Stumpp MT, Forrer P, Amstutz P, Plückthun A. 2003. Designing repeat proteins: Well-expressed, soluble and stable proteins from combinatorial libraries of consensus ankyrin repeat proteins. *J Mol Biol* **332**: 489–503.

Chen YQ, Ghosh S, Ghosh G. 1998b. A novel DNA recognition mode by the NF-κB p65 homodimer. *Nat Struct Biol* **5**: 67–73.

Chen FE, Huang DB, Chen YQ, Ghosh G. 1998a. Crystal structure of p50/p65 heterodimer of transcription factor NF-κB bound to DNA *Nature* **391**: 410–413.

Chen YQ, Sengchanthalangsy LL, Hackett A, Ghosh G. 2000. NF-κB p65 (RelA) homodimer uses distinct mechanisms to recognize DNA targets. *Structure* **8**: 419–428.

Connelly MA, Marcu KB. 1995. CHUK, a new member of the helix-loop-helix and leucine zipper families of interacting proteins, contains a serine-threonine kinase catalytic domain. *Cell Mol Biol Res* **41**: 537–549.

Cordier F, Grubisha O, Traincard F, Véron M, Delepierre M, Agou F. 2009. The zinc finger of NEMO is a functional ubiquitin-binding domain. *J Biol Chem* **284**: 2902–2907.

Cordier F, Vinolo E, Véron M, Delepierre M, Agou F. 2008. Solution structure of NEMO zinc finger and impact of an anhidrotic ectodermal dysplasia with immunodeficiency-related point mutation. *J Mol Biol* **377**: 1419–1432.

Cramer P, Larson CJ, Verdine GL, Müller CW. 1997. Structure of the human NF-κB p52 homodimer-DNA complex at 2.1 Å resolution. *EMBO J* **16**: 7078–7090.

Cramer P, Varrot A, Barillas-Mury C, Kafatos FC, Müller CW. 1999. Structure of the specificity domain of the Dorsal homologue Gambif1 bound to DNA *Structure* **7**: 841–852.

Croy CH, Bergqvist S, Huxford T, Ghosh G, Komives EA. 2004. Biophysical characterization of the free IκBα

ankyrin repeat domain in solution. *Protein Sci* **13**: 1767–1777.

DiDonato JA, Hayakawa M, Rothwarf DM, Zandi E, Karin M. 1997. A cytokine-responsive IκB kinase that activates the transcription factor NF-κB *Nature* **388**: 548–554.

Drew D, Shimada E, Huynh K, Bergqvist S, Talwar R, Karin M, Ghosh G. 2007. Inhibitor κB kinase β binding by inhibitor κB kinase γ. *Biochemistry* **46**: 12482–12490.

Ernst MK, Dunn LL, Rice NR. 1995. The PEST-like sequence of IκBα is responsible for inhibition of DNA binding but not for cytoplasmic retention of c-Rel or RelA homodimers. *Mol Cell Biol* **15**: 872–882.

Escalante CR, Shen L, Thanos D, Aggarwal AK. 2002. Structure of NF-κB p50/p65 heterodimer bound to the PRDII DNA element from the interferon-β promoter. *Structure* **10**: 383–391.

Eto A, Muta T, Yamazaki S, Takeshige K. 2003. Essential roles for NF-κB and a Toll/IL-1 receptor domain-specific signal(s) in the induction of IκBζ. *Biochem Biophys Res Commun* **301**: 495–501.

Fiorini E, Schmitz I, Marissen WE, Osborn SL, Touma M, Sasada T, Reche PA, Tibaldi EV, Hussey RE, Kruisbeek AM, et al. 2002. Peptide-induced negative selection of thymocytes activates transcription of an NF-κB inhibitor. *Mol Cell* **9**: 637–648.

Franzoso G, Bours V, Park S, Tomita-Yamaguchi M, Kelly K, Siebenlist U. 1992. The candidate oncoprotein Bcl-3 is an antagonist of p50/NF-κB-mediated inhibition. *Nature* **359**: 339–342.

Fusco AJ, Huang DB, Miller D, Wang VY, Vu D, Ghosh G. 2009. NF-κB p52:RelB heterodimer recognizes two classes of κB sites with two distinct modes. *EMBO Rep* **10**: 152–159.

Ganchi PA, Sun SC, Greene WC, Ballard DW. 1993. A novel NF-κB complex containing p65 homodimers: Implications for transcriptional control at the level of subunit dimerization. *Mol Cell Biol* **13**: 7826–7835.

Ge B, Li O, Wilder P, Rizzino A, McKeithan TW. 2003. NF-κB regulates BCL3 transcription in T lymphocytes through an intronic enhancer. *J Immunol* **171**: 4210–4218.

Ghosh S, Karin M. 2002. Missing pieces in the NF-κB puzzle. *Cell* **109 Suppl**: S81–96.

Ghosh S, May MJ, Kopp EB. 1998. NF-κB AND REL PROTEINS: Evolutionarily Conserved Mediators of Immune Responses. *Annu Rev Immunol* **16**: 225–260.

Ghosh G, van Duyne G, Ghosh S, Sigler PB. 1995. Structure of NF-κB p50 homodimer bound to a κB site. *Nature* **373**: 303–310.

Groves MR, Barford D. 1999. Topological characteristics of helical repeat proteins. *Curr Opin Struct Biol* **9**: 383–389.

Haruta H, Kato A, Todokoro K. 2001. Isolation of a novel interleukin-1-inducible nuclear protein bearing ankyrin-repeat motifs. *J Biol Chem* **276**: 12485–12488.

Hatada EN, Nieters A, Wulczyn FG, Naumann M, Meyer R, Nucifora G, McKeithan TW, Scheidereit C. 1992. The ankyrin repeat domains of the NF-κB precursor p105 and the protooncogene bcl-3 act as specific inhibitors of NF-κB DNA binding. *Proc Natl Acad Sci* **89**: 2489–2493.

Hayden MS, Ghosh S. 2008. Shared principles in NF-κB signaling. *Cell* **132:** 344–362.

Herscovitch M, Comb W, Ennis T, Coleman K, Yong S, Armstead B, Kalaitzidis D, Chandani S, Gilmore TD. 2008. Intermolecular disulfide bond formation in the NEMO dimer requires Cys54 and Cys347. *Biochem Biophys Res Commun* **367:** 103–108.

Hirotani T, Lee PY, Kuwata H, Yamamoto M, Matsumoto M, Kawase I, Akira S, Takeda K. 2005. The nuclear IκB protein IκBNS selectively inhibits lipopolysaccharide-induced IL-6 production in macrophages of the colonic lamina propria. *J Immunol* **174:** 3650–3657.

Hoffmann A, Natoli G, Ghosh G. 2006. Transcriptional regulation via the NF-κB signaling module. *Oncogene* **25:** 6706–6716.

Hoffmann A, Levchenko A, Scott ML, Baltimore D. 2002. The IκB-NF-κB signaling module: Temporal control and selective gene activation. *Science* **298:** 1241–1245.

Hu Y, Baud V, Oga T, Kim KI, Yoshida K, Karin M. 2001. IKKα controls formation of the epidermis independently of NF-κB *Nature* **410:** 710–714.

Huang DB, Vu D, Ghosh G. 2005. NF-κB RelB forms an intertwined homodimer. *Structure* **13:** 1365–1373.

Huang DB, Huxford T, Chen YQ, Ghosh G. 1997. The role of DNA in the mechanism of NF-κB dimer formation: Crystal structures of the dimerization domains of the p50 and p65 subunits. *Structure* **5:** 1427–1436.

Huang TT, Kudo N, Yoshida M, Miyamoto S. 2000. A nuclear export signal in the N-terminal regulatory domain of IκBα controls cytoplasmic localization of inactive NF-κB/IκBα complexes. *Proc Natl Acad Sci* **97:** 1014–1019.

Huxford T, Malek S, Ghosh G. 2000. Preparation and crystallization of dynamic NF-κB:IκB complexes. *J Biol Chem* **275:** 32800–32806.

Huxford T, Huang DB, Malek S, Ghosh G. 1998. The crystal structure of the IκBα/NF-κB complex reveals mechanisms of NF-κB inactivation. *Cell* **95:** 759–770.

Ikeda F, Hecker CM, Rozenknop A, Nordmeier RD, Rogov V, Hofmann K, Akira S, Dotsch V, Dikic I. 2007. Involvement of the ubiquitin-like domain of TBK1/IKK-i kinases in regulation of IFN-inducible genes. *EMBO J* **26:** 3451–3462.

Jacobs MD, Harrison SC. 1998. Structure of an IκBα/NF-κB complex. *Cell* **95:** 749–758.

Jacque E, Tchenio T, Piton G, Romeo PH, Baud V. 2005. RelA repression of RelB activity induces selective gene activation downstream of TNF receptors. *Proc Natl Acad Sci* **102:** 14635–14640.

Jaffray E, Wood KM, Hay RT. 1995. Domain organization of IκBα and sites of interaction with NF-κB p65. *Mol Cell Biol* **15:** 2166–2172.

Johnson C, Van Antwerp D, Hope TJ. 1999. An N-terminal nuclear export signal is required for the nucleocytoplasmic shuttling of IκBα. *EMBO J* **18:** 6682–6693.

Kearns JD, Basak S, Werner SL, Huang CS, Hoffmann A. 2006. IκBε provides negative feedback to control NF-κB oscillations, signaling dynamics, and inflammatory gene expression. *J Cell Biol* **173:** 659–664.

Kitamura H, Kanehira K, Okita K, Morimatsu M, Saito M. 2000. MAIL, a novel nuclear IκB protein that potentiates LPS-induced IL-6 production. *FEBS Lett* **485:** 53–56.

Lawrence T, Bebien M, Liu GY, Nizet V, Karin M. 2005. IKKα limits macrophage NF-κB activation and contributes to the resolution of inflammation. *Nature* **434:** 1138–1143.

Li Z, Nabel GJ. 1997. A new member of the IκB protein family, IκBε, inhibits RelA (p65)-mediated NF-κB transcription. *Mol Cell Biol* **17:** 6184–6190.

Li Z-W, Chu W, Hu Y, Delhase M, Deerinck T, Ellisman M, Johnson R, Karin M. 1999c. The IKKβ subunit of IκB Kinase (IKK) is essential for nuclear factor κB activation and prevention of apoptosis. *J Exp Med* **189:** 1839–1845.

Li Y, Kang J, Friedman J, Tarassishin L, Ye J, Kovalenko A, Wallach D, Horwitz MS. 1999b. Identification of a cell protein (FIP-3) as a modulator of NF-κB activity and as a target of an adenovirus inhibitor of tumor necrosis factor α-induced apoptosis. *Proc Natl Acad Sci* **96:** 1042–1047.

Li Q, Van Antwerp D, Mercurio F, Lee KF, Verma IM. 1999a. Severe liver degeneration in mice lacking the IκB kinase 2 gene. *Science* **284:** 321–325.

Lo YC, Lin SC, Rospigliosi CC, Conze DB, Wu CJ, Ashwell JD, Eliezer D, Wu H. 2009. Structural basis for recognition of diubiquitins by NEMO *Mol Cell* **33:** 602–615.

Makris C, Godfrey VL, Krahn-Senftleben G, Takahashi T, Roberts JL, Schwarz T, Feng L, Johnson RS, Karin M. 2000. Female mice heterozygous for IKKγ/NEMO deficiencies develop a dermatopathy similar to the human X-linked disorder incontinentia pigmenti. *Mol Cell* **5:** 969–979.

Malek S, Chen Y, Huxford T, Ghosh G. 2001. IκBβ, but not IκBα, functions as a classical cytoplasmic inhibitor of NF-κB dimers by masking both NF-κB nuclear localization sequences in resting cells. *J Biol Chem* **276:** 45225–45235.

Malek S, Huang DB, Huxford T, Ghosh S, Ghosh G. 2003. X-ray crystal structure of an IκBβ:NF-κB p65 homodimer complex. *J Biol Chem* **278:** 23094–23100.

Manning G, Whyte DB, Martinez R, Hunter T, Sudarsanam S. 2002. The protein kinase complement of the human genome. *Science* **298:** 1912–1934.

Marienfeld RB, Palkowitsch L, Ghosh S. 2006. Dimerization of the IκB kinase-binding domain of NEMO is required for tumor necrosis factor α-induced NF-κB activity. *Mol Cell Biol* **26:** 9209–9219.

Marienfeld R, Berberich-Siebelt F, Berberich I, Denk A, Serfling E, Neumann M. 2001. Signal-specific and phosphorylation-dependent RelB degradation: A potential mechanism of NF-κB control. *Oncogene* **20:** 8142–8147.

Marienfeld R, May MJ, Berberich I, Serfling E, Ghosh S, Neumann M. 2003. RelB forms transcriptionally inactive complexes with RelA/p65. *J Biol Chem* **278:** 19852–19860.

Mathes E, O'Dea EL, Hoffmann A, Ghosh G. 2008. NF-κB dictates the degradation pathway of IκBα. *EMBO J* **27:** 1357–1367.

May MJ, Marienfeld RB, Ghosh S. 2002. Characterization of the IκB-kinase NEMO binding domain. *J Biol Chem* **277:** 45992–46000.

May MJ, D'Acquisto F, Madge LA, Glockner J, Pober JS, Ghosh S. 2000. Selective inhibition of NF-κB activation by a peptide that blocks the interaction of NEMO with the IκB kinase complex. *Science* **289:** 1550–1554.

Mercurio F, Murray BW, Shevchenko A, Bennett BL, Young DB, Li JW, Pascual G, Motiwala A, Zhu H, Mann M, et al. 1999. IκB kinase (IKK)-associated protein 1, a common component of the heterogeneous IKK complex. *Mol Cell Biol* **19:** 1526–1538.

Mercurio F, Zhu H, Murray BW, Shevchenko A, Bennett BL, Li J, Young DB, Barbosa M, Mann M, Manning A, et al. 1997. IKK-1 and IKK-2: Cytokine-activated IκB kinases essential for NF-κB activation. *Science* **278:** 860–866.

Michel F, Soler-Lopez M, Petosa C, Cramer P, Siebenlist U, Müller CW. 2001. Crystal structure of the ankyrin repeat domain of Bcl-3: A unique member of the IκB protein family. *EMBO J* **20:** 6180–6190.

Moorthy AK, Huang DB, Wang VY, Vu D, Ghosh G. 2007. X-ray structure of a NF-κB p50/RelB/DNA complex reveals assembly of multiple dimers on tandem κB sites. *J Mol Biol* **373:** 723–734.

Motoyama M, Yamazaki S, Eto-Kimura A, Takeshige K, Muta T. 2005. Positive and negative regulation of nuclear factor-κB-mediated transcription by IκBζ, an inducible nuclear protein. *J Biol Chem* **280:** 7444–7451.

Müller CW, Rey FA, Harrison SC. 1996. Comparison of two different DNA-binding modes of the NF-κB p50 homodimer. *Nat Struct Biol* **3:** 224–227.

Müller CW, Rey FA, Sodeoka M, Verdine GL, Harrison SC. 1995. Structure of the NF-κB p50 homodimer bound to DNA *Nature* **373:** 311–317.

Mura C, McCammon JA. 2008. Molecular dynamics of a κB DNA element: Base flipping via cross-strand intercalative stacking in a microsecond-scale simulation. *Nucleic Acids Res* **36:** 4941–4955.

Muta T, Yamazaki S, Eto A, Motoyama M, Takeshige K. 2003. IκBζ, a new anti-inflammatory nuclear protein induced by lipopolysaccharide, is a negative regulator for nuclear factor-κB *J Endotoxin Res* **9:** 187–191.

O'Dea EL, Barken D, Peralta RQ, Tran KT, Werner SL, Kearns JD, Levchenko A, Hoffmann A. 2007. A homeostatic model of IκB metabolism to control constitutive NF-κB activity. *Mol Syst Bio* **3:** 111.

Ohno H, Takimoto G, McKeithan TW. 1990. The candidate proto-oncogene bcl-3 is related to genes implicated in cell lineage determination and cell cycle control. *Cell* **60:** 991–997.

Pando MP, Verma IM. 2000. Signal-dependent and -independent Degradation of Free and NF-κB-bound IκBα *J Biol Chem* **275:** 21278–21286.

Panne D, Maniatis T, Harrison SC. 2007. An atomic model of the interferon-β enhanceosome. *Cell* **129:** 1111–1123.

Peters RT, Maniatis T. 2001. A new family of IKK-related kinases may function as IκB kinase kinases. *Biochim Biophys Acta* **1471:** M57–62.

Rahighi S, Ikeda F, Kawasaki M, Akutsu M, Suzuki N, Kato R, Kensche T, Uejima T, Bloor S, Komander D, et al. 2009. Specific recognition of linear ubiquitin chains by NEMO is important for NF-κB activation. *Cell* **136:** 1098–1109.

Regnier CH, Song HY, Gao X, Goeddel DV, Cao Z, Rothe M. 1997. Identification and characterization of an IκB kinase. *Cell* **90:** 373–383.

Riemann M, Endres R, Liptay S, Pfeffer K, Schmid RM. 2005. The IκB protein Bcl-3 negatively regulates transcription of the IL-10 gene in macrophages. *J Immunol* **175:** 3560–3568.

Rogers S, Wells R, Rechsteiner M. 1986. Amino acid sequences common to rapidly degraded proteins: The PEST hypothesis. *Science* **234:** 364–368.

Rothwarf DM, Zandi E, Natoli G, Karin M. 1998. IKKγ is an essential regulatory subunit of the IκB kinase complex. *Nature* **395:** 297–300.

Rudolph D, Yeh WC, Wakeham A, Rudolph B, Nallainathan D, Potter J, Elia AJ, Mak TW. 2000. Severe liver degeneration and lack of NF-κB activation in NEMO/IKKγ-deficient mice. *Genes Dev* **14:** 854–862.

Rushe M, Silvian L, Bixler S, Chen LL, Cheung A, Bowes S, Cuervo H, Berkowitz S, Zheng T, Guckian K, et al. 2008. Structure of a NEMO/IKK-associating domain reveals architecture of the interaction site. *Structure* **16:** 798–808.

Savinova OV, Hoffmann A, Ghosh G. 2009. The NFKB1 and NFKB2 proteins p105 and p100 function as the core of high molecular-weight heterogenous complexes. *Mol Cell* **12:** 591–602.

Schmidt-Supprian M, Bloch W, Courtois G, Addicks K, Israël A, Rajewsky K, Pasparakis M. 2000. NEMO/IKKγ-deficient mice model incontinentia pigmenti. *Mol Cell* **5:** 981–992.

Sedgwick SG, Smerdon SJ. 1999. The ankyrin repeat: A diversity of interactions on a common structural framework. *Trends Biochem Sci* **24:** 311–316.

Sen R, Baltimore D. 1986. Multiple nuclear factors interact with the immunoglobulin enhancer sequences. *Cell* **46:** 705–716.

Senftleben U, Cao Y, Xiao G, Greten FR, Krahn G, Bonizzi G, Chen Y, Hu Y, Fong A, Sun SC, et al. 2001. Activation by IKKα of a second, evolutionary conserved, NF-κB signaling pathway. *Science* **293:** 1495–1499.

Shaul JD, Farina A, Huxford T. 2008. The human IKKβ subunit kinase domain displays CK2-like phosphorylation specificity. *Biochem Biophys Res Commun* **374:** 592–597.

Shih VF, Kearns JD, Basak S, Savinova OV, Ghosh G, Hoffmann A. 2009. Kinetic control of negative feedback regulators of NF-κB/RelA determines their pathogen- and cytokine-receptor signaling specificity. *Proc Natl Acad Sci* **106:** 9619–9624.

Sil AK, Maeda S, Sano Y, Roop DR, Karin M. 2004. IκB kinase-α acts in the epidermis to control skeletal and craniofacial morphogenesis. *Nature* **428:** 660–664.

Simeonidis S, Liang S, Chen G, Thanos D. 1997. Cloning and functional characterization of mouse IκBε. *Proc Natl Acad Sci* **94:** 14372–14377.

Sue SC, Cervantes C, Komives EA, Dyson HJ. 2008. Transfer of flexibility between ankyrin repeats in IκBα upon formation of the NF-κB complex. *J Mol Biol* **380:** 917–931.

Tam WF, Sen R. 2001. IκB family members function by different mechanisms. *J Biol Chem* **276:** 7701–7704.

Tanaka M, Fuentes ME, Yamaguchi K, Durnin MH, Dalrymple SA, Hardy KL, Goeddel DV. 1999. Embryonic lethality, liver degeneration, and impaired NF-κB activation in IKKβ-deficient mice. *Immunity* **10:** 421–429.

Tegethoff S, Behlke J, Scheidereit C. 2003. Tetrameric oligomerization of IκB kinase γ (IKKγ) is obligatory for IKK complex activity and NF-κB activation. *Mol Cell Biol* **23:** 2029–2041.

Thompson JE, Phillips RJ, Erdjument-Bromage H, Tempst P, Ghosh S. 1995. IκBβ regulates the persistent response in a biphasic activation of NF-κB *Cell* **80:** 573–582.

Tojima Y, Fujimoto A, Delhase M, Chen Y, Hatakeyama S, Nakayama K, Kaneko Y, Nimura Y, Motoyama N, Ikeda K, et al. 2000. NAK is an IκB kinase-activating kinase. *Nature* **404:** 778–782.

Trinh DV, Zhu N, Farhang G, Kim BJ, Huxford T. 2008. The nuclear I κB protein IκBζ specifically binds NF-κB p50 homodimers and forms a ternary complex on κB DNA *J Mol Biol* **379:** 122–135.

Whiteside ST, Epinat JC, Rice NR, Isräel A. 1997. IκBε, a novel member of the IκB family, controls RelA and cRel NF-κB activity. *EMBO J* **16:** 1413–1426.

Woronicz JD, Gao X, Cao Z, Rothe M, Goeddel DV. 1997. IκB kinase-β: NF-κB activation and complex formation with IκB kinase-a and NIK *Science* **278:** 866–869.

Yamamoto M, Yamazaki S, Uematsu S, Sato S, Hemmi H, Hoshino K, Kaisho T, Kuwata H, Takeuchi O, Takeshige K, et al. 2004. Regulation of Toll/IL-1-receptor-mediated gene expression by the inducible nuclear protein IκBζ. *Nature* **430:** 218–222.

Yamamoto Y, Verma UN, Prajapati S, Kwak YT, Gaynor RB. 2003. Histone H3 phosphorylation by IKKα is critical for cytokine-induced gene expression. *Nature* **423:** 655–9.

Yamaoka S, Courtois G, Bessia C, Whiteside ST, Weil R, Agou F, Kirk HE, Kay RJ, Isräel A. 1998. Complementation cloning of NEMO, a component of the IκB kinase complex essential for NF-κB activation. *Cell* **93:** 1231–1240.

Yamazaki S, Muta T, Takeshige K. 2001. A novel IκB protein, IκB-ζ, induced by proinflammatory stimuli, negatively regulates nuclear factor-κB in the nuclei. *J Biol Chem* **276:** 27657–27662.

Zandi E, Rothwarf DM, Delhase M, Hayakawa M, Karin M. 1997. The IκB kinase complex (IKK) contains two kinase subunits, IKKα and IKKβ, necessary for IκB phosphorylation and NF-κB activation. *Cell* **91:** 243–252.

Zandi E, Chen Y, Karin M. 1998. Direct phosphorylation of IκB by IKKα and IKKβ: Discrimination between free and NF-κB-bound substrate. *Science* **281:** 1360–1363.

The IKK Complex, a Central Regulator of NF-κB Activation

Alain Israël

Unite de Signalisation Moleculaire et Activation Cellulaire, URA 2582 CNRS, Institut Pasteur, 25 rue du Dr Roux, 75724 Paris Cedex 15, France

Correspondance: aisrael@pasteur.fr

The IKK kinase complex is the core element of the NF-κB cascade. It is essentially made of two kinases (IKKα and IKKβ) and a regulatory subunit, NEMO/IKKγ. Additional components may exist, transiently or permanently, but their characterization is still unsure. In addition, it has been shown that two separate NF-κB pathways exist, depending on the activating signal and the cell type, the canonical (depending on IKKβ and NEMO) and the noncanonical pathway (depending solely on IKKα). The main question, which is still only partially answered, is to understand how an NF-κB activating signal leads to the activation of the kinase subunits, allowing them to phosphorylate their targets and eventually induce nuclear translocation of the NF-κB dimers. I will review here the genetic, biochemical, and structural data accumulated during the last 10 yr regarding the function of the three IKK subunits.

NF-κB represents a family of transcription factors that are normally kept inactive in the cytoplasm through interaction with inhibitory molecules of the IκB family. In response to multiple stimuli such as inflammatory cytokines, bacterial or viral products, or various types of stress, the IκB molecules become phosphorylated on two critical serine residues. This modification allows their polyubiquitination and destruction by the proteasome. As a consequence, free NF-κB enters the nucleus and activates transcription of a variety of genes participating in the immune and inflammatory response, cell adhesion, growth control, and protection against apoptosis.

The kinase(s) responsible for the phosphorylation of the IκB inhibitors remained elusive for many years, until the biochemical purification of a cytoplasmic high-molecular-weight complex migrating around 700–900 kDa and containing two related catalytic subunits, IKKα and IKKβ (Chen et al. 1996; Didonato et al. 1997). An additional component of the so-called IKK complex, NEMO/IKKγ (thereafter referred to as NEMO), has subsequently been identified through genetic complementation of an NF-κB activation-defective cell line (Yamaoka et al. 1998) and sequencing of IKK-associated polypeptides (Rothwarf et al. 1998; Mercurio et al. 1999). Although NEMO does not have catalytic properties, cell lines defective for NEMO do not activate NF-κB in response to many stimuli, demonstrating the key role of this protein in activating the NF-κB pathway.

THE KINASE SUBUNITS

The two kinase subunits of the IKK complex, IKKα and IKKβ, have been purified and cloned on the basis of their ability to phosphorylate the IκB inhibitors following biochemical purification of the complex (Mercurio et al. 1997). In addition, IKKα was cloned as an interactor of another kinase, NIK, which was later shown to be located upstream of IKKα in the noncanonical NF-κB activation pathway (Regnier et al. 1997). IKKα and IKKβ show a similar structure (50% sequence identity), and include an amino-terminal kinase domain, a helix-loop-helix (HLH) that functions in modulating IKK kinase activity, and a leucine zipper (LZ), whose role is to allow homo- or heterodimerization of the kinases (Fig. 1). In addition, IKKα shows a putative nuclear localization signal (Sil et al. 2004), possibly linked to its nuclear activity. IKKβ contains a ubiquitin-like domain whose function is currently unknown, but which does not seem

to be recognized by the ubiquitin binding domains that have been tested (May et al. 2004).

The interaction domain between the kinase subunits and NEMO was identified as a small peptide at the carboxyl terminus of IKKα and IKKβ. Interestingly, a cell-permeable 11-amino-acid peptide derived from the carboxyl terminus of IKKβ (aa 735–745) behaves as an effective and specific inhibitor of NF-κB by inhibiting the IKK–NEMO interaction and proved to be effective in two distinct experimental mouse models of acute inflammation (May et al. 2000).

The exact mechanism by which the kinase subunits become activated remains obscure. However, it is clear that to become active they need to be phosphorylated on two serine residues (Ser 177 and 181 for IKKβ, and Ser 176 and 180 for IKKα) located in an activation loop, similar to a large number of other kinases (Mercurio et al. 1997; Delhase et al. 1999). This phosphorylation probably leads to a conformational change and to kinase activation. Mutation of these Ser to Ala prevents activation of the kinases whereas mutation to Glu (phosphomimetic) renders them constitutively active. The search for an upstream kinase of IKKs has been extremely lengthy. It now

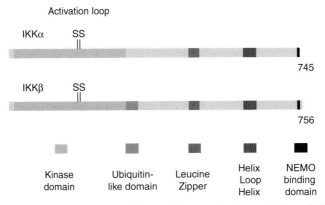

Figure 1. The kinase subunits. The domains of the two kinase subunits are indicated: The kinase domain is located at the amino-terminus (the activation loop is also shown: Amino acids 176–180 of IKKα and 177–181 of IKKβ). An ubiquitin-like domain (aa 307–384 in IKKβ) is located carboxy-terminal to the kinase domain of IKKβ (but not IKKα), and seems to be involved in the catalytic activity of IKKβ. The function of the leucine zipper domain is to allow homo- or heterodimerization of the kinases. The role of the helix loop helix domain is less clear, but it seems to be involved in the modulation of the kinase activity. Finally a ~40 amino-acid region at the extreme carboxyl terminus of the kinases (aa 705–743) is required for their interaction with NEMO.

seems clear that TAK1 behaves as an IKKK, at least in response to certain signals (Ninomiya-Tsuji et al. 1999; Wang et al. 2001). Indeed it has been shown that TAK1, which is also an upstream kinase for the JNK pathway, can phosphorylate IKKβ in the activation loop, that its down-regulation interferes with IKK activation, and that mutations in the TAK1 gene in *Drosophila* interfere with the NF-κB and JNK pathways (Vidal et al. 2001; Silverman et al. 2003). However it has also been shown that tissue-specific deletion of the TAK1 gene in the mouse does not lead to a defect in the NF-κB response to B-cell antigen (Sato et al. 2005). TAK1 is normally associated with the cofactors TAB1 and TAB2 (or TAB3). Although TAB2 or TAB3 seems to be involved in NF-κB activation, this does not seem to be the case for TAB1 (Shim et al. 2005). Another kinase, MEKK3, has also been suggested to act upstream of the IKK complex, as cells lacking MEKK3 are partially defective in NF-κB activation in response to certain stimuli (Yang et al. 2001; Huang et al. 2004). In addition, it has been suggested that two pathways, dependent on either TAK1 or MEKK3, diverge downstream of IRAK1 in the NF-κB response to IL1 (Qin et al. 2006; Yao et al. 2007).

Finally, another possibility of activation of the kinases, which might be involved in NF-κB activation by certain viral proteins, such as HTLV1 Tax, is that there would be no specific requirement for an upstream kinase and that the ability of the IKKs to autophosphorylate and phosphorylate their dimerization partner might be activated by stimulus-dependent conformational changes or oligomerization, something that might in part depend on the presence of NEMO.

Although phosphorylation of IKKβ is necessary for activation of the canonical pathway, phosphorylation of IKKα is not, though it is required for activation of the alternative pathway. As is often the case, the identity of the phosphatases that deactivate the kinases is relatively unclear. PP2A has been shown to dephosphorylate IKKβ in vitro (Didonato et al. 1997), whereas PP2Cβ seems to be associated with IKKβ when overexpressed (Prajapati et al. 2004).

The similarity between the two kinase subunits, as well as their presence in the same complex, suggested that they would probably have overlapping functions. Unexpectedly, subsequent biochemical and genetic approaches have indicated that they have relatively distinct substrates and functions. This in turn relates to the existence of two NF-κB pathways: The so-called canonical pathway is turned on by proinflammatory stimuli, such as TNF, IL1, or TLR ligands such as LPS, and is therefore involved in the innate response. This pathway essentially requires the IKKβ and NEMO subunit and leads to the rapid degradation of the IκB inhibitory molecules. More recently, an alternative or noncanonical NF-κB pathway has been identified (Senftleben et al. 2001) and shown to be involved in the response to ligands such as BAFF, CD40 ligand, and LTβ, and to be associated with lymphoid organogenesis. This pathway relies on the IKKα subunit and its upstream kinase NIK (Park et al. 2005), and does not require IKKβ nor NEMO. It leads to the slow partial degradation of the NF-κB precursor subunit p100 and the association of the processing product p52 with relB, another member of the NF-κB family. In addition to its role in the alternative pathway, IKKα shows functions that are independent of the NF-κB pathway. For example, it has been shown to bind promoters of estrogen responsive genes, such as cyclin D1 and c-myc, and to activate their transcription by forming a transcription complex with the estrogen receptor ERα and the coactivator AIB1/SRC-3 (Park et al. 2005).

Besides the IκBα inhibitors, other substrates of the IKK complex have been identified, including components of the NF-κB cascade such as NEMO itself, relA, c-rel, CylD, and Bcl10. The latter is a component of the NF-κB response to antigens in T cells (Thome and Tschopp 2002). Interestingly, IKK-mediated Bcl10 phosphorylation induces its degradation, resulting in down-regulation of NF-κB signaling (Lobry et al. 2007). This indicates that the IKK complex can behave both as a positive and a negative regulator of the cascade. Other substrates do not belong to the NF-κB cascade. They include IRS-1, a component of insulin signaling (Nakamori et al. 2006). IRS-1 phosphorylation has been shown to participate in the mechanism by which TNF attenuates insulin signaling through activation of the IKK complex. Another substrate is TSC1, a component of a tumor suppressor complex (Lee et al. 2007). Phosphorylation of TSC1 by IKKβ activates the mTOR pathway, enhances angiogenesis, and results in tumor development. Finally, IKK has also been shown to phosphorylate and induce the degradation of the transcription factor FOXO3a, which plays an important role in controlling cell proliferation and survival, therefore promoting tumorigenesis (Hu et al. 2004).

THE NEMO/IKKγ REGULATORY SUBUNIT, AND THE FIRST STRUCTURAL DATA

The third subunit of the IKK complex is a noncatalytic 48 kDa protein, called NEMO/IKKγ (Fig. 2). Although devoid of catalytic activity, NEMO is absolutely required for the canonical NF-κB activation pathway. Structure prediction, confirmed by recent X-ray crystallography data, indicates that it is essentially a long parallel dimeric intermolecular coiled coil, except for the carboxyl terminus. The amino-terminal part of NEMO (aa 47–120 in human NEMO) is responsible for interaction with the kinase subunits (May et al. 2000). The X-ray structure

of amino acids 44–111 of NEMO bound to amino acids 701–746 of IKKβ has been reported recently (Rushe et al. 2008): It forms an asymmetrical four-helix bundle made of a parallel NEMO dimer, each monomer being a crescent shape α-helix associated with two mainly helical IKKβ peptides, which do not interact with each other.

Interestingly, replacement of a phosphoacceptor Ser at position 68 by a phosphomimetic Glu decreases NEMO dimerization and reduces IKKβ binding (Palkowitsch et al. 2008). However, the kinase responsible for targeting Ser68 has not been unambiguously identified.

The rest of the molecule contains a first coiled coil (CC1, aa 100–196) and a second one (CC2, aa 255–291), followed by an LZ (LZ, aa 300–343) (see Fig. 1) and a carboxy-terminal ZF (aa 394–419).

Another recent report describes the X-ray structure of the central region of NEMO associated with a fragment of the viral protein vFLIP (Bagnéris et al. 2008). vFLIP is a transactivator encoded by the Kaposi's sarcoma herpes virus (KSHV) whose interaction with NEMO activates NF-κB. The structure indicates that the region encompassing aa 192–252 is another parallel intermolecular coiled coil. How vFLIP activates NF-κB cannot easily be deduced from this costructure.

A breakthrough in understanding the function of NEMO came when it was realized

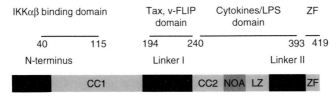

Figure 2. The NEMO molecule. Human NEMO is a 419–amino-acid dimeric molecule essentially structured under the form of a series of parallel intermolecular coiled coils (based on the available structural data). CC1, coiled coil 1; CC2, coiled coil 2; NOA, ubiquitin binding domain; ZF, Zinc Finger (and a second ubiquitin binding domain). The determination of the structure of linker 1 indicated that it is also structured as an intermolecular coiled coil. The structure of CC1 and linker II has not been determined yet. It must be stressed that the dimeric structure of NEMO is relatively unstable in the absence of interacting partners (kinases, polyubiquitin, …). The region of interaction with some of these partners has been indicated: The amino terminus is involved in the interaction with the two kinases. Linker 1 is involved in the interaction with viral transactivators such as HTLV1 Tax and KSHV v-FLIP. The entire carboxy-terminal region is required for transmission of the signal, and the NOA and the ZF domains bind polyubiquitin chains.

that nondegradative polyubiquitination of the Lys63-linked type played an important role in the NF-κB cascade. It was in 1996 that the requirement of nondegradative protein ubiquitination for IKK activation was first shown (Chen et al. 1996). Later on, biochemical purification and a reconstituted in vitro IKK activation system enabled the demonstration that TRAF6, an E3 ligase and a component of the NF-κB cascade downstream of proinflammatory molecules such as IL1 or LPS, associates with a dimeric E2 complex (Ubc13/Uev1a) to generate K63-linked polyubiquitin chains (Deng et al. 2000). This activity is necessary for IKK activation. Another necessary component of TRAF6-dependent NF-κB activation turned out to be a complex containing the kinase TAK1 and its two cofactors TAB1 and TAB2 (Wang et al. 2001). This complex is able to phosphorylate IKK in a manner dependent on TRAF6 and Ubc13/Uev1a (although the role of Ubc13 in IKK activation has been challenged [Habelhah et al. 2004; Yamamoto et al. 2006]). Later on it was shown that Lys63-linked polyubiquitination of several components of the cascade seems to be a general feature of the response to different types of stimuli (Chen 2005). This is in particular the case for RIP1, a kinase and adaptor in the NF-κB response to TNF. This modification serves to recruit proteins or protein complexes to polyubiquitinated substrates. Recently, it was shown that NEMO can specifically recognize Lys63-linked polyubiquitin chains and becomes itself ubiquitinated by the same type of chains following activation of the NF-κB cascade; these two properties seem to be required for NEMO activity (Tang et al. 2003; Ea et al. 2006; Wu et al. 2006a). As an example, the TNF cascade leading to NF-κB activation is shown in Figure 3: TNF induces trimerization of the TNF receptor and leads to the recruitment of TRADD, the E3 ubiquitin-ligase TRAF2 (and/or TRAF5), and the kinase RIP1 (other molecules may be recruited but will not be discussed here). Lys63-linked polyubiquitination of RIP1 on Lys 377, possibly mediated by TRAF2/5, leads to the recruitment of the TAK1/TAB1/TAB2 complex through

the ubiquitin-binding Zinc Finger of TAB2. Through an unknown mechanism this leads to the activation of the TAK1 kinase. The IKK complex is also recruited to these K63 polyubiquitin chains through the specific ubiquitin-binding domain of NEMO. Alternatively, NEMO ubiquitination might directly recruit the TAK1 complex, allowing it to phosphorylate and activate the IKKs. Irrespective of the precise mechanism, activated TAK1 phosphorylates IKKβ on its activation loop, leading to its activation and to phosphorylation of the inhibitor IκBα. IκBα is then polyubiquitinated through Lys48-linked polyubiquitin chains and degraded by the proteasome.

The importance of the ability of NEMO to bind ubiquitin has been confirmed by mutagenesis, as well as by the analysis of the mutations found associated with human pathologies (see the following discussion). Indeed, mutations in the domain of NEMO allowing recognition of K63-linked polyubiquitin chains (at aa 311, 315, 319) have been found in patients affected with EDA-ID. Database search allowed us to narrow down the region of NEMO involved in binding to ubiquitin to a small 30–40 aa domain located in the middle of the LZ domain, through identification of four other proteins that contain a similar region and also bind K63 linked polyubiquitin chains: optineurin and ABIN-1,2,3 (Sebban et al. 2006; Wagner et al. 2008). This region has been termed NOA/UBAN/NUB. Mutations in conserved residues interfere with binding of these proteins to ubiquitin, but the functional consequences are unclear in the case of ABINs as their actual function is still a matter of debate. Regarding optineurin, it has been shown recently that it acts as a negative regulator of TNF-induced NF-κB activation by competing with NEMO for binding to polyubiquitin chains (Zhu et al. 2007).

Very recently, a combination of X-ray and NMR analysis revealed the structural basis for recognition of polyubiquitin by the NOA domain (Lo et al. 2009). The data show that the region encompassing aa 265–330 forms a parallel intermolecular coiled coil with a kink at residue Pro299, which marks the boundary

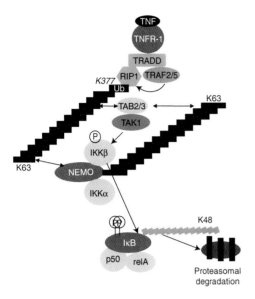

Figure 3. The NF-κB response to TNF. TNF induces trimerization of the TNF receptor, leading to the recruitment of TRADD, the E3 ubiquitin-ligase TRAF2 (and/or TRAF5), and the kinase RIP1 (other recruited molecules have been omitted for simplicity). K63-linked polyubiquitination of RIP1 on Lys 377, possibly mediated by TRAF2/5 (curved arrow), leads to the recruitment of the TAK1/TAB1/TAB2 complex through the ubiquitin-binding zinc finger of TAB2 (TAB2 can be replaced by TAB3, and TAB1 has been omitted for clarity). Through an unknown mechanism, this leads to the activation of the TAK1 kinase. The IKK complex is also recruited to these K63 polyubiquitin chains through the ubiquitin-binding domain of NEMO, allowing TAK1 to phosphorylate and activate the IKKs. K63-linked polyubiquitination of NEMO has been observed, but its actual role is currently unclear. One possibility is that it allows recruitment of the TAK1 complex in close proximity to the IKK kinase subunits, allowing their activation by TAK1; alternatively it might allow NEMO oligomerization through cross-recognition by its own ubiquitin binding domain. Whatever the exact mechanism of activation of the IKKs, they eventually phosphorylate the IκBα inhibitory subunit of NF-κB. IκBα is then polyubiquitinated through Lys48-linked polyubiquitin chains by the β-TrCP E3 ubiquitin ligase, leading to its degradation by the proteasome and to nuclear translocation of free NF-κB dimers, ultimately ending in activation of NF-κB target genes.

between the CC2 and LZ domains. A combination of mutagenesis and NMR analysis of a NOA-K63 diubiquitin complex reveals that diubiquitin is positioned perpendicular to the coiled coil and that each ubiquitin contacts both chains of a NEMO dimer. In addition, each ubiquitin interacts asymmetrically with NEMO. Interestingly, NEMO can also bind linear polyubiquitin with a much higher affinity and with a slightly different surface of contact. The physiological significance of this observation is currently unclear. However, it must be noted that a recent report (Tokunaga et al. 2009) describes the characterization of a dimeric E3 ligase made of two ring finger proteins, HOIP and HOIL-1L, which is able to generate linear polyubiquitin chains and to attach them to NEMO, onto Lys 285 and 309. This E3 ligase activity seems to be necessary for an NF-κB pathway which is independent of Ubc13.

Unpublished results from the group of F. Agou and M. Veron (Institut Pasteur) propose a different structural model whereby a dimeric NEMO molecule binds two K63-linked polyubiquitination chains that run parallel to the NEMO coiled coil (submitted). In parallel, the same group has shown that the carboxy-terminal zinc finger of NEMO is also able to bind ubiquitin (Cordier et al. 2009). A further collaboration with the group of F. Agou allowed us to demonstrate that the specificity for K63 chains of NEMO does not depend on NEMO or ZF alone, but requires both domains (submitted). Therefore, the carboxy-terminal half of NEMO seems to represent a new type of bipartite K63-specific ubiquitin binding domain. Determination of the X-ray structure of a complex between this bipartite module and a K63-linked polyubiquitin chain will be necessary to understand how the specificity is achieved. Interestingly, this bipartite module is conserved in two other proteins, ABIN2 and optineurin. These results suggest that the main function of NEMO is to bring its associated kinases to polyubiquitinated targets. This, however, requires some level of specificity and a basal level of affinity between NEMO and its targets, as has been shown for RIP1 (Zhang et al. 2000).

Although the role of the ubiquitin binding ability of NEMO is reasonably clear, the role of NEMO polyubiquitination is far less obvious. NEMO ubiquitination through K63-linked chains in response to multiple stimuli has been observed (see for example Tang et al. 2003; Zhou et al. 2004; Yamamoto et al. 2006), but the kinetics of this ubiquitination with regard to IKK activation remains unclear: Is it an early event required for activation of the IKK complex, or a late event involved in turning down NF-κB activation? It could, for example, be imagined that NEMO ubiquitination allows direct recruitment of the TAK1 complex (instead of IKK and TAK1 complexes being corecruited to polyubiquitinated RIP1) (see Fig. 3), which could then phosphorylate and activate the NEMO-associated kinases. Alternatively, ubiquitinated NEMO might allow the recruitment of deubiquitinases such as CylD or A20, which have been described as negative regulators of NF-κB, through the use of adaptor proteins such as the ABINs or TAX1BP1, which recognize K63-linked polyubiquitin chains (Mauro et al. 2006). From that point of view, one interesting observation is the fact that Ubc13-deficient cells are only mildly affected in NF-κB activation, whereas NEMO ubiquitination is strongly impaired (Yamamoto et al. 2006). The identification of the lysine residues which are targeted would probably help, but so far only a single site has been unambiguously identified by mass spectrometry: Lys285 is ubiquitinated in response to the NOD2-RIP2 pathway (Abbott et al. 2004). A Lys residue located in the zinc finger at position 399 has also been postulated to be an important site of ubiquitination, essentially on the basis of mutagenesis experiments, but has never been confirmed by mass

spectrometry analysis. Interestingly, this site is located in the middle of the ubiquitin binding surface of the zinc finger and its ubiquitination would interfere with NEMO binding to ubiquitin.

NEMO ubiquitination probably implies that it somehow needs to be deubiquitinated. A K63-specific deubiquitinase, CylD, has been found to interact with NEMO and to behave as a negative regulator of the NF-κB pathway (Sun 2008). CylD can deubiquitinate NEMO in vitro. Therefore it was postulated that CylD inhibits the NF-κB cascade by deubiquitinating specific substrates such as NEMO, RIP1, or others. Unfortunately, at the moment the identity of the physiological substrates of CylD is still unclear. Incidentally, other deubiquitinases such as A20 or Cezanne show the same K63 specificity and down-regulate the NF-κB pathway, but their exact substrates have not been identified yet (Sun 2008).

Although NEMO seems to be essentially dedicated to the NF-κB pathway, it has been shown recently that it also seems to bridge the NF-κB and the interferon regulatory factor (IRF) signaling pathways (Zhao et al. 2007). The IRF3/IRF7 pathway of the innate response to virus infection relies on two IKK-related kinases (TBK1 and IKKε), which phosphorylate the IRF3 and IRF7 transcription factors and allow their nuclear translocation and subsequent activation of their target genes. These two kinases belong to a complex that has been described to include one of three known additional subunits, TANK, NAP1, or SINTBAD (Chau et al. 2008). Zhao et al. showed that virus-induced activation of IRF3 and IRF7 requires NEMO, which acts upstream of the kinases TBK1 and IKKε (Zhao et al. 2007). The authors also demonstrate that NEMO does not directly interact with the two kinases, but binds TANK. The exact role of NEMO in this cascade remains, however, unclear.

OTHER POST-TRANSLATIONAL MODIFICATIONS OF NEMO

Although NEMO ubiquitination has been extensively discussed, few reports deal with

NEMO phosphorylation. IKK-dependent phosphorylation of NEMO on Ser 31, 43, 68, and 376 has been suggested to regulate NF-κB activity, but these phosphorylation events have never been confirmed by mass spectrometric analysis (Carter et al. 2003; Palkowitsch et al. 2008). On the other hand, some other phosphorylation sites have been identified by global proteome-wide mass spectrometric analysis, but their actual significance remains unclear.

In parallel, a specific NF-κB activating signal, DNA damage, has been studied in detail and has revealed the role of a further modification of NEMO, sumoylation. It was first shown that double-strand breaks induce NF-κB through an ATM–IKK pathway, raising the question of how a nuclear signal could trigger a cytoplasmic response (Wu and Miyamoto 2007). The conclusion of a series of studies is now that NF-κB activation by DNA-damaging agents is probably mediated by the conjunction of two convergent pathways: DNA damage-induced ATM activation and a parallel stress pathway that causes SUMO-1 modification of NEMO to permit NF-κB activation. Indeed it was shown that a sumoylated form of NEMO accumulated in the nucleus in response to genotoxic stress, and that this form of NEMO was apparently free (Huang et al. 2003). The sites of sumoylation have been identified as Lys 277 and 309. Additional proteins, PIDD (p53 induced protein with a death domain) and RIP1, were found to associate with NEMO in the nucleus, and to favor its sumoylation, although their exact role remains unclear (Janssens et al. 2005; Wu et al. 2006b). NEMO sumoylation is ATM-independent, but to allow NF-κB activation, NEMO must then be phosphorylated by ATM on Lys 85, leading to its monoubiquitination and to the export of a NEMO–ATM complex out of the nucleus (Huang et al. 2003). How this leads to activation of the IKK complex remains unclear, although it might involve the ELKS protein (see the following discussion). More recently the SUMO ligase responsible for NEMO sumoylation has been identified as PIASy (Mabb et al. 2006). The interaction between PIASy and NEMO is increased by genotoxic stress, and occurs in

the nucleus; NEMO–PIASy and NEMO–IKK interactions are mutually exclusive. One may wonder what is the role of NF-κB activation in response to DNA damaging agents. A series of reports suggest that NF-κB activation inhibits cell death induced by DNA-damaging anticancer drugs and radiation and thus facilitates malignant cell survival and growth.

NEMO AND THE FIRST NF-κB LINKED HUMAN GENETIC DISEASES

In 2000, the first human genetic diseases associated with mutations in the NF-κB pathway were identified (Smahi et al. 2000). Two distinct X-linked human diseases, incontinentia pigmenti (IP) and anhidrotic ectodermal dysplasia associated with immunodeficiency (EDA-ID), have been linked to NEMO dysfunction, providing a unique view of the role that NF-κB plays in human development, skin homeostasis, and innate/acquired immunity.

Loss of Function Mutations of NEMO Results in IP

IP is a severe X-linked (the IKBKG gene encoding NEMO is located in Xq28) genodermatosis that is lethal for males early during development. In females, the most characteristic feature of the disease is a dermatosis that usually begins after birth and evolves according to a stereotyped sequence. In addition to the manifestations at the epidermis, IP patients also suffer from ophthalmologic, odontological (missing or deformed teeth) and, in rare cases, neurological problems. Interestingly, 85% of the patients show the same deletion that results in an inactive NEMO truncation product and a complete loss of NF-κB activation by most known stimuli.

The complex skin phenotype of female IP patients is difficult to interpret, and it is impossible to cover all the possible explanations in this review (Smahi et al. 2002). NEMO knockout mice have been engineered by several groups (Makris et al. 2000; Rudolph et al. 2000; Schmidt-Supprian et al. 2000) and the phenotype of these mice is apparently similar to that of IP patients. Male mice die very early during embryogenesis (E12.5) from massive liver apoptosis, which is characteristic of a global defect in NF-κB activation. It is, however, not known whether liver apoptosis is also responsible for male lethality in IP patients.

Hypomorphic Mutations of NEMO Results in EDA-ID

As indicated above, complete NEMO loss of function is embryonic lethal for males but females can survive, showing a complex phenotype because of their mosaic character regarding X-inactivation. More interestingly in terms of understanding the role played by NF-κB in humans, hypomorphic NEMO mutations allow affected males to survive, and they show a pathology directly caused by a reduced response of the NF-κB cascade. This pathology, anhidrotic EDA-ID (also known as HED-ID, OMIM # 300291), associates the previously known symptoms of EDA (absence of sweat glands, sparce scalp hair, and missing teeth) with immunodeficiency (Doffinger et al. 2001). EDA is caused by mutations in several components of the EDA/EDAR cascade (EDA is a member of the TNF family and EDAR is its receptor), which is involved in the formation of ectodermal derivatives, and signals through NEMO and NF-κB. The mutations affecting NEMO interfere with both the EDA/EDAR cascade and the NF-κB-dependent pathways involved in the innate and acquired immune response. Mutations leading to EDA-ID have been shown to cover the entire NEMO coding region, and are essentially missense mutations and small deletions (Fusco et al. 2008; Hanson et al. 2008). Some patients carrying NEMO mutations show infectious susceptibility to specific pathogens, whereas others associate EDA-ID with osteopetrosis and lymphedema. The relevance to NF-κB of these two last symptoms is, however, relatively unclear.

An intriguing question is whether IP and EDA-ID represent the extreme boundaries of a genetic continuum. Indeed, depending on the severity of the mutation in terms of NF-κB response, male patients will show more or less severe pathologies linked to defective immune response and EDA/EDAR pathway, or will die before birth in the case of complete loss of function. However, caution should be taken when translating the severity of the mutation and its position in the molecule into a dissection of the functional domains of NEMO, as two independently isolated patients carrying the same mutation show different symptoms, emphasizing the influence of the genetic background (Doffinger et al. 2001; Jain et al. 2001).

ARE THERE OTHER COMPONENTS IN THE IKK COMPLEX?

First, it is important to mention that several types of complexes most likely exist in cells, although this has not been systematically studied: It has already been mentioned that IKKα dimers exist in the absence of NEMO and IKKβ, but is it likely that complexes containing NEMO and a dimer of IKKβ also exist. In addition, free NEMO is also present in cells, independently of the DNA damage response (Fontan et al. 2007; Wu and Miyamoto 2007).

The exact stoichiometry of the IKK complex has not been unambiguously determined, and its apparent molecular weight in sizing columns (600–800 kDa) does not help much, as it is essentially caused by the elongated shape of NEMO. Based on the most recent structural data, the stoichiometry is probably one NEMO dimer for two kinase subunits, but higher order structures may exist. As the IKK complex needs relatively harsh conditions to be purified, one may wonder whether other components exist, besides NEMO and the kinases. The literature has described a large number of potential candidates (see Table 1 in Sebban et al. 2006) identified through two-hybrid screens or coimmunoprecipitation, but it is unclear whether any of these proteins is a *bona fide* permanent component of the IKK complex. Some of these proteins might just transiently interact with IKK, which would still make them important components of the cascade, but precludes their detailed description within the restricted space of this article. It is, however, worth mentioning the ELKS protein (Ducut Sigala et al. 2004). Identified as a component of the IKK complex by purification and mass spectrometry, ELKS, a 105 kDa protein, is necessary for full NF-κB activation, and seems to be involved in recruiting IκBα to the IKK complex. As discussed previously, ELKS also seems to be involved in the NF-κB response to DNA damaging agents (Wu et al. 2006b).

Chaperones such as hsp90 and hsp70 have also been described as components of the IKK complex (Salminen et al. 2008). Hsp70 seems to behave as a NEMO-interacting inhibitor of NF-κB signaling, while hsp90 associated with its co-chaperone cdc37 behaves as a stabilizing factor of IKK through interaction between cdc37 and the kinase domains of IKKα and IKKβ. However, hsp90 also interacts with other kinases and seems to be a general stabilizer of kinase domain folding.

WHAT DO KO MICE TELL US ABOUT THE FUNCTION OF THE IKK SUBUNITS?

This section does not intend to cover the detailed phenotypes observed in mice carrying null or mutant alleles of the components of the IKK complex, but to briefly mention how in vivo studies have advanced our knowledge of the function of these proteins. The consequences of NEMO inactivation or mutation have been essentially studied in human patients affected with IP or EDA-ID (see previous discussion). Regarding the kinase subunits, it rapidly became clear that they differ in their substrate specificities and, as a result, have distinct biological functions. Although IKKβ is a true IκB kinase, IKKα is not, and shows other, not necessary NF-κB-related activities. IKKβ was originally considered as the essential subunit responsible for NF-κB response to cytokines and various pathogen-derived antigens, which was confirmed by subsequent studies; KO of IKKβ results in embryonic lethality because of massive hepatocyte apoptosis (Li et al. 1999b; Li et al. 1999c; Tanaka et al. 1999), a mark of complete NF-κB inactivation. This confirms the major role of IKKβ in the NF-κB response. However, compared for example with a NEMO

KO, death occurs a few days later, suggesting the possibility that IKKα can partially compensate for the absence of IKKβ. Compared to IKKβ, the IKKα subunit seems to show much more diverse functions. As mentioned above, it does play a role, although probably minor, in the canonical NF-κB cascade. Unexpectedly, it plays a prominent role in RANK-induced classical NF-κB activation in mammary epithelial cells (Cao et al. 2001), through NF-κB-mediated optimal cyclin D1 induction. It is, however, unclear why it is specifically IKKα that regulates these events. Based on IKKα specificity for the p100 precursor of p52 and its prominent role in the noncanonical cascade (Senftleben et al. 2001), one might have expected specific immune defects affecting secondary lymphoid organs and B cells in IKKα KO mice. Unexpectedly, these KO mice die a few days after birth and show defects affecting multiple morphogenetic events, including limb and skeletal patterning, and show keratinocyte hyperproliferation without differentiation (Hu et al. 1999; Li et al. 1999a; Takeda et al. 1999). Further studies concluded that IKKα controls keratinocyte differentiation, but that this pathway does not rely on NF-κB nor on the kinase activity of IKKα (Hu et al. 2001). More recently, the crucial nuclear role of IKKα in this phenomenon was shown through the identification of a functional NLS in the IKKα kinase domain (Sil et al. 2004). Inactivation of this NLS represses keratinocyte differentiation, indicating that IKKα exerts its function within the nucleus of basal keratinocytes in the epidermis. It must be noted that the expected B-cell phenotype associated with a defect in IKKα could indeed be observed in experiments involving establishment of bone marrow chimera through transfer of IKKα-deficient fetal liver cells (Kaisho et al. 2001).

Later on it was shown that IKKα can also be found in the nucleus in the absence of NEMO and IKKβ, where it acts at different levels to regulates NF-κB-dependent and -independent gene expression (Gloire et al. 2006). It was first shown that TNF induces the recruitment of IKKα, together with relA and CBP, onto the promoter of NF-κB target genes (such as

IκBα, IL-8, or IL-6), where it phosphorylates histone H3 on Ser10, triggering its subsequent CBP-mediated acetylation on Lys14, a crucial step in modulating chromatin accessibility (Anest et al. 2003; Yamamoto et al. 2003). A similar situation was observed following LPS treatment of macrophages (Park et al. 2006). IKKα was later shown to regulate additional steps of NF-κB-dependent gene transcription. First it allows derepression of NF-κB target genes by phosphorylating the SMRT repressor, which is recruited by p50 and p52 homodimers, and inducing its nuclear export (together with HDAC3) and degradation (Hoberg et al. 2004). Then it phosphorylates chromatin-bound p65 on Ser536, leading to the displacement of the SMRT-HDAC3 repressor activity and allowing p300 to acetylate p65 at Lys310, an event necessary for full transcription (Hoberg et al. 2006).

Unexpectedly, a similar type of activity of IKKα has been shown to be involved in the negative regulation of macrophage activation and inflammation and to represent a crucial element in limiting the inflammatory response to Gram-negative infection (Lawrence et al. 2005). The molecular basis of this activity of IKKα was its ability to act as a chromatin-bound relA carboxy-terminal kinase responsible for p65 turnover, in turn limiting macrophage activation and inflammation.

REFERENCES

Abbott DW, Wilkins A, Asara JM, Cantley LC. 2004. The Crohn's disease protein, NOD2, requires RIP2 in order to induce ubiquitinylation of a novel site on NEMO. *Curr Biol* **14:** 2217–2227.

Anest V, Hanson JL, Cogswell PC, Steinbrecher KA, Strahl BD, Baldwin AS. 2003. A nucleosomal function for IkB kinase a (IKKa) is essential for NF-kB dependent gene expression. *Nature* **423:** 659–663.

Bagnéris C, Ageichik AV, Cronin N, Wallace B, Collins M, Boshoff C, Waksman G, Barrett T. 2008. Crystal structure of a vFlip-IKKγ complex: Insights into viral activation of the IKK signalosome. *Molecular Cell* **30:** 620–631.

Cao Y, Bonizzi G, Seagroves TN, Greten FR, Johnson R, Schmidt EV, Karin M. 2001. IKKα Provides an Essential Link between RANK Signaling and Cyclin D1 Expression during Mammary Gland Development. *Cell* **107:** 763–775.

Carter RS, Pennington KN, Ungurait BJ, Ballard DW. 2003. In Vivo Identification of Inducible Phosphoacceptors in the IKKγ/NEMO Subunit of Human IκB Kinase. *J Biol Chem* **278:** 19642–19648.

Chau TL, Gioia R, Gatot JS, Patrascu F, Carpentier I, Chapelle JP, O'Neill LAJ, Beyaert R, Piette J, Chariot A. 2008. Are the IKKs and IKK-related kinases TBK1 and IKK-ε similarly activated? *Trends Biochem Sci* **33:** 171–180.

Chen ZJ. 2005. Ubiquitin signalling in the NF-κB pathway. *Nat Cell Biol* **7:** 758–765.

Chen ZJ, Parent L, Maniatis T. 1996. Site-specific phosphorylation of IkBa by a novel ubiquitination-dependent protein kinase activity. *Cell* **84:** 853–862.

Cordier F, Grubisha O, Traincard F, Véron M, Delepierre M, Agou F. 2009. The zinc finger of NEMO is a functional ubiquitin-binding domain. *J Biol Chem* **284:** 2902–2907.

Delhase M, Hayakawa M, Chen Y, Karin M. 1999. Positive and negative regulation of I κ B kinase activity through IKK β subunit phosphorylation. *Science* **284:** 309–313.

Deng L, Wang C, Spencer E, Yang LY, Braun A, You JX, Slaughter C, Pickart C, Chen ZJ. 2000. Activation of the I κ B kinase complex by TRAF6 requires a dimeric ubiquitin-conjugating enzyme complex and a unique polyubiquitin chain. *Cell* **103:** 351–361.

Didonato JA, Hayakawa M, Rothwarf DM, Zandi E, Karin M. 1997. A cytokine-responsive IkB kinase that activates the transcription factor NF-kb. *Nature* **388:** 548–554.

Doffinger R, Smahi A, Bessia C, Geissmann F, Feinberg J, Durandy A, Bodemer C, Kenwrick S, Dupuis-Girod S, Blanche S, et al. 2001. X-linked anhidrotic ectodermal dysplasia with immunodeficiency is caused by impaired NF-κB signaling. *Nat Genet* **27:** 277–285.

Ducut Sigala JL, Bottero V, Young DB, Shevchenko A, Mercurio F, Verma IM. 2004. Activation of transcription factor NF-κB requires ELKS, an IκB kinase regulatory subunit. *Science* **304:** 1963–1967.

Ea CK, Deng L, Xia ZP, Pineda G, Chen ZJ. 2006. Activation of IKK by TNFα Requires Site-Specific Ubiquitination of RIP1 and Polyubiquitin Binding by NEMO. *Mol Cell* **22:** 1–13.

Fontan E, Traincard F, Levy SG, Yamaoka S, Veron M, Agou F. 2007. NEMO oligomerization in the dynamic assembly of the IκB kinase core complex. *Febs J* **274:** 2540–2551.

Fusco F, Pescatore A, Bal E, Ghoul A, Paciolla M, Lioi MB, D'Urso M, Rabia SH, Bodemer C, Bonnefont JP, et al. 2008. Alterations of the IKBKG locus and diseases: An update and a report of 13 novel mutations. *Hum Mutat* **29:** 595–604.

Gloire G, Dejardin E, Piette J. 2006. Extending the nuclear roles of IκB kinase subunits. *Biochem Pharmacol* **72:** 1081–1089.

Habelhah H, Takahashi S, Cho SG, Kadoya T, Watanabe T, Ronai Z. 2004. Ubiquitination and translocation of TRAF2 is required for activation of JNK but not of p38 or NF-κB. *Embo J* **23:** 322–332.

Hanson EP, Monaco-Shawver L, Solt LA, Madge LA, Banerjee PP, May MJ, Orange JS. 2008. Hypomorphic nuclear factor-κB essential modulator mutation database and reconstitution system identifies phenotypic and immunologic diversity. *J Allergy Clin Immunol* **122:** 1169–1177.e1116.

Hoberg JE, Yeung F, Mayo MW. 2004. SMRT derepression by the IκB kinase α: A prerequisite to NF-κB transcription and survival. *Mol Cell* **16:** 245–255.

Hoberg JE, Popko AE, Ramsey CS, Mayo MW. 2006. IκB kinase α-mediated derepression of SMRT potentiates acetylation of RelA/p65 by p300. *Mol Cell Biol* **26:** 457–471.

Hu YL, Baud V, Delhase M, Zhang PL, Deerinck T, Ellisman M, Johnson R, Karin M. 1999. Abnormal morphogenesis but intact IKK activation in mice lacking the IKKa subunit of IkB kinase. *Science* **284:** 316–320.

Hu Y, Baud V, Oga T, Kim KI, Yoshida K, Karin M. 2001. IKKa controls formation of the epidermis independently of NF-kB. *Nature* **410:** 710–714.

Hu MC, Lee DF, Xia W, Golfman LS, Ou-Yang F, Yang JY, Zou Y, Bao S, Hanada N, Saso H, et al. 2004. IκB kinase promotes tumorigenesis through inhibition of forkhead FOXO3a. *Cell* **117:** 225–237.

Huang TT, Wuerzberger-Davis SM, Wu ZH, Miyamoto S. 2003. Sequential modification of NEMO/IKKγ by SUMO-1 and ubiquitin mediates NF-κB activation by genotoxic stress. *Cell* **115:** 565–576.

Huang Q, Yang J, Lin Y, Walker C, Cheng J, Liu ZG, Su B. 2004. Differential regulation of interleukin 1 receptor and Toll-like receptor signaling by MEKK3. *Nat Immunol* **5:** 98–103.

Jain A, Ma CA, Liu S, Brown M, Cohen J, Strober W. 2001. Specific missense mutations in NEMO result in hyper-IgM syndrome with hypohydrotic ectodermal dysplasia. *Nature Immunol* **2:** 223–228.

Janssens S, Tinel A, Lippens S, Tschopp J. 2005. PIDD mediates NF-κB activation in response to DNA damage. *Cell* **123:** 1079–1092.

Kaisho T, Takeda K, Tsujimura T, Kawai T, Nomura F, Terada N, Akira S. 2001. IκB kinase α is essential for mature B cell development and function. *J Exp Med* **193:** 417–426.

Lawrence T, Bebien M, Liu GY, Nizet V, Karin M. 2005. IKKα limits macrophage NF-κB activation and contributes to the resolution of inflammation. *Nature* **434:** 1138–1143.

Lee DF, Kuo HP, Chen CT, Hsu JM, Chou CK, Wei Y, Sun HL, Li LY, Ping B, Huang WC, et al. 2007. IKK β suppression of TSC1 links inflammation and tumor angiogenesis via the mTOR pathway. *Cell* **130:** 440–455.

Li QT, Lu QX, Hwang JY, Buscher D, Lee KF, Izpisua-Belmonte JC, Verma IM. 1999a. IKK1-deficient mice exhibit abnormal development of skin and skeleton. *Genes Dev* **13:** 1322–1328.

Li QT, Van Antwerp D, Mercurio F, Lee KF, Verma IM. 1999b. Severe liver degeneration in mice lacking the IkB kinase 2 gene. *Science* **284:** 321–325.

Li ZW, Chu WM, Hu YL, Delhase M, Deerinck T, Ellisman M, Johnson R, Karin M. 1999c. The IKKb subunit of IkB kinase (IKK) is essential for Nuclear Factor κ B activation and prevention of apoptosis. *J Exp Med* **189:** 1839–1845.

Lo YC, Lin SY, Rospigliosi CC, Conze DB, Wu CJ, Ashwell JD, Eliezer D, Wu H. 2009. Structural Basis for Recognition of Diubiquitins by NEMO. *Mol Cell*: doi:10.1016/j.mol- cel.2009.1001.1012.

Lobry C, Lopez T, Israel A, Weil R. 2007. Negative feedback loop in T cell activation through IκB kinase-induced phosphorylation and degradation of Bcl10. *Proc Natl Acad Sci* **104:** 908–913.

Mabb AM, Wuerzberger-Davis SM, Miyamoto S. 2006. PIASy mediates NEMO sumoylation and NF-κB activation in response to genotoxic stress. *Nat Cell Biol* **8:** 986–993.

Makris C, Godfrey VL, Krahn-Senftleben G, Takahashi T, Roberts JL, Schwarz T, Feng LL, Johnson RS, Karin M. 2000. Female mice heterozygous for IKK γ/NEMO deficiencies develop a dermatopathy similar to the human X-linked disorder incontinentia pigmenti. *Mol Cell* **5:** 969–979.

Mauro C, Pacifico F, Lavorgna A, Mellone S, Iannetti A, Acquaviva R, Formisano S, Vito P, Leonardi A. 2006. ABIN-1 binds to NEMO/IKKγ and co-operates with A20 in inhibiting NF-κB. *J Biol Chem* **281:** 18482–18488.

May MJ, D'Acquisto F, Madge LA, Glockner J, Pober JS, Ghosh S. 2000. Selective inhibition of NF-κ B activation by a peptide that blocks the interaction of NEMO with the I κ B kinase complex. *Science* **289:** 1550–1554.

May MJ, Larsen SE, Shim JH, Madge LA, Ghosh S. 2004. A novel ubiquitin-like domain in IκB kinase β is required for functional activity of the kinase. *J Biol Chem* **279:** 45528–45539.

Mercurio F, Zhu H, Murray BW, Shevchenko A, Bennett BL, Li J, Young DB, Barbosa M, Mann M, Manning A, et al. 1997. IKK-1 and IKK-2: Cytokine-activated IκB kinases essential for NF-κB activation. *Science* **278:** 860–866.

Mercurio F, Murray BW, Shevchenko A, Bennett BL, Young DB, Li JW, Pascual G, Motiwala A, Zhu H, Mann M, et al. 1999. IkB kinase (IKK)-associated protein 1, a common component of the heterogeneous IKK complex. *Mol Cell Biol* **19:** 1526–1538.

Nakamori Y, Emoto M, Fukuda N, Taguchi A, Okuya S, Tajiri M, Miyagishi M, Taira K, Wada Y, Tanizawa Y. 2006. Myosin motor Myo1c and its receptor NEMO/IKK-γ promote TNF-α-induced serine307 phosphorylation of IRS-1. *J Cell Biol* **173:** 665–671.

Ninomiya-Tsuji J, Kishimoto K, Hiyama A, Inoue J, Cao ZD, Matsumoto K. 1999. The kinase TAK1 can activate the NF-κB as well as the MAP kinase cascade in the IL-1 signalling pathway. *Nature* **398:** 252–256.

Palkowitsch L, Leidner J, Ghosh S, Marienfeld RB. 2008. Phosphorylation of serine 68 in the IκB kinase (IKK)-binding domain of NEMO interferes with the structure of the IKK complex and tumor necrosis factor-α-induced NF-κB activity. *J Biol Chem* **283:** 76–86.

Park KJ, Krishnan V, O'Malley BW, Yamamoto Y, Gaynor RB. 2005. Formation of an IKK[α]-Dependent Transcription Complex Is Required for Estrogen Receptor-Mediated Gene Activation. *Mol Cell* **18:** 71–82.

Park GY, Wang X, Hu N, Pedchenko TV, Blackwell TS, Christman JW. 2006. NIK is involved in nucleosomal regulation by enhancing histone H3 phosphorylation by IKKα. *J Biol Chem* **281:** 18684–18690.

Prajapati S, Verma U, Yamamoto Y, Kwak YT, Gaynor RB. 2004. Protein phosphatase 2Cβ association with the IκB kinase complex is involved in regulating NF-κB activity. *J Biol Chem* **279:** 1739–1746.

Cite this article as *Cold Spring Harb Perspect Biol* 2010;2:a000158

Qin J, Yao J, Cui G, Xiao H, Kim TW, Fraczek J, Wightman P, Sato S, Akira S, Puel A, et al. 2006. TLR8-mediated NF-κB and JNK activation are TAK1-independent and MEKK3-dependent. *J Biol Chem* **281**: 21013–21021.

Regnier CH, Song HY, Gao X, Goeddel DV, Cao Z, Rothe M. 1997. Identification and characterization of an IκB kinase. *Cell* **90**: 373–383.

Rothwarf DM, Zandi E, Natoli G, Karin M. 1998. IKK-g is an essential regulatory subunit of the IkB kinase complex. *Nature* **395**: 297–300.

Rudolph D, Yeh WC, Wakeham A, Rudolph B, Nallainathan D, Elia A, Potter J, Mak TW. 2000. Severe liver degeneration and lack of NF-kB activation in NEMO/IKK-γ deficient mice. *Genes Dev* **14**: 854–862.

Rushe M, Silvian L, Bixler S, Chen LL, Cheung A, Bowes S, Cuervo H, Berkowitz S, Zheng T, Guckian K, et al. 2008. Structure of a NEMO/IKK-Associating Domain Reveals Architecture of the Interaction Site. *Structure* **16**: 798–808.

Salminen A, Paimela T, Suuronen T, Kaarniranta K. 2008. Innate immunity meets with cellular stress at the IKK complex: Regulation of the IKK complex by HSP70 and HSP90. *Immunology Letters* **117**: 9–15.

Sato S, Sanjo H, Takeda K, Ninomiya-Tsuji J, Yamamoto M, Kawai T, Matsumoto K, Takeuchi O, Akira S. 2005. Essential function for the kinase TAK1 in innate and adaptive immune responses. *Nat Immunol* **6**: 1087–1095.

Schmidt-Supprian M, Bloch W, Courtois G, Addicks K, Israël A, Rajewsky K, Pasparakis M. 2000. NEMO/IKK γ-deficient mice model Incontinentia Pigmenti. *Mol Cell* **5**: 981–992.

Sebban H, Yamaoka S, Courtois G. 2006. Posttranslational modifications of NEMO and its partners in NF-κB signaling. *Trends Cell Biol* **16**: 569–577.

Senftleben U, Cao Y, Xiao G, Greten FR, Krahn G, Bonizzi G, Chen Y, Hu Y, Fong A, Sun SC, et al. 2001. Activation by IKKα of a second, evolutionary conserved, NF-κ B signaling pathway. *Science* **293**: 1495–1499.

Shim JH, Xiao C, Paschal AE, Bailey ST, Rao P, Hayden MS, Lee KY, Bussey C, Steckel M, Tanaka N, et al. 2005. TAK1, but not TAB1 or TAB2, plays an essential role in multiple signaling pathways in vivo. *Genes Dev* **19**: 2668–2681.

Sil AK, Maeda S, Sano Y, Roop DR, Karin M. 2004. IκB kinase-α acts in the epidermis to control skeletal and craniofacial morphogenesis. *Nature* **428**: 660–664.

Silverman N, Zhou R, Erlich RL, Hunter M, Bernstein E, Schneider D, Maniatis T. 2003. Immune activation of NF-κB and JNK requires Drosophila TAK1. *J Biol Chem* **278**: 48928–48934.

Smahi A, Courtois G, Vabres P, Yamaoka S, Heuertz S, Munnich A, Israël A, Heiss NS, Klauck S, Kioschis P, et al. 2000. Genomic rearrangement in NEMO impairs NF-kB activation and is a cause of Incontinentia Pigmenti. *Nature* **405**: 466–472.

Smahi A, Courtois G, Rabia SH, Doffinger R, Bodemer C, Munnich A, Casanova JL, Israël A. 2002. The NF-κB signalling pathway in human diseases: From incontinentia pigmenti to ectodermal dysplasias and immune-deficiency syndromes. *Hum Mol Genet* **11**: 2371–2375.

Sun SC. 2008. Deubiquitylation and regulation of the immune response. *Nat Rev Immunol* **8**: 501–511.

Takeda K, Takeuchi O, Tsujimura T, Itami S, Adachi O, Kawai T, Sanjo H, Yoshikawa K, Terada N, Akira S. 1999. Limb and skin abnormalities in mice lacking IKKa. *Science* **284**: 313–316.

Tanaka M, Fuentes ME, Yamaguchi K, Durnin MH, Dalrymple SA, Hardy KL, Goeddel DV. 1999. Embryonic lethality, liver degeneration, and impaired NF-kB activation in IKKb-deficient mice. *Immunity* **10**: 421–429.

Tang ED, Wang CY, Xiong Y, Guan KL. 2003. A role for NF-κB essential modifier/IκB kinase-γ (NEMO/IKKγ) ubiquitination in the activation of the IκB kinase complex by tumor necrosis factor-α. *J Biol Chem* **278**: 37297–37305.

Thome M, Tschopp J. 2002. Bcl10. *Curr Biol* **12**: R45.

Tokunaga F, Sakata S, Saeki Y, Satomi Y, Kirisako T, Kamei K, Nakagawa T, Kato M, Murata S, Yamaoka S, et al. 2009. Involvement of linear polyubiquitylation of NEMO in NF-κB activation. *Nat Cell Biol* **11**: 123–132.

Vidal S, Khush RS, Leulier F, Tzou P, Nakamura M, Lemaitre B. 2001. Mutations in the Drosophila dTAK1 gene reveal a conserved function for MAPKKKs in the control of rel/NF-κB-dependent innate immune responses. *Genes Dev* **15**: 1900–1912.

Wagner S, Carpentier I, Rogov V, Kreike M, Ikeda F, Lohr F, Wu CJ, Ashwell JD, Dotsch V, Dikic I, et al. 2008. Ubiquitin binding mediates the NF-κB inhibitory potential of ABIN proteins. *Oncogene* **27**: 3739–3745.

Wang C, Deng L, Hong M, Akkaraju GR, Inoue J, Chen ZJ. 2001. TAK1 is a ubiquitin-dependent kinase of MKK and IKK. *Nature* **412**: 346–351.

Wu Miyamoto S. 2007. Many faces of NF-κB signaling induced by genotoxic stress. *J Mol Med* **85**: 1187–1202.

Wu CJ, Conze DB, Li T, Srinivasula SM, Ashwell JD. 2006a. NEMO is a sensor of Lys 63-linked polyubiquitination and functions in NF-κB activation. *Nat Cell Biol* **8**: 398–406.

Wu ZH, Shi Y, Tibbetts RS, Miyamoto S. 2006b. Molecular Linkage Between the Kinase ATM and NF-κB Signaling in Response to Genotoxic Stimuli. *Science* **311**: 1141–1146.

Yamamoto Y, Verma U, Prajapati S, Kwak YT, Gaynor RB. 2003. Histone H3 phosphorylation by IKKα is critical for cytokine-induced gene expression. *Nature* **423**: 655–659.

Yamamoto M, Okamoto T, Takeda K, Sato S, Sanjo H, Uematsu S, Saitoh T, Yamamoto N, Sakurai H, Ishii KJ, et al. 2006. Key function for the Ubc13 E2 ubiquitin-conjugating enzyme in immune receptor signaling. *Nat Immunol* **7**: 962–970.

Yamaoka S, Courtois G, Bessia C, Whiteside ST, Weil R, Agou F, Kirk HE, Kay RJ, Israël A. 1998. Complementation cloning of NEMO, a component of the IkB kinase complex essential for NF-kB activation. *Cell* **93**: 1231–1240.

Yang J, Lin Y, Guo Z, Cheng J, Huang J, Deng L, Liao W, Chen Z, Liu Z, Su B. 2001. The essential role of MEKK3 in TNF-induced NF-κB activation. *Nat Immunol* **2**: 620–624.

Yao J, Kim TW, Qin J, Jiang Z, Qian Y, Xiao H, Lu Y, Qian W, Gulen MF, Sizemore N, et al. 2007. Interleukin-1

(IL-1)-induced TAK1-dependent Versus MEKK3-dependent NFκB activation pathways bifurcate at IL-1 receptor-associated kinase modification. *J Biol Chem* **282:** 6075–6089.

Zhang SQ, Kovalenko A, Cantarella G, Wallach D. 2000. Recruitment of the IKK signalosome to the p55 TNF receptor: RIP and A20 bind to NEMO (IKKγ) upon receptor stimulation. *Immunity* **12:** 301–311.

Zhao T, Yang L, Sun Q, Arguello M, Ballard DW, Hiscott J, Lin R. 2007. The NEMO adaptor bridges the nuclear factor-κB and interferon regulatory factor signaling pathways. *Nat Immunol* **8:** 592–600.

Zhou H, Wertz I, O'Rourke K, Ultsch M, Seshagiri S, Eby M, Xiao W, Dixit VM. 2004. Bcl10 activates the NF-κB pathway through ubiquitination of NEMO. *Nature* **427:** 167–171.

Zhu G, Wu CJ, Zhao Y, Ashwell JD. 2007. Optineurin negatively regulates TNFα- induced NF-κB activation by competing with NEMO for ubiquitinated RIP. *Curr Biol* **17:** 1438–1443.

Ubiquitination and Degradation of the Inhibitors of NF-κB

Naama Kanarek, Nir London, Ora Schueler-Furman, and Yinon Ben-Neriah

Departments of Immunology and Genetics and Biotechnology, Hebrew University-Hadassah Medical School, Institute of Medical Research Israel-Canada, Jerusalem, 91120, Israel

Correspondence: yinon@cc.huji.ac.il

The key step in NF-κB activation is the release of the NF-κB dimers from their inhibitory proteins, achieved via proteolysis of the IκBs. This irreversible signaling step constitutes a commitment to transcriptional activation. The signal is eventually terminated through nuclear expulsion of NF-κB, the outcome of a negative feedback loop based on IκBα transcription, synthesis, and IκBα-dependent nuclear export of NF-κB (Karin and Ben-Neriah 2000). Here, we review the process of signal-induced IκB ubiquitination and degradation by comparing the degradation of several IκBs and discussing the characteristics of IκBs' ubiquitin machinery.

Retrospectively, one of the most remarkable milestones in NF-κB research was the identification of its cytoplasmic inhibitor IκB (Inhibitor of NF-κB), soon after the discovery of the DNA binding activity of the factor (Sen and Baltimore 1986a; Sen and Baltimore 1986b; Baeuerle and Baltimore 1988). It immediately focused the research of NF-κB activation on the mechanism of liberation of the factor from the inhibitory effects of IκB. Using a simple detergent treatment of cell extracts, Baeuerle and Baltimore provided the first proof-of-concept for a model entailing the dissociation of the inhibitor as a major step in NF-κB activation (Baeuerle and Baltimore 1988). Subsequent experiments in cell lines suggested that stimulus-induced IκB phosphorylation triggered release of associated NF-κB, which could account for physiological activation of NF-κB (Ghosh and Baltimore 1990). Yet, later experiments showed that IκB phosphorylation was insufficient for NF-κB activation (Alkalay et al. 1995a; DiDonato et al. 1995), and IκB degradation must precede NF-κB activation (Beg et al. 1993; Brown et al. 1993; Mellits et al. 1993; Sun et al. 1993). Baeuerle and colleagues showed that blockade of IκB degradation prevented NF-κB activation (Henkel et al. 1993), but elucidation of the mechanism of IκB proteolysis awaited further studies by Maniatis, Goldberg, Ben-Neriah, Ciechanover, and colleagues who showed that signal-induced ubiquitination and proteasomal degradation of IκB were required for NF-κB activation (Palombella et al. 1994; Alkalay et al. 1995b; Chen et al. 1995). These findings implicated for the first time ubiquitin-dependent proteolysis as an integral step of signal induced transcriptional activation, yet the specificity basis for IκB degradation remained to be shown. Sequence homology comparisons and site-directed mutagenesis revealed a conserved amino-terminal sequence containing two serine residues that were found to be phosphorylated after phorbol ester stimulation and are necessary for signal-induced IκBα ubiquitination and degradation by the proteasome (Brown et al. 1995; Chen et al. 1995). This information was critical for the subsequent molecular characterization of both the IκB kinase and ubiquitin ligase (E3). Using a set of specific phosphopeptides, Yaron et al. showed that the phosphorylation-based motif (DpSGXXpS) of IκBα is sufficient for IκB recognition by components of the ubiquitin-system in vitro (Yaron et al. 1997). Subsequently, the specificity component of the E3, β-TrCP, was identified by mass spectroscopy as a protein that specifically interacts with the IκB phosphopeptide motif (Yaron et al. 1998). Significantly, a *Drosophila* homolog of β-TrCP, *slimb*, had previously been identified by Jiang and Struhl in a genetic screen as a likely E3 for β-catenin and cubitus interruptus (Ci) (Jiang and Struhl 1998). Several labs (Spencer et al. 1999; Tan et al. 1999; Winston et al. 1999) then showed that β-TrCP is the substrate binding (receptor) subunit of an SCF-type E3 ligase (Deshaies 1999). The structural basis for recognition of phospho-IκBα by the IκB E3 ligase was then established by Pavletich and his colleagues, who solved the crystal structure of the substrate interacting pocket of β-TrCP in complex with the β-catenin phosphopeptide (sharing with IκBα the destruction motif, i.e., degron) (Wu et al. 2003). β-TrCP was subsequently shown to control the proteasome-mediated degradation of other IκBs (IκBβ and IκBε) (Shirane et al. 1999; Wu and Ghosh 1999), and of NF-κB1/p105 (Orian et al. 2000; Heissmeyer et al. 2001; Lang et al. 2003), as well as the signal-induced processing of NF-κB2/p100 to p52 by the proteasome (Fong and Sun 2002).

THE IκB E3 UBIQUITIN LIGASE

The half-life of IκBα in resting cells is 138 min, two orders of magnitude longer than the half-life of IκBα in stimulated cells—1.5 min (Henkel et al. 1993). The signal that transforms the stable IκBα in resting cells to the fast degrading form in stimulated cells is the phosphorylation of residues 32 and 36 (Chen et al. 1995; Yaron et al. 1997). Rapid degradation of the phosphorylated IκBα demands an E3 with features allowing the speed, specificity, and inducibility that characterize IκBα degradation. An SCF-type E3 possesses the required characteristics: (1) It recognizes mainly phosphorylated substrates, allowing recognition only after signal induced phosphorylation of the substrate, and (2) it has a variable receptor subunit, the F-box protein, which recognizes a wide spectrum of different substrates, permitting the necessary specificity (Deshaies 1999). The SCF complex forms a bridge between the E2 ubiquitin-conjugating enzyme and the different substrates. This interaction increases the effective concentration of the target lysine residues near the active site of the E2, thus catalyzing the ubiquitination of the attached substrate (Wu et al. 2003). The SCF complex is composed of the adapter protein Skp1 (Feldman et al. 1997), the scaffold protein cullin-1 (Feldman et al. 1997), the RING domain protein Rbx1 (Roc or Hrt1) (Skowyra et al. 1999) (an E2 adapter), and a variable F-box protein (FBP), which allows the broad specificity (Skowyra et al. 1997). All the FBPs share a conserved 40-amino-acid domain called the F-box domain, which connects to the rest of the SCF complex through Skp1 (Bai et al. 1996). The substrate-binding domain of the FBP is almost invariably positioned directly carboxy-terminal to the F-box domain in the sequence, and no FBP has more than one F-box domain (Cardozo and Pagano 2004). The FBPs are classified according to their substrate-binding motif. The FBW family that includes β-TrCP, which recognize IκB, has a WD-40 repeats domain, structured as a β-propeller (Deshaies 1999). The WD-40 domain recognizes motifs containing phosphorylated Serines or Threonines, for example, the DpSGXXpS degron recognized by β-TrCP (Yaron et al. 1997; Smith et al. 1999). The FBL family has a Leucine-rich repeat in an arc-shaped α-β-repeat structure. A nonobligatory preference of substrate recognition by FBL is the prephosphorylation of the substrate or, at times, its attachment to other proteins in a complex (Kobe and Kajava 2001). Other FBPs are termed FBX and contain a variety of protein–protein interaction domains. It is very common for proteins in this family to recognize their substrates only following different post-translational modifications, such as glycosylation, hydroxylation, and others (Jaakkola et al. 2001; Yoshida et al. 2002; Yoshida et al. 2003). The three FBP families share the requirement for a substrate change before recognition, by either post-translational modification or binding of other proteins. This mode of regulation enables individual recognition of many substrates by the same FBP over different cellular conditions.

One of the best characterized FBPs is the FBW β-TrCP. This protein is located mainly in the nucleus (Davis et al. 2002), yet may show some activity also in the cytoplasm (Jiang and Struhl 1998). It binds an impressive list of phosphorylated substrates, including regulators of inflammation and cell fate (the IκBs), development and tissue organization (β-catenin, Snail), many cell cycle regulators (Emi1, Wee1), and DNA damage responders (Claspin, Cdc25A) (Jiang and Struhl 1998; Busino et al. 2003; Guardavaccaro et al. 2003; Watanabe et al. 2004; Zhou et al. 2004; Peschiaroli et al. 2006). In spite of β-TrCP's pivotal contribution to many oncogenic cellular pathways, there are not many known cases of mutations in the β-TrCP gene in human malignancies. This might reflect the essential physiological role of this E3 or redundancy of the two β-TrCP paralogues encoded by two different genes. The two paralogues are identical in their biochemical qualities because the "pocket" responsible for substrate recognition is identical (Fuchs et al. 1999). The redundancy may explain to the mild testicular phenotype observed in the in vivo ablation of β-TrCP1 (Guardavaccaro et al. 2003). The two best characterized substrates of β-TrCP, β-catenin and IκB, do not accumulate in β-TrCP-deficient mouse embryonic fibroblasts (MEF$^{\beta\text{-TrCP1}-/-}$). The

phenotype detected in the knockout mice might stem from differential regulation of the two paralogues: For instance, β-TrCP2 may not be expressed in all cells of the testes at which the knockout phenotype has been noticed. The best way to validate the differential contribution of the two paralogs in vivo is probably a combination of inducible deletion of β-TrCP2 and β-TrCP1, which is currently underway in some laboratories.

REGULATION OF β-TrCP

β-TrCP is responsible for the degradation of many proteins in response to various stimuli, among which are cell cycle regulators, some with opposing functions. Each substrate demands the attention of β-TrCP in a different temporal, spatial, and physiological context. Some of the substrates, such as the IκBs, must be degraded quickly following stimuli, whereas others, like β-catenin, are continuously ubiquitinated by β-TrCP and must be stabilized on a stimulus. Therefore, β-TrCP must be available at all times and the main mode of its functional regulation has to be the recognition and binding to the different substrates in the right context. This feature is provided by the signal dependent phosphorylation of all β-TrCP substrates (Deshaies 1999; Karin and Ben-Neriah 2000; Nakayama and Nakayama 2005). Any regulation of β-TrCP itself must therefore be bounded to these limitations. There is, however, only limited data concerning the transcriptional regulation of β-TrCP. It was reported that β-TrCP1 is regulated by the Wnt pathway at the mRNA level, both by modulation of the rate of transcription and the stabilization of mRNA (Spiegelman et al. 2000; Ballarino et al. 2004). β-TrCP1 mRNA stability is also increased following Jun amino-terminal kinase (Jnk) activation in a transcription-independent manner (Spiegelman et al. 2002). The transcriptional regulation of β-TrCP2, however, is probably different, with Wnt stimulus leading to its transcriptional inhibition and mitogen-activated signaling resulting in a β-TrCP2 transcriptional increase (Spiegelman et al. 2002). There still remains the question of the physiological relevance of these controls modes, as unless the isoforms have different substrate specificity, opposing transcriptional regulation of the two may offset each other.

Taking into account the entire SCF complex as one functional E3 unit, it is difficult to predict the impact of the change in the level of one subunit on E3 function. A comparison between two transgenic mice, one overexpressing the WT β-TrCP1 and the other expressing a dominant negative mutant of β-TrCP1, exemplifies the problematics of up-regulation of the β-TrCP subunit by itself. The dominant negative mutant that lacks the F-box is unable to assemble an SCF complex, but rather sequesters β-TrCP substrates and hinders their recognition by WT β-TrCP. Yet, surprisingly, overexpression of WT β-TrCP caused the same degree of β-catenin accumulation and tumorigenesis rate (Belaidouni et al. 2005). Therefore, when WT β-TrCP is expressed in excess, the stoichiometry of the other components in the SCF complex is inadequate, resulting in the apparent dominant-negative function of WT β-TrCP transgene. The need for balanced expression of the SCF subunits likely rebuts the notion of modulating β-TrCP expression levels as a major mode of regulation of the E3 activity.

Acting as a part of the SCF complex, β-TrCP activity is also subject to regulation directed at other components of the complex. The Nedd8 ubiquitin-like molecule was found to regulate the assembly of the SCF complex (Wei and Deng 2003; Wolf et al. 2003). Nedd8 modification of Cul1 stabilizes the SCF complex by preventing the binding of Cand1. Cand1 binds to cullin-Rbx1 complex and occludes Skp1, thereby preventing SCF assembly (Liu et al. 2002). Cleavage of Nedd8 by the COP9/signalosome (CSN) will allow the binding of Cand1 and decrease the SCF activity (Lyapina et al. 2001). Thus, complex assembly provides another level of E3 regulation, preventing spurious activity of the E3 complex.

Another case of SCF regulation is evident in the interaction of β-TrCP with the abundant nuclear protein hnRNP-U, reminiscent of a pseudo-substrate control, as hnRNP-U binds to the E3 but is not ubiquitinated (Davis et al. 2002). This may affect β-TrCP by maintaining it in the nucleus and raising its substrate-binding threshold. Raising the recognition threshold is an effective tool to avoid degradation of low affinity substrates, for example, none-, or partially phosphorylated substrates. Examples of a low affinity substrate are IκBα molecules that are phosphorylated only at one of the two serines of the β-TrCP degron (Ser36 by IKKε, and Ser32 by Rsk1) (Ghoda et al. 1997; Schouten et al. 1997; Peters et al. 2000): These will not compete effectively with saturating levels of hnRNP-U and will therefore escape β-TrCP ubiquitination.

These examples probably represent only the tip of the iceberg, as the intricate SCF structure provides ample room for activity regulation through multiple intrinsic (e.g., subunit modification) and extrinsic (accessory molecules) factors, mostly yet to be discovered.

IκB RECOGNITION BY β-TrCP

Rarely there is as detailed a mechanistic understanding of the E3-substrate interaction as with β-TrCP. This understanding emerged from peptide inhibition studies, defining the degron sequence and the necessity for dual phosphorylation for ubiquitination of phosphorylated IκBα (Yaron 1997). Yet a clearer interaction picture awaited the crystallography analysis of Pavletich and colleagues who pinpointed the critical features of substrate binding by β-TrCP (Wu 2003). This has been achieved by the cocrystallography of β-TrCP with a phosphorylated β-catenin peptide, sharing with IκB the core degron motif DpSGΦXpS (GΦX referred to as a "spacer," Φ stands for a hydrophobic, and X is any residue). The following binding details are observed in the structure: Phosphopeptide binding is mediated by one face of the β-propeller, a structure created by the seven WD repeat domain of β-TrCP; other domains of the F-box protein in the WD vicinity (e.g., the F-box, the F-box-WD linker, and the dimerization domain) or other SCF components do not participate in substrate binding; a central groove running through the middle of the WD propeller structure accommodates the degron moiety of the substrate. Wu et al. noticed that all seven WD repeats of β-TrCP contribute contacts to the bound peptide. Nearly all of the β-TrCP contacts are made by the six-residue degron; the side chains of the Asp, the hydrophobic residue (Φ), the backbone of the Gly, and the spacer residues (X) insert the farthest into the groove, making intermolecular contacts in a mostly buried environment. The phosphate groups of the two serines bind sites at the rim of the groove and along with the Asp make the largest number of contacts with β-TrCP residues surrounding the groove through hydrogen bonds and electrostatic interactions. Whereas Pavletich's data explains well the interaction of β-TrCP with IκBs and other canonical degron motif substrates, a remaining question is how the E3 can accommodate peptides containing atypical degrons, such as p105, which contains a longer spacer. To that end, we used the Rosetta software (Das and Baker 2008) to model a loop of varying length and sequence onto the scaffold of the two phosphorylated serine anchors of the β-catenin peptide bound to β-TrCP (Wu et al. 2003). A comparison of the structural models of phospho-peptides derived from β-catenin (DpSG**IH**pS), IκBα (DpSG**LD**pS), p105 (DpSG**VET**pS), and CDC25B (DpSG**FCLD**pS) suggests that these peptides of different spacer length can indeed all be accommodated into the groove (Fig. 1): Although the two phosphorylated serines are oriented similarly to those of β-catenin, the spacer between them—although of different length and sequence—displays similar characteristics. In all of the motifs, a glycine is followed by a hydrophobic residue (Ile, Leu, Val, and Phe) which is packed against a hydrophobic patch in the groove. The longer the binding motifs, the deeper into the groove will this hydrophobic residue be inserted. This advances our understanding of the principles that underlie substrate specificity, and in addition is invaluable for attempts to design β-TrCP blocking agents as NF-κB inhibitors (see the following discussion).

DEGRADATION KINETICS OF THE DIFFERENT IκBs

The three IκBs (IκBα, IκBβ, and IκBε) are degraded with different kinetics, IκBα the fastest (approximately 15 min in Jurkat and THP-1cells) and IκBε the slowest (120 min in Jurkat cells) (Whiteside et al. 1997). It is hard to define the factor that influences mostly these kinetics, and there might be a variation of the limiting factor in different physiological contexts. The first candidate is the stimulus itself, but even when examining one of the most effective stimuli, TNFα, there is hardly any correlation between the applied dose and the kinetics of IKK phosphorylation activity or IκB degradation (Cheong et al. 2006). IKK itself can be the factor that distinguishes between the three IκBs. Mathematical models and experimental data indicated that NF-κB dynamics are sensitive to the timing and duration of the IKK activity (Cheong et al. 2008). Indeed, the affinity between IKK and the different IκBs correlates with their degradation rate, implicating IKK as the main factor that shapes the differential kinetics (Heilker et al. 1999).

Another factor that might influence the degradation kinetics is the ubiquitination rate. Because the three IκBs share the same DSG degron, it is not plausible that the recognition by β-TrCP is discriminative, but perhaps other domains in the IκB structure can influence the processivity of ubiquitination as with the different substrates of another E3 (Rape et al. 2006). The processivity of ubiquitination is influenced by the affinity to the E3 ligase and by the associated activity of deubiquitinating enzymes (DUBs) (Rape et al. 2006). Deubiquitination is often a major means of terminating ubiquitination-dependent signaling response, and could also function in reducing the basal, signal-independent ubiquitination of several proteins in NF-κB signaling, including IκBs. CYLD and A20 are the two key DUBs in this regulation, but neither of them targets IκBs (Sun and Ley 2008). So far, no DUB has been found to negatively regulate the degradation of IκB, but considering equivalent signaling systems (e.g., the DNA damage

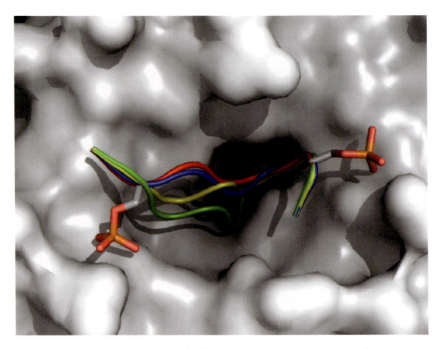

Figure 1. Accommodation of peptide degrons of different lengths into the β-TrCP binding groove. The β-TrCP WD40 domain is shown in white surface representation from a top view (1P22:A); the two phosphorylated Serine residues that are conserved among all the peptides are shown in stick representation; and different peptide backbones are colored according to the peptide: The β-catenin peptide is shown in red (1P22:C); a model of the IκBα peptide (DSG**LD**S) is shown in blue; p105 peptide (DSG**VET**S) in yellow; and CDC25B peptide (DSG**FCLD**S) in green. Note that with increasing length of the spacer between the two phosphorylated serine residues, the hydrophobic part of the peptide binds deeper into the pocket.

response) (Zhang et al. 2006; Bassermann et al. 2008), it may exist, providing an important regulatory aspect to NF-κB signaling.

The unique degradation characteristics of the IκBs fit their nonredundant roles in the regulation of the NF-κB response. It is best exemplified in the computational model generated by Hoffmann et al. (2002). Using three mouse models in which each IκB was deleted separately, it was shown that IκBα is inimitable in inducing a rapid, strong negative feedback regulation, resulting in an oscillatory NF-κB activation profile. IκBβ, and particularly IκBε, respond more slowly to IKK activation and act to dampen the long-term oscillations of the NF-κB response. The interplay between these isoforms dictates the onset and termination of NF-κB activation, allowing a relatively stable NF-κB response during long-term stimulation. Consistent with the major role of IκBα, deletion

of IκBβ and IκBε in animals is mostly harmless, indicating that other IκBs can compensate for their loss (Memet et al. 1999; Hoffmann et al. 2002). In contrast, deletion of IκBα resulted in constitutive NF-κB activation and embryonic lethality (Beg et al. 1995), yet, once IκBβ is expressed under the promoter of IκBα, it provided full compensation for the absence of IκBα (Cheng et al. 1998). This finding highlighted the importance of unique transcriptional regulation of IκBα rather than its unique activity or its own degradation rate, in controlling NF-κB activity.

In addition to their distinctive degradation kinetics, the three IκBs show different quantitative contribution to the NF-κB response. Unfortunately, a quantitative model based on the experimental data cited above failed to distinguish between the three IκBs (Cho et al. 2003). Thus, this aspect still awaits the formulation of a proper underlying model. In particular, the relative stoichiometry of association of each IκB with NF-κB is not known. Nevertheless, there is a negative correlation between the relative concentration of nuclear NF-κB and the overall cellular IκBα levels, indicating a major role of IκBα in controlling the NF-κB localization (Pogson et al. 2008). It will be interesting to study NF-κB localization with respect to the relative concentrations of the other IκBs, and thus incorporate further kinetic and quantitative aspects into a unified model of NF-κB activation.

UBIQUITINATION OF p100 AND p105 CONTROLS THEIR IκB FUNCTION

p100 and p105 also function as IκBs (Rice et al. 1992; Mercurio et al. 1993) but do not share the main features of a canonical degradation pathway. Their primary function is to serve as precursors for the mature NF-κB proteins p52 and p50, a function regulated by proteasomal processing (Fan and Maniatis 1991). Yet, the ankyrin repeat structure of p100 and p105 binds NF-κB/Rel proteins, similarly to the homologous region of the "professional IκBs" (α, β, and ε), and can therefore fulfill an IκB function, assuming that the fraction of NF-κB bound to these precursor proteins is sometimes free to dissociate, enter the nucleus, and induce transcriptional activity. In this respect, precursor processing regulates the NF-κB pathway in two planes: producing new and releasing old NF-κB molecules for the benefit of the transcriptional response. p100 is phosphorylated by IKKα that is recruited and activated by NF-κB-inducing kinase (NIK) as part of the noncanonical NF-κB signaling pathway (Xiao et al. 2001; Xiao et al. 2004). This phosphorylation promotes the binding of β-TrCP, polyubiquitination, and proteasomal processing, which generates active, IκB-free p52 (Fong and Sun 2002). In contrast, p105 is constitutively processed both in resting and stimulated cells (Weil et al. 1999; Orian et al. 2000), a course of action mediated by an independent, yet to be identified E3 that induces mono-ubiquitination rather than the polyubiquitination mediated by

β-TrCP (Kravtsova-Ivantsiv et al. 2009). Still, the inhibitory function of p105 is compliant with canonical IKKβ phosphorylation, and p105 proteasomal processing, which is facilitated by signaling (Mercurio et al. 1993) reduces the inhibition (Heissmeyer et al. 1999). In addition, a fraction of p105 is targeted for complete degradation following certain cell stimuli, such as TNFα (Lang et al. 2003).

Overexpressed p105 was found to sequester c-rel and p65 in the cytoplasm of COS-7 and CV-1 cells (Rice et al. 1992). Endogenous p105 and p100 were shown to associate with c-rel and p65 in Jurkat cells (Mercurio et al. 1993) and in WT MEFs in vivo (Basak et al. 2007). Unfortunately, in these studies, the question of whether the binding of p100 or p105 is quantitatively significant has not been addressed and the stoichiometry of the NF-κB fraction that is bound to these proteins is unknown. The RelA fraction that is bound to p100 or p105 could vary from cell to cell and be subject to cell stimulation or other cellular conditions. Thus, its quantification might provide a better understanding of the IκB role played by the NF-κB precursors.

A study focusing on the noncanonical NF-κB signaling downstream to the LTβR provided evidence that p100 functions as an inhibitor in IκBα$^{-/-}$, IκBβ$^{-/-}$, and IκBε$^{-/-}$ MEFs (Basak et al. 2007). This work also showed that the fraction of p100-bound, dissociable p65 was increased three- to fourfold following TNFα stimulation, proposing a specific time window for IκB function of p100 after canonical signaling in WT cells. This suggestion was further supported by transcriptional synergism between LTβR and TNFα. However, some relevant questions were left unanswered: whether the overall stoichiometry of p100-bound NF-κB is significantly changed following canonical stimulation, and whether the transcriptional synergism between LTβR and TNFα stimuli brought as evidence for an IκB function of p100 could in fact be caused by IKK activation with LTβR (Chang et al. 2002).

An in vivo research of the inhibitory role of p100 and p105 is hard to conduct because most genetic manipulations will also influence p50 and p52 and thus, it will be hard to distinguish the IκB-relevant effects. One way around this point is to stabilize a precursor by way of site directed mutagenesis, which will spare processing. Indeed this was recently achieved in the case of p105 in T cells, showing that substitution of the modifiable serine residues controlling p105 stability resulted in a moderate "IκB super-repressor"-like molecule, affecting T-cell development (Sriskantharajah et al. 2009). However, again the stoichiometry of p105-bound NF-κB is nonphysiological, thus not necessarily attesting to a physiological IκB function of p105 in T cells. As a heterozygous mutation that substitutes human Ser32 of IκBα is sufficient to cause T-cell immunodeficiency and ectodermal dysplasia (Courtois et al. 2003; Janssen et al. 2004), it is possible that equivalent serine substituting mutations in the *NFKB1* and *NFKB2* genes will cause similar human pathologies, thereby supporting the IκB function of NF-κB precursor protein.

To conclude, better understanding of the true IκB role of p100 or p105 requires quantitative and kinetic studies, with comparison to the "professional" IκBs, preferentially under physiologic conditions (i.e., in WT cells in response to various stimuli) (see Fig. 2). This will probably be best achieved with the help of mathematical modeling. Only then may we understand the role of each IκB in physiological NF-κB response.

SUBCELLULAR LOCATION OF IκB UBIQUITINATION AND DEGRADATION

A large missing piece in the NF-κB activation scheme is the site of IκBα ubiquitination and degradation. This issue is underscored by the discordant subcellular localization of the IκBs, both the "professionals" and the NF-κB precursors, which are mostly cytoplasmic, and the E3 β-TrCP, which is mostly nuclear. A major fraction of β-TrCP is sequestered in the nucleus by the pseudo-substrate hnRNP-U, and co-immunoprecipitation studies show that this hnRNP-U-β-TrCP complex is available for ubiquitinating IκBα on exchange of the low affinity hnRNP-U pseudosubstrate by the high affinity pIκBα (Davis et al. 2002). Apart from

the β-TrCP, other components of the SCF, as well as the proteasome, are localized and active in the nucleus (Nakayama and Nakayama 2005; von Mikecz 2006). Nuclear proteasome substrates include the p53 inhibitor Mdm2, the CDK inhibitor Far1, and others (von Mikecz 2006). How can we therefore envision the interaction between a nuclear E3 and cytoplasmic substrates? One way is to assume that the location of the components is not stable, and at least one component, the substrate or the E3, may switch its subcellular location randomly or on demand. It was therefore interesting to note that IκBα-NF-κB complexes shuttle continuously between the cytoplasm and the nucleus of unstimulated cells. The shuttling is mediated by a nuclear export signal (NES), located near the amino terminus of IκBα, and a likely unmasked nuclear localization signal (NLS) of the p50 proteins (Malek et al. 2001). IκBβ, which lacks the NES, is probably more efficient in masking the NLS because IκBβ-NF-κB complexes are found exclusively in the cytoplasm (Huang and Miyamoto 2001). IκBα shuttling is apparently unperturbed as long as the NF-κB/IκBα complex is stable. Once IκBα is phosphorylated via activated IKK, it is recognized by β-TrCP and degraded, following which NF-κB is retained in the nucleus. Whereas phosphorylation of IκBα is thought to occur in the cytoplasm (Karin and Ben-Neriah 2000), the subcellular site of its ubiquitination and degradation is unknown.

It is important to note that although the IκBα/NF-κB shuttling is a reasonable solution for introducing a nuclear E3 to a cytoplasmic substrate, most of the relevant experiments have relied on LMB treatment as an inhibitor of IκBα nuclear export. Other influences of this drug that may affect NF-κB signaling have never been taken into account. Therefore, any activation model entailing steady state shuttling must be taken with a grain of salt.

THE ROLE OF THE IκBs IN TERMINATION OF NF-κB SIGNALING

Temporal accuracy is required for induction and for termination of NF-κB activity. One of

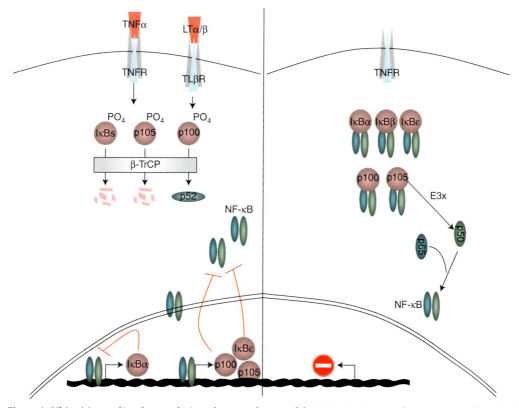

Figure 2. Ubiquitin-mediated proteolysis at the central stage of the NF-κB. Many regulatory steps in the NF-κB pathway involve ubiquitin-regulated proteolysis. In resting cells (*right* panel), basal level processing of the NF-κB1 precursor p105 to p50, mediated by a yet unidentified E3(s) (E3x), is the major proteolytic event. The processing product p50 associates with p65 to generate the NF-κB heterodimer, maintained inactive in the cytosol through association with an inhibitor: one of the "professional IκBs" or an NF-κB precursor (both p105 and p100). On cell stimulation (*left* panel), TNFα and LTα/β signal-mediated degradation of the IκBs and processing NF-κB2 (p100) to the mature p52 subunit, respectively, are the key proteolytic steps. All these ubiquitin-proteolysis events are controlled by a single E3 ubiquitin ligase, β-TrCP. A prerequisite for β-TrCP-induced ubiquitination is phosphorylation of its target proteins, mediated by IKK (not shown). Following IκB degradation, NF-κB accumulates in the nucleus and is free to induce transcriptional activation. Another major feature of the NF-κB pathways is the negative feedback control (red lines) by which first the newly transcribed IκBα and then the NF-κB precursors function to shorten signal duration by inhibiting NF-κB-controlled transcription. Pairing IκBα transcription and resynthesis to its rapid degradation shapes an oscillatory NF-κB response, which might have a particular regulatory significance, yet to be discovered. IκBε transcription by NF-κB function to reduce the oscillatory magnitude of the response and stabilize it during longer stimulations.

the major factors that controls the lifetime of the NF-κB response is a negative feedback loop based on NF-κB-dependent IκB synthesis (Le Bail et al. 1993). Following a near-complete signal-induced degradation of IκB, the newly synthesized IκBα enters the nucleus and pulls NF-κB away from the chromatin and back to the cytoplasm (Arenzana-Seisdedos et al. 1995). This process should allow a sufficient period of NF-κB activity to induce relevant target genes and prevent the extension of the transcriptional activity beyond the necessary time, to avoid deleterious overactivation effects (Hoffmann et al. 2002). It has been argued that NF-κB modification is one of the key factors that prevents premature termination of the transcriptional activity. For instance, RelA acetylation could prohibit its interaction with IκB, whereas its deacetylation by HDAC3 increases its affinity to IκB, thus allowing the eviction of NF-κB from the chromatin by the newly synthesized IκBα. Indeed, a timely acetylation–deacetylation cycle was shown to contribute to signal maintenance for 45 min (Chen et al. 2001). Following TNFα stimulus, NF-κB is activated for approximately 45 min (Hoffmann et al. 2002), whereas IκBα is transcribed as early as 15 min and peaks within 1 h (Scott et al. 1993). However, IKK activity peaks after 15 min and is maintained for more than an hour (Cheong et al. 2006), a period during which all the newly synthesized IκBα could be immediately phosphorylated by IKK and subsequently degraded. How then can IκB exert its termination role in the presence of an active IKK?

One possibility is that both IKKα and IKKβ phosphorylate NF-κB-bound IκB more efficiently than free IκB (Zandi et al. 1998), explaining how the free IκB can accumulate in IKK-active cells. Nevertheless, a major fraction of the newly synthesized IκB appears phosphorylated (Werner et al. 2005), suggesting that phosphorylation would not be the rate-limiting step in the destruction of this newly synthesized molecule. Another possibility is reduced ubiquitination of the unbound IκB by β-TrCP. Unlike many SCF substrates, most lysine residues of IκB are not targeted for ubiquitination in vivo, nor in the bound state in vitro, and the only degradation-relevant residues are Lys 21 and 22 of IκBα (Scherer et al. 1995). Lys 21 is also the target residue for sumoylation, a modification that might antagonize proteolysis-marking ubiquitination of IκB and could spare the newly synthesized IκB if happening soon after synthesis. However, sumoylation was shown to happen only in the nucleus (Rodriguez et al. 2001), which similarly to, or in parallel with RelA acetylation, might imply a role in terminating the NF-κB response, merely by stabilizing nuclear IκBα. Overexpressed Sumo-1 blocked TNFα and IL-1 dependent ubiquitination and degradation of IκBα in COS7 cells (Franzoso et al. 1992), yet S32/36

phosphorylated IκBα was resistant to sumoylation in vitro, possibly indicating the relevance of this modification only for the newly synthesized, nonphosphorylated IκB.

One should bear in mind that an unbound IκB could also be targeted by E3s other than β-TrCP, which do not require IKK phosphorylation (Yaron et al. 1997; O'Dea et al. 2008) and might be degraded as efficiently as phosphorylated IκB (O'Dea et al. 2008). IκBα can also be phosphorylated independently of a signal on its carboxyl terminus by the kinase CK2 (Barroga et al. 1995) or in response to a noncanonical stimulus, such as UV (Kato et al. 2003). Knockdown of the CK2β subunit or mutation of the phosphorylated residues in the carboxy-terminal fragment of IκBα abrogated the NF-κB response following UV (Kato et al. 2003). In addition, in response to doxorubicin-induced DNA damage, IκBα undergoes proteasomal degradation, both in WT and in IKKα$^{-/-}$ and IKKβ$^{-/-}$ MEFs (Wu and Miyamoto 2007). If this is physiologically relevant, it basically means that we do not understand how the newly synthesized IκB, whether phosphorylated or not, escapes ubiquitination and degradation, and there must be a mechanism to seclude it from the ubiquitination machinery or reverse its ubiquitination effectively (i.e., by deubiquitination).

THERAPEUTIC IMPLICATIONS OF IκB DEGRADATION

Aberrant NF-κB regulation has been implicated in autoimmune diseases and in certain types of cancer. Given their central role of regulating NF-κB activation, the proteins involved in ubiquitination and degradation of the IκBs are attractive targets for drug development. The successful treatment of a major hematological malignancies using a proteasome inhibitor has validated the ubiquitin–proteasome system as therapeutic target (Rajkumar et al. 2005). Bortezomib (formerly known as PS-341), a boronic acid dipeptide that binds directly with and inhibits the proteasome enzymatic complex, has a significant therapeutic activity in the treatment of advanced multiple myeloma

(MM) (Hideshima et al. 2002) and was approved for MM treatment by the FDA in 2003. MM is often characterized by NF-κB pathway mutations (Annunziata et al. 2007; Keats et al. 2007) and Bortezomib's effects on cells are partly mediated through NF-κB inhibition (Russo et al. 2001; Hideshima et al. 2002), resulting in apoptosis, decreased angiogenic cytokine expression, and inhibition of tumor cell adhesion to stroma. However, continuous proteasomal inhibition is toxic and therefore, bortezomib cannot be administered more than twice weekly, with proteasome inhibition lasting approximately 24 h after each injection (Adams 2002). An alternative approach to inhibiting NF-κB is to block IκB ubiquitination. This would be expected to have less undesired side effects than global proteasome inhibition and may be particularly attractive for cancer therapy (Fuchs et al. 2004). Here, too, one could believe in blocking a class of relevant E3s, or targeting specifically β-TrCP. Recent studies indicate the feasibility of blocking SCF-type E3s through interference with their activation requiring cullin modification by the ubiquitin-like molecule Nedd8 (Soucy et al. 2009). A more specific strategy for NF-κB inhibition would be selective IκB stabilization. The validity of this approach has been established in numerous experiments, both in vitro and in vivo, using a dominant IκB "super-repressor," which cannot be inducibly phosphorylated and ubiquitinated but maintains its NF-κB inhibitory capacity (Cusack et al. 2000; Lavon et al. 2000). For example, the expression of such an inhibitor in liver cells has been shown to halt tumor growth in a murine model of hepatitis-associated cancer in vivo (Pikarsky et al. 2004). A clinically relevant method for that purpose is to prevent IκB ubiquitination or to impair β-TrCP recruitment to IκB. Cell-penetrating IκB phosphopeptides have been used for this purpose in cell lines (Yaron et al. 1997), but their efficacy has yet to be shown in vivo. It might also be possible to develop small-molecule inhibitors that structurally mimic the ligase recognition motif or specifically inactivate $SCF^{\beta\text{-}TrCP}$ (through allosteric inhibition or disassembly). One obvious

drawback with targeting β-TrCP is the potential to affect $SCF^{\beta\text{-}TrCP}$ targets other than IκBs (Spiegelman et al. 2002). A critical issue is likely to be whether $SCF^{\beta\text{-}TrCP}$ is the only E3 that contributes to the degradation of any given substrate. One obvious concern is stabilization of β-catenin, an important factor in colorectal tumorigenesis and other cancer types (van Es et al. 2003). One possibility, however, is that whereas IκB degradation relies exclusively on β-TrCP (see Fig. 2), other substrates like β-catenin have alternative ubiquitination and degradation routes, operating on β-TrCP inhibition. Consistent with this hypothesis, a distinct E3 complex containing the Siah RING finger protein has been shown to recognize and promote the degradation of β-catenin independently of $SCF^{\beta\text{-}TrCP}$ (Liu et al. 2001; Matsuzawa and Reed 2001). Future analysis of β-TrCP-deficient mice will air this issue, proving whether the NF-κB pathway is affected in these mice far more than other signaling pathways. Furthermore, even if the selective inhibition of IκB ubiquitination is a technically feasible goal, systemic inhibition of NF-κB is not without risk because of its crucial role in regulating immune responses (Caamano and Hunter 2002). This is a particularly important consideration when NF-κB blockade is considered in combination with standard chemotherapy, which itself compromises the immune system and exacerbates tissue damage (Lavon et al. 2000). Furthermore, NF-κB inhibition may in fact be a "double-edged sword," as in the presence of a carcinogen, and perhaps even under chemotherapeutic poisoning, NF-κB inhibition may facilitate, rather than prevent, tumor development (Kamata et al. 2005).

CONCLUDING REMARKS

Inducible degradation of IκB has become a paradigm of targeting a regulatory protein by the ubiquitin proteasome system as a major step in the activation of a signaling pathway. A central feature of this process is the coupling of two protein modification events, substrate bi-phosphorylation by a dedicated kinase,

IKK, and ubiquitination, based on recognition of the phosphorylated degron by the E3. However, NF-κB is regulated by five different inhibitory proteins (IκBα, β, and ε and the two IκB-like NF-κB precursors), each holding some portion of NF-κB in an inactive state. Therefore, maximal activation of the pathway requires simultaneous elimination of all five IκBs. This may be achieved on costimulation of both the canonical and alternative branches of the pathway, inducing degradation and processing of different IκBs, respectively. No wonder that the coordination of the degradation response relies on a similar, IKK-induced phospho-degron that is shared among all IκBs and on a single E3, β-TrCP, which evolved to fit the nuances of the degron in the different IκBs.

Despite of the major advances achieved in understanding ubiquitin-signaling processes of NF-κB activation, several key questions remain unanswered, for example: What is the relative role of each IκB in controlling NF-κB activation in response to different stimuli? How is the NF-κB-transcribed, newly synthesized IκB protected from degradation as long as IKK is still activated? Which parameters other than phosphorylation affect the rate of degradation and processing of the different IκB? Where in the cell are IκB ubiquitination and proteolysis taking place? How important is the regulation of the ubiquitin–proteasome machinery itself in NF-κB signaling? Is IκB ubiquitination also regulated by specific de-ubiquitination enzymes? Are there any specific proteasome adaptor molecules that ensure degradation of IκBs, while protecting associated NF-κB from proteolysis by the proteasome? How frequently are mutations in the IκBs or their destruction machinery encountered in human disease?

Considering these and other outstanding issues, the regulation and function of ubiquitination are likely to remain a major focus of research in the NF-κB field.

ACKNOWLEDGMENTS

Relevant research in the authors' laboratories was supported by the Israel Science Foundation, the RUBICON Network of Excellence of the European Commission (FP6), The German Israeli Foundation, and the Adelson Medical Research Foundation.

REFERENCES

Adams J. 2002. Proteasome inhibition: A novel approach to cancer therapy. *Trends Mol Med* **8:** S49–54.

Alkalay I, Yaron A, Hatzubai A, Jung S, Avraham A, Gerlitz O, Pashut-Lavon I, Ben-Neriah Y. 1995a. In vivo stimulation of I κ B phosphorylation is not sufficient to activate NF-κ B. *Mol Cell Biol* **15:** 1294–1301.

Alkalay I, Yaron A, Hatzubai A, Orian A, Ciechanover A, Ben-Neriah Y. 1995b. Stimulation-dependent I κ B α phosphorylation marks the NF-κB inhibitor for degradation via the ubiquitin-proteasome pathway. *Proc Natl Acad Sci* **92:** 10599–10603.

Annunziata CM, Davis RE, Demchenko Y, Bellamy W, Gabrea A, Zhan F, Lenz G, Hanamura I, Wright G, Xiao W, et al. 2007. Frequent engagement of the classical and alternative NF-κB pathways by diverse genetic abnormalities in multiple myeloma. *Cancer Cell* **12:** 115–130.

Arenzana-Seisdedos F, Thompson J, Rodriguez MS, Bachelerie F, Thomas D, Hay RT. 1995. Inducible nuclear expression of newly synthesized IκB α negatively regulates DNA-binding and transcriptional activities of NF-κ B. *Mol Cell Biol* **15:** 2689–2696.

Baeuerle PA, Baltimore D. 1988. Activation of DNA-binding activity in an apparently cytoplasmic precursor of the NF-κ B transcription factor. *Cell* **53:** 211–217.

Bai C, Sen P, Hofmann K, Ma L, Goebl M, Harper JW, Elledge SJ. 1996. SKP1 connects cell cycle regulators to the ubiquitin proteolysis machinery through a novel motif, the F-box. *Cell* **86:** 263–274.

Ballarino M, Fruscalzo A, Marchioni M, Carnevali F. 2004. Identification of positive and negative regulatory regions controlling expression of the *Xenopus laevis* βTrCP gene. *Gene* **336:** 275–285.

Barroga CF, Stevenson JK, Schwarz EM, Verma IM. 1995. Constitutive phosphorylation of IκB α by casein kinase II. *Proc Natl Acad Sci* **92:** 7637–7641.

Basak S, Kim H, Kearns JD, Tergaonkar V, O'Dea E, Werner SL, Benedict CA, Ware CF, Ghosh G, Verma IM et al. 2007. A fourth IκB protein within the NF-κB signaling module. *Cell* **128:** 369–381.

Bassermann F, Frescas D, Guardavaccaro D, Busino L, Peschiaroli A, Pagano M. 2008. The Cdc14B-Cdh1-Plk1 axis controls the G2 DNA-damage-response checkpoint. *Cell* **134:** 256–267.

Beg AA, Finco TS, Nantermet PV, Baldwin AS Jr. 1993. Tumor necrosis factor and interleukin-1 lead to phosphorylation and loss of IκBα: A mechanism for NF-κ B activation. *Mol Cell Biol* **13:** 3301–3310.

Beg AA, Sha WC, Bronson RT, Baltimore D. 1995. Constitutive NF-κB activation, enhanced granulopoiesis, and neonatal lethality in IκB α-deficient mice. *Genes Dev* **9:** 2736–2746.

Belaidouni N, Peuchmaur M, Perret C, Florentin A, Benarous R, Besnard-Guerin C. 2005. Overexpression of human β-TrCP1 deleted of its F box induces tumorigenesis in transgenic mice. *Oncogene* **24:** 2271–2276.

Brown K, Park S, Kanno T, Franzoso G, Siebenlist U. 1993. Mutual regulation of the transcriptional activator NF-κ B and its inhibitor, IκBα. *Proc Natl Acad Sci* **90:** 2532–2536.

Brown K, Gerstberger S, Carlson L, Franzoso G, Siebenlist U. 1995. Control of IκBα proteolysis by site-specific, signal-induced phosphorylation. *Science* **267:** 1485–1488.

Busino L, Donzelli M, Chiesa M, Guardavaccaro D, Ganoth D, Dorrello NV, Hershko A, Pagano M, Draetta GF. 2003. Degradation of Cdc25A by β-TrCP during S phase and in response to DNA damage. *Nature* **426:** 87–91.

Caamano J, Hunter CA. 2002. NF-κB family of transcription factors: Central regulators of innate and adaptive immune functions. *Clin Microbiol Rev* **15:** 414–429.

Cardozo T, Pagano M. 2004. The SCF ubiquitin ligase: Insights into a molecular machine. *Nat Rev Mol Cell Biol* **5:** 739–751.

Chang YH, Hsieh SL, Chen MC, Lin WW. 2002. Lymphotoxin β receptor induces interleukin 8 gene expression via NF-κB and AP-1 activation. *Exp Cell Res* **278:** 166–174.

Chen Z, Hagler J, Palombella VJ, Melandri F, Scherer D, Ballard D, Maniatis T. 1995. Signal-induced site-specific phosphorylation targets IκBα to the ubiquitin-proteasome pathway. *Genes Develop* **9:** 1586–1597.

Chen L, Fischle W, Verdin E, Greene WC. 2001. Duration of nuclear NF-κB action regulated by reversible acetylation. *Science* **293:** 1653–1657.

Cheng JD, Ryseck RP, Attar RM, Dambach D, Bravo R. 1998. Functional redundancy of the nuclear factor κ B inhibitors IκBα and IκBβ. *J Exp Med* **188:** 1055–1062.

Cheong R, Bergmann A, Werner SL, Regal J, Hoffmann A, Levchenko A. 2006. Transient IκB kinase activity mediates temporal NF-κB dynamics in response to a wide range of tumor necrosis factor-α doses. *J Biol Chem* **281:** 2945–2950.

Cheong R, Hoffmann A, Levchenko A. 2008. Understanding NF-κB signaling via mathematical modeling. *Mol Syst Biol* **4:** 192.

Cho KH, Shin SY, Lee HW, Wolkenhauer O. 2003. Investigations into the analysis and modeling of the TNFα-mediated NF-κB-signaling pathway. *Genome Res* **13:** 2413–2422.

Courtois G, Smahi A, Reichenbach J, Doffinger R, Cancrini C, Bonnet M, Puel A, Chable-Bessia C, Yamaoka S, Feinberg J, et al. 2003. A hypermorphic IκBα mutation is associated with autosomal dominant anhidrotic ectodermal dysplasia and T cell immunodeficiency. *J Clin Invest* **112:** 1108–1115.

Cusack JC Jr, Liu R, Baldwin AS Jr. 2000. Inducible chemoresistance to 7-ethyl-10-[4-(1-piperidino)-1-piperidino]-carbonyloxycamptothe cin (CPT-11) in colorectal cancer cells and a xenograft model is overcome by inhibition of nuclear factor-κB activation. *Cancer Res* **60:** 2323–2330.

Das R, Baker D. 2008. Macromolecular modeling with rosetta. *Annu Rev Biochem* **77:** 363–382.

Davis M, Hatzubai A, Andersen JS, Ben-Shushan E, Fisher GZ, Yaron A, Bauskin A, Mercurio F, Mann M, Ben-Neriah Y. 2002. Pseudosubstrate regulation of the SCF(β-TrCP) ubiquitin ligase by hnRNP-U. *Genes Develop* **16:** 439–451.

Deshaies RJ. 1999. SCF and Cullin/Ring H2-based ubiquitin ligases. *Ann Rev Cell Develop Biol* **15:** 435–467.

DiDonato JA, Mercurio F, Karin M. 1995. Phosphorylation of IκBα precedes but is not sufficient for its dissociation from NF-κB. *Mol Cell Biol* **15:** 1302–1311.

Fan CM, Maniatis T. 1991. Generation of p50 subunit of NF-κB by processing of p105 through an ATP-dependent pathway. *Nature* **354:** 395–398.

Feldman RM, Correll CC, Kaplan KB, Deshaies RJ. 1997. A complex of Cdc4p, Skp1p, and Cdc53p/cullin catalyzes ubiquitination of the phosphorylated CDK inhibitor Sic1p. *Cell* **91:** 221–230.

Fong A, Sun SC. 2002. Genetic evidence for the essential role of β-transducin repeat-containing protein in the inducible processing of NF-κB2/p100. *J Biol Chem* **277:** 22111–22114.

Franzoso G, Bours V, Park S, Tomita-Yamaguchi M, Kelly K, Siebenlist U. 1992. The candidate oncoprotein Bcl-3 is an antagonist of p50/NF-κB-mediated inhibition. *Nature* **359:** 339–342.

Fuchs SY, Chen A, Xiong Y, Pan ZQ, Ronai Z. 1999. HOS, a human homolog of Slimb, forms an SCF complex with Skp1 and Cullin1 and targets the phosphorylation-dependent degradation of IκB and β-catenin. *Oncogene* **18:** 2039–2046.

Fuchs SY, Spiegelman VS, Kumar KG. 2004. The many faces of β-TrCP E3 ubiquitin ligases: Reflections in the magic mirror of cancer. *Oncogene* **23:** 2028–2036.

Ghoda L, Lin X, Greene WC. 1997. The 90-kDa ribosomal S6 kinase (pp90rsk) phosphorylates the N-terminal regulatory domain of IκBα and stimulates its degradation in vitro. *J Biol Chem* **272:** 21281–21288.

Ghosh S, Baltimore D. 1990. Activation in vitro of NF-κB by phosphorylation of its inhibitor IκB. *Nature* **344:** 678–682.

Guardavaccaro D, Kudo Y, Boulaire J, Barchi M, Busino L, Donzelli M, Margottin-Goguet F, Jackson PK, Yamasaki L, Pagano M. 2003. Control of meiotic and mitotic progression by the F box protein β-Trcp1 in vivo. *Dev Cell* **4:** 799–812.

Heilker R, Freuler F, Vanek M, Pulfer R, Kobel T, Peter J, Zerwes HG, Hofstetter H, Eder J. 1999. The kinetics of association and phosphorylation of IκB isoforms by IκB kinase 2 correlate with their cellular regulation in human endothelial cells. *Biochemistry* **38:** 6231–6238.

Heissmeyer V, Krappmann D, Hatada EN, Scheidereit C. 2001. Shared pathways of IκB kinase-induced SCF(βTrCP)-mediated ubiquitination and degradation for the NF-κB precursor p105 and IκBα. *Mol Cell Biol* **21:** 1024–1035.

Heissmeyer V, Krappmann D, Wulczyn FG, Scheidereit C. 1999. NF-κB p105 is a target of IκB kinases and controls signal induction of Bcl-3-p50 complexes. *EMBO J* **18:** 4766–4778.

Henkel T, Machleidt T, Alkalay I, Kronke M, Ben-Neriah Y, Baeuerle PA. 1993. Rapid proteolysis of IκBα is necessary for activation of transcription factor NF-κB. *Nature* **365:** 182–185.

Hideshima T, Chauhan D, Richardson P, Mitsiades C, Mitsiades N, Hayashi T, Munshi N, Dang L, Castro A, Palombella V, et al. 2002. NF-κB as a therapeutic target in multiple myeloma. *J Biol Chem* **277:** 16639–16647.

Hoffmann A, Levchenko A, Scott ML, Baltimore D. 2002. The IκB-NF-κB signaling module: Temporal control and selective gene activation. *Science* **298:** 1241–1245.

Huang TT, Miyamoto S. 2001. Postrepression activation of NF-κB requires the amino-terminal nuclear export signal specific to IκBα. *Mol Cell Biol* **21:** 4737–4747.

Jaakkola P, Mole DR, Tian YM, Wilson MI, Gielbert J, Gaskell SJ, Kriegsheim A, Hebestreit HF, Mukherji M, Schofield CJ, et al. 2001. Targeting of HIF-α to the von Hippel-Lindau ubiquitylation complex by O2-regulated prolyl hydroxylation. *Science* **292:** 468–472.

Janssen R, van Wengen A, Hoeve MA, ten Dam M, van der Burg M, van Dongen J, van de Vosse E, van Tol M, Bredius R, Ottenhoff TH, et al. 2004. The same IκBα mutation in two related individuals leads to completely different clinical syndromes. *J Exp Med* **200:** 559–568.

Jiang J, Struhl G. 1998. Regulation of the Hedgehog and Wingless signalling pathways by the F-box/WD40-repeat protein Slimb. *Nature* **391:** 493–496.

Kamata H, Honda S, Maeda S, Chang L, Hirata H, Karin M. 2005. Reactive oxygen species promote TNFα-induced death and sustained JNK activation by inhibiting MAP kinase phosphatases. *Cell* **120:** 649–661.

Karin M, Ben-Neriah Y. 2000. Phosphorylation meets ubiquitination: The control of NF-κB activity. *Annu Rev Immunol* **18:** 621–663.

Kato T Jr, Delhase M, Hoffmann A, Karin M. 2003. CK2 Is a C-terminal IκB kinase responsible for NF-κB activation during the UV response. *Mol Cell* **12:** 829–839.

Keats JJ, Fonseca R, Chesi M, Schop R, Baker A, Chng WJ, Van Wier S, Tiedemann R, Shi CX, Sebag M, et al. 2007. Promiscuous mutations activate the noncanonical NF-κB pathway in multiple myeloma. *Cancer cell* **12:** 131–144.

Kobe B, Kajava AV. 2001. The leucine-rich repeat as a protein recognition motif. *Curr Opin Struct Biol* **11:** 725–732.

Kravtsova-Ivantsiv Y, Cohen S, Ciechanover A. 2009. Modification by single ubiquitin moieties rather than polyubiquitination is sufficient for proteasomal processing of the p105 NF-κB precursor. *Mol Cell* **33:** 496–504.

Lang V, Janzen J, Fischer GZ, Soneji Y, Beinke S, Salmeron A, Allen H, Hay RT, Ben-Neriah Y, Ley SC. 2003. βrCP-mediated proteolysis of NF-κB1 p105 requires phosphorylation of p105 serines 927 and 932. *Mol Cell Biol* **23:** 402–413.

Lavon I, Goldberg I, Amit S, Landsman L, Jung S, Tsuberi BZ, Barshack I, Kopolovic J, Galun E, Bujard H, et al. 2000. High susceptibility to bacterial infection, but no liver dysfunction, in mice compromised for hepatocyte NF-κB activation. *Nature medicine* **6:** 573–577.

Le Bail O, Schmidt-Ullrich R, Israël A. 1993. Promoter analysis of the gene encoding the I κ B-α/MAD3 inhibitor of NF-κB: Positive regulation by members of the rel/NF-κB family. *EMBO J* **12:** 5043–5049.

Liu J, Stevens J, Rote CA, Yost HJ, Hu Y, Neufeld KL, White RL, Matsunami N. 2001. Siah-1 mediates a novel β-catenin degradation pathway linking p53 to the adenomatous polyposis coli protein. *Mol Cell* **7:** 927–936.

Liu J, Furukawa M, Matsumoto T, Xiong Y. 2002. NEDD8 modification of CUL1 dissociates p120(CAND1), an inhibitor of CUL1-SKP1 binding and SCF ligases. *Mol Cell* **10:** 1511–1518.

Lyapina S, Cope G, Shevchenko A, Serino G, Tsuge T, Zhou C, Wolf DA, Wei N, Shevchenko A, Deshaies RJ. 2001. Promotion of NEDD-CUL1 conjugate cleavage by COP9 signalosome. *Science* **292:** 1382–1385.

Malek S, Chen Y, Huxford T, Ghosh G. 2001. IκBβ, but not IκBα, functions as a classical cytoplasmic inhibitor of NF-κB dimers by masking both NF-κB nuclear localization sequences in resting cells. *J Biol Chem* **276:** 45225–45235.

Matsuzawa SI, Reed JC. 2001. Siah-1, SIP, and Ebi collaborate in a novel pathway for β-catenin degradation linked to p53 responses. *Mol Cell* **7:** 915–926.

Mellits KH, Hay RT, Goodbourn S. 1993. Proteolytic degradation of MAD3 (I κ B α) and enhanced processing of the NF-κ B precursor p105 are obligatory steps in the activation of NF-κ B. *Nucleic Acids Res* **21:** 5059–5066.

Memet S, Laouini D, Epinat JC, Whiteside ST, Goudeau B, Philpott D, Kayal S, Sansonetti PJ, Berche P, Kanellopoulos J, et al. 1999. IκBε-deficient mice: Reduction of one T cell precursor subspecies and enhanced Ig isotype switching and cytokine synthesis. *J Immunol* **163:** 5994–6005.

Mercurio F, DiDonato JA, Rosette C, Karin M. 1993. p105 and p98 precursor proteins play an active role in NF-κ B-mediated signal transduction. *Genes Dev* **7:** 705–718.

Nakayama KI, Nakayama K. 2005. Regulation of the cell cycle by SCF-type ubiquitin ligases. *Semin Cell Dev Biol* **16:** 323–333.

O'Dea EL, Kearns JD, Hoffmann A. 2008. UV as an amplifier rather than inducer of NF-κB activity. *Mol Cell* **30:** 632–641.

Orian A, Gonen H, Bercovich B, Fajerman I, Eytan E, Israël A, Mercurio F, Iwai K, Schwartz AL, Ciechanover A. 2000. SCF(β-TrCP) ubiquitin ligase-mediated processing of NF-κB p105 requires phosphorylation of its C-terminus by IκB kinase. *Embo J* **19:** 2580–2591.

Palombella VJ, Rando OJ, Goldberg AL, Maniatis T. 1994. The ubiquitin-proteasome pathway is required for processing the NF-κB1 precursor protein and the activation of NF-κ B. *Cell* **78:** 773–785.

Peschiaroli A, Dorrello NV, Guardavaccaro D, Venere M, Halazonetis T, Sherman NE, Pagano M. 2006. SCF(βTrCP)-mediated degradation of Claspin regulates recovery from the DNA replication checkpoint response. *Mol Cell* **23:** 319–329.

Peters RT, Liao SM, Maniatis T. 2000. IKKε is part of a novel PMA-inducible IκB kinase complex. *Mol Cell* **5:** 513–522.

Pikarsky E, Porat RM, Stein I, Abramovitch R, Amit S, Kasem S, Gutkovich-Pyest E, Urieli-Shoval S, Galun E, Ben-Neriah Y. 2004. NF-κB functions as a tumour

Cite this article as *Cold Spring Harb Perspect Biol* 2010;2:a000166

promoter in inflammation-associated cancer. *Nature*
431: 461–466.

Pogson M, Holcombe M, Smallwood R, Qwarnstrom E.
2008. Introducing spatial information into predictive
NF-κB modelling–an agent-based approach. *PLoS
ONE* **3:** e2367.

Rajkumar SV, Richardson PG, Hideshima T, Anderson KC.
2005. Proteasome inhibition as a novel therapeutic target
in human cancer. *J Clin Oncol* **23:** 630–639.

Rape M, Reddy SK, Kirschner MW. 2006. The processivity of
multiubiquitination by the APC determines the order of
substrate degradation. *Cell* **124:** 89–103.

Rice NR, MacKichan ML, Israël A. 1992. The precursor of
NF-κ B p50 has IκB-like functions. *Cell* **71:** 243–253.

Rodriguez MS, Dargemont C, Hay RT. 2001. SUMO-1
conjugation in vivo requires both a consensus modification motif and nuclear targeting. *J Biol Chem* **276:**
12654–12659.

Russo SM, Tepper JE, Baldwin AS Jr, Liu R, Adams J, Elliott
P, Cusack JC Jr. 2001. Enhancement of radiosensitivity
by proteasome inhibition: Implications for a role of
NF-κB. *Int J Rad Oncol, Biol, Phys* **50:** 183–193.

Scherer DC, Brockman JA, Chen Z, Maniatis T, Ballard DW.
1995. Signal-induced degradation of IκB α requires
site-specific ubiquitination. *Proc Natl Acad Sci* **92:**
11259–11263.

Schouten GJ, Vertegaal AC, Whiteside ST, Israël A, Toebes
M, Dorsman JC, van der Eb AJ, Zantema A. 1997. IκB
α is a target for the mitogen-activated 90 kDa ribosomal
S6 kinase. *EMBO J* **16:** 3133–3144.

Scott ML, Fujita T, Liou HC, Nolan GP, Baltimore D. 1993.
The p65 subunit of NF-κB regulates IκB by two distinct
mechanisms. *Genes Dev* **7:** 1266–1276.

Sen R, Baltimore D. 1986a. Inducibility of κ immunoglobulin enhancer-binding protein Nf-κ B by a posttranslational mechanism. *Cell* **47:** 921–928.

Sen R, Baltimore D. 1986b. Multiple nuclear factors interact
with the immunoglobulin enhancer sequences. *Cell* **46:**
705–716.

Shirane M, Hatakeyama S, Hattori K, Nakayama K. 1999.
Common pathway for the ubiquitination of IκBα,
IκBβ, and IκBε mediated by the F-box protein FWD1.
J Biol Chem **274:** 28169–28174.

Skowyra D, Craig KL, Tyers M, Elledge SJ, Harper JW. 1997.
F-box proteins are receptors that recruit phosphorylated
substrates to the SCF ubiquitin-ligase complex. *Cell* **91:**
209–219.

Skowyra D, Koepp DM, Kamura T, Conrad MN, Conaway
RC, Conaway JW, Elledge SJ, Harper JW. 1999.
Reconstitution of G1 cyclin ubiquitination with complexes containing SCFGrr1 and Rbx1. *Science* **284:**
662–665.

Smith TF, Gaitatzes C, Saxena K, Neer EJ. 1999. The WD
repeat: A common architecture for diverse functions.
Trends Biochem Sci **24:** 181–185.

Soucy TA, Smith PG, Milhollen MA, Berger AJ, Gavin JM,
Adhikari S, Brownell JE, Burke KE, Cardin DP,
Critchley S, et al. 2009. An inhibitor of NEDD8-activat-
ing enzyme as a new approach to treat cancer. *Nature*
458: 732–736.

Spencer E, Jiang J, Chen ZJ. 1999. Signal-induced ubiquitination of IκBα by the F-box protein Slimb/β-TrCP.
Genes Dev **13:** 284–294.

Spiegelman VS, Slaga TJ, Pagano M, Minamoto T, Ronai Z,
Fuchs SY. 2000. Wnt/β-catenin signaling induces the
expression and activity of β-TrCP ubiquitin ligase receptor. *Mol Cell* **5:** 877–882.

Spiegelman VS, Tang W, Katoh M, Slaga TJ, Fuchs SY. 2002.
Inhibition of HOS expression and activities by Wnt
pathway. *Oncogene* **21:** 856–860.

Sriskantharajah S, Belich MP, Papoutsopoulou S, Janzen J,
Tybulewicz V, Seddon B, Ley SC. 2009. Proteolysis of
NF-κB1 p105 is essential for T cell antigen receptor-
induced proliferation. *Nat Immunol* **10:** 38–47.

Sun SC, Ley SC. 2008. New insights into NF-κB regulation
and function. *Trends Immunol* **29:** 469–478.

Sun SC, Ganchi PA, Ballard DW, Greene WC. 1993. NF-κ B
controls expression of inhibitor IκBα: Evidence for
an inducible autoregulatory pathway. *Science* **259:**
1912–1915.

Tan P, Fuchs SY, Chen A, Wu K, Gomez C, Ronai Z, Pan ZQ.
1999. Recruitment of a ROC1-CUL1 ubiquitin ligase by
Skp1 and HOS to catalyze the ubiquitination of IκB α.
Mol Cell **3:** 527–533.

van Es JH, Barker N, Clevers H. 2003. You Wnt some, you
lose some: Oncogenes in the Wnt signaling pathway.
Current Opinion Gen Develop **13:** 28–33.

von Mikecz A. 2006. The nuclear ubiquitin-proteasome
system. *J Cell Sci* **119:** 1977–1984.

Watanabe N, Arai H, Nishihara Y, Taniguchi M, Hunter T,
Osada H. 2004. M-phase kinases induce phospho-
dependent ubiquitination of somatic Wee1 by
SCF(β-TrCP). *Proc Natl Acad Sci* **101:** 4419–4424.

Wei N, Deng XW. 2003. The COP9 signalosome. *Ann Rev
Cell Develop Biol* **19:** 261–286.

Weil R, Sirma H, Giannini C, Kremsdorf D, Bessia C,
Dargemont C, Brechot C, Israël A. 1999. Direct association and nuclear import of the hepatitis B virus X
protein with the NF-κB inhibitor IκBα. *Mol Cell Biol*
19: 6345–6354.

Werner SL, Barken D, Hoffmann A. 2005. Stimulus specificity of gene expression programs determined by temporal control of IKK activity. *Science* **309:** 1857–1861.

Whiteside ST, Epinat JC, Rice NR, Israël A. 1997. IκB ε, a
novel member of the IκB family, controls RelA and cRel
NF-κB activity. *EMBO J* **16:** 1413–1426.

Winston JT, Strack P, Beer-Romero P, Chu CY, Elledge SJ,
Harper JW. 1999. The SCF(β-TRCP)-ubiquitin ligase
complex associates specifically with phosphorylated
destruction motifs in IκBα and β-catenin and stimulates
IκBα ubiquitination in vitro. *Genes Dev* **13:** 270–283.

Wolf DA, Zhou C, Wee S. 2003. The COP9 signalosome: An
assembly and maintenance platform for cullin ubiquitin
ligases? *Nature cell biology* **5:** 1029–1033.

Wu C, Ghosh S. 1999. βTrCP mediates the signal-induced
ubiquitination of IκBβ. *J Biol Chem* **274:** 29591–29594.

Wu ZH, Miyamoto S. 2007. Many faces of NF-κB
signaling induced by genotoxic stress. *J Mol Med* **85:**
1187–1202.

Wu G, Xu G, Schulman BA, Jeffrey PD, Harper JW, Pavletich
NP. 2003. Structure of a β-TrCP1-Skp1-β-catenin

complex: Destruction motif binding and lysine specificity of the SCF(β-TrCP1) ubiquitin ligase. *Molecular cell* **11**: 1445–1456.

Xiao G, Harhaj EW, Sun SC. 2001. NF-κB-inducing kinase regulates the processing of NF-κB2 p100. *Mol Cell* **7**: 401–409.

Xiao G, Fong A, Sun SC. 2004. Induction of p100 processing by NF-κB-inducing kinase involves docking IκB kinase α (IKKα) to p100 and IKKα-mediated phosphorylation. *J Biol Chem* **279**: 30099–30105.

Yaron A, Gonen H, Alkalay I, Hatzubai A, Jung S, Beyth S, Mercurio F, Manning AM, Ciechanover A, Ben-Neriah Y. 1997. Inhibition of NF-κB cellular function via specific targeting of the IκB-ubiquitin ligase. *Embo J* **16**: 6486–6494.

Yaron A, Hatzubai A, Davis M, Lavon I, Amit S, Manning AM, Andersen JS, Mann M, Mercurio F, Ben-Neriah Y. 1998. Identification of the receptor component of the IκBα-ubiquitin ligase. *Nature* **396**: 590–594.

Yoshida Y, Chiba T, Tokunaga F, Kawasaki H, Iwai K, Suzuki T, Ito Y, Matsuoka K, Yoshida M, Tanaka K, et al. 2002. E3 ubiquitin ligase that recognizes sugar chains. *Nature* **418**: 438–442.

Yoshida Y, Tokunaga F, Chiba T, Iwai K, Tanaka K, Tai T. 2003. Fbs2 is a new member of the E3 ubiquitin ligase family that recognizes sugar chains. *J Biol Chem* **278**: 43877–43884.

Zandi E, Chen Y, Karin M. 1998. Direct phosphorylation of IκB by IKKα and IKKβ: Discrimination between free and NF-κB-bound substrate. *Science* **281**: 1360–1363.

Zhang D, Zaugg K, Mak TW, Elledge SJ. 2006. A role for the deubiquitinating enzyme USP28 in control of the DNA-damage response. *Cell* **126**: 529–542.

Zhou BP, Deng J, Xia W, Xu J, Li YM, Gunduz M, Hung MC. 2004. Dual regulation of Snail by GSK-3β-mediated phosphorylation in control of epithelial-mesenchymal transition. *Nat Cell Biol* **6**: 931–940.

Signaling to NF-κB: Regulation by Ubiquitination

Ingrid E. Wertz[1] and Vishva M. Dixit[2]

[1]Department of Protein Engineering, Genentech, Inc., South San Francisco, California 94080

[2]Department of Physiological Chemistry, Genentech, Inc., South San Francisco, California 94080

Correspondence: ingrid@gene.com

The NF-κB pathway is a ubiquitous stress response that activates the NF-κB family of transcription factors. Antigen receptors, receptors of the innate immune system, and certain intracellular stressors are potent activators of this pathway. The transcriptional program that is activated is both antiapoptotic and highly proinflammatory. Indeed, any compromise in engagement of the pathway results in immunodeficiency, whereas constitutive activation generates a sustained inflammatory response that may promote malignancy. As such, NF-κB activation is under tight regulation by a number of post-translational modifications, including phosphorylation and ubiquitination. This article attempts to synthesize our current knowledge regarding the regulation of NF-κB signaling by ubiquitination, specifically highlighting the biochemical basis for both positive and negative feedback loops that function in unison to generate coordinated signals that are essential for the viability of metazoan animals.

INTRODUCTION TO NF-κB SIGNALING

The NF-κB family of transcription factors includes critical regulators of proinflammatory and antiapoptotic gene transcription programs. As such, they play essential homoeostatic roles in the development and orchestration of the host immune response. NF-κB-mediated transcription is the endpoint of a complex series of reactions that are initiated by a vast array of stimuli, ranging from cellular stress to the engagement of receptors that mediate innate and adaptive immunity. In this article, we review the unique aspects of the pathways that culminate in NF-κB activation and also highlight the common components that are shared by these diverse signaling cascades (see Table 1).

Each NF-κB-activating stimulus converges on the activation of one of two kinase complexes termed inhibitors of -κB (I-κB) kinase (I-κK) complexes. The heterotrimeric I-κK complexes are composed of a regulatory subunit I-κK-γ, or NF-κB essential modulator (NEMO), and two kinases I-κK-α or I-κK-β. Canonical NF-κB signaling pathways assemble proximal signaling complexes on polyubiquitin chain scaffolds; assembly of these complexes promotes the activation of I-κK-β, which in turn phosphorylates I-κB inhibitory proteins. These inhibitors mask the nuclear localization signals (NLS) within NF-κB transcription

Table 1. Classes of signaling components in NF-κB transduction pathways

Stimuli	TNFα	IL1β, LPS	CD40L	MHC/ antigen	Viral RNA	Bacterial PAMPs	DNA damage
Receptors	TNFR1	IL1R1, TLR4	CD40	TCR	**RIG-I***	NOD1/2	PIDD?
Adaptors	TRADD	MyD88 TRIF MAL TRAM	**TRAF3***	CARMA **Bcl10*** MALT1	MAVS FADD	?	PIDD
Proximal (adaptor) kinases	**RIP1***	**IRAK1*** IRAK4 RIP1	?	Src/Syk PI3K PDK1 PKCθ	RIP1	RIP2	RIP1
Non- degradative ligases	TRAF2 TRAF5 cIAP1/2 LUBAC	TRAF6 Pellino	**TRAF2***	TRAF2 TRAF6	TRIM25 TRAF2 TRAF6	TRAF2 TRAF5 TRAF6 cIAP1/2	PIASy
Degradative ligases	A20 RNF11 ITCH	Pellino?	**cIAP1/2***	NEDD4 ITCH	RNF125	?	?
I-κK- activating kinases	TAK1 MEKK3 aPKC	TAK1 MEKK3 aPKC	NIK	TAK1	TAK1	TAK1	ATM?
DUBs	A20 CYLD Cezanne	A20 CYLD	?	A20 CYLD	A20 CYLD	A20 CYLD	?

Representative stimuli and the cognate receptors (where known) are listed for each signaling pathway. Degradative ligases assemble polyubiquitin chains that promote substrate degradation, whereas nondegradative ligases polymerize polyubiquitin chains that promote assembly of proximal signaling complexes and subsequent activation of NF-κB signaling. PIASy, though a SUMO ligase, can therefore be classified as a nondegradative ligase in the DNA damage pathway since NEMO SUMOylation does not promote NEMO degradation and is a proximal event that is required for NF-κB activation. See text for details.

*Indicates these proteins are targets of ubiquitin editing.

factors, thereby retaining them in the cytosol. Phosphorylation of I-κB by I-κK-β permits the recognition of I-κB by the SCF^βTrCP ubiquitin ligase complex, which subsequently targets I-κB for proteasomal degradation. Thus, heterodimeric NF-κB transcription factors such as p50/p65 are released from I-κB inhibition and enter the nucleus to activate transcription of NF-κB target genes (Hoffmann and Baltimore 2006). Alternatively, stimuli that activate noncanonical signaling pathways promote stabilization of NF-κB-inducing kinase (NIK), a labile proximal kinase. When sufficient levels accumulate, NIK phosphorylates and activates I-κK-α within the I-κK complex, which in turn phosphorylates p100, a precursor NF-κB subunit. Phosphorylated p100 is also ubiquitinated by the SCF^βTrCP ubiquitin ligase complex and is subsequently processed by the proteaseome to p52, which is a transcriptionally competent

NF-kB subunit in conjunction with RelB (Vallabhapurapu and Karin 2009). Thus, ubiquitination is a critical regulatory mechanism at multiple steps within NF-κB signaling cascades. We therefore provide a brief review of the ubiquitin/ proteasome system.

THE UBIQUITIN/PROTEASOME SYSTEM (UPS)

Ubiquitin is a 76-amino-acid protein that when covalently linked to target proteins may alter their half-life, localization, or function. An enzymatic cascade composed of three types of proteins—the E1 ubiquitin-activating enzyme, the E2 ubiquitin-conjugating enzyme, and the E3 ubiquitin ligase—mediates the conjugation of ubiquitin to substrate proteins (Fig. 1). The E1 enzyme hydrolyses ATP and transfers a thio-ubiquitin intermediate to the

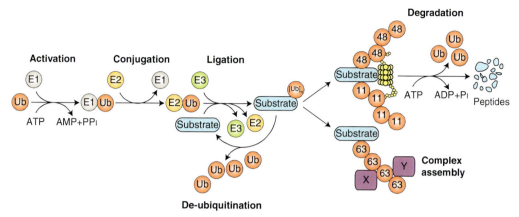

Figure 1. Enzymes and reactions of the ubiquitin/proteasome system.

active site cysteine of one of over 30 E2 enzymes. The E2 enzyme "charged" with a thio-ubiquitin may transfer the ubiquitin to a HECT domain of an E3 ligase, or instead, may bind to a RING domain (or a related variant domain) of an E3 ligase, which number in the hundreds. Importantly, the E3 also binds the substrate and orients the reactants for facile ligation of the ubiquitin carboxyl terminus to the ε-amino group or the amino terminus of a substrate lysine (Lys) residue (Schwartz and Ciechanover 2009).

Multiple rounds of ubiquitination, with ubiquitin itself serving as a substrate, generates polyubiquitin chains. Ubiquitin has seven Lys residues, Lys-6, Lys-11, Lys-27, Lys-29, Lys-33, Lys-48, and Lys-63, and any one of these can participate in polyubiquitin chain formation. The amino terminus of ubiquitin may also be conjugated to another ubiquitin carboxyl terminus in which case linear polyubiquitin chains are generated. Whereas the HECT domain-containing E3 ligases are intrinsically responsible for dictating ubiquitin linkage specificity, it is the E2 associated with RING-type ligases that determines the type of polyubiquitin chain formed. For example, the E2 UBC13, in cooperation with the E2 variant UEV1 and TRAF RING domains, promotes the formation of Lys-63-linked ubiquitin chains. However, the E2 UBC5, together with the RING-containing protein ROC1/RBX1 of the SCF$^{\beta TrCP}$ ligase, promotes Lys-48-linked polyubiquitination (Vallabhapurapu and Karin 2009). Embedded

in the topology of the various ubiquitin chains is information that dictates signaling outcome. For example, Lys-48- and Lys-11-linked polyubiquitination usually, but not always, targets substrates for proteosomal degradation (Fig. 1). In contrast, Lys-63-linked polyubiquitin chains function as scaffolds to assemble signaling complexes and thereby participate in diverse cellular processes ranging from DNA repair to activation of NF-κB signaling (Fig. 1) (Pickart and Fushman 2004; Ikeda and Dikic 2008). Distinct ubiquitin-binding domains (UBDs) are the receptors that bind different ubiquitin chain conformations and dictate the fate of ubiquitinated substrates. For example, the ubiquitin-associated (UBA) domains of proteasome subunits Rpn13/ARM1 and Rpn10/S5a preferentially bind Lys-48-linked chains, thereby recruiting substrates for degradation. Alternatively, the UBAN (UBD in ABIN and NEMO) motif of NEMO binds linear and Lys-63-linked polyubiquitin chains, thereby shuttling NEMO and the associated I-κK proteins, to activated signaling complexes (Ikeda and Dikic 2008).

Ubiquitination can be reversed by proteases termed deubiquitinases (DUBs) (Fig. 1). The human genome encodes approximately 100 DUBs that fall into five families: Four are papain-like cysteine proteases (ubiquitin carboxy-terminal hydrolase, UCH; ubiquitin specific protease, USP; ovarian tumor domain, OTU; and Machado-Joseph disease proteases, MJD) and the fifth is a metalloprotease (JAMM). In addition to protease domains, DUBs often contain protein interaction motifs, including UBDs, that direct their recruitment to particular signaling complexes and promote preferential binding to specific poly-Ub chain linkages (Reyes-Turcu and Wilkinson 2009). DUBs are critical regulators of NF-κB signaling pathways. For example, genetic ablation of the OTU DUB A20 causes rampant inflammation in multiple organ systems because of unchecked NF-κB activity, resulting in perinatal lethality (Lee et al. 2000). Furthermore, mutation or deletion of *TNFAIP3*, which encodes A20, is associated with deregulation of NF-κB signaling in a variety of lymphomas (Compagno et al. 2009; Kato et al. 2009; Novak et al. 2009; Schmitz et al. 2009) and *TNFAIP3* polymorphisms are associated with the autoimmune disorder systemic lupus erythematosus (Musone et al. 2008). The USP DUB CYLD is also a tumor suppressor that negatively regulates NF-κB signaling by de-ubiquitinating a number of critical pathway components. Somatic mutations are linked to familial cylindromatosis and loss of CYLD expression has been implicated in a variety of malignancies including colonic, hepatocellular, and renal carcinomas, as well as multiple myeloma (Sun 2009). Thus, cycles of ubiquitination and deubiquitination, each mediated by distinct enzymes, control cellular processes that are essential for the maintenance of cellular homeostasis.

UPS REGULATION OF NF-κB SIGNALING: GENERAL CONCEPTS

Translating Polyubiquitination into I-κK Activation

Having introduced the NF-κB transcription factors and the UPS, we can now address some general concepts in the signaling pathways that activate NF-κB. The first is how the polymerization of nondegradative ubiquitin chains promotes I-κK-β activation. There are several pieces of evidence suggesting that polyubiquitination plays an essential role in the activation of I-κK-β. First, NEMO is both a ubiquitin binding protein and a target of ubiquitination in classical NF-κB signaling pathways. Genetic

ablation studies have shown that NEMO is essential for activation of I-κK-β (Makris et al. 2000; Rudolph et al. 2000). Indeed, mutations in NEMO that disrupt ubiquitin binding cause anhidrotic ectodermal dysplasia and immunodeficiency caused by improper activation of NF-κB signaling (Chen and Sun 2009). Finally, as indicated previously, mutation or deletion of the DUBs A20 and CYLD deregulate NF-κB signaling, thereby promoting unchecked inflammation and tumorigenesis.

Several models of how ubiquitination might promote I-κK-β activation have been proposed. These include: (1) conformational changes in NEMO induced by polyubiquitin binding that promote I-κK-β activation, (2) induced proximity of I-κK-β proteins within the I-κK complex that activates *trans*-phosphorylation, and (3) induced proximity of I-κK kinases, such as TAK1 or MEKK3, with the I-κK complex. Structural studies have indicated that NEMO binds Lys-63- and linear polyubiquitin chains, which is critical for NF-κB activation in vivo (Rahighi et al. 2009; Lo et al. 2009; Ivins et al. 2009). Furthermore, it was suggested that NEMO association with polyubiquitin could induce sufficient conformational changes to activate associated I-κK-β (Rahighi et al. 2009). The role of NEMO polyubiquitination in I-κK-β activation is less certain. Several polyubiquitination sites have been mapped on NEMO but only Lys-392, the equivalent of human Lys-399, has been tested in a murine knock-in model. These mice were more resistant to lipopolysaccharide (LPS)-induced endotoxic shock but had no defect in T-cell receptor (TCR)-induced proliferation, both of which are mediated by NF-κB signaling (Ni et al. 2008). This finding could suggest that ubiquitination of Lys-392 on NEMO is not important for TCR-induced NF-κB activation. However, it is also possible that alternative sites on NEMO may be sufficiently ubiquitinated following TCR activation if Lys-392 is not available for modification. Thus, NEMO ubiquitination on specific residues may be more important in certain NF-κB signaling pathways than in others. Knockout studies have established a role for MEKK3 and TAK1 as I-κK kinases in certain pathways. MEKK3 is thought to associate with RIP1 in the TNFR1 pathway and activate I-κK-β that is recruited to proximal signaling complexes via NEMO (Vallabhapurapu and Karin 2009). Similarly, the TAB2 and -3 regulatory proteins within the TAK1/TAB complex have UBDs that facilitate binding to Lys-63 ubiquitinated substrates. It is thought that corecruitment of TAK1/TAB and I-κK complexes to polyubiquitinated proteins within proximal signaling complexes may also allow TAK1 to phosphorylate I-κK-β via induced proximity (Chiu et al. 2009).

Thus, there is ample evidence for each of the three models to explain how polyubiquitination of proteins that participate in classical NF-κB signaling pathways could promote I-κK activation. In theory, none of the three models are mutually exclusive and all could contribute to I-κK activation. For example, polyubiquitin chains on one NEMO subunit could bind the UBD of another NEMO subunit to promote the assembly of a larger I-κK complex. This could promote low-level I-κK activity via *trans*-phosphorylation, and subsequent recruitment of this oligomeric I-κK complex to activated signaling complexes via polyubiquitin/UBD interactions could further enhance I-κK activity by induced proximity with I-κK kinases such as TAK1. Such a scenario is all the more plausible given that NEMO can be modified by and can associate with both Lys-63-linked and linear polyubiquitin chains: One type of chain could promote assembly of oligomeric I-κK complexes, whereas the other type of chain could promote the recruitment of I-κK complexes to activated signaling networks.

Inactivation of NF-κB Signaling

Down-regulation of NF-κB signaling is as important as NF-κB activation. Unchecked NF-κB activity may promote rampant inflammation or tumorigenesis, as revealed by mutations or deletions of the DUBs CYLD and A20. In addition to DUBs, a number of inhibitory proteins down-regulate NF-κB signaling, including RIP3, MyD88s, IRAK-M, SARM, and I-κB. RIP3, MyD88s, and SARM competitively block

the recruitment of key signaling proteins to activated complexes (Barton and Medzhitov 2004; O'Neill and Bowie 2007); IRAK-M blocks the dissociation of IRAK1 and -4 from MyD88, which is a critical step to activate NF-κB downstream of IL1R1 and TLR (O'Neill and Bowie 2007); and I-κB masks the NLS on NF-κB hetrerodimers to block nuclear translocation (Hoffmann and Baltimore 2006). DUBs may also collaborate with ubiquitin ligases to inactivate and degrade critical mediators of NF-κB signaling. This form of regulation is termed ubiquitin editing and usually regulates the ubiquitination status of essential positive regulators of NF-κB signaling complexes. First, Lys-63 polyubiquitin chains are depolymerized by DUBs to disassemble the platform on which signaling complexes are organized. Then, Lys-48 polyubiquitin chains are polymerized on the same protein to target it for proteasomal degradation (Wertz et al. 2004; Heyninck and Beyaert 2005; Newton et al. 2008). However, it is important to note that ubiquitin editing need not be restricted to regulation of Lys-48 and Lys-63-linked polyubiquitination or even to ubiquitination in general. In fact, ubiquitin editing can be broadly conceptualized as the concerted removal of any modification that promotes signaling complex assembly, such as Lys-63-linked polyubiquitination, linear ubiquitin chains, or even ligation with the ubiquitin-like protein SUMO, combined with modifications that promote substrate degradation, such as Lys-11 or Lys-48 polyubiquitination. Table 1 includes a summary of proteins regulated by ubiquitin editing in NF-κB signaling pathways.

UPS REGULATION OF NF-κB SIGNALING: SPECIFIC PATHWAYS

Introduction

As mentioned previously, an array of stimuli may activate host receptors to initiate the upstream signaling events that culminate in the activation of NF-κB transcription factors. Although each signaling cascade is activated by unique stimuli and contains highly specialized components, the pathways share some functionally common elements, including: (1) adaptors that link the activated receptor to downstream effector proteins, (2) proximal kinases that propagate signaling, sometimes in a kinase-independent manner by serving as adaptor proteins, (3) nondegradative ligases, (4) degradative E3 ligases, (5) distal I-κK kinases, and (6) DUBs. These components, and the pathways that they participate in, are listed in Table 1.

TNFR1

Tumor necrosis factor-α (TNF-α) was first described in 1975 as a factor that caused tumor necrosis in a murine sarcoma model (Carswell et al. 1975). Since then, TNF-α has been characterized as a cytokine that activates a variety of cellular responses including NF-κB signaling, as depicted in Figure 2. TNFR1 is trimerized on TNF-α binding, thereby promoting recruitment of the adaptor protein TNF receptor associated protein with a death domain (TRADD). TRADD assembles at least two distinct signaling complexes that initiate opposing signaling pathways: Complex 1-mediated activation of antiapoptotic and proinflammatory mediators, including NF-κB, or complex 2-mediated activation of apoptosis. Activation of apoptosis by complex 2 has been extensively reviewed elsewhere (Varfolomeev and Vucic 2008); in this article, we focus on NF-κB activation by complex 1.

In addition to TRADD, complex 1 includes receptor interacting protein-1 (RIP1), cellular inhibitor of apoptosis protein-1 (cIAP1), cIAP2,

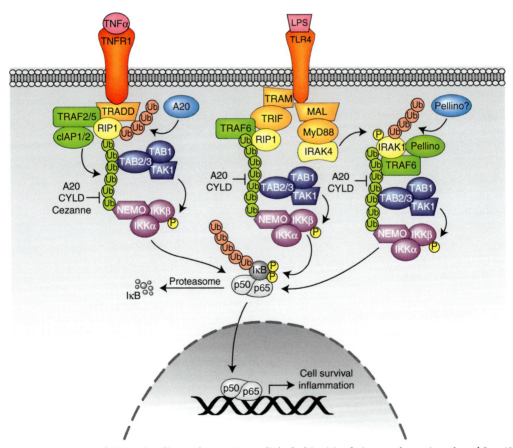

Figure 2. TNFR1 and TLR4 signaling pathways. Lys-48-linked ubiquitin chains are shown in red, and Lys-63-linked ubiquitin chains are shown in green. See text for additional details.

and TNF receptor associated factor-2 (TRAF2). RIP1 is a kinase that is essential for TNF-α-mediated NF-κB signaling in a kinase-independent manner (Ting et al. 1996; Kelliher et al. 1998). On TNF stimulation, TNFR1-associated RIP1 is rapidly modified by Lys-63-linked polyubiquitin chains. These Lys-63-linked chains create a scaffold to recruit the I-κK complex via NEMO. Mutations in NEMO that abolish binding to Lys-63-linked polyubiquitin chains also disrupt association of NEMO with RIP in TNF-α-stimulated cells, block the recruitment of I-κK to TNFR1, and inhibit I-κK activation (Ea et al. 2006; Wu et al. 2006). The UBDs in the TAB2/3 regulatory proteins of the TAK1 kinase complex also bind Lys-63-linked polyubiquitin chains, thus TAB2/3 is recruited to polyubiquitinated RIP on TNF-α stimulation (Kanayama et al. 2004). The subsequent activation of the TAK1 complex promotes I-κK activation (Kovalenko and Wallach 2006). Lys-377 is likely the primary residue that is ubiquitinated on RIP and appears to be important for NF-κB activation, given that a Lys-377 mutation to arginine attenuates RIP ubiquitination, prevents the recruitment of TAK1 and I-κK complexes to TNFR1, and inhibits I-κK activation (Ea et al. 2006). TNFR1-stimulated signaling pathways may also activate MEKK3 and aPKC that in turn phosphorylate the I-κK complex.

Given the importance of RIP modification with Lys-63 polyubiquitin chains in propagating TNFR1 signaling, the definitive identification of RIP E3 ligases is of significant interest. The initial candidate was TRAF2, a RING domain E3 that is essential for TNF-α-stimulated and NF-κB signaling (Yeh et al. 1997; Tada et al. 2001), possibly in collaboration with the heterodimeric E2 UBC13/UEV1 (Shi and Kehrl 2003). Evidence for TRAF2-mediated RIP ubiquitination is based on TRAF2 RNAi (Wertz et al. 2004) and gene ablation (Lee et al. 2004) experiments, which decreased RIP ubiquitination in the absence of TRAF2. However, these data could simply reflect that TRAF2 is an adaptor that recruits another ligase to RIP1. Thus, formal proof demonstrating direct ubiquitination of RIP1 by TRAF2 is still

needed. TRAF5 is another RING domain E3 that is thought to participate in TNFR1-induced NF-κB activation because TRAF2 and -5 are functionally redundant in promoting NF-κB activation (Tada et al. 2001). However, formal proof of TRAF5 ligase activity is missing, and TRAF5 recruitment to complex 1 has not been reported.

The cIAP1 and cIAP2 (cIAP1/2) proteins are another pair of RING E3s that associate with activated TNFR1 (Shu et al. 1996; Srinivasula and Ashwell 2008). Because cIAP1/2 are recruited to TNFR1 via TRAF2, they are candidate TRAF2-associated RIP1 ligases. Indeed, cIAP1/2 promote RIP1 polyubiquitination in vivo, and in vitro experiments revealed that cIAP1/2 couple with the E2 UBCH5 to catalyze Lys-63-linked RIP polyubiquitination. Furthermore, deletion of both cIAP1 and -2 revealed their essential and redundant roles in TNF-α-induced NF-κB activation (Varfolomeev et al. 2008; Mahoney et al. 2008; Bertrand et al. 2008). It is therefore unclear what specific functions TRAF2 and -5 and cIAP1/2 have in RIP1 ubiquitination and NF-κB activation. It is possible that TRAF2 (and perhaps TRAF5) simply recruits cIAP1/2 to TNFR1 but has no additional role in RIP1 ubiquitination. Alternatively, TRAF2 and -5 could regulate cIAP1/2 ubiquitin ligase activity, possibly via ubiquitination, or perhaps TRAF2 and -5 ubiquitinate RIP1 on different sites and/or in different contexts than cIAP1/2 to fine-tune TNFR1-induced NF-κB activation.

The E2 UBC13 was first implicated in TNFR1 signaling when a dominant–negative version of UBC13 was reported to block TNF-α- and TRAF2-induced NF-κB activity (Deng et al. 2000). However, UBC13 genetic ablation experiments performed by two separate groups have generated conflicting results. In each set of studies, homozygous UBC13 ablation was embryonic lethal, thus necessitating study of hemizygous UBC13$^{-/+}$ mice (Fukushima et al. 2007) or mice conditionally deficient in UBC13 (Yamamoto et al. 2006a). Although UBC13$^{+/-}$ macrophages and splenocytes displayed blunted activation of NF-κB in response to TNF-α (Fukushima et al. 2007), UBC13$^{-/-}$ MEFs displayed no alteration in

TNF-α-induced NF-κB signaling relative to wild-type MEFs (Yamamoto et al. 2006a). These discrepancies may be attributable to other E2 enzymes or E2/ligase pairs that can substitute for UBC13 in certain cell lines. For example, it was shown in a cell-free system that UBC4/5 can promote I-κB-α phosphorylation with an unidentified ligase (Chen et al. 1996). Furthermore, TRAF2 and -5 fail to bind UBC13 in vitro (Yin et al. 2009), implicating another E2 and/or E2/ligase pair in TNFR1-induced NF-κB activation. The recent identification of a distinct E2/E3 enzyme complex that modifies NEMO with linear polyubiquitin chains and is essential for TNF-α-activated NF-κB signaling may explain some of the discrepancies revealed in the previous studies (Tokunaga et al. 2009).

Genetic ablation experiments established A20 as a critical negative regulator of TNF-α-induced NF-κB signaling (Lee et al. 2000), and functional studies later revealed that A20 contains both an OTU DUB domain and a C_2/C_2 ZnF E3 ligase domain. As such, A20 is a dual-function ubiquitin editing enzyme for RIP1: The A20 DUB domain first depolymerizes Lys-63-linked ubiquitin chains from RIP1, and the A20 E3 ligase motif then promotes the ligation of Lys-48-linked ubiquitin chains on RIP1 (Wertz et al. 2004). Several modulators of A20 ubiquitin ligase activity have also been identified. TAXBP1, a binding protein of the hTLV TAX protein, is an A20-binding protein and cooperates with A20 to attenuate TNF-α signaling by recruiting the HECT-domain ubiquitin ligase Itch (Shembade et al. 2008). The RING domain E3 RNF11 also collaborates with TAXBP1 and Itch to promote RIP1 degradation after TNF-α treatment (Shembade et al. 2009), although the precise roles for each E3 are unknown. A number of additional RIP1 DUBs have also been reported. The A20-like protein cellular zinc finger antiNF-κB (Cezanne) also has an OTU domain and promotes RIP1 deubiquitination on TNF-α stimulation (Enesa et al. 2008), and CYLD is also proposed to deubiquitinate RIP1, among other critical targets, in TNF-α-induced NF-κB signaling pathways (Sun 2009). In vivo evidence of RIP1 ubiquitin editing

was recently revealed by ubiquitin linkage-specific antibodies (Newton et al. 2008), thus the interplay between the various ubiquitin modifying enzymes in orchestrating RIP1 ubiquitination and degradation will be interesting to elucidate.

IL1R1/TLR4

Interleukin-1 (IL1) receptor-1 (IL1R1) and Toll-like receptors (TLR) are transmembrane proteins that share a common intracellular Toll and IL1 receptor (TIR) domain. As such, they recruit related complexes of signaling proteins with distinct variations that fine-tune regulation and mediate signaling specificity. Like TNFR1, IL1R1-activated signaling pathways may also activate MEKK3 and aPKC that in turn activate the I-κK complex. IL1R1 is activated by the potent inflammatory cytokine IL1-β, whereas TLRs, of which there are at least 10 in humans, recognize pathogen-associated molecular patterns (PAMPs) such as LPS and viral nucleic acids. On activation, most TLRs and IL1R1 recruit the adaptor protein MyD88 either directly or via MAL (also known as TIRAP) through TIR/TIR interactions. TLR4 can also recruit TRIF via TRAM, and TLR3 directly recruits TRIF (for more extensive reviews, see O'Neill and Bowie 2007; Verstrepen et al. 2008).

Here, we focus on TLR4 as a prototypical TIR domain-containing receptor. A wealth of studies have investigated the mechanistic details leading to I-κK activation, thus the model presented here highlights general concepts (Fig. 2). LPS binding activates two primary pathways downstream of TLR4 that culminate in TAK1 activation. In one branch, TRAM and TRIF are recruited to TLR4, and TRIF recruits both TRAF6 and RIP1 to the proximal receptor signaling complex. It is thought that TRAF6 and RIP1 Lys-63-linked polyubiquitination both facilitate TAK1 activation (Schauvliege et al. 2007; Vallabhapurapu and Karin 2009). In the other branch, recruitment of MyD88 via MAL promotes the assembly of a proximal signaling complex that includes IRAK1 and -4 and TRAF6. IRAK4 phosphorylates IRAK1, which promotes dissociation of IRAK1 and bound TRAF6 from TLR4. This complex subsequently associates with the TAK1/TAB complex, perhaps via TAB2/3 binding to TRAF6 Lys-63-linked chains. Pellino RING ligases are also recruited and may ubiquitinate IRAK1 with Lys-63-linked chains. IRAK1 and Pellino proteins are ubiquitinated and targeted for proteasomal degradation, perhaps by Pellino ligases that promote Lys-48 and Lys-11 polyubiquitination. IRAK1 degradation and TAK/TAB phosphorylation may facilitate the release of TAK1/TAB complex and associated TRAF6, but the role of such translocation in NF-κB activation is unclear (Moynagh 2009). Thus, most evidence indicates that Lys-63-linked ubiquitination is critical for TAK1 and IKK activation via MyD88 and MAL even though the precise mechanisms are unknown.

Indeed, TRAF6, IRAK1, NEMO, and the Pellino proteins are all reported targets of Lys-63-linked ubiquitination, and the ubiquitination sites that are critical for propagating NF-κB signaling have been mapped on IRAK1 (Conze et al. 2008) and TRAF6 (Lamothe et al. 2007). Nevertheless, the E2 and E3 enzymes responsible for activating NF-κB signaling downstream of IL1R1 and TLR are not completely characterized. For example, conditional UBC13$^{-/-}$ B cells, bone marrow-derived macrophages, and MEFs have no defects in NF-κB signaling after stimulation with IL1-β or TLR agonists including LPS, CpG DNA, or bacterial lipopeptide (Yamamoto et al. 2006a). These results suggest that UBC13 in collaboration with TRAF ligases may not be responsible for catalyzing Lys-63 polyubiquitination, and/or that Lys-63 polyubiquitination is not important for IL1R1 or TLR signaling pathways. In contrast, UBC13$^{+/-}$ macrophages and splenocytes showed decreased NF-κB activity in response to LPS (Fukushima et al. 2007). Adding to the complexity of interpreting these studies, it was also reported that LPS treatment promoted less TRAF6 ubiquitination in UBC13$^{+/-}$ spleen lysates and splenocytes relative to wild-type controls (Fukushima et al. 2007). However, it is not clear whether the lysates were denatured before TRAF6 immunoprecipitation to dissociate noncovalently bound proteins; if not, the polyubiquitination detected in TRAF6 immunoprecipitates could reflect the ubiquitination status of TRAF6-associated proteins (including UBC13), rather than TRAF6 itself. Finally, the role of the TRAF6 RING in TLR- and IL1R1 signaling is unclear. One study showed that the TRAF6 RING and first ZnF were dispensable for NF-κB activation in complementation experiments with TRAF6-deficient cells (Kobayashi et al. 2001), whereas similar complementation studies (Lamothe et al. 2007; Conze et al. 2008; Walsh et al. 2008) and structure/function studies (Yin et al. 2009) indicate that the TRAF6 RING domain is essential. Similarly, structure/function studies have shown that TRAF6 interaction with UBC13 is essential for IL1-induced NF-κB activation (Yin et al. 2009). It is therefore possible that in certain conditions alternate E2s, ligases, and/or even polyubiquitin chains can propagate IL1R1- and TLR-induced NF-κB signaling, such as the LUBAC ligase complex that polymerizes linear ubiquitin chains that are essential for IL1-β-induced NF-κB signaling (Tokunaga et al. 2009).

Ubiquitin editing also appears to be an important regulatory mechanism for NF-κB activation in IL1R1 and TLR4 signaling pathways. IRAK ubiquitination and degradation on receptor activation was first reported 12 years ago (Yamin and Miller 1997), but it was only recently shown that IRAK is ubiquitinated with Lys-63 polyubiquitination before degradation (Ordureau et al. 2008). Furthermore, ubiquitin-linkage-specific antibodies revealed that IRAK1 is a target of ubiquitin editing after IL1-β stimulation (Newton et al. 2008). A20 (Boone et al. 2004) and CYLD (Sun 2009) have been shown to remove Lys-63 polyubiquitin chains from TRAF6 and thereby regulate TLR4 and IL1R1 signaling, but the enzymes responsible for IRAK1 ubiquitin editing are largely unknown. Certain Pellino proteins are reported to synthesize Lys-11, Lys-48, and Lys-63 polyubiquitin chains and are therefore attractive candidates for IRAK ligases in addition to TRAF6 (Moynagh 2009), and Pellino genetic ablation studies will likely further elucidate the physiological targets.

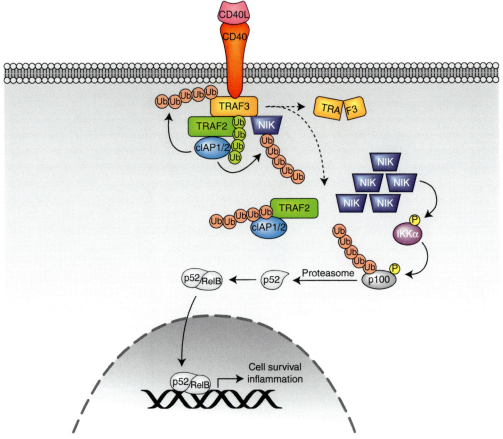

Figure 3. CD40 signaling pathways. Lys-48-linked ubiquitin chains are shown in red, and Lys-63-linked ubiquitin chains are shown in green. See text for additional details.

Noncanonical NF-κB Signaling: CD40

Most TNFR family members can activate both canonical and noncanonical NF-κB signaling pathways. A subset of receptors including CD40 and BAFF-R on B cells, LTB-R on stromal cells, TWEAK-R on endothelial cells, and RANK on osteoclasts primarily activate the noncanonical pathway, making these receptors key systems for study (Hacker and Karin 2006). Here, we focus on CD40 as a model system of NF-κB activation by noncanonical signaling pathways (Fig. 3).

A key feature of noncanonical NF-κB signaling is the proteasomal processing of p100 to p52. Generation of p52 allows the active NF-κB heterodimer to enter the nucleus and promote transcription. Processing of p100 is initiated by IκK-α-mediated phosphorylation, which is activated by NIK. NIK is normally maintained at low levels in the cytosol as a result of efficient degradation by the ubiquitin/proteasome system (Vallabhapurapu and Karin 2009). Thus, the identification and mechanistic characterization of the ligase(s) that promote NIK degradation has been the subject of intense research.

Several clues to the identity and function of NIK ligases have been provided by studies using knockout mice, RNAi, and small molecule antagonists of signaling components of noncanonical NF-κB signaling pathways. In unstimulated cells, NIK is constitutively bound to TRAF2 and -3, which are nonredundant negative regulators: Deficiency in TRAF2 or -3 promotes constitutive NIK activation, and crossing TRAF2 or -3 deficient mice with NIK$^{-/-}$ mice resolves the pathology (Wallach and Kovalenko 2008). Furthermore, inhibition of cIAP1 or -2 expression using RNAi or small molecule antagonists also activates NIK (Wu et al. 2007). Because TRAF2 and -3 are thought to catalyze nondegradative Lys-63 polyubiquitination, it was proposed that TRAF2 and -3 participate in NIK degradation by serving as a molecular link to cIAP1/2, which could assemble degradative polyubiquitin chains on NIK. Additional evidence to support this hypothesis includes degradation of endogenous cIAP1/2 on activation with ligands that activate noncanonical NF-κB signaling (CD40, BAFF, and TWEAK) and genetic deletion of cIAP1/2 in multiple myeloma cells, which is correlated with NIK stabilization, p100 processing, and chronic NF-κB activation (Varfolomeev and Vucic 2008).

Additional studies have investigated the coordinated activities of TRAF2 and -3 with cIAP1/2 in NIK degradation in response to CD40 activation. These studies suggested that on receptor ligation, TRAF3 serves as an adaptor to recruit NIK, as well as TRAF2 and associated cIAP1/2, to receptors. TRAF2 then modifies cIAP1/2 with Lys-63-linked polyubiquitination, thereby enhancing the ability of cIAP1/2 to catalyze Lys-48-linked polyubiquitination of TRAF3. The resultant degradation of TRAF3 dissociates NIK from the cIAPs, thereby permitting NIK accumulation, p100 processing, and NF-κB activiation (Wallach and Kovalenko 2008). Collectively, these studies point to cIAP1/2, and possibly to TRAF2 and -3, as targets of ubiquitin editing: On receptor activation, these proteins are initially modified with Lys-63-linked chains (Wallach and Kovalenko 2008; Varfolomeev and Vucic 2008), and are subsequently modified with degradative polyubiquitin linkages that target them for proteasomal destruction (Wu et al. 2007; Wallach and Kovalenko 2008). Notably, in the case of noncanonical NF-κB signaling, ubiquitin editing promotes NF-κB activity, as opposed to canonical NF-κB signaling in which ubiquitin editing attenuates NF-κB activity.

These reports may explain why TRAF2 and -3 have nonredundant functions in CD40

signaling but, as with any new information in complex systems, the findings prompt further investigation. For example, the role of the TRAF3 RING domain is unclear. Complementation experiments of TRAF3 null MEFs suggested that a TRAF3 RING mutant is unable to destabilize endogenous NIK or inhibit endogenous p100 processing, implicating the importance of the TRAF3 RING domain in regulating noncanonical NF-κB signaling (He et al. 2007). Conversely, experiments in HEK 293T cells with RING-deleted TRAF3 (Vallabhapurapu et al. 2008) and in TRAF2$^{-/-}$ or TRAF3$^{-/-}$ MEFs reconstituted with TRAF2/TRAF3 chimeras suggest that TRAF3 serves as an adaptor only, and the TRAF3 RING is not required to regulate p100 processing (Zarnegar et al. 2008). Additionally, how TRAF2 and -3 generate polyubiquitin chains is uncertain. Structure/function studies suggest that only TRAF6, but not TRAF2 or -3 RING domains, interact with UBC13. Thus, the E2 enzymes that cooperate with TRAF2 and -3 are unknown (Yin et al. 2009). This could explain why conditional ablation of UBC13 in B cells does not alter p100 processing in response to anti-CD40 or BAFF stimulation (Yamamoto et al. 2006a), and suggests that another ligase and/or E2 that catalyze Lys-63 polyubiquitination participate in noncanonical NF-κB signaling. Finally, it is unclear whether NIK requires additional regulation for full kinase activity, or whether simple accumulation of critical protein levels is sufficient for activation (Wallach and Kovalenko 2008).

Antigen Receptors: TCR

Antigen receptors are activated on ligation with MHC-bound antigenic peptides from pathogens that are presented on the surface of antigen-presenting cells. Receptor ligation activates a prototypical signaling cascade that is common to an array of receptors, including angiotensin II- or lysophosphatidic acid-activated G-protein coupled receptors as well as a number of receptors containing immunoreceptor tyrosine-based activation motifs (ITAMs). ITAM-containing receptors include TCR, BCR, the natural killer

(NK) cell receptor NKG2D, and the osteoclast receptor OSCAR (Thome 2008). In each case, receptor activation initiates a phosphorylation cascade that promotes the assembly of a complex that is composed of CARMA1, Bcl10, and MALT1 (the CBM signalosome), which ultimately activates I-κK to promote NF-κB transcription.

Here, we focus on TCR as a prototype of antigen receptor signaling (Fig. 4). TCR ligation with a peptide/MHC complex activates TCR-associated Src/Syc family kinases. A subsequent phosphorylation cascade that includes PI3K and PDK1 culminates in PKC-Θ phosphorylation. Activated PKC-Θ then phosphorylates CARMA1, resulting in membrane recruitment and assembly of the CBM signalosome. CBM assembly promotes polymerization of Lys-63-

linked ubiquitin chains and subsequent activation of TAK1 to stimulate NF-κB-mediated transcription (Bhoj and Chen 2009).

TRAF2 and -6 are candidate ligases that may polymerize Lys-63-linked Ub chains downstream of activated TCRs. Although TRAF2 or -6 knockout animals have no deficiency in TCR-induced NF-κB activation, RNAi of both TRAF2 and -6 almost completely abolished I-κK activation downstream of TCR. Because MALT1 has TRAF2 and -6 binding motifs, it was proposed that TRAF oligomerization by assembly of the CBM complex activates TRAF E3 activity and subsequent generation of Lys-63 polyubiquitin chains (Sun et al. 2004). It will be important to confirm these findings with TRAF2/6 double knockout mice. Given that TRAF proteins have been shown in other

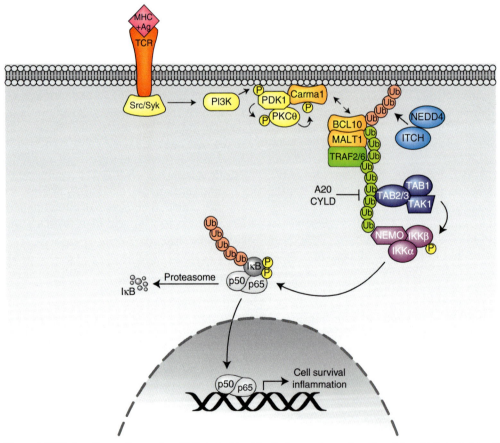

Figure 4. TCR signaling pathways. Lys-48-linked ubiquitin chains are shown in red, and Lys-63-linked ubiquitin chains are shown in green. See text for additional details.

pathways to recruit additional ligases that catalyze Lys-63-linked polyUb chains, such as cIAP1/2 and Pellinos, additional studies with TRAF2 and -6 mutants that are deficient in ligase recruitment motifs will also be required to determine the relative contribution of each ligase in TCR-induced NF-κB activation. Indeed, UBC13-deficient thymocytes are severely deficient in TAK1 activation, but IκB-α phosphorylation and degradation are only modestly affected (Yamamoto et al. 2006b). These findings suggest that other upstream ligases that collaborate with alternate E2s and that activate additional I-κK-activating kinases could be involved in TCR-mediated NF-κB activation.

TRAF6, MALT1, NEMO, and Bcl10 have all been proposed as targets for Lys-63 polyubiquitination, and primary ubiquitination sites have been mapped in most cases, revealing their importance in propagating NF-κB signaling (Sun et al. 2004; Zhou et al. 2004; Oeckinghaus et al. 2007; Wu and Ashwell 2008). Interestingly, Bcl10 is also destabilized following TCR activation. It has been shown that Bcl10 is ubiquitinated and targeted for lysosomal degradation by NEDD4 and Itch, ligases that are important for regulation of immune responses (Scharschmidt et al. 2004). Thus, Bcl10 is another example of a critical signaling component that is regulated by ubiquitin editing: Bcl10 is rapidly modified by Lys-63 polyubiquitination to facilitate TCR activation, and is subsequently tagged with degradative polyubiquitin modifications to down-regulate Bcl10 and thereby attenuate TCR signaling.

Recent reports have also implicated Bcl10, as well as A20, as substrates of the caspase-like activity of MALT1. In both cases, MALT1 proteolytic activity is not required for TCR-induced NF-κB activation, in agreement with previous reports showing that substitution of the putative MALT1 catalytic Cys residue only modestly reduces NF-κB activity (Uren et al. 2000) (Lucas et al. 2001). Rather, MALT1-mediated proteolysis seems to be important for fine-tuning TCR-induced NF-κB activation. Bcl10 cleavage promotes T-cell adhesion to fibronectin following TCR activation, which is important for T-cell stimulation, migration, and extravasation (Rebeaud et al. 2008). Unlike most cells in which A20 is induced by NF-κB as a negative-feedback mechanism to prevent protracted NF-κB signaling, A20 is constitutively expressed in lymphoid cells. Thus, MALT1-induced A20 cleavage is proposed to inactivate A20 and thereby release the constitutive brake on TCR-induced NF-κB signaling. These studies provide an additional mechanism for regulation of TCR-induced NF-κB signaling and motivate the search for additional caspase substrates. This is especially significant given the genetic (Hacker and Karin 2006) and biochemical (Sun et al. 2008) evidence that caspase-8 is critically required for NF-κB activation in response to TCR ligation, yet the mechanistic details of how caspase-8 achieves these effects remain unclear.

NF-κB Activation by Intracellular Stimuli: NOD2, RIG-I, and DNA Damage

Although the previous examples of NF-κB-activating signaling pathways are all initiated by extracellular ligands that activate transmembrane receptors, a number of stimuli may also initiate NF-κB signaling from within the cytosol or nucleus. These include microbe-derived PAMPs that bind to nucleotide-binding domain leucine-rich repeat (NLR) proteins or RIG-I-like receptors (RLR), as well as DNA damage (Fig. 5). Other noxious stimuli, such as oxidation and endoplasmic reticulum stress, may also activate NF-κB signaling. However, the signaling components are less well characterized, thus these pathways will not be discussed here and the reader is instead referred to several excellent reviews (Brzoska and Szumiel 2009; Zhang and Kaufman 2008).

The NLR proteins are defined by a tripartite domain organization that includes an amino-terminal protein/protein interaction domain (CARD, pyrin, or BIR), a central nucleotide-binding oligomerization domain (NOD) that promotes activation-induced oligomerization, and a carboxy-terminal leucine-rich repeat (LRR) domain that is important for detection of PAMPs. Here, we discuss NOD1 and -2, the most extensively studied members of the NLR

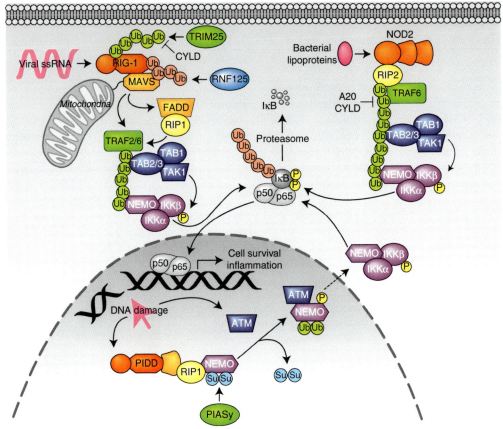

Figure 5. NF-κB activation by intracellular receptors. Lys-48-linked ubiquitin chains are shown in red, Lys-63-linked ubiquitin chains are shown in green, SUMOylation is indicated by blue circles, and monoubiquitination is indicated by green circles. See text for additional details.

family (Shaw et al. 2008). NOD1 is widely expressed, whereas NOD2 expression is limited to intestinal Paneth cells, dendritic cells, and monocytes and macrophages. Both are activated by distinct components of bacterial cell walls that induce homo-oligomerization. These activation-induced conformational changes promote the recruitment of RIP2, followed by RIP2 ubiquitination with Lys-63-linked chains. Several ligases have been proposed for RIP2. Some reports suggest that TRAF2 and -5 promote RIP2 polyubiquitination in the NOD1 signaling pathway, whereas TRAF6 may be important for polyubiquitination of targets downstream of NOD2 (Vallabhapurapu and Karin 2009). However, another group showed that, like RIP1 in the TNFR signaling pathway, cIAP1 and -2 promote RIP2 Lys-63

polyubiquitination in the NOD1/2 pathways. Interestingly, cIAP1/2 did not seem to be redundant as in the TNFR1 signaling pathway, because ablation of either cIAP1 or cIAP2 attenuated NF-κB signaling (Bertrand et al. 2009). Lys-63 polyubiquitinated RIP2 then recruits TAK1 via associated TAB proteins, leading to I-κK activation (Reardon and Mak 2009). A20 ablation results in exaggerated NOD2 signaling, suggesting that A20 also edits polyubiquitination of NOD2 signaling components. Indeed, RIP2 polyubiquitination is markedly enhanced in A20 null cells, and A20 deubiquitinates RIP2 in vitro. Notably, mutations in A20 and in NOD2 are associated with Crohn's disease (Hitotsumatsu et al. 2008). NEMO is also ubiquitinated with Lys-63-linked chains following NOD2 activation, thereby activating I-κK.

Interestingly, the NOD2 mutations that are found in Crohn's disease inhibit NEMO ubiquitination and NF-κB signaling, underscoring the importance of proper regulation of ubiquitination in NLR signaling pathways (Chen et al. 2009).

RLR family members include RIG-I, MDA5, and LGP2, all of which bind cytosolic viral RNA via helicase domains. Here, we focus on RIG-I as the prototypical RLR. Binding of viral RNA promotes association of the CARD motif of RIG-I with the CARD of MAVS, a mitochondrial-localized adaptor protein. The RIG-I/MAVS interaction is also facilitated by TRIM25, which polyubiquitinates RIG-I with Lys-63-linked chains. More specifically, a RIG-I Lys-172-Arg point mutant that cannot be ubiquitinated by TRIM25 neither interacts with MAVS nor activates NF-κB. MAVS also contains TRAF3 binding sites, which are important for IRF3 activation, and TRAF2 and -6 binding sites, which may promote IKK activation (Kawai and Akira 2008). FADD and RIP1 may also participate in RIG-I-mediated NF-κB activation downstream of MAVS (Hacker and Karin 2006). RIG-I activity is down-regulated by ubiquitin editing: CYLD depolymerizes Lys-63-linked polyubiquitin chains on RIG-I (Sun 2009) and the ligase RNF125 targets RIG-I and MAVS for proteasomal degradation (Chiu et al. 2009).

DNA damage is a potent activator of NF-κB activity, but many of the pathway components and mechanistic details are still being elucidated. Indeed, the "receptor" for DNA damage is unknown, although the leucine-rich repeats of PIDD are thought to play a role in detecting DNA damage. Modification of nuclear NEMO by PIASy with the ubiquitin-like protein SUMO also appears to be an early event in transducing DNA damage signals. NEMO SUMOylation also requires RIP and PIDD, which associate via their respective death domains, and complex with NEMO on DNA damage. SUMOylated NEMO accumulates in the nucleus and is subsequently phosphorylated by ATM, a protein kinase that is activated by DNA damage. NEMO is then monoubiquitinated by unknown enzymes on the same residues that were SUMOylated and is exported

with ATM to the cytosol, where it associates with I-κK. Both NEMO monoubiquitination and ATM are required for I-κK activation, but the mechanistic details are unknown (Janssens and Tschopp 2006; Brzoska and Szumiel 2009).

CONCLUSIONS AND FUTURE PERSPECTIVES

The study of NF-κB signaling has opened up a Pandora's box, highlighting the interplay between phosphorylation and the various forms of ubiquitination that work in a concerted manner to both activate and extinguish a signaling cascade of central importance to the livelihood of metazoan animals. As we learn from the cornucopia of information that characterization of this pathway has yielded and will continue to yield, it will be important to bear in mind a number of questions. To begin, are mechanisms like ubiquitin editing more generally applicable to other signaling pathways? With the advent of ubiquitin chain-specific antibodies and improved mass spectroscopic techniques, this pressing question should be answerable. Secondly, what determines whether ubiquitin ligases such as cIAP 1/2 synthesize Lys-48- or Lys-63-linked chains? We suspect that linkage specificity is determined by the nature of the E2 enzyme, but how is docking of various E2s to the same ligase regulated? Thirdly, what are the respective roles of the RING domains in TRAFs and cIAP1/2, and why are these ligases often recruited simultaneously to proximal signaling complexes? Are the TRAFs bona fide ubiquitin ligases and if so, do they exclusively synthesize Lys-63 polyubiquitin chains? Can TRAFs polymerize Lys-63-linked chains in conjunction with E2 enzymes other than UBC13? Is it possible that the purpose of simultaneously recruiting multiple ubiquitin ligases in proximal signaling complexes is that each type of ligase predominantly polymerizes one type of ubiquitin chain? Addressing these questions by knocking down components in transformed cells or by overexpression studies may be misleading because of ineffective knockdown and the propensity of the TRAFs to promiscuously oligomerize on overexpression. Definitive answers to these important questions in a physiological context will likely require the generation of RING-mutant knock-in mice and careful analysis of any compromise in ubiquitin chain generation and in NF-κB signaling. Finally, what regulates the negative regulators like A20 and CYLD? Both DUBs may be modulated by phosphorylation (Reyes-Turcu et al. 2009; Sun 2009), but is this their only mode of regulation? Whatever the answers, the field promises to teach us much about how NF-κB, and cellular signaling in general, is regulated. Most importantly, much of what we learn is likely to have therapeutic benefits.

ACKNOWLEDGMENTS

The authors would like to thank Eugene Varfolomeev, Domagoj Vucic, and Nobuhiko Kayagaki for critical reading of the manuscript, and Allison Bruce for graphics assistance. Our apologies to our colleagues whose important contributions are not cited due to space constraints.

REFERENCES

Barton GM, Medzhitov R. 2004. Toll signaling: RIPping off the TNF pathway. *Nat Immunol* **5:** 472–474.

Bertrand MJ, Doiron K, Labbe K, Korneluk RG, Barker PA, Saleh M. 2009. Cellular inhibitors of apoptosis cIAP1 and cIAP2 are required for innate immunity signaling by the pattern recognition receptors NOD1 and NOD2. *Immunity* **30:** 789–801.

Bertrand MJ, Milutinovic S, Dickson KM, Ho WC, Boudreault A, Durkin J, Gillard JW, Jaquith JB, Morris SJ, Barker PA. 2008. cIAP1 and cIAP2 facilitate cancer cell survival by functioning as E3 ligases that promote RIP1 ubiquitination. *Mol Cell* **30:** 689–700.

Bhoj VG, Chen ZJ. 2009. Ubiquitylation in innate and adaptive immunity. *Nature* **458:** 430–437.

Boone DL, Turer EE, Lee EG, Ahmad RC, Wheeler MT, Tsui C, Hurley P, Chien M, Chai S, Hitotsumatsu O, et al. 2004. The ubiquitin-modifying enzyme A20 is required for termination of Toll-like receptor responses. *Nat Immunol* **5:** 1052–1060.

Brzoska K, Szumiel I. 2009. Signalling loops and linear pathways: NF-κB activation in response to genotoxic stress. *Mutagenesis* **24:** 1–8.

Carswell EA, Old LJ, Kassel RL, Green S, Fiore N, Williamson B. 1975. An endotoxin-induced serum factor that causes necrosis of tumors. *Proc Natl Acad Sci* **72:** 3666–3670.

Chen ZJ, Sun LJ. 2009. Nonproteolytic functions of ubiquitin in cell signaling. *Mol Cell* **33:** 275–286.

Chen ZJ, Parent L, Maniatis T. 1996. Site-specific phosphorylation of IκBα by a novel ubiquitination-dependent protein kinase activity. *Cell* **84:** 853–862.

Chen G, Shaw MH, Kim YG, Nunez G. 2009. NOD-like receptors: Role in innate immunity and inflammatory disease. *Ann Rev Pathol* **4:** 365–398.

Chiu YH, Zhao M, Chen ZJ. 2009. Ubiquitin in NF-κB signaling. *Chem Rev* **109:** 1549–1560.

Compagno M, Lim WK, Grunn A, Nandula SV, Brahmachary M, Shen Q, Bertoni F, Ponzoni M, Scandurra M, Califano A, et al. 2009. Mutations of multiple genes cause deregulation of NF-κB in diffuse large B-cell lymphoma. *Nature* **459:** 717–721.

Conze DB, Wu CJ, Thomas JA, Landstrom A, Ashwell JD. 2008. Lys63-linked polyubiquitination of IRAK-1 is required for interleukin-1 receptor- and toll-like receptor-mediated NF-κB activation. *Mol Cell Biol* **28:** 3538–3547.

Deng L, Wang C, Spencer E, Yang L, Braun A, You J, Slaughter C, Pickart C, Chen ZJ. 2000. Activation of the IκB kinase complex by TRAF6 requires a dimeric ubiquitin-conjugating enzyme complex and a unique polyubiquitin chain. *Cell* **103:** 351–361.

Ea CK, Deng L, Xia ZP, Pineda G, Chen ZJ. 2006. Activation of IKK by TNFα requires site-specific ubiquitination of RIP1 and polyubiquitin binding by NEMO. *Mol Cell* **22:** 245–257.

Enesa K, Zakkar M, Chaudhury H, Luong LA, Rawlinson L, Mason JC, Haskard DO, Dean JL, Evans PC. 2008. NF-κB suppression by the deubiquitinating enzyme cezanne: A novel negative feedback loop in pro-inflammatory signaling. *J Biol Chem* **283:** 7036–7045.

Fukushima T, Matsuzawa S, Kress CL, Bruey JM, Krajewska M, Lefebvre S, Zapata JM, Ronai Z, Reed JC. 2007. Ubiquitin-conjugating enzyme Ubc13 is a critical component of TNF receptor-associated factor (TRAF)-mediated inflammatory responses. *Proc Natl Acad Sci* **104:** 6371–6376.

Hacker H, Karin M. 2006. Regulation and function of IKK and IKK-related kinases. *Sci STKE* **2006:** re13.

He JQ, Saha SK, Kang JR, Zarnegar B, Cheng G. 2007. Specificity of TRAF3 in its negative regulation of the noncanonical NF-κB pathway. *J Biol Chem* **282:** 3688–3694.

Heyninck K, Beyaert R. 2005. A20 inhibits NF-κB activation by dual ubiquitin-editing functions. *Trends Biochem Sci* **30:** 1–4.

Hitotsumatsu O, Ahmad RC, Tavares R, Wang M, Philpott D, Turer EE, Lee BL, Shiffin N, Advincula R, Malynn BA, et al. 2008. The ubiquitin-editing enzyme A20 restricts nucleotide-binding oligomerization domain containing 2-triggered signals. *Immunity* **28:** 381–390.

Hoffmann A, Baltimore D. 2006. Circuitry of nuclear factor κB signaling. *Immunol Rev* **210:** 171–186.

Ikeda F, Dikic I. 2008. Atypical ubiquitin chains: New molecular signals. 'Protein Modifications: Beyond the Usual Suspects' review series. *EMBO reports* **9:** 536–542.

Ivins FJ, Montgomery MG, Smith SJ, Morris-Davies AC, Taylor IA, Rittinger K. 2009. NEMO oligomerisation and its ubiquitin-binding properties. *Biochem J* **421**: 243–251.

Janssens S, Tschopp J. 2006. Signals from within: The DNA-damage-induced NF-κB response. *Cell Death Differentiation* **13**: 773–784.

Kanayama A, Seth RB, Sun L, Ea CK, Hong M, Shaito A, Chiu YH, Deng L, Chen ZJ. 2004. TAB2 and TAB3 activate the NF-κB pathway through binding to polyubiquitin chains. *Mol Cell* **15**: 535–548.

Kato M, Sanada M, Kato I, Sato Y, Takita J, Takeuchi K, Niwa A, Chen Y, Nakazaki K, Nomoto J, Asakura Y, et al. 2009. Frequent inactivation of A20 in B-cell lymphomas. *Nature* **459**: 712–716.

Kawai T, Akira S. 2008. Toll-like receptor and RIG-I-like receptor signaling. *Ann New York Acad Sci* **1143**: 1–20.

Kelliher MA, Grimm S, Ishida Y, Kuo F, Stanger BZ, Leder P. 1998. The death domain kinase RIP mediates the TNF-induced NF-κB signal. *Immunity* **8**: 297–303.

Kobayashi N, Kadono Y, Naito A, Matsumoto K, Yamamoto T, Tanaka S, et al. 2001. Segregation of TRAF6-mediated signaling pathways clarifies its role in osteoclastogenesis. *EMBO* **20**: 1271–1280.

Kovalenko A, Wallach D. 2006. If the prophet does not come to the mountain: Dynamics of signaling complexes in NF-κB activation. *Mol Cell* **22**: 433–436.

Lamothe B, Besse A, Campos AD, Webster WK, Wu H, Darnay BG. 2007. Site-specific Lys-63-linked tumor necrosis factor receptor-associated factor 6 auto-ubiquitination is a critical determinant of I κ B kinase activation. *J Biol Chem* **282**: 4102–4112.

Lee EG, Boone DL, Chai S, Libby SL, Chien M, Lodolce JP, Ma A. 2000. Failure to regulate TNF-induced NF-κB and cell death responses in A20-deficient mice. *Science* **289**: 2350–2354.

Lee TH, Shank J, Cusson N, Kelliher MA. 2004. The kinase activity of Rip1 is not required for tumor necrosis factor-α-induced IκB kinase or p38 MAP kinase activation or for the ubiquitination of Rip1 by Traf2. *J Biol Chem* **279**: 33185–33191.

Lo YC, Lin SC, Rospigliosi CC, Conze DB, Wu CJ, Ashwell JD, Eliezer D, Wu H. 2009. Structural basis for recognition of diubiquitins by NEMO. *Mol Cell* **33**: 602–615.

Lucas PC, Yonezumi M, Inohara N, McAllister-Lucas LM, Abazeed ME, Chen FF, Yamaoka S, Seto M, Nunez G. 2001. Bcl10 and MALT1, independent targets of chromosomal translocation in malt lymphoma, cooperate in a novel NF-κB signaling pathway. *J Biol Chem* **276**: 19012–19019.

Mahoney DJ, Cheung HH, Mrad RL, Plenchette S, Simard C, Enwere E, Arora V, Mak TW, Lacasse EC, Waring J, Korneluk RG. 2008. Both cIAP1 and cIAP2 regulate TNFα-mediated NF-κB activation. *Proc Natl Acad Sci* **105**: 11778–11783.

Makris C, Godfrey VL, Krahn-Senftleben G, Takahashi T, Roberts JL, Schwarz T, Feng L, Johnson RS, Karin M. 2000. Female mice heterozygous for IKK γ/NEMO deficiencies develop a dermatopathy similar to the human X-linked disorder incontinentia pigmenti. *Mol Cell* **5**: 969–979.

Moynagh PN. 2009. The Pellino family: IRAK E3 ligases with emerging roles in innate immune signalling. *Trends Immunol* **30**: 33–42.

Musone SL, Taylor KE, Lu TT, Nititham J, Ferreira RC, Ortmann W, Shifrin N, Petri MA, Kamboh MI, Manzi S, et al. 2008. Multiple polymorphisms in the TNFAIP3 region are independently associated with systemic lupus erythematosus. *Nat Gen* **40**: 1062–1064.

Newton K, Matsumoto ML, Wertz IE, Kirkpatrick DS, Lill JR, Tan J, Dugger D, Gordon N, Sidhu SS, Fellouse FA, et al. 2008. Ubiquitin chain editing revealed by polyubiquitin linkage-specific antibodies. *Cell* **134**: 668–678.

Ni CY, Wu ZH, Florence WC, Parekh VV, Arrate MP, Pierce S, Schweitzer B, Van Kaer L, Joyce S, Miyamoto S, Ballard DW, Oltz EM. 2008. Cutting edge: K63-linked polyubiquitination of NEMO modulates TLR signaling and inflammation in vivo. *J Immunol* **180**: 7107–7111.

Novak U, Rinaldi A, Kwee I, Nandula SV, Rancoita PM, Compagno M, Cerri M, Rossi D, Murty VV, Zucca E, et al. 2009. The NF-κB negative regulator TNFAIP3 (A20) is inactivated by somatic mutations and genomic deletions in marginal zone lymphomas. *Blood* **113**: 4918–4921.

O'Neill LA, Bowie AG. 2007. The family of five: TIR-domain-containing adaptors in Toll-like receptor signalling. *Nat Rev Immunol* **7**: 353–364.

Oeckinghaus A, Wegener E, Welteke V, Ferch U, Arslan SC, Ruland J, Scheidereit C, Krappmann D. 2007. Malt1 ubiquitination triggers NF-κB signaling upon T-cell activation. *EMBO J* **26**: 4634–4645.

Ordureau A, Smith H, Windheim M, Peggie M, Carrick E, Morrice N, Cohen P. 2008. The IRAK-catalysed activation of the E3 ligase function of Pellino isoforms induces the Lys63-linked polyubiquitination of IRAK1. *Biochem J* **409**: 43–52.

Pickart CM, Fushman D. 2004. Polyubiquitin chains: Polymeric protein signals. *Curr Opin Chem Biol* **8**: 610–616.

Rahighi S, Ikeda F, Kawasaki M, Akutsu M, Suzuki N, Kato R, Kensche T, Uejima T, Bloor S, Komander D, et al. 2009. Specific recognition of linear ubiquitin chains by NEMO is important for NF-κB activation. *Cell* **136**: 1098–1109.

Reardon C, Mak TW. 2009. cIAP Proteins: Keystones in NOD Receptor Signal Transduction. *Immunity* **30**: 755–756.

Rebeaud F, Hailfinger S, Posevitz-Fejfar A, Tapernoux M, Moser R, Rueda D, Gaide O, Guzzardi M, Iancu EM, Rufer N, et al. 2008. The proteolytic activity of the paracaspase MALT1 is key in T cell activation. *Nat Immunol* **9**: 272–281.

Reyes-Turcu FE, Wilkinson KD. 2009. Polyubiquitin binding and disassembly by deubiquitinating enzymes. *Chem Rev* **109**: 1495–1508.

Reyes-Turcu FE, Ventii KH, Wilkinson KD. 2009. Regulation and cellular roles of ubiquitin-specific deubiquitinating enzymes. *Ann Rev Biochem* **78**: 363–397.

Rudolph D, Yeh WC, Wakeham A, Rudolph B, Nallainathan D, Potter J, Elia AJ, Mak TW. 2000. Severe liver degeneration and lack of NF-κB activation in NEMO/IKKγ-deficient mice. *Genes Develop* **14**: 854–862.

Scharschmidt E, Wegener E, Heissmeyer V, Rao A, Krappmann D. 2004. Degradation of Bcl10 induced by T-cell activation negatively regulates NF-κ B signaling. *Mol Cell Biol* **24:** 3860–3873.

Schauvliege R, Janssens S, Beyaert R. 2007. Pellino proteins: Novel players in TLR and IL-1R signalling. *J Cell Mol Med* **11:** 453–461.

Schmitz R, Hansmann ML, Bohle V, Martin-Subero JI, Hartmann S, Mechtersheimer G, Klapper W, Vater I, Giefing M, Gesk S, et al. 2009. TNFAIP3 (A20) is a tumor suppressor gene in Hodgkin lymphoma and primary mediastinal B cell lymphoma. *J Exp Med* **206:** 981–989.

Schwartz AL, Ciechanover A. 2009. Targeting proteins for destruction by the ubiquitin system: Implications for human pathobiology. *Ann Rev Pharmacol Toxicol* **49:** p73-96.

Shaw MH, Reimer T, Kim YG, Nunez G. 2008. NOD-like receptors (NLRs): Bona fide intracellular microbial sensors. *Current Opinion Immunol* **20:** 377–382.

Shembade N, Harhaj NS, Parvatiyar K, Copeland NG, Jenkins NA, Matesic LE, Harhaj EW. 2008. The E3 ligase Itch negatively regulates inflammatory signaling pathways by controlling the function of the ubiquitin-editing enzyme A20. *Nat Immunol* **9:** 254–262.

Shembade N, Parvatiyar K, Harhaj NS, Harhaj EW. 2009. The ubiquitin-editing enzyme A20 requires RNF11 to downregulate NF-κB signalling. *EMBO J* **28:** 513–522.

Shi CS, Kehrl JH. 2003. Tumor necrosis factor (TNF)-induced germinal center kinase-related (GCKR) and stress-activated protein kinase (SAPK) activation depends upon the E2/E3 complex Ubc13-Uev1A/TNF receptor-associated factor 2 (TRAF2). *J Biol Chem* **278:** 15429–15434.

Shu HB, Takeuchi M, Goeddel DV. 1996. The tumor necrosis factor receptor 2 signal transducers TRAF2 and c-IAP1 are components of the tumor necrosis factor receptor 1 signaling complex. *Proc Natl Acad Sci* **93:** 13973–13978.

Srinivasula SM, Ashwell JD. 2008. IAPs: What's in a name? *Mol Cell* **30:** 123–135.

Sun SC. 2009. CYLD: A tumor suppressor deubiquitinase regulating NF-κB activation and diverse biological processes. *Cell Death Differentiation* doi: 10.1038/cdd.2009.43.

Sun H, Gong S, Carmody RJ, Hilliard A, Li L, Sun J, Kong L, Xu L, Hilliard B, Hu S, et al. 2008. TIPE2, a negative regulator of innate and adaptive immunity that maintains immune homeostasis. *Cell* **133:** 415–426.

Sun L, Deng L, Ea CK, Xia ZP, Chen ZJ. 2004. The TRAF6 ubiquitin ligase and TAK1 kinase mediate IKK activation by BCL10 and MALT1 in T lymphocytes. *Mol Cell* **14:** 289–301.

Tada K, Okazaki T, Sakon S, Kobarai T, Kurosawa K, Yamaoka S, Hashimoto H, Mak TW, Yagita H, Okumura K, et al. 2001. Critical roles of TRAF2 and TRAF5 in tumor necrosis factor-induced NF-κ B activation and protection from cell death. *J Biol Chem* **276:** 36530–36534.

Thome M. 2008. Multifunctional roles for MALT1 in T-cell activation. *Nat Rev Immunol* **8:** 495–500.

Ting AT, Pimentel-Muinos FX, Seed B. 1996. RIP mediates tumor necrosis factor receptor 1 activation of NF-κB but not Fas/APO-1-initiated apoptosis. *EMBO J* **15:** 6189–6196.

Tokunaga F, Sakata S, Saeki Y, Satomi Y, Kirisako T, Kamei K, Nakagawa T, Kato M, Murata S, Yamaoka S, et al. 2009. Involvement of linear polyubiquitylation of NEMO in NF-κB activation. *Nat Cell Biol* **11:** 123–132.

Uren AG, O'Rourke K, Aravind LA, Pisabarro MT, Seshagiri S, Koonin EV, Dixit VM. 2000. Identification of paracaspases and metacaspases: Two ancient families of caspase-like proteins, one of which plays a key role in MALT lymphoma. *Mol Cell* **6:** 961–967.

Vallabhapurapu S, Karin M. 2009. Regulation and function of NF-κB transcription factors in the immune system. *Ann Rev Immunol* **27:** 693–733.

Vallabhapurapu S, Matsuzawa A, Zhang W, Tseng PH, Keats JJ, Wang H, Vignali DA, Bergsagel PL, Karin M. 2008. Nonredundant and complementary functions of TRAF2 and TRAF3 in a ubiquitination cascade that activates NIK-dependent alternative NF-κB signaling. *Nat immunol* **9:** 1364–1370.

Varfolomeev E, Vucic D. 2008. (Un)expected roles of c-IAPs in apoptotic and NFκB signaling pathways. *Cell Cycle* **7:** 1511–1521.

Varfolomeev E, Goncharov T, Fedorova AV, Dynek JN, Zobel K, Deshayes K, Fairbrother WJ, Vucic D. 2008. c-IAP1 and c-IAP2 are critical mediators of tumor necrosis factor α (TNFα)-induced NF-κB activation. *J Biol Chem* **283:** 24295–24299.

Verstrepen L, Bekaert T, Chau TL, Tavernier J, Chariot A, Beyaert R. 2008. TLR-4, IL-1R and TNF-R signaling to NF-κB: Variations on a common theme. *Cell Mol Life Sci* **65:** 2964–2978.

Wallach D, Kovalenko A. 2008. Self-termination of the terminator. *Nat Immunol* **9:** 1325–1327.

Walsh MC, Kim GK, Maurizio PL, Molnar EE, Choi Y. 2008. TRAF6 autoubiquitination-independent activation of the NFκB and MAPK pathways in response to IL-1 and RANKL. *PloS one* **3:** e4064.

Wertz IE, O'Rourke KM, Zhou H, Eby M, Aravind L, Seshagiri S, Wu P, Wiesmann C, Baker R, Boone DL, et al. 2004. De-ubiquitination and ubiquitin ligase domains of A20 downregulate NF-κB signalling. *Nature* **430:** 694–699.

Wu CJ, Ashwell JD. 2008. NEMO recognition of ubiquitinated Bcl10 is required for T cell receptor-mediated NF-κB activation. *Proc Natl Acad Sci* **105:** 3023–3028.

Wu H, Tschopp J, Lin SC. 2007. Smac mimetics and TNFα: A dangerous liaison? *Cell* **131:** 655–658.

Wu CJ, Conze DB, Li T, Srinivasula SM, Ashwell JD. 2006. Sensing of Lys 63-linked polyubiquitination by NEMO is a key event in NF-κB activation [corrected]. *Nat Cell Biol* **8:** 398–406.

Yamamoto M, Okamoto T, Takeda K, Sato S, Sanjo H, Uematsu S, Saitoh T, Yamamoto N, Sakurai H, Ishii KJ, et al. 2006a. Key function for the Ubc13 E2 ubiquitin-conjugating enzyme in immune receptor signaling. *Nat Immunol* **7:** 962–970.

Yamamoto M, Sato S, Saitoh T, Sakurai H, Uematsu S, Kawai T, Ishii KJ, Takeuchi O, Akira S. 2006b. Cutting Edge: Pivotal function of Ubc13 in thymocyte TCR signaling. *J Immunol* **177:** 7520–7524.

Cite this article as *Cold Spring Harb Perspect Biol* 2010;2:a003350

Yamin TT, Miller DK. 1997. The interleukin-1 receptor-associated kinase is degraded by proteasomes following its phosphorylation. *J Biol Chem* **272:** 21540–21547.

Yeh WC, Shahinian A, Speiser D, Kraunus J, Billia F, Wakeham A, de la Pompa JL, Ferrick D, Hum B, Iscove N, et al. 1997. Early lethality, functional NF-κB activation, and increased sensitivity to TNF-induced cell death in TRAF2-deficient mice. *Immunity* **7:** 715–725.

Yin Q, Lin SC, Lamothe B, Lu M, Lo YC, Hura G, Zheng L, Rich RL, Campos AD, Myszka DG, et al. 2009. E2 interaction and dimerization in the crystal structure of TRAF6. *Nat Struct Mol Biol* **16:** 658–666.

Zarnegar BJ, Wang Y, Mahoney DJ, Dempsey PW, Cheung HH, He J, Shiba T, Yang X, Yeh WC, Mak TW, et al. 2008. Noncanonical NF-κB activation requires coordinated assembly of a regulatory complex of the adaptors cIAP1, cIAP2, TRAF2 and TRAF3 and the kinase NIK. *Nat Immunol* **9:** 1371–1378.

Zhang K, Kaufman RJ. 2008. From endoplasmic-reticulum stress to the inflammatory response. *Nature* **454:** 455–462.

Zhou H, Wertz I, O'Rourke K, Ultsch M, Seshagiri S, Eby M, Xiao W, Dixit VM. 2004. Bcl10 activates the NF-κB pathway through ubiquitination of NEMO. *Nature* **427:** 167–171.

Selectivity of the NF-κB Response

Ranjan Sen[1] and Stephen T. Smale[2]

[1]Laboratory of Cellular and Molecular Biology, National Institute on Aging, National Institutes of Health, Baltimore, Maryland 21225

[2]Department of Microbiology, Immunology, and Molecular Genetics, David Geffen School of Medicine at UCLA, Los Angeles, California 90095-1662

Correspondence: ranjan.sen@nih.gov, smale@mednet.ucla.edu

NF-κB is activated by many stimuli and NF-κB binding sites have been identified in a wide variety of genes. Yet, NF-κB-dependent gene expression must be stimulus- and cell-type-specific. In others words, the cellular response to different NF-κB activating stimuli, such as TNFα, IL-1, and LPS, must be different; and the response of different cell types, such as lymphocytes, fibroblasts, or epithelial cells, to the same NF-κB-inducing stimulus must also be different. Finally, kinetics of gene expression must be accounted for, so that all NF-κB-dependent genes are not activated simultaneously even if cell type and stimulus are constant. Here, we explore the mechanistic framework in which such regulatory aspects of NF-κB-dependent gene expression have been analyzed because they are likely to form the basis for physiological responses.

NF-κB was first reported in 1986 as a DNA-binding protein that recognizes, in a DNA sequence-specific manner, an important motif within the intronic enhancer of the immunoglobulin (Ig) κ light chain gene (Sen and Baltimore 1986). NF-κB DNA-binding activity was observed following stimulation of a pre-B cell line and represented the first example of inducible DNA binding as a primary response to cell stimulation. Subsequent studies revealed that NF-κB activity is induced in most cell types in response to a wide variety of stimuli, with major roles in cell activation, survival, and differentiation (Hayden and Ghosh 2004; Hoffmann et al. 2006; Ghosh and Hayden 2008; Vallabhapurapu and Karin 2009). In the classical model, NF-κB is found in the cytoplasm of unstimulated or resting cells in association with an inhibitory IκB protein. In response to cell stimulation, IκB is phosphorylated, ubiquitinated, and degraded, freeing NF-κB to translocate to the nucleus, bind its recognition sites in promoters and enhancers, and activate gene transcription.

Although this classical model of NF-κB has been well-documented and widely discussed, it has long been known that NF-κB function and regulation involves considerable complexity and diversity. This complexity and diversity, which are thought to be important for facilitating the differential and highly selective regulation of NF-κB target genes, are apparent at a number of levels, beginning with the existence of multiple NF-κB family members (Ghosh et al. 1998). Most vertebrates that have been studied contain five genes encoding the NF-κB family members RelA (p65), c-Rel, RelB, p50, and p52. These family members can bind DNA in a variety of heterodimeric species and all but RelB can bind as homodimers. The=existence of multiple family members and numerous dimeric species provides strategies for the selective regulation of target genes through differential expression of specific family members and dimeric species, differential DNA recognition, and differential transactivation mechanisms.

A second level of complexity arises from the existence of carboxy-terminal ankyrin repeat domains within the p105 and p100 precursors of p50 and p52, respectively. The ankyrin repeats are often removed immediately by proteolytic processing, but they are retained in a subset of stable dimers in the cytoplasm, providing an opportunity to regulate NF-κB activity via the regulation of proteolytic processing (Ghosh et al. 1998; Hoffmann et al. 2006; Ghosh and Hayden 2008; Vallabhapurapu and Karin 2009). The existence of multiple IκB proteins in vertebrates provides a third level of complexity. A primary function of classical IκBs is to sequester NF-κB dimers in the cytoplasm, but other IκBs appear to regulate NF-κB activity in the nucleus by functioning as transcriptional coregulators (Ghosh and Hayden 2008).

Although the existence of multiple NF-κB and IκB family members with variable properties provides intrinsic mechanisms for facilitating the selective regulation of NF-κB target genes, selectivity is further achieved through extrinsic mechanisms. In particular, variations in the kinetics of NF-κB activation can strongly influence target gene selection in response to a stimulus. In addition, the differential organization of chromatin at NF-κB target genes can regulate susceptibility to activation by NF-κB complexes (Natoli et al. 2006; Natoli 2009). The developmental history of a cell can further dictate which target genes are susceptible to activation, because of differences in chromatin structure established during development and differences

in the expression of other transcription factors and coregulatory proteins that influence the expression of NF-κB target genes. Together, these intrinsic and extrinsic mechanisms allow NF-κB to regulate distinct but often overlapping sets of genes in different cell types and in response to different stimuli.

In this article, we highlight progress toward understanding the mechanisms by which the selective regulation of NF-κB target genes can be achieved. Rather than attempting to summarize the enormous body of literature relevant to this topic, we instead focus on a few key topics and on specific challenges that must be overcome to fully understand the determinants of selectivity. A primary focus is on the mechanisms by which closely related NF-κB family members and differences in the duration of activation can contribute to selectivity.

TARGET GENE SELECTIVITY OF DISTINCT NF-κB DIMERS

One fundamental contributor to the selectivity of an NF-κB response is the regulation of distinct sets of target genes by different NF-κB heterodimeric and homodimeric species (Fig. 1). The variable phenotypes of mice lacking different NF-κB family members provides strong evidence that different dimers regulate distinct sets of genes (Gerondakis et al. 1999, 2006). Some of the phenotypic differences observed in mice lacking specific NF-κB family members are caused by differences in the expression patterns of the family members. For example, RelA appears to be ubiquitously expressed, whereas c-Rel is largely restricted to hematopoietic tissues (Ghosh et al. 1998; Gerondakis et al. 1999, 2006; Hoffmann et al. 2006). Although expression differences are important to consider, phenotypic differences are apparent in cell types that express all family members, such as hematopoietic cells (Gerondakis et al. 1999, 2006). In these cells, many NF-κB target genes appear to be expressed normally in the absence of individual family members, suggestive of redundancy or compensation, but some target genes show strong

Cite this article as *Cold Spring Harb Perspect Biol* 2009;2:a000257

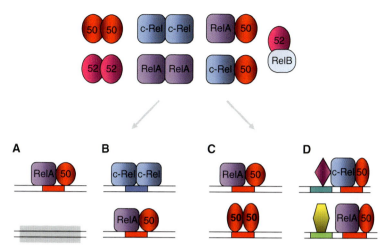

Figure 1. Mechanisms that contribute to the specificity of gene activation by NF-κB proteins. Homo- and heterodimerization between Rel-homology region (RHR) containing proteins generates a diverse range of κB sequence element binding proteins (*top*). Typically, several of these may be present in the cell nucleus after an activating stimulus, although the composition varies according to cell type, stimulus, and duration of signaling. The spectrum of genes activated is determined by several factors discussed in the text. (*A*) Tissue- or signal-specific marking may allow NF-κB binding to some genes (*top*) but not others (*bottom*). Gray box represents chromatin constraints that may preclude NF-κB binding. (*B*) The sequence of the κB element in promoters, represented by blue and red boxes, within DNA may bind specific RHR family proteins. (*C*) The same κB element (red box) may bind more than one RHR dimer under different circumstances, leading to different transcriptional outcomes. (*D*) The proximity of κB elements (red box) to other transcription factor binding sites (blue and green boxes) may differentially regulate gene expression by combinatorial mechanisms.

dependence on a particular family member for expression.

The differential regulation of specific target genes by closely related NF-κB family members has been studied in greatest depth with RelA and c-Rel. These two family members are of particular interest because, in addition to the amino-terminal Rel homology region (RHR), which defines the NF-κB/Rel family and consists of the DNA-binding domain and dimerization domain, RelA and c-Rel contain potent transactivation domains at their carboxyl terminus (Ghosh et al. 1998; Ghosh and Hayden 2008; Vallabhapurapu and Karin 2009). (RelB also contains a transactivation domain, but RelB-containing heterodimers appear to be activated primarily by the nonclassical pathway [Vallabhapurapu and Karin 2009].) The RHRs of RelA and c-Rel are more closely related than are the RHRs of any other pair of NF-κB family members. Furthermore,

structural studies have revealed that the residues that contact DNA are identical in these two family members (Chen and Ghosh 1999). Many NF-κB target genes in nonhematopoietic cell types are thought to require RelA for activation because c-Rel is absent. However, in hematopoietic cells that express RelA and c-Rel homodimers, as well as various heterodimeric species, both redundant and selective functions appear to exist.

Both c-Rel- and RelA-deficient mice show immune defects, demonstrating that the functions of the two proteins are not entirely redundant (Gerondakis et al. 1999, 2006). However, considerable redundancy appears to exist, as only a small number of NF-κB target genes have been identified that show strong dependence on either c-Rel or RelA. Microarray experiments performed in T cells revealed only a few genes that show strong dependence on c-Rel, although weak dependence was broadly observed (Bunting et al. 2007). In murine macrophages, *Il12b* and *Il23a* represent two of fewer than two dozen inducible genes identified in microarray experiments that show substantially reduced expression in the absence of c-Rel (Sanjabi et al. 2000, 2005; Carmody et al. 2007; K.J.W. and S.T.S., unpubl. data). RelA-dependent target genes have rarely been described in hematopoietic cells. However, one study revealed RelA-dependence in dendritic cells of several inflammatory cytokine genes, including *Il6* (Wang et al. 2007).

A major unanswered question is whether genes that require either RelA or c-Rel for expression are regulated by homodimers of these family members or heterodimers. For example, c-Rel-dependent genes may require either c-Rel:c-Rel homodimers or any of a variety of heterodimeric species, such as c-Rel:p50 or c-Rel:p52. Because the functions of p50 and p52 appear to be partially or largely redundant, it is difficult to rely on analyses of p50- or p52-deficient mice for determining whether heterodimeric species are critical for the activation of c-Rel-dependent genes. *Il12b*, which is c-Rel-dependent, appears to be expressed normally in $p50^{-/-}p52^{-/-}$ mice, consistent with the hypothesis that a c-Rel:c-Rel homodimer is responsible for selective regulation (Franzoso et al. 1997). However, because mechanisms can be envisioned that compensate for the simultaneous loss of these two family members, additional mechanistic insights are needed to determine whether c-Rel homodimers are indeed responsible for the selective regulation of *Il12b*. Unfortunately, chromatin immunoprecipitation (ChIP) experiments may not be informative, as both functional and nonfunctional dimers may associate with a given motif. For example, RelA, c-Rel, and p50 all appear to associate with the *Il12b* promoter, despite the c-Rel-dependence of transcription and the evidence that p50 and p52 may play no functional role (Sanjabi et al. 2005). These findings illustrate the challenge of understanding the selective regulation of NF-κB target genes by distinct dimeric species.

One strategy for distinguishing between the functions of homodimers and heterodimers may

be to analyze substitution mutations that selectively disrupt the formation of homodimers or heterodimers. Mutations in v-Rel have been described that prevent the formation of v-Rel:v-Rel homodimers but not v-Rel:p50 heterodimers (Liss and Bose 2002). Analogous mutations in c-Rel or RelA may prove to be highly beneficial for distinguishing the targets of heterodimers from those of homodimers.

MECHANISMS OF TARGET GENE SELECTIVITY BY DIFFERENT NF-κB FAMILY MEMBERS

Although several examples of target genes that require specific NF-κB family members have been reported, surprisingly little is known about the range of mechanisms by which selective regulation is achieved. As mentioned previously, the RHRs of RelA and c-Rel are closely related but their transactivation domains show little similarity. At first glance, this observation suggests that the transactivation domains are likely to be responsible for each protein's unique functions. However, the transactivation domains of both c-Rel and RelA are poorly conserved through vertebrate evolution, suggesting that, like transactivation domains in other transcription factors, their mechanisms of activation may not be highly specific. This lack of conservation is difficult to evaluate because structural features that are not reflected in sequence conservation may be conserved through evolution and may support highly specific protein–protein interactions. A number of post-translational modifications have been reported within the transactivation domains of both RelA and c-Rel, but the relevance of these modifications for the selective functions of the proteins have been examined in physiological assays in only a few instances.

One post-translational modification of RelA that has been examined extensively is the phosphorylation of serine 276 (S276) (Zhong et al. 1998, 2002; Dong et al. 2008). However, this residue is located within the RHR rather than the transactivation domain. Susceptibility to S276 phosphorylation is influenced by the

RelA transactivation domain. S276 phosphory-lation leads to the recruitment of the p300 coactivator, which is necessary for the activation of a subset of NF-κB target genes (Zhong et al. 1998, 2002). Interestingly, although the S276 residue is conserved in c-Rel, p300 overex-pression does not enhance the transactivation capacity of c-Rel, in contrast to the strong enhancement observed with RelA (Wang et al. 2007). Indeed, an analysis of RelA-dependent genes expressed in dendritic cells suggested that the RelA requirement was because of RelA's uni-que ability to associate with p300 (Wang et al. 2007). Thus, genes that require the function of the p300 coactivator for their expression may show a selective requirement for RelA-contain-ing dimers.

Although differential interactions with co-regulatory proteins may contribute to target gene selectivity of RelA and c-Rel, differences in binding properties may be of equal impor-tance. Although the RHRs of RelA and c-Rel are highly conserved, and although the residues that contact DNA are identical, critical differ-ences in binding properties appear to exist. Surprisingly, however, knowledge of these dif-ferences and of their functional relevance is limited. One major reason for our limited knowledge is that detailed studies of the range of potential recognition sites for each dimeric species have not been performed. A second reason is that the studies that have been completed suggest that, although recognition sequence differences can be identified, the differences are subtle and quantitative. A third and final reason for our limited knowledge is that it is very difficult to determine which of the many DNA motifs capable of binding each dimer with a range of affinities can support the regulation of endogenous target genes.

Remarkably, one of the only studies to char-acterize consensus DNA recognition motifs for NF-κB dimers was published almost two dec-ades ago (Kunsch et al. 1992). In this study, an in vitro binding site selection analysis was performed to identify preferred recognition sequences for p50:p50, RelA:RelA, and c-Rel:c-Rel dimers. Fewer than two dozen high-affinity recognition sequences were described for each

complex, revealing considerable similarity, but with clear differences. In particular, a few of the sequences that stably associated with c-Rel:c-Rel dimers were unable to bind to either RelA:RelA or p50:p50 dimers in electrophoretic mobility shift assays (EMSAs).

Although these results raised the possibility that c-Rel:c-Rel dimers may selectively bind some DNA motifs and selectively regulate target genes containing these motifs, these initial find-ings have not been significantly extended since their publication in 1992. In particular, com-parable studies of multiple heterodimeric spe-cies have not been performed and detailed studies examining the precise binding affinities of homodimers and heterodimers for a range of sequences (in side-by-side experiments) have not been performed. Knowledge of binding affini-ties would not by itself reveal the degree to which selective binding can contribute to the selective regulation of target genes by various NF-κB dimers, but this knowledge would rep-resent a critical step toward this goal. It is important to add that, although biochemical experiments can be used to identify intrinsic preferences, NF-κB, like most transcription factors, has the potential to bind DNA coopera-tively with other factors. Cooperative binding can lead to functional activity at low affinity sites, because the specificity and affinity of the NF-κB-DNA interaction can be enhanced by the binding partner.

As an alternative strategy for understanding the selective regulation of NF-κB target genes, one of our laboratories analyzed c-Rel/RelA chimeric proteins to identify the residues of c-Rel that are responsible for its unique ability to activate the endogenous *Il12b* gene in LPS-stimulated murine macrophages (Sanjabi et al. 2005). By using retroviral vectors to express a series of chimeric proteins in c-Rel$^{-/-}$ macro-phages, an 86-residue region of the c-Rel RHR was found to be sufficient for rescuing *Il12b* activation when inserted in place of the corre-sponding residues of RelA. DNA-binding analy-sis of these chimeric proteins revealed that this 86-residue region, which contains only 46 residues that differ between RelA and c-Rel, is responsible for the ability of c-Rel homodimers to bind with high affinity to a broader range of nonconsensus NF-κB recognition sequences than RelA homodimers, consistent with the earlier studies of Kunsch et al. (1992). Thus, these results provide strong evidence that differential DNA binding plays a major role in the selective activation of *Il12b* by c-Rel. Importantly, no other regions of c-Rel were important for *Il12b* activation in the chimeric protein analysis, suggesting that the unique DNA-binding properties of c-Rel may be solely responsible for the selective activation of at least one c-Rel target gene.

Another mechanism of selective regulation by different NF-κB family members emerged from the initial observation that NF-κB recognition sequences in specific promoters are strictly conserved through evolution, despite evidence that nucleotide substitutions could easily be tolerated while maintaining high-affinity binding (Leung et al. 2004). Furthermore, the NF-kB recognition sites were often found in pairs within promoters. Dependence versus independence on the p50 or p52 subunits of NF-κB, as shown by an analysis of $p50^{-/-}p52^{-/-}$ MEFs, revealed that subunit dependence was dictated by the specific recognition sequences and by cooperation between the pairs of sites. Most significantly, alterations in the highly conserved nucleotides within these DNA motifs altered subunit dependence, but apparently without altering binding affinity. Instead, the different recognition sequences and pairs of recognition sequences varied in their responsiveness to specific coactivator proteins. Although much remains to be learned about the mechanisms by which different NF-κB dimers, recognition sequences, and coactivator proteins can contribute to selectivity of the NF-κB response, this study illustrates the degree of complexity that is likely to be involved.

DURATION OF NF-κB ACTIVATION

The preceding sections focused on the mechanisms by which the differential expression patterns of NF-κB family members, as well as their variable binding and transactivation properties, may contribute to the selective regulation

of NF-κB target genes. Another fundamental regulatory feature of NF-κB activation via the canonical pathway is its transience. The duration of NF-κB activation may contribute to the selectivity of the transcriptional response by dictating which other factors can cooperate with NF-κB to activate gene transcription, by regulating the relative levels of different NF-κB proteins, and by changes in association with transcriptional coactivators and corepressors (Fig. 2). NF-κB duration may also regulate physiological responses by maintaining cell viability or cytokine and chemokine gene expression. Thus, the kinetics of NF-κB activation and inactivation is a central regulator of its function. Several pathways contribute to transient NF-κB activation, as discussed below.

IκB-dependent Export

One major pathway is via nuclear export of IκBα and IκBε-associated nuclear proteins. IκBα and IκBε are NF-kB target genes that are rapidly activated at the transcriptional level in response to NF-kB induction in all cell types tested. This leads to de novo IκBα and IκBε protein synthesis to replenish IκBs degraded in response to the NF-κB-inducing stimulus. Newly synthesized IκB proteins translocate to the nucleus, associate with nuclear NF-κB proteins, and export the transcription factors to the cytoplasm via nuclear export sequences present at their amino-termini (Tam et al. 2000; Huang and Miyamoto 2001; Johnson et al. 1999; Kearns et al. 2006). Early observations that IκBα can disrupt DNA-bound NF-κB complexes in vitro suggested that this mechanism may apply to promoter-bound NF-κB in the nucleus (Zabel and Baeuerle 1990). Although disruption of promoter complexes would be sufficient to stop NF-κB-dependent transcription, it is believed that nuclear export serves the essential function of recreating a pool of cytoplasmic NF-κB/IκB complexes to respond to ongoing or subsequent stimuli.

Three aspects of IκB-dependent transcription termination are noteworthy with respect to the selectivity issue. First, IκBα gene transcription precedes IκBε transcription (Kearns

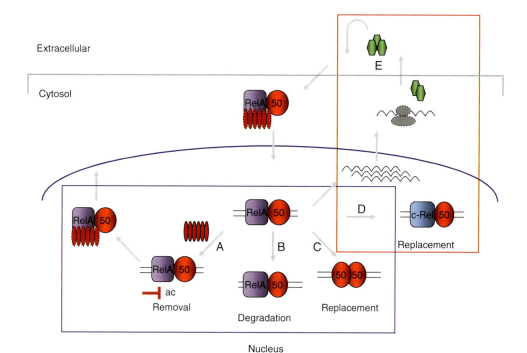

Figure 2. Mechanisms that regulate duration of NF-κB activation. NF-κB activation via the classical pathway is initiated by degradation of IκB protein (cytosol center), resulting in NF-κB (schematized by a p50/RelA heterodimer here, but could involve many of the homo-/heterodimeric pairs shown in Fig. 1) translocation to the nucleus and gene activation after DNA binding. Mechanisms summarized within the blue box reduce the duration of NF-κB-mediated gene expression. (*A*) Newly synthesized IκB proteins can remove DNA-bound NF-κB and export the complex out to the cytosol; (*B*) post-translational modifications of RelA, such as methylation and phosphorylation, can target it for proteasome-mediated degradation within the nucleus. The fate of the heterodimeric partner in the complex is not known; (*C*) transcription-activating heterodimers (such as p50/RelA or p50/c-Rel) can be replaced by p50 homodimers that have been implicated in repressing transcription (note however, that there is also evidence that p50 homodimers may activate transcription in association with nonclassical IκB-like proteins such as Bcl3 and IκBξ). Mechanisms summarized in the red box can extend the duration of NF-κB-mediated gene expression. (*D*) De novo c-Rel gene transcription and translation can lead to long-term induction of c-Rel-containing heterodimers, and (*E*) NF-κB target genes such as TNFα (green hexagon) activated during the initial stimulus may feed back in an autocrine fashion to maintain nuclear NF-κB.

et al. 2006); therefore, IκBα protein is resynthe-sized first and is likely to be the major player in NF-κB down-regulation. However, the kinetics of IκBα and IκBε synthesis may vary depending on the cell type and stimulus, and should there-fore be evaluated in specific circumstances. Second, IκBβ does not contain a nuclear ex-port sequence (NES) and has been proposed to be the only IκB that truly sequesters NF-κB in the cytosol (Tam and Sen 2001). However, newly synthesized IκBβ can translocate to the nucleus and associate with promoter-bound

NF-κB proteins without disrupting DNA-protein interactions (Suyang et al. 1996). This mechanism has been proposed to "protect" DNA-bound NF-κB from IκBα-mediated dis-ruption, thereby prolonging NF-κB-dependent gene expression. Third, acetylation of RelA by CBP/p300 complexes has been shown to pre-vent IκBα association, and thereby proposed to extend the duration of functional RelA complexes on some promoters (Chen et al. 2001; Chen et al. 2002; Kiernan et al. 2003). However, the proportion of RelA that gets acetylated is quite low; thus, the contribution of this pathway to the duration of NF-κB activation needs to be evaluated in specific situations. It is also likely that this mechanism applies to only a subset of promoters, such as those in which independently recruited p300/CBP is available to acetylate Rel proteins after cell stimulation. In our view, acetylation may provide protection to a subset of NF-κB-dependent promoters from rapid shut down by de novo synthesized IκBs.

Finally, NF-κB proteins themselves can contribute to their differential dwell time in the nucleus. RelA contains an NES in its carboxy-terminal domain, whereas c-Rel does not (Tam et al. 2001; Harhaj and Sun 1999). Therefore, different NF-κB homo- and heterodimers vary in their propensity to be nuclear. Specifically, RelA homodimers carry two NESs even in the absence of associated IκBα that can lead to their export from the nucleus. In contrast, c-Rel homodimers have no NESs and therefore require help to be exported from the nucleus. Although the RelA NES can mediate its export from the nucleus, IκBα and ε help reduce nuclear re-entry of RelA containing complexes by partially obscuring the nuclear localization signal. However, IκBα association does not completely prevent nuclear entry of NF-κB complexes; consequently, IκBα-associated complexes shuttle between the nucleus and the cytosol (Tam et al. 2000). The preponderance of IκBα-associated complexes in the cytosol in biochemical or immunohistochemical analyses therefore reflects a "snapshot" of a dynamic state in which the rate of export greatly exceeds the rate of import of these complexes. In this view, the NF-κB complexes vary in their nuclear propensity in approximate proportion to the number and strength of NLSs and NESs.

Nuclear Degradation of RelA

Transient NF-κB-dependent gene expression even in the absence of IκBα first suggested that other mechanisms may contribute to the duration of nuclear NF-κB (Saccani et al. 2004). Natoli and colleagues showed that a proportion of RelA in cells is degraded by the proteasome while bound to gene promoters. Since then, the combined effort of several laboratories has shown that ubiquitination is mediated by a SOCS1-containing E3 ligase complex in conjunction with COMMD1, a ubiquitously expressed negative regulator of NF-κB (Maine et al. 2007; Mao et al. 2009). Another E3 ligase, PDLIM2, has also been shown to ubiquitinate and target RelA to PML bodies where it is degraded by the proteasome (Tanaka et al. 2007). The relative contribution of each pathway for targeting RelA degradation may vary according to cell type and stimulus.

Recently, methylation of lysines 314 and 315 in RelA by the methyl transferase Set 7/9 has been implicated in marking promoter-bound RelA for degradation. Yang et al. (2009) found that knockdown of Set 7/9 in the U206 cell line resulted in higher levels of IL-6 and IL-8 gene expression in response to TNFα; in contrast, expression of a different NF-κB target gene, A20, was unaffected. Using chromatin immunoprecipitation, they located Set 7/9 at the IL-8 promoter in unactivated cells, but not at the A20 promoter regardless of the state of cell activation. This led to the hypothesis that prebound Set 7/9 at certain promoters is one mechanism by which promoters are marked for down-regulation by RelA degradation. At present, it is not clear why certain promoters are cleared in this manner, or whether this mechanism varies depending on the cell type or stimulus.

Degradation of nuclear RelA has also been proposed to be the basis for extended NF-κB-dependent gene expression in macrophages that lack IKK1 (Lawrence et al. 2005). In this case, Karin and colleagues showed that phosphorylation of nuclear RelA was reduced, as was proteasome-mediated nuclear degradation (Vallabhapurapu and Karin 2009). The generality of these observations to other cell types, or to other NF-κB-inducing stimuli in macrophages, remains to be determined. Taken together, these observations indicate that multiple mechanisms ensure that the duration of the NF-κB response is strictly regulated.

Secondary Stimulation Controls NF-κB Duration

Bacterial LPS is a well known inducer of canonical NF-κB via activation of Toll-like receptor 4 (TLR4). Typically, LPS induces long-lasting nuclear NF-κB compared with induction by TNFα or IL-1. The basis for this in mouse embryo fibroblasts (MEFs) appears to be that LPS treatment induces production of endogenous TNFα, which acts in an autocrine fashion to maintain long-term nuclear NF-κB activity. Werner et al. (2005) found that long-term IKK activation and nuclear NF-κB expression was significantly reduced in TNFα-deficient MEFs. Concomitantly, transcription of a subset of late-activated NF-κB-dependent genes, such as IL-12β and IL-1α and β, was also attenuated in these cells.

De Novo c-Rel Synthesis as a Means of Prolonging the NF-κB Response

Despite nuclear degradation of RelA after an activating stimulus, the level of RelA protein for the most part does not vary significantly in activated cells. Presumably, this is because the bulk of induced protein is not targeted for nuclear degradation. The small decrease in RelA is likely made up by constitutive protein synthesis from pre-existing RelA mRNA. Thus, post-translational mechanisms account for most of RelA dynamics in response to cellular activation. However, this is not the case for c-Rel. In T lymphocytes activated via the T-cell receptor (or its pharmacologic analog phorbol ester plus calcium ionophore), c-Rel gene transcription is induced, leading to de novo synthesis of c-Rel protein (Venkataraman et al. 1996). This newly synthesized c-Rel accumulates in the nucleus with delayed kinetics compared with post-translationally induced NF-κB via the canonical pathway. This is likely to be the basis for early observations in Jurkat T cells, that "late" NF-κB DNA binding activity consisted of c-Rel containing complexes. The immunosuppressive drugs cyclosporin A (CsA) and FK506 inhibit inducible c-Rel gene transcription, which has been proposed to be mediated by the transcription factor NFAT (Grumont

et al. 2004). Because this kind of c-Rel expression is controlled at the level of transcription, it does not occur with all forms of NF-κB-inducting stimuli. Notably, TNFα, IL-1, or LPS treatment, which do not induce long-term calcium mobilization, do not induce de novo c-Rel synthesis. Therefore, the nature of the signal determines when cells adopt this mode of "long-term" NF-κB (specifically, c-Rel) activation.

These observations have been extended to B lymphocytes activated via the B-cell antigen receptor (BCR). Surface immunoglobulin cross-linking induces two waves of nuclear NF-κB activity (B. Damdinsuren and R. Sen, unpubl. data). The first phase consists of both RelA and c-Rel and is mediated via the canonical post-translational pathway. This phase is transient, lasting between 4 and 6 hours, and coincides with reduced cellular IκB levels. It is followed by a slower second phase, composed mainly of c-Rel protein that is mediated by de novo c-Rel transcription and translation. As found in T cells, this long-term c-Rel induction is suppressed by CsA. Gene expression analyses in BCR-activated B lymphocytes indicate that phase I NF-κB induces auto-regulatory genes such as *NFKBIZ* and *NFKBIE*, and genes important for G1 progression such as *c-myc*. Phase II c-Rel is essential for cell survival and entry into S phase of the cell cycle. These observations show that early and late NF-κB induces distinct sets of genes and emphasize the importance of the duration of NF-κB activation as a key determinant of cellular responses.

PATTERNS OF NF-κB-DEPENDENT GENE EXPRESSION

In the preceding sections, we summarized mechanisms by which NF-κB target genes can be selectively regulated by different NF-κB family members and dimeric species, and by variations in the duration of classical NF-κB activation. Here, we briefly consider the impact of lineage-specific and stimulus-specific marking of target genes on the selectivity of an NF-κB response, and we further discuss the kinetics of NF-κB-dependent gene expression. A recurring theme is the central role played by the specific architecture of the control regions for each NF-κB target gene in defining the combination of inducible transcription factors required for gene activation, and in orchestrating changes in chromatin structure that modulate susceptibility to activation.

Lineage-specific Marking

The interleukin 2 (Il2) gene serves as a good example to explore the mechanisms that distinguish which NF-κB-dependent genes will be induced in specific cell types in response to NF-κB-inducing signals. Il2 is a T lymphocyte-specific gene that is activated in response to antigen receptor (CD3) plus costimulatory (CD28) signals. Its promoter has been extensively studied and contains functionally important binding sites for inducible transcription factors NF-κB, NF-AT and AP-1. The most important NF-κB binding site comprises the so-called CD28 response element (CD28RE) that is essential for high level gene induction (Himes et al. 1996; Shapiro et al. 1997; Rao et al. 2003). Two properties determine importance of the CD28RE. First, CD28 cross-linking, via its ability to activate the PI3 kinase pathway in T cells, leads to efficient IKK2 activation (Park et al. 2009). Consequently, IκB proteins are phosphorylated and degraded, releasing IκB-bound NF-κB protein to translocate to the nucleus. Second, the CD28RE preferentially binds c-Rel containing NF-κB proteins; therefore, CD28 costimulation is largely ineffective in c-Rel-deficient T cells. Because most of the cytoplasmic c-Rel in naïve CD4 T cells is bound to IκBβ (Banerjee et al. 2005), efficient CD28-dependent degradation of IκBβ is essential for c-Rel translocation to the nucleus to activate IL-2 gene transcription.

Although TCR signals are sufficient to induce NFAT and AP-1, the requirement for CD28 indicates that c-Rel is essential for promoter activation, particularly in naïve CD4+ T cells. However, these three transcription factors are also induced in BCR-activated B cells. Yet, Il2 gene transcription is T-cell specific. These observations emphasize that tissue-specific NF-κB target genes must be preselected during

differentiation to respond to NF-κB inducing stimuli. This preselection is likely to consist of the changes in chromatin structure during T-cell differentiation that confer susceptibility to activation on TCR engagement. These developmental changes in chromatin structure and the mechanisms by which they are regulated remain poorly understood for the Il2 gene and other NF-κB target genes. However, in simplistic terms, because different loci become susceptible to NF-κB activation in different cell lineages, the determinants of susceptibility must consist of specific DNA motifs within each locus that bind developmentally regulated transcription factors capable of catalyzing changes in chromatin structure and susceptibility to activation.

It is important to add that the combinatorial activation of the Il2 gene by multiple inducible transcription factors illustrates that stimuli that induce only a subset of these proteins will not be sufficient to activate the Il2 gene in T cells. For example, TNFα stimulation of T cells activates NF-κB, but not NFAT; consequently, TNFα does not activate Il2 gene transcription.

Stimulus-specific Marking

In addition to lineage-specific marking (such as Il2 in T vs. B cells), genes can also be marked for stimulus-specific gene expression. One of the best examples of this kind of marking comes from the work of Natoli and colleagues (Saccani et al. 2002). They showed that certain NF-κB-dependent promoters, such as Il6, Il12b, and MCP-1, were marked by p38-MAP kinase-dependent phosphorylation of histone H3 in human dendritic cells treated with LPS. Pharmacologic inhibition of p38 MAP kinase reduced H3S10 phosphorylation, and delayed transcription activation of these genes, despite normal NF-κB induction. Not all NF-κB-dependent inducible genes in these cells were marked in this way, suggesting multiple regulatory mechanisms in play. Importantly, TNFα treatment induced NF-κB but not p38 in these cells, resulting in reduced expression of p38-marked NF-κB-dependent genes without affecting other NF-κB-dependent inducible transcription. How p38-dependent H3S10 phosphorylation helps to selectively confer NF-κB responsiveness at certain promoters remains to be clarified; nor is it clear how certain genes come to be marked in this fashion.

Stimulus-specific activation can also be regulated by the remodeling of promoter-encompassing nucleosomes catalyzed by stimulus-specific transcription factors. Many rapidly induced NF-κB target genes can be activated without a requirement for nucleosome remodeling by ATP-dependent nucleosome remodeling complexes of the SWI/SNF family (Ramirez-Carrozzi et al. 2006, 2009). These genes usually contain CpG-island promoters, which may be directly responsible for remodeling-independent activation through their intrinsic assembly into unstable nucleosomes (Ramirez-Carrozzi et al. 2009). Importantly, remodeling-independent activation is generally associated with genes known to be activated by a broad and diverse range of stimuli. In contrast, remodeling-dependent activation is generally observed at genes that are activated more selectively. These genes contain non-CpG-island promoters that assemble into stable nucleosomes, leading to a requirement for nucleosome remodeling by SWI/SNF complexes. In response to LPS stimulation of macrophages via TLR4, for example, genes that depend on both NF-κB and IRF3 for activation almost always require nucleosome remodeling. IRF3 is essential for nucleosome remodeling, thereby restricting activation of these genes to stimuli capable of inducing IRF3 in addition to NF-κB.

In this stimulus specificity example, as in the lineage-specific regulation of *Il2* transcription, promoter architecture plays a central role in selective regulation. In this case, the presence or absence of a CpG-island largely dictates whether activation will be remodeling-independent or remodeling-dependent. Then, for remodeling-dependent genes, the presence of DNA recognition motifs for specific transcription factors capable of promoting the recruitment of nucleosome remodeling complexes dictates which NF-κB-inducing stimuli will be capable of inducing gene transcription.

Kinetics of NF-κB-dependent Gene Expression

Appropriate expression of NF-κB-dependent genes can be broadly categorized in terms of three parameters: (a) the rate of gene induction, (b) the duration of functional promoter-bound complexes, and (c) the rate of mRNA degradation. A rapidly induced gene will generate mRNA quickly but accumulation of mRNA will depend on its stability and the rate of continued RNA synthesis. Conversely, a gene may be induced slowly, but its expression level may be maintained by stability of the promoter complex and the mRNA that is generated. We shall briefly consider what is known about each of these parameters.

Rate of Gene Induction

Promoter architecture is the key determinant of the rate of transcription induction by induced NF-κB. For example, the promoter of the well-accepted NF-κB target gene IκBα contains 6 NF-κB binding sites located relatively close to the transcription start site (Rupec et al. 1999). To date, no other inducible factors have been implicated in IκBα transcription. Simplistically, we can therefore consider the IκBα promoter to be activated purely based on the presence or absence of NF-κB in the nucleus. This is consistent with the biology of IκBα, which is induced in response to all NF-κB-inducing stimuli, and in all cell types examined, to initiate feedback inhibition of NF-κB activity. In contrast, most other NF-κB-inducible genes are subject to more complex regulation, such as lineage- and/or stimulus-specificity discussed previously.

Therefore, most NF-κB-inducible promoters are more complex. This complexity may be reflected in the number of NF-κB binding sites present, their disposition relative to each other as well as the transcription initiation site, and most importantly in the requirement for other constitutive, or inducible, transcription factors to coordinate with inducible NF-κB to activate gene expression. These variables unfortunately have not been systematically investigated in enough genes to draw general conclusions. Nevertheless, it is easy to see how a promoter that requires a "late" inducible factor plus NF-κB could be activated later than a promoter that requires only NF-κB, or NF-κB plus a constitutively present transcription factor. Finally, some NF-κB-inducible genes may be regulated by distal NF-κB-dependent transcriptional enhancers, such as the immunoglobulin κ light chain gene.

Duration of Gene Transcription

The duration of gene transcription will also be determined by the duration of a functional promoter complex. As described previously, transient presence of NF-κB components in the nucleus is one way in which the complex can be disrupted. Mechanistically, this can occur by IκBα or IκBε proteins migrating into the nucleus and actively disrupting promoter complexes by removing DNA-bound NF-κB. Alternately, IκBα proteins may only remove predissociated NF-κB. It is very likely that the nature of the promoter complex determines the pathway for NF-κB eviction and thereby the effectiveness of transcription termination. For example, NF-κB interactions with its nearest neighbors, which will vary from promoter to promoter, may determine the stability of the promoter complex to IκB-mediated disruption. Thereby, newly synthesized IκBs will disrupt some promoters earlier than others. Another way to destabilize promoters could be based on the nuclear half-lives of other inducible transcription factors, or their post-translational modifications.

NF-κB-dependent transcription may also be terminated by replacing transcriptionally active NF-κB proteins by transcriptionally inactive ones. In particular, p50 homodimers have been proposed to repress transcription, because they lack transcription activation domains. Additionally, DNA-bound p50 homodimers have been shown to recruit histone deacetylases to promoters to epigenetically inactivate transcriptionally permissive chromatin state (Zhong et al. 2002; Williams et al. 2006). Because the NF-κB1 gene, which encodes the

p105 precursor to p50, is thought to be an NF-κB target gene, this route of gene suppression is yet another form of feedback inhibition mediated by induced NF-κB.

Post-transcriptional Regulation of NF-κB Target Genes

The variety of ways by which NF-κB is downregulated from the nucleus indicates that long-term NF-κB activation is detrimental. To effectively terminate NF-κB-dependent gene expression, however, post-transcriptional regulatory mechanisms are also in place. For example, if NF-κB target gene mRNA persists significantly after termination of transcription, the effect of NF-κB down-regulation will not be obvious. Similarly, NF-κB transcription factor down-regulation will have little impact if proteins expressed from NF-κB target genes remain functionally active in cells. One of the earliest characterized examples of the functional consequences of dysregulating NF-κB target genes at the post-transcriptional level is that of TNFα. The 3′ untranslated region of the TNFα mRNA contains AU-rich elements (AREs) that determine mRNA stability and translational control (Kontoyiannis et al. 1999; Hitti et al. 2006). Germline deletion of the ARE results in increased TNFα production and development of chronic inflammatory pathologies (Kontoyiannis et al. 1999).

Recently, Hao and Baltimore reported the first systematic analysis of the role of mRNA stability in regulating NF-κB-dependent gene expression (Hao and Baltimore 2009). They found that the expression patterns of TNFα-inducible genes could be grouped on the basis of mRNA stability. Specifically, mRNAs that were rapidly induced and rapidly down-regulated had short half-lives (t1/2), mRNAs that were rapidly induced and then maintained had intermediate t1/2, and those that were induced gradually over the time course of their studies had long t1/2. The t1/2s were determined by the 3′ UTRs of these genes, with shorter t1/2 correlating with higher numbers of AREs within the 3′ UTR. Interestingly, genes within each category were functionally related with regard to their roles during inflammation. Thus, post-transcriptional mechanisms constitute a critical component of NF-κB-dependent responses.

CONCLUDING REMARKS

NF-κB dimers can be induced in virtually all cell types by a remarkably broad range of stimuli, yet the set of target genes induced in a cell type by a given stimulus is highly variable. Much of this selectivity can be attributed to differences in the set of signaling pathways induced by each stimulus, which lead to the induction of multiple transcription factors that act in concert with NF-κB dimers through combinatorial mechanisms to activate specific target genes. Although considerable effort has been devoted to the elucidation of signaling pathways, and associated transcription factors, induced by NF-κB inducing stimuli, there exists a considerable gap in our mechanistic understanding of how these factors cooperate with NF-κB to coordinate cellular responses. Given the growing importance of NF-κB dysregulation in human disease, it is likely that such combinatorial specificities will provide additional targets for therapeutic intervention.

In this article, we attempted to minimize documenting the considerable advances that have been made in NF-κB-dependent transcriptional control. Instead, we highlight a subset of regulatory strategies that reveal important goals for the future. Amongst these are the elaboration of different functions of closely related NF-κB family members as homo- or heterodimeric species, and the mechanism by which NF-κB complexes work with other transcription factors to activate transcription. Similarly, although the duration of NF-κB response is now known to be tightly regulated and contributes to selective gene activation, much remains to be learned about the coordination between various levels of regulation that impact NF-κB-dependent transcription. Additionally, it is likely that NF-κB regulation itself, as well as its transcriptional cross talk with other factors, will vary considerably between tissue types. Further studies in this area will be particularly interesting. Finally, the control regions

associated with each NF-κB target gene must, by necessity, contain all of the information required to regulate cell type and stimulus-specific activation, yet we lack a complete understanding of how genes are selected to become NF-κB responsive. The availability of complete genome sequences and of an increasing repertoire of molecular and bioinformatic techniques for analyzing mechanisms of selective gene transcription will undoubtedly lead to rapid progress.

ACKNOWLEDGMENTS

Research in the NF-κB field performed in the authors' laboratories was supported by the Intramural Research Program of the National Institute of Aging (R.S.) and by NIH grant R01 AI073868 (S.T.S.).

REFERENCES

Banerjee D, Liou HC, Sen R. 2005. c-Rel-dependent priming of naïve T cells by inflammatory cytokines. *Immunity* **4:** 445–458.

Bunting K, Rao S, Hardy K, Woltring D, Denyer GS, Wang J, Gerondakis S, Shannon MF. 2007. Genome-wide analysis of gene expression in T cells to identify targets of the NF-κ B transcription factor c-Rel. *J Immunol* **178:** 7097–7109.

Carmody RJ, Ruan Q, Liou HC, Chen YH. 2007. Essential roles of c-Rel in TLR-induced IL-23 19 gene expression in dendritic cells. *J Immunol* **178:** 186–191.

Chen FE, Ghosh G. 1999. Regulation of DNA binding by Rel/NF-κB transcription factors: Structural views. *Oncogene* **18:** 6845–6852.

Chen LF, Mu Y, Greene WC. 2002. Acetylation of RelA at discrete sites regulates distinct nuclear functions of NF-κB. *EMBO J* **21:** 6539–6548.

Chen LF, Fischle W, Verdin E, Greene WC. 2001. Duration of nuclear NF-κB action regulated by reversible acetylation. *Science* **293:** 1653–1657.

Dong J, Jimi E, Zhong H, Hayden MS, Ghosh S. 2008. Repression of gene expression by unphosphorylated NF-κB 65 through epigenetic mechanisms. *Genes Dev* **22:** 1159–1173.

Franzoso G, Carlson L, Xing L, Poljak L, Shores EW, Brown KD, Leonardi A, Tran T, Boyce BF, Siebenlist U. 1997. Requirement for NF-κB in osteoclast and B-cell development. *Genes Dev* **11:** 3482–3496.

Gerondakis S, Grossmann M, Nakamura Y, Pohl T, Grumont R. 1999. Genetic approaches in mice to understand Rel/NF-κB and IκB function: Transgenics and knockouts. *Oncogene* **18:** 6888–6895.

Gerondakis S, Grumont R, Gugasyan R, Wong L, Isomura I, Ho W, Banerjee A. 2006. Unravelling the complexities of the NF-κB signalling pathway using mouse knockout and transgenic models. *Oncogene* **25:** 6781–6799.

Ghosh S, Hayden MS. 2008. New regulators of NF-κB in inflammation. *Nat Rev Immunol* **8:** 837–848.

Ghosh S, May MJ, Kopp EB. 1998. NF-κB and Rel proteins: Evolutionarily conserved mediators of immune responses. *Annu Rev Immunol* **16:** 225–260.

Grumont R, Lock P, Mollinari M, Shannon FM, Moore A, Gerondakis S. 2004. The mitogen-induced increase in T cell size involves PKC and NFAT activation of Rel/NF-κB-dependent c-mye expression. *Immunity* **21:** 19–30.

Hao S, Baltimore D. 2009. The stability of mRNA influences the temporal order of the induction of genes encoding inflammatory molecules. *Nature Immunology* **10:** 281–288.

Harhaj EW, Sun SC. 1999. Regulation of RelA subcellular localization by a putative nuclear export signal and 50. *Mol Cell Biol* **19:** 7088–7095.

Hayden MS, Ghosh S. 2004. Signaling to NF-κB. *Genes Dev* **18:** 2195–2224.

Himes SR, Coles LS, Reeves R, Shannon MF. 1996. High mobility group protein I(Y) is required for function and for c-Rel binding to CD28 response elements within the Gm-CSF and IL-2 promoters. *Immunity* **5:** 479–489.

Hitti E, Iakovieve T, Brook M, Deppenmeier S, Gruber AD, Radzioch D, Clark AR, Blackshear PJ, Kotlyarov A, Gaestel M. 2006. Mitogen-activated protein kinase-activated protein kinase 2 regulates tumor necrosis factor mRNA stability and translation mainly by altering tristetraprolin expression, stability, and binding to adenine/uridine-rich element. *Mol Cell Biol* **26:** 2399–2407.

Hoffmann A, Natoli G, Ghosh G. 2006. Transcriptional regulation via the NF-κB signaling module. *Oncogene* **25:** 6706–6716.

Huang TT, Miyamoto S. 2001. Postrepression activation of NF-κB requires the amino-terminal nuclear export signal specific to IκBα. *Mol Cell Biol* **21:** 4737–4747.

Johnson C, Van Antwerp D, Hope TJ. 1999. An N-terminal nuclear export signal is required for the nucleocytoplasmic shuttling of IκBα. *EMBO J* **18:** 6682–6693.

Kearns JD, Basak S, Werner SL, Huang CS, Hoffmann A. 2006. IκBε provides negative feedback to control NF-κB oscillations, signaling dynamics, and inflammatory gene expression. *J Biol Chem* **173:** 659–664.

Kiernan R, Bres V, Ng RW, Coudart MP, El Messaoudi S, Sardet C, Jin DY, Emiliani S, Benkirane M. 2003. Post-activation turn-off of NF-κ B-dependent transcription is regulated by acetylaltion of 65. *J Biol Chem* **278:** 2758–2766.

Kontoyiannis D, Pasparakis M, Pizarro TT, Cominelli F, Kollias G. 1999. Impaired on/off regulation of TNF biosynthesis in mice lacking TNF AU-rich elements: Implications for joint and gut-associated immunopathologies. *Immunity* **10:** 387–398.

Kunsch C, Ruben SM, Rosen CA. 1992. Selection of optimal κ B/Rel DNA-binding motifs: Interaction of both subunits of NF-κ B with DNA is required for transcriptional activation. *Mol Cell Biol* **12:** 4412–4421.

Lawrence T, Bebien M, Liu GY, Nizet V, Karin M. 2005. IKKα limits macrophage NF-κB activation and contributes to the resolution of inflammation. *Nature* **434:** 1138–1143.

Leung TH, Hoffmann A, Baltimore D. 2004. One nucleotide in a κB site can determine cofactor specificity for NF-κB dimers. *Cell* **20:** 453–464.

Liss AS, Bose HR Jr. 2002. Mutational analysis of the v-Rel dimerization interface reveals a critical role for v-Rel homodimers in transformation. *J Virol* **76:** 4928–4939.

Maine GN, Mao X, Komarck CM, Burstein E. 2007. COMMD1 promotes the ubiquitination of NF-κB subunits through a cullin-containing ubiquitin ligase. *EMBO J* **26:** 436–477.

Mao X, Gluck N, Li D, Maine GN, Li H, Zaidi IW, Repaka A, Mayo MW, Burstein E. 2009. GCN5 is a required cofactor for a ubiquitin ligase that targets NF-κB/RelA. *Genes Dev* **23:** 849–861.

Natoli G, Saccani S, Bosisio D, Marazzi I. 2005. Interactions of NF-κB with chromatin: The art of being at the right place at the right time. *Nat Immunol* **6:** 439–445.

Natoli G. 2009. Control of NF-κB-dependent transcriptional responses by chromatin organization. *Cold Spring Harb Perspect Biol* **1:** a000224.

Park SG, Schulze-Luehrman J, Hayden MS, Hashimoto N, Ogawa W, Kasuga M, Ghosh S. 2009. The kinase PDK1 integrates T cell antigen receptor and CD28 coreceptor signaling to induce NF-κB and activate T cells. *Nat Immunol* **10:** 158–166.

Ramirez-Carrozzi VR, Braas D, Bhatt DM, Cheng CM, Hong C, Doty KR, Black JC, Hoffmann A, Carey M, Smale ST. 2009. A unifying model for the selective regulation of inducible transcription by CpG islands and nucleosome remodeling. *Cell* **138:** 114–128.

Ramirez-Carrozzi VR, Nazarian AA, Li CC, Gore SL, Sridharan R, Imbalzano AN, Smale ST. 2006. Selective and antagonistic functions of SWI/SNF and Mi-2β nucleosome remodeling complexes during an inflammatory response. *Genes Dev* **20:** 282–296.

Rao S, Gerondakis S, Woltring D, Shannon MF. 2003. c-Rel is required for chromatin remodeling across the IL-2 gene promoter. *J Immunol* **170:** 3724–3731.

Rupec RA, Poujol D, Grosgeorge J, Carle GF, Livolsi A, Peyron JF, Schmid RM, Baeuerle PA, Messer G. 1999. Structural analysis, expression, and chromosomal localization of themouse ikba gene. *Immunogenetics* **49:** 395–403.

Saccani S, Pantano S, Natoli G. 2002. 38-dependent marking of inflammatory genes for increased NF-κB recruitment. *Nat Immunol* **3:** 69–75.

Saccani S, Marazzi I, Beg AA, Natoli G. 2004. Degradation of promoter-bound 65/RelA is essential for the prompt termination of the nuclear factor in κB response. *J Exp Med* **200:** 107–113.

Sanjabi S, Hoffmann A, Liou HC, Baltimore D, Smale ST. 2000. Selective requirement for c-Rel during IL-12 40 gene induction in macrophages. *Proc Natl Acad Sci USA* **97:** 12705–12710.

Sanjabi S, Williams KJ, Saccani S, Zhou L, Hoffmann A, Ghosh G, Gerondakis S, Natoli G, Smale ST. 2005. A c-Rel subdomain responsible for enhanced DNA-binding affinity and selective gene activation. *Genes Dev* **19:** 2138–2151.

Sen R, Baltimore D. 1986. Inducibility of κ immunoglobulin enhancer-binding protein Nf-κ B by a posttranslational mechanism. *Cell* **47:** 921–928.

Shapiro VS, Truitt KE, Imboden JB, Weiss A. 1997. CD28 mediates transcriptional upregulation of the interleukin-2 (IL-2) promoter through a composite element containing the CD28RE and NF-IL-2B AP-1 sites. *Mol Cell Biol* **17:** 4051–4058.

Suyang H, Phillips R, Douglas I, Ghosh S. 1996. Role of unphosphorylated newly synthesized I κ B β in persistent activation of NF-κB. *Mol Cell Biol* **16:** 5444–5449.

Tam WF, Sen R. 2001. IκB family members function by different mechanisms. *J Biol Chem* **276:** 7701–7704.

Tam WF, Wang W, Sen R. 2001. Cell-specific association and shuttling of IκBα provides a mechanism for nuclear NF-κB in B lymphocytes. *Mol Cell Biol* **21:** 4837–4846.

Tam WF, Lee LH, Davis L, Sen R. 2000. Cytoplasmic sequestration of rel proteins by IκBα requires CRM1-dependent nuclear export. *Mol Cell Biol* **20:** 2269–2284.

Tanaka T, Grusby MJ, Kaisho T. 2007. PDLIM2-mediated termination of transcription factor NF-κB activation by intranuclear sequestration and degradation of the 65 subunit. *Nat Immunol* **8:** 584–591.

Vallabhapurapu S, Karin M. 2009. Regulation and function of NF-κB transcription factors in the immune system. *Annu Rev Immunol* **27:** 693–733.

Venkataraman L, Wang W, Sen R. 1996. Differential regulation of c-Rel translocation in activated B and T cells. *J Immunol* **157:** 1149–1155.

Wang J, Wang X, Hussain S, Zheng Y, Sanjabi S, Ouaaz F, Beg AA. 2007. Distinct roles of different NF-κ B subunits in regulating inflammatory and T cell stimulatory gene expression in dendritic cells. *J Immunol* **178:** 6777–6788.

Werner SL, Barken D, Hoffmann A. 2005. Stimulus specificity of gene expression programs determined by temporal control of IKK activity. *Science* **309:** 1857–1861.

Williams SA, Chen LF, Kwon H, Ruiz-Jarabo CM, Verdin E, Greene WC. 2006. NF-κB 50 promotes HIV latency through HDAC recruitment and repression of transcriptional initiation. *EMBO J* **25:** 139–149.

Yang XD, Huang B, Li M, Lamb A, Kelleher NL, Chen LF. 2009. Negative regulation of NF-κ B action by Set9-mediated lysine methylation of the RelA subunit. *EMBO J* **28:** 1055–1066.

Zabel U, Baeuerle PA. 1990. Purified human I κ B can rapidly dissociate the complex of the NF-κ B transcription factor with its cognate DNA. *Cell* **61:** 255–265.

Zhong H, Voll RE, Ghosh S. 1998. Phosphorylation of NF-κ B 65 by PKA stimulates transcriptional activity by promoting a novel bivalent interaction with the coactivator CBP/p300. *Mol Cell* **1:** 661–671.

Zhong H, May MJ, Jimi E, Ghosh S. 2002. The phosphorylation status of nuclear NF-κ B determines its association with CBP/p300 or HDAC-1. *Mo. Cell* **9:** 625–636.

Control of NF-κB-dependent Transcriptional Responses by Chromatin Organization

Gioacchino Natoli

Department of Experimental Oncology, European Institute of Oncology (IEO), IFOM-IEO Campus, Via Adamello 16, 20139 Milan, Italy

Correspondence: gioacchino.natoli@ifom-ieo-campus.it

A large number of genes have been positively selected and recruited to participate in various phases of the inflammatory response triggered by microbial stimuli. Because of the complexity of the response, the many phases in which it is deployed, and the many "flavors" in which it appears (depending on quality and intensity of the stimulus as well as the target organ), very elaborated mechanisms evolved to ensure that the expression of the induced genes is carefully and precisely organized so that each gene is expressed in response to specific stimuli and with kinetics and intensities that suit the peculiar function of its product(s). Data accumulated in recent years have strengthened the concept that chromatin is an essential substrate at which multiple signals are integrated to promote a correctly choreographed expression of the genes involved in inflammatory transcriptional responses. Although the current level of understanding of these mechanisms is far from complete, some concepts and ideas have resisted experimental challenges and now represent accepted paradigms that are the subject of this article.

NF-κB AND THE COMPLEXITY OF INFLAMMATORY TRANSCRIPTIONAL RESPONSES

The ability to efficiently cope with microbial infections is an essential requisite of complex life on earth. Indeed, prototypes of the innate immune system and rudimentary inflammatory responses already appeared in rather primitive multicellular eukaryotes, and extremely more complex and evolved versions of them are now found throughout not only the animal but also the plant kingdom (Janeway and Medzhitov 2002).

The sources of complexity in inflammatory responses, and the ensuing need to efficiently cope with them, include the great diversity of microbial stimuli; the need to adjust the intensity of the response to that of its inducer; the need for innate immune cells to properly interact with and control the activation of the adaptive immune system; the presence of multiple organs with individual properties in terms of exposure to microbial sources, tissue organization, repair capacity, blood, and lymphatic supply; and finally the need to create an efficient interface between inflammatory responses and complex networks of metabolic pathways, so that even massive recruitment of energy sources to fight microbes will not jeopardize metabolic homeostasis. Moreover, several of the molecules and mechanisms invented during evolution to fight microbes were also recruited by nonmicrobial inflammatory responses, such as those triggered by tissue damage. Because of this complexity and of the many contexts in which an inflammatory response may unfold, the products of a huge number of genes have been selected during evolution to participate in one or more phases of inflammation. Although some genes, like the one encoding TNFα, are activated with rather monotonous patterns in almost every type of inflammatory response, several others are induced with complex kinetics and in a highly stimulus-selective and cell-type- or tissue-restricted fashion (Nau et al. 2002). In simple words, even if thousands of genes are regulated at the level of transcription to participate in inflammatory responses, kinetic, spatial, and quantitative parameters of activation are highly gene-specific, as they reflect the specific function of each gene product in the unfolding or control of the response.

The transcription factors of the NF-κB/Rel family directly control the induction of a very large fraction of the inflammatory transcriptome. Moreover, they contribute to other transcriptional programs, like those induced by antigen-receptor triggering in lymphocytes. The almost universal requirement for NF-κB/Rel in the induction of inflammatory genes poses a conundrum: How can we explain such a complexity in transcriptional patterns if only a few master controllers are involved? Data accumulated in recent years showed an essential role of chromatin organization in integrating information from multiple sources to control NF-κB recruitment to the underlying genomic sequences and generate complex and diverse NF-κB-regulated gene expression programs (Natoli et al. 2005).

TWO CLASSES OF NF-κB-REGULATED GENES DEFINED BY DIFFERENTIAL CHROMATIN CONFIGURATIONS AND INDUCTION KINETICS

Genes activated in response to inflammatory cues are a highly heterogeneous set whose products participate in several biological processes occurring sequentially or in parallel during inflammation (Nau et al. 2002), including the following: Chemoattractants (chemokines) promoting the recruitment of inflammatory cells to the peripheral sites of inflammation or recruiting cells of the adaptive immune system for antigen presentation; general inflammatory cytokines (like TNFα and interleukin 1β) that mainly contribute to response amplification and immune cell activation; antiapoptotic and antioxidant proteins that enable cells to survive both to the infection and to toxic products accumulating in the infiltrate as a consequence of cell activation and death (mainly of neutrophils); enzymes producing mediators that control vascular permeability and endothelial growth factors controlling inflammatory neoangiogenesis; proteins or enzymes involved in microbial recognition and killing; and cytokines instructing the differentiation of T lymphocytes (like IL12, which promotes Th1 differentiation of CD4+ T lymphocytes). Although all of these genes are required for different components of the response, they are not activated simultaneously. For instance, in LPS-stimulated human dendritic cells (DCs), expression of chemokines attracting T cells in the T-dependent areas of the lymph node (like ELC/CCL19) is postponed some hours as compared to the induction of a chemokine-like IL8/Cxcl8, which recruits neutrophils to peripheral inflammatory sites (Saccani et al. 2002). Therefore, DCs express IL8 in the periphery shortly after the encounter with the microbe and ELC/Ccl19 only after migration to the lymph node.

An initial clue on the mechanistic bases of kinetic complexity in NF-κB-dependent gene expression came from the analysis of NF-κB recruitment to target genes using chromatin immunoprecipitation (ChIP) (Saccani et al. 2001). Although entry of NF-κB in the nucleus after a stimulus like TNFα or LPS is massive, fast, and synchronous, kinetics of recruitment to target genes was found to be asynchronous and complex. Although some genes were bound by NF-κB without any detectable delay after its nuclear entry ("fast"

Cite this article as *Cold Spring Harb Perspect Biol* 2009;1:a000224

genes), others were bound (and activated) with much slower kinetics, in some cases even hours after NF-κB nuclear translocation ("slow" genes). Slow genes include primary genes activated with slow kinetics but not requiring new protein synthesis, and secondary genes (requiring new protein synthesis for activation) (Ramirez-Carrozzi et al. 2006). Delayed recruitment of NF-κB to slow target genes does not depend on an overall lower affinity of their promoters for NF-κB, because both fast and slow genes bear high-affinity DNA-binding sites (Saccani et al. 2001). Conversely, chromatin configuration at the two groups of genes is clearly different. Before stimulation, fast genes display a chromatin landscape typical of genes poised for immediate activation, including high levels of histone H3 and H4 acetylation, and trimethylation of lysine 4 of histone H3 (H3K4me3, a histone modification specifically enriched at active or poised transcription start sites, TSS) (Saccani

et al. 2001). Moreover, the promoters of fast NF-κB targets are constitutively accessible to nucleases, thus indicating an overall "open" and accessible organization (Fig. 1) (Ramirez-Carrozzi et al. 2006). Conversely, slow genes in unstimulated cells are associated with hypoacetylated histones and are negative for H3K4me3. In response to activation, both acetylation and H3K4 trimethylation progressively increase, with a kinetics that apparently precedes NF-κB recruitment (Fig. 1) (Saccani et al. 2001; Kayama et al. 2008). In any case, the final levels of acetylation and H3K4me3 usually do not reach those of fast genes. Importantly, changes in histone modifications are paralleled by an increase in accessibility to nucleases, thus suggesting a progressive decompaction of the locus. Decompaction seems to be an overall inefficient process, occurring in only a fraction of the cells, thus often leading to a mosaic expression of the corresponding gene (Weinmann et al. 2001). According to these data, the kinetic behavior

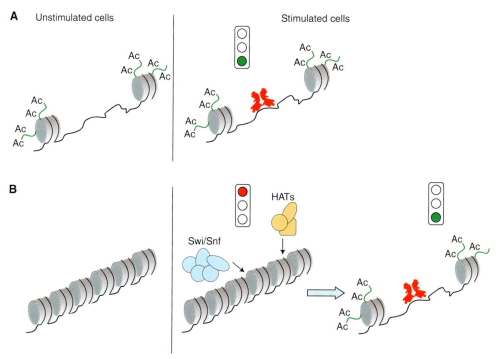

Figure 1. Differential chromatin states at NF-κB-dependent genes. In the *upper* part of the figure, a gene is shown with an open and accessible chromatin structure (*A*), whereas in the *bottom* part, a gene is depicted with a closed chromatin configuration requiring Swi/Snf-dependent chromatin remodeling to enhance accessibility to NF-κB and transcriptional activation (*B*). NF-κB is shown in red.

of NF-κB target genes reflects the chromatin configuration of the nucleosomes in the vicinity of/overlapping the NF-κB binding sites. On these premises, we have proposed a classification of NF-κB target genes based on chromatin configuration-dependent differential kinetics of NF-κB recruitment. According to this classification, we indicate the "fast" genes as those with *constitutive and immediate accessibility* (CIA) and the "slow" ones as those with *regulated and late accessibility* (RLA). The properties of these two groups of genes are summarized in Table 1 (some of them are further discussed later).

An essential and still open question is whether these two different chromatin configurations are written in the underlying DNA sequence or are epigenetically determined.

GENETIC DISSECTION OF CHROMATIN REMODELING EVENTS AT NF-κB-REGULATED GENES

The data described previously established a clear correlation between alternative chromatin configurations and differential kinetics of induction of NF-κB target genes. Subsequent data from Stephen Smale and colleagues provided formal evidence that chromatin remodeling is required for the activation of essentially all RLA genes (Ramirez-Carrozzi et al. 2006). "Chromatin remodeling" generically refers to all of the alterations of histone/DNA contacts performed by multimolecular ATP-dependent machines (Becker and Horz 2002). The Swi/

Snf chromatin remodeling complexes contain the essential ATPase subunits Brg1 and Brm, which are largely interchangeable: Simultaneous depletion of both ATPases by RNA interference abrogated the induction of all RLA genes, including both the late primary genes and the secondary genes (Ramirez-Carrozzi et al. 2006). Impaired induction was associated with lack of remodeling, as indicated by the inability of LPS stimulation to increase accessibility of chromatin at their promoters to restriction enzymes.

Interestingly, the basal and repressed state was actively enforced by an opposing chromatin remodeling complex, Mi2/Nurd. Cells depleted of the helicase component of Mi2/Nurd (Mi2β/Chd4) showed normal kinetics of gene activation but greatly enhanced levels of induction, possibly because of unopposed Swi/Snf-mediated remodeling. In the interpretation of these data, it should be considered that apparently remodeling is an incomplete event, often occurring in only a fraction of the stimulated cells: One possibility is that the relative levels or activity of Swi/Snf versus Mi2/Nurd in individual cells tilt the balance towards activation or repression and that in the absence of Mi2/Nurd, the actual fraction of cells expressing RLA genes is increased.

Additional genetic analyses of NF-κB-dependent gene induction in LPS-stimulated cells have uncovered further regulatory layers (Kayama et al. 2008). LPS signals through Toll-like receptor 4 (TLR4) via two main adapters, MyD88 and Trif (Kawai and Akira

Table 1. Properties of NF-κB regulated genes with different chromatin configurations

	Constitutively and immediately accessible genes	Genes with regulated and late accessibility
Timing of mRNA induction after stimulus	<30 min	>60 min (to several hours)
Kinetics of NF-κB recruitment	Immediate (minutes after nuclear entry)	Delayed (>60 min)
Requirement for chromatin remodeling	No	Yes
Requirement for new protein synthesis	No	Yes/No depending on the gene
Constitutive H3/H4 acetylation	Yes	No
Constitutive H3K4me3	Yes	No

2007). In LPS-stimulated Myd88$^{-/-}$ macrophages NF-κB activation is only marginally impaired, because of compensation by TRIF. These minimal alterations of NF-κB activation are surprisingly associated with a profound defect in the expression of most or all secondary genes (Kayama et al. 2008). Defective inducibility is associated with an absolute loss of nucleosome remodeling, absent increase in H3K4 trimethylation or histone acetylation, and finally impaired or absent recruitment of NF-κB. Therefore, NF-κB recruitment to secondary LPS-induced genes is directly controlled by remodeling events triggered by MyD88. Conversely, activation of all fast/CIA genes occurs normally in MyD88-deficient cells.

IkBζ, a distant paralog of IkBα encoded by the Nfkbiz gene, is a transcriptional coregulator whose expression is rapidly induced by NF-κB, thus behaving as an early primary response gene (Haruta et al. 2001).

In the absence of IkBζ, the process leading to the activation of some secondary genes (like IL12b and IL6) is aborted at a specific stage after chromatin remodeling but before recruitment of NF-κB and loading of the preinitiation complex (Yamamoto et al. 2004; Kayama et al. 2008). In IkBζ-deficient cells, accessibility of the DNA associated with promoter nucleosomes is normally increased after stimulation, but NF-κB and Pol II are not loaded. The increase in H3K4me3 that accompanies the induction of these genes is also completely blocked in the absence of IkBζ, consistent with the idea that H3K4 methyltransferases are recruited to TSSs by association with initiating or early elongating forms of RNA Pol II (Kayama et al. 2008). Therefore, chromatin remodeling and downstream events (including NF-κB recruitment and loading of the preinitiation complex) can be genetically dissected. The simplest and most intuitive model that can be proposed on the basis of the data described previously is that remodeling exposes NF-κB binding sites and makes them available for binding. However, the fact that remodeling is not followed by NF-κB recruitment in the absence of IkBζ suggests the existence of additional intermediate events, whose

molecular nature is unknown, occurring between remodeling and factor recruitment.

Overall, the following steps can be identified as part of the process leading to the activation of late primary and secondary inflammatory NF-κB-dependent genes. The first step is the induction or activation of factors that promote chromatin remodeling. The identity of such factors has remained elusive until now: A likely possibility is that they are transcription factors rapidly activated by stimulation and acting mainly or only to prepare the target DNA for NF-κB landing. Remodeling itself is operated mainly or exclusively by Swi/Snf complexes. However, although obviously required, remodeling is not sufficient to precipitate downstream events, as it can be dissociated from NF-κB recruitment and gene activation in cells lacking IkBζ. Remodeling is then followed by NF-κB recruitment, Pol II loading, and H3K4 trimethylation of a few nucleosomes surrounding the TSS. It should be considered that some transcription factors have a marked preference for H3K4me3-positive genomic regions (Guccione et al. 2006); therefore, it is possible that initial Pol II recruitment and H3K4 trimethylation trigger a feed-forward mechanism promoting further NF-κB recruitment and maximal gene induction. In any case, NF-κB recruitment to hundreds of inflammatory genes is the last step of a complex multistep program whose final aim is to relieve NF-κB binding sites from the negative control exerted by chromatin. Coordinated control of chromatin remodeling and NF-κB persistence in the nucleus in response to stimulation may also underlie selectivity in gene induction in response to alternative stimuli or to stimuli of different duration (Hoffmann et al. 2002). For instance, a short-lasting NF-κB activation is by itself incompatible with the induction of RLA genes, as NF-κB will be extruded from the nucleus before remodeling is completed and kB sites made accessible. Moreover, stimulus specificity in gene activation may arise as a consequence of the differential ability of alternative stimuli with similar NF-κB activation potential to trigger the signaling pathways leading to chromatin remodeling.

THE MOLECULAR BASES OF DIFFERENTIAL CHROMATIN ACCESSIBILITY TO NF-κB PROTEINS

The data discussed in the previous sections indicate that chromatin remodeling precedes, and is required for, NF-κB recruitment to a large fraction of NF-κB-dependent genes, which basically include all target genes except the early primary ones. However, although in vivo biochemical data demonstrate that chromatin accessibility to nucleases at regions surrounding NF-κB binding sites increases in response to stimulation and before NF-κB binding, they do not provide any clue about the nature of the chromatin configurations that are nonpermissive for NF-κB binding. In this section, I discuss how the fundamental unit of chromatin organization, the nucleosome, impacts on NF-κB interaction with cognate binding sites.

The general paradigm is that the packaging of DNA with core histones to generate nucleosomes, and the further folding of the nucleosomal chain into higher order fibers with various degrees of compaction, imposes a steric hindrance to transcription factor binding (Kornberg and Lorch 1999). However, transcription factors bind DNA using a large variety of structural motifs (Garvie and Wolberger 2001): Therefore, their ability to bind nucleosomal sites shows a continuous distribution, with some factors completely unable to contact a site contained in a nucleosome, and some others able to bind nucleosomal and naked sites with a comparable affinity. The TATA-box binding protein (TBP) lies at one end of this distribution, as it has no measurable affinity towards nucleosomal sites (Imbalzano et al. 1994). The heat shock factor (HSF1) can bind nucleosomal elements but the affinity is 100-fold lower than that for naked sites (Taylor et al. 1991). Finally, the glucocorticoid receptor is representative of those factors able to bind with similar affinity naked and nucleosomal DNA (Li and Wrange 1995).

Available structural data are substantially incompatible with the formation of a complex between any NF-κB dimer and high affinity binding sites incorporated in a nucleosome. The Rel homology domain (RHD) of different NF-κB dimers crystallized onto a κB site has a butterfly-like shape, with the wings connected to a cylindrical body of DNA (Chen et al. 1998; Huxford et al. 1999). The amino-terminal domains of the two monomers make base-specific contacts, whereas the carboxy-terminal domains interact with each other and generate the interface between the two subunits. Moreover, they both make nonspecific contacts with the sugar-phosphate backbone of the DNA. The NF-κB dimer does not completely encircle the DNA; however, the DNA-binding cleft delimited by the two amino-terminal domains is clearly too narrow to accommodate the surface of a nucleosome (Fig. 2) (Natoli et al. 2005). Therefore, the ensuing prediction is that nucleosomal sites cannot be bound by NF-κB and that the incorporation of a κB site in a nucleosome is completely sufficient to prevent binding. Surprisingly, experimental data contradict this prediction. Incorporation of κB

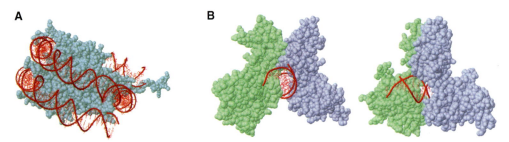

Figure 2. Structural views of the nucleosome and NF-κB. (*A*) Nucleosome structure showing the DNA surrounding the histone octamer. (*B*) The structure of an NF-κB dimer (p65/p50) crystallized onto an NF-κB binding site with front (*left*) and side (*right*) views. p50 is shown in green, p65 in gray, and DNA in red. (From Natoli et al. Nature Immunol. 6, 439, 2005; Nature Publishing Group.)

sites in a nucleosome causes no detectable reduction in the affinity for NF-κB, suggesting that nucleosomes by themselves are transparent to NF-κB (Angelov et al. 2004). Importantly, high affinity binding to nucleosomal sites is independent of the position of the NF-κB binding site relative to the dyad axis of the nucleosome (Angelov et al. 2004). This observation is very relevant in terms of understanding the mechanisms of NF-κB binding to nucleosomes. In fact, when a transcription factor binding site is placed close to the dyad axis of the nucleosome, inhibition of binding is stronger than when the same site is placed in a peripheral location. This difference is because of the spontaneous and transient (on a millisecond time scale) uncoiling of DNA from the periphery of the nucleosome, followed by its quick reassociation with histones (Widom 1998). These "breathing motions" of mono-nucleosomal DNA occur almost exclusively at the peripheral edge of the nucleosome and are strongly reduced or completely absent close to the dyad axis. Although a transient "site exposure" may occur at peripherally located sites and explain their accessibility even to transcription factors encircling DNA, this is not the case for internally located sites. Moreover, it is unclear if and how these breathing motions occur in vivo in the content of nucleosomal chains. In any case, it is safe to conclude that NF-κB binding to nucleosomal sites does not occur as a consequence of the spontaneous uncoiling of DNA from the surface of the nucleosome, as in this case a marked reduction or complete loss of affinity for internally located sites and a marginal or no reduction for peripheral sites would be observed.

Since in vitro mono-nucleosomal kB sites are as accessible to NF-κB proteins as the naked ones, the requirement for chromatin remodeling for NF-κB binding and activation of all RLA genes in vivo cannot be accounted for by the simple wrapping of kB sites around histone octamers. Two possibilities can be envisioned: First, although nucleosomes by themselves are transparent to NF-κB, compacted nucleosomal chains may create an efficient barrier to binding. In this case, remodeling would not result in the simple movement of one or a few nucleosomes or in the release of DNA from the nucleosomal surface, but it would consist in a more extensive reconfiguration of chromatin at the promoters and TSSs of RLA genes. The second possibility is that the nucleosomes controlling NF-κB recruitment have special properties that make them able to efficiently block the interaction of NF-κB with cognate sites. Nucleosomes in the vicinity of TSSs are extensively modified by a plethora of post-translational modifications (Kouzarides 2007). Some of them alter the intrinsic ability of nucleosomes to make inter-nucleosomal interactions and generate compact fibers. For instance, acetylation of the amino-terminal histone tails neutralizes their positive charge and reduces the attractive interactions with the negatively charged DNA (both on the same and on adjacent nucleosomes) (Kornberg and Lorch 1999). Severely hypoacetylated nucleosomes bearing tails folded back on the nucleosomal DNA may be efficient at occluding kB sites, and nucleosome remodeling may act to release DNA from the nucleosomal surface. An alternative possibility is that the incorporation of histone variants in place of canonical core histones may affect the accessibility of NF-κB binding sites. When incorporated in nucleosomes, the macroH2A variant (which contains an extended carboxy-terminal macro domain), can completely block access to centrally located but not peripheral kB sites in vitro (Angelov et al. 2003). This effect depends on the presence of the macro domain, which is also required for inhibition of chromatin remodeling and transcriptional activation by macroH2A. In a specific case, this macroH2A-dependent mechanism of kB site occlusion has been shown to occur in vivo (Agelopoulos and Thanos 2006). However, the genomic distribution of macroH2A-containing nucleosomes is still debated, and it is unknown whether macroH2A is commonly incorporated in nucleosomes in the vicinity of TSSs and used for dynamic regulatory purposes.

A piece of information that is still unavailable and is likely necessary to address several of the issues discussed previously, is a high-resolution map of nucleosome distribution (and modifications associated with individual nucleosomes) at the regulatory regions of NF-κB-dependent genes, including promoters, TSS, and distant enhancers. The current understanding of the rules dictating nucleosome positions is rather primitive and therefore algorithms with a strong predictive ability are not available yet (Segal et al. 2006). However, it is easy to predict that the exponential progress in this area, mainly fueled by high-throughput multiparallel next-generation sequencing technologies, will lead to the solution of this problem in a relatively short time, in spite of the complexity and high content in repeats of mammalian genomes. The main issue is to understand to what extent the underlying genomic sequence controls the position of individual nucleosomes and how well-positioned nucleosomes impact on the accessibility of DNA binding sites. Definitive maps at nucleotide resolution have just been reported in yeast (Mavrich et al. 2008a) and *Drosophila* (Mavrich et al. 2008b), and in both cases they are consistent with the idea that only nucleosomes in the vicinity of/surrounding TSSs are well-positioned, in a manner possibly dictated by the DNA sequence. Positioned nucleosomes at these locations then act as physical barriers imposing the position to adjacent nucleosomes (Mavrich et al. 2008a). Such barrier effect is progressively lost, moving away from the positioned nucleosomes. In the specific case of the transcriptional responses controlled by NF-κB, only anecdotic reports are available to date, and in some cases they are compatible with the idea that nucleosomes may physically occlude NF-κB binding sites. For instance, in macrophages, a positioned nucleosome encompassing the *cis*-regulatory elements required for the induction of Il12b gene transcription is quickly remodeled in response to activation (Weinmann et al. 1999). Although Il12b induction requires cRel, remodeling is cRel independent (Weinmann et al. 2001). Therefore, it is possible to envision a nucleosome-mediated sequential mechanism of cooperation between early acting transcription factors and NF-κB: Factors binding to repressive chromatin in the promoter (or at a distant regulatory site in an accessible conformation) drive chromatin decompaction and/or nucleosomal changes, eventually making the promoter accessible to NF-κB.

RESTRICTING THE SEARCH SPACE FOR NF-κB: ANOTHER ROLE FOR NUCLEOSOMAL INCORPORATION OF NF-κB SITES?

A major and long-standing issue in transcriptional regulation is to understand how transcription factors locate relevant binding sites in the genome (von Hippel and Berg 1989; Halford and Marko 2004). This issue is particularly relevant for inducible factors like NF-κB, which must be able to contact binding sites in a very short time after nuclear entry: Induction of some NF-κB target genes occurs almost instantly after NF-κB nuclear translocation, even if the genomic search space is immense. The kB site is conventionally represented as $G_{-5}G_{-4}G_{-3}R_{-2}N_{-1}N_0Y_{+1}Y_{+2}C_{+3}C_{+4}$ (R = purine, N = any nucleotide, Y = pyrimidine) or by position weight matrixes showing the probability that each position in the site is occupied by a given nucleotide. A definitive estimate of how frequently kB sites appear in mammalian genomes is unavailable and it is expected that a fraction of them will be species-specific. The 256 variants of the most restrictive kB site consensus sequence in which the first ($G_{-5}G_{-4}G_{-3}$) and the last ($C_{+3}C_{+4}$) nucleotides are fixed, are found at over 11,000 locations on the smallest human chromosome (chr22), accounting for about 1/67th of the human genome (Udalova et al. 2002). This leads to an estimate of a minimal number of human genomic kB sites in the order of magnitude of almost 10^6. The requirement for nucleosome remodeling machines for NF-κB binding and activation of some NF-κB target genes suggests, as described above, that incorporation in chromatin may efficiently prevent NF-κB binding. It can be inferred that a simple way to reduce the number of kB sites competing for available nuclear NF-κB molecules is to wrap a fraction of them (including nonfunctional sites and sites belonging to NF-κB target genes with restricted/tissue-specific expression) in nonaccessible chromatin contexts.

Whether this is really the case is not yet known. However, the advent of ChIP sequencing should allow this point to be directly addressed, and specifically, it should allow the fraction of identifiable genomic sites that are in fact contacted by NF-κB in a given cell type and/or experimental condition to be detected. Initial data obtained with high throughput approaches (Schreiber et al. 2006; Lim et al. 2007) have not specifically addressed this point and therefore no definitive conclusion can be safely drawn at this stage.

It should also be considered that the average residence time of NF-κB on high affinity binding sites in vivo is in the time scale of seconds (Bosisio et al. 2006): Therefore, binding to nonfunctional genomic sites will not lead to the sequestration of NF-κB molecules, and exploiting the insulating properties of chromatin to reduce the representation of binding-competent kB sites may in fact not be necessary.

FUTURE PERSPECTIVES

Our current understanding of the interplay between NF-κB, chromatin organization of the genomic information, and transcriptional output allows the definition of some features of the NF-κB-dependent transcriptional responses that are in part (or largely) chromatin-regulated:

- the kinetic complexity of induction of NF-κB target genes;

- the stimulus-specific induction of target genes;

- the cooperative effects (mediated by nucleosomes) between NF-κB and partner transcription factors.

The definitive understanding of the mechanistic bases of the interaction between chromatin and NF-κB will require:

- the generation of comprehensive nucleosome occupancy maps at NF-κB target genes and the definition of the underlying sequence determinants;

- the identification of the transcription factor synergizing with NF-κB at removing the chromatin-dependent hindrance to gene activation;

- the identification of the nuclear enzymatic activities that are required to overcome the block imposed by chromatin to the activation of NF-κB target genes. The possibility that some of these enzymes may represent drug targets is rather obvious and a scenario in which inhibition of these enzymes may lead to therapeutic down-regulation of selected NF-κB target genes is less futuristic than it may seem.

REFERENCES

Agelopoulos M, Thanos D. 2006. Epigenetic determination of a cell-specific gene expression program by ATF-2 and the histone variant macroH2A. *EMBO J* **25**: 4843–4853.

Angelov D, Lenouvel F, Hans F, Muller CW, Bouvet P, Bednar J, Moudrianakis EN, Cadet J, Dimitrov S. 2004. The histone octamer is invisible when NF-κB binds to the nucleosome. *J Biol Chem* **279**: 42374–42382.

Angelov D, Molla A, Perche PY, Hans F, Cote J, Khochbin S, Bouvet P, Dimitrov S. 2003. The histone variant macroH2A interferes with transcription factor binding and SWI/SNF nucleosome remodeling. *Mol Cell* **11**: 1033–1041.

Becker PB, Horz W. 2002. ATP-dependent nucleosome remodeling. *Annu Rev Biochem* **71**: 247–273.

Bosisio D, Marazzi I, Agresti A, Shimizu N, Bianchi ME, Natoli G. 2006. A hyper-dynamic equilibrium between promoter-bound and nucleoplasmic dimers controls NF-κB-dependent gene activity. *Embo J* **25**: 798–810.

Chen FE, Huang DB, Chen YQ, Ghosh G. 1998. Crystal structure of p50/p65 heterodimer of transcription factor NF-κB bound to DNA. *Nature* **391**: 410–413.

Garvie CW, Wolberger C. 2001. Recognition of specific DNA sequences. *Mol Cell* **8**: 937–946.

Guccione E, Martinato F, Finocchiaro G, Luzi L, Tizzoni L, Dall' Olio V, Zardo G, Nervi C, Bernard L, Amati B. 2006. Myc-binding-site recognition in the human genome is determined by chromatin context. *Nat Cell Biol* **8**: 764–770.

Halford SE, Marko JF. 2004. How do site-specific DNA-binding proteins find their targets? *Nucleic Acids Res* **32**: 3040–3052.

Haruta H, Kato A, Todokoro K. 2001. Isolation of a novel interleukin-1-inducible nuclear protein bearing ankyrin-repeat motifs. *J Biol Chem* **276**: 12485–12488.

Hoffmann A, Levchenko A, Scott ML, Baltimore D. 2002. The IκB-NF-κB signaling module: Temporal control and selective gene activation. *Science* **298:** 1241–1245.

Huxford T, Malek S, Ghosh G. 1999. Structure and mechanism in NF-κB/IκB signaling. *Cold Spring Harb Symp Quant Biol* **64:** 533–540.

Imbalzano AN, Kwon H, Green MR, Kingston RE. 1994. Facilitated binding of TATA-binding protein to nucleosomal DNA. *Nature* **370:** 481–485.

Janeway CA Jr, Medzhitov R. 2002. Innate immune recognition. *Annu Rev Immunol* **20:** 197–216.

Kawai T, Akira S. 2007. TLR signaling. *Semin Immunol* **19:** 24–32.

Kayama H, Ramirez-Carrozzi VR, Yamamoto M, Mizutani T, Kuwata H, Iba H, Matsumoto M, Honda K, Smale ST, Takeda K. 2008. Class-specific regulation of pro-inflammatory genes by MyD88 pathways and IκBζ. *J Biol Chem* **283:** 12468–12477.

Kornberg RD, Lorch Y. 1999. Twenty-five years of the nucleosome, fundamental particle of the eukaryote chromosome. *Cell* **98:** 285–294.

Kouzarides T. 2007. Chromatin modifications and their function. *Cell* **128:** 693–705.

Li Q, Wrange O. 1995. Accessibility of a glucocorticoid response element in a nucleosome depends on its rotational positioning. *Mol Cell Biol* **15:** 4375–4384.

Lim CA, Yao F, Wong JJ, George J, Xu H, Chiu KP, Sung WK, Lipovich L, Vega VB, Chen J, et al. 2007. Genome-wide mapping of RELA(p65) binding identifies E2F1 as a transcriptional activator recruited by NF-κB upon TLR4 activation. *Mol Cell* **27:** 622–635.

Mavrich TN, Ioshikhes IP, Venters BJ, Jiang C, Tomsho LP, Qi J, Schuster SC, Albert I, Pugh BF. 2008a. A barrier nucleosome model for statistical positioning of nucleosomes throughout the yeast genome. *Genome Res* **18:** 1073–1083.

Mavrich TN, Jiang C, Ioshikhes IP, Li X, Venters BJ, Zanton SJ, Tomsho LP, Qi J, Glaser RL, Schuster SC, et al. 2008b. Nucleosome organization in the *Drosophila* genome. *Nature* **453:** 358–362.

Natoli G, Saccani S, Bosisio D, Marazzi I. 2005. Interactions of NF-κB with chromatin: The art of being at the right place at the right time. *Nat Immunol* **6:** 439–445.

Nau GJ, Richmond JF, Schlesinger A, Jennings EG, Lander ES, Young RA. 2002. Human macrophage activation programs induced by bacterial pathogens. *Proc Natl Acad Sci* **99:** 1503–1508.

Ramirez-Carrozzi VR, Nazarian AA, Li CC, Gore SL, Sridharan R, Imbalzano AN, Smale ST. 2006. Selective and antagonistic functions of SWI/SNF and Mi-2β nucleosome remodeling complexes during an inflammatory response. *Genes Dev* **20:** 282–296.

Saccani S, Pantano S, Natoli G. 2001. Two waves of nuclear factor κB recruitment to target promoters. *J Exp Med* **193:** 1351–1359.

Saccani S, Pantano S, Natoli G. 2002. p38-Dependent marking of inflammatory genes for increased NF-κB recruitment. *Nat Immunol* **3:** 69–75.

Schreiber J, Jenner RG, Murray HL, Gerber GK, Gifford DK, Young RA. 2006. Coordinated binding of NF-κB family members in the response of human cells to lipopolysaccharide. *Proc Natl Acad Sci* **103:** 5899–5904.

Segal E, Fondufe-Mittendorf Y, Chen L, Thastrom A, Field Y, Moore IK, Wang JP, Widom J. 2006. A genomic code for nucleosome positioning. *Nature* **442:** 772–778.

Taylor IC, Workman JL, Schuetz TJ, Kingston RE. 1991. Facilitated binding of GAL4 and heat shock factor to nucleosomal templates: Differential function of DNA-binding domains. *Genes Dev* **5:** 1285–1298.

Udalova IA, Mott R, Field D, Kwiatkowski D. 2002. Quantitative prediction of NF-κB DNA-protein interactions. *Proc Natl Acad Sci* **99:** 8167–8172.

von Hippel PH, Berg OG. 1989. Facilitated target location in biological systems. *J Biol Chem* **264:** 675–678.

Weinmann AS, Plevy SE, Smale ST. 1999. Rapid and selective remodeling of a positioned nucleosome during the induction of IL-12 p40 transcription. *Immunity* **11:** 665–675.

Weinmann AS, Mitchell DM, Sanjabi S, Bradley MN, Hoffmann A, Liou HC, Smale ST. 2001. Nucleosome remodeling at the IL-12 p40 promoter is a TLR-dependent, Rel-independent event. *Nat Immunol* **2:** 51–57.

Widom J. 1998. Structure, dynamics, and function of chromatin in vitro. *Annu Rev Biophys Biomol Struct* **27:** 285–327.

Yamamoto M, Yamazaki S, Uematsu S, Sato S, Hemmi H, Hoshino K, Kaisho T, Kuwata H, Takeuchi O, Takeshige K, et al. 2004. Regulation of Toll/IL-1-receptor-mediated gene expression by the inducible nuclear protein IκBζ. *Nature* **430:** 626–630.

The Regulatory Logic of the NF-κB Signaling System

Ellen O'Dea and Alexander Hoffmann

Signaling Systems Laboratory, Department of Chemistry and Biochemistry, University of California, San Diego, La Jolla, California 92093

Correspondence: ahoffmann@ucsd.edu

NF-κB refers to multiple dimers of Rel homology domain (RHD) containing polypeptides, which are controlled by a stimulus-responsive signaling system that mediates the physiological responses to inflammatory intercellular cytokines, pathogen exposure, and developmental signals. The NF-κB signaling system operates on transient or short timescales, relevant to inflammation and immune responses, and on longer-term timescales relevant to cell differentiation and organ formation. Here, we summarize our current understanding of the kinetic mechanisms that allow for NF-κB regulation at these different timescales. We distinguish between the regulation of NF-κB dimer formation and the regulation of NF-κB activity. Given the number of regulators and reactions involved, the NF-κB signaling system is capable of integrating a multitude of signals to tune NF-κB activity, signal dose responsiveness, and dynamic control. We discuss the prevailing mechanisms that mediate signaling cross talk.

How can the regulatory logic of a signaling system be understood? The regulatory logic refers to the properties of a system that are not evident by studying one molecular component in isolation, or even the interaction between two; hence, they are sometimes referred to as emergent system properties. Straightforward emergent properties of a signaling system are signal-dose responses (which may be linear or sinusoidal) and dynamic control of the response (which may be fast or slow ramping, transient or oscillatory). More complex emergent properties may pertain to the integration of different signals (synergistically or antagonistically), memory functions, or contingencies for prior stimulus exposures to signal transduction. Those properties are mediated by the molecular components arranged in a particular network topology. Quantitative measurements of the relevant biochemical reactions is a prerequisite for a molecular understanding of the regulatory logic, and tracking a multitude of reactions via a computational model is an effective and practical strategy.

Understanding the function of a network must begin with comprehensive accounting of the parts list, the list of molecular components. The NF-κB signaling system consists of two protein families, the NF-κBs (activators) and the IκBs (inhibitors) (Fig. 1). The NF-κB transcription factors are the result of combinatorial dimerization of five monomers that can

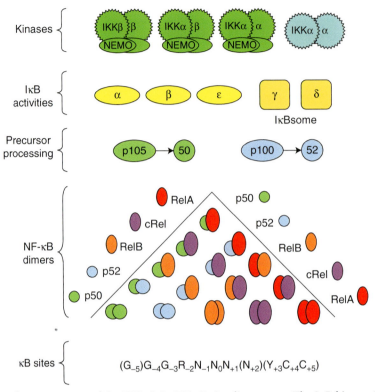

Figure 1. Molecular components of the IKK–IκB–NF-κB signaling system. The IκB kinases form canonical NEMO-containing (green) complexes and noncanonical IKKα complexes (blue), which control the degradation of IκB proteins as well as precursor processing. IκBα, IκBβ, IκBε, p105 (IκBγ), and p100 (IκBδ) are able to bind and sequester NF-κB dimers ("IκB activity"). The p50 and p52 NF-κB proteins are initially synthesized as the precursor proteins p105 and p100, respectively. The five NF-κB family members can potentially form 15 possible dimers, which may bind to a large family of κB sites in DNA.

produce 15 possible dimers. IκB activities act as stoichiometric inhibitors of the DNA binding activities of these dimers, and all combinations are, in principle, possible. Two classes of IκB kinase (IKK) complexes (the canonical and noncanonical IKK complexes) control the half-life and therefore abundances of the IκB activities. The canonical IKK are defined as those complexes bound and regulated by NF-κB essential modulator (NEMO), and the noncanonical complexes require both IKKα and NF-κB inducing kinase (NIK) activities but are independent of NEMO (Scheidereit 2006). The canonical IKK complexes act on the classical IκB proteins IκBα, -β, -ε, and the IκBγ activity mediated by p105, whereas the noncanonical IKK complex acts on the IκBδ activity mediated by p100. Both of the latter

nonclassical IκB activities (IκBγ and IκBδ) reside in a high-molecular weight complex (Savinova et al. 2009), which we term the IκBsome. In addition, both IKK activities can affect NF-κB dimer generation by controlling the processing of precursor proteins during or shortly after their translational synthesis. Canonical IKK may enhance processing of p105 to p50, thereby increasing the availability of p50-containing dimers, whereas noncanonical IKK controls processing of p100 to p52, thereby generating p52-containing dimers.

Considering the regulatory logic of the NF-κB signaling system, we distinguish regulatory mechanisms on two different timescales at which they operate: NF-κB dimer formation via NF-κB protein synthesis and dimerization, and the regulation of NF-κB dimer activity via

IκB degradation and resynthesis. The former (NF-κB protein synthesis and dimerization) is largely associated with longer-term cell differentiation processes, whereas the latter (IκB degradation and resynthesis) controls reversible responses to often transient inflammatory stimuli. Hence, we first consider the regulatory logic of NF-κB dimer generation and then discuss the mechanisms that regulate dimer activity.

REGULATION OF NF-κB DIMER GENERATION

NF-κB refers to homo- and heterodimeric DNA binding complexes that consist of Rel homology domain (RHD) containing polypeptides. Of the 15 potential dimers (Fig. 1), three do bind DNA but lack transcriptional activity (p50:p50, p52:p52, and p50:p52), and three are not known to bind DNA (RelA:RelB, cRel:RelB, and RelB:RelB), leaving nine potential transcriptional activators (Hoffmann and Baltimore 2006). Dimer formation and DNA binding occur through the RHD. Upon dimerization, NF-κB dimers can interact with IκB proteins, as the ankyrin repeat domain (ARD) of IκB forms a large interaction surface around the dimeric interface (Hayden and Ghosh 2004; Hoffmann et al. 2006). IκB binding biases cellular localization of the NF-κB dimer to the cytoplasm, thereby inhibiting the DNA binding activity of the NF-κB dimers and allowing for stimulus-responsive activation by cytoplasmic IKK complexes.

Several biochemical reactions control the generation and therefore availability of NF-κB dimers. Synthesis of the RHD polypeptides is the first step, and is regulated at the level of transcription in a cell-type-specific manner. Furthermore, RHD polypeptide genes (except *rela*) are inducible by NF-κB activity (see nf-kb.org for a list of NF-κB target genes), potentially resulting in positive feedback as well as functional interdependence between them. In addition, the generation of p50 and p52, the two RHD polypeptides that lack transcriptional activation domains (TADs) but act as binding partners for cRel, RelB, and RelA, is

controlled by proteolysis of p105 and p100. These proteolytic events may be responsive to canonical and noncanonical signals.

RHD polypeptides require dimerization to avoid degradation, presumably via unfolded protein degradation pathways. NF-κB dimer formation and stability may not only be determined by intrinsic association and dissociation rate constants between RHD polypeptides, but may also involve interactions with IκB proteins. One intriguing possibility is that association with the IκBs stabilizes the NF-κB dimers by slowing their dissociation into monomers, and possibly also the degradation of the intact dimers. If this hypothesis is found to be valid, IκB interactions with NF-κB dimers would not only inhibit NF-κB function, but would also contribute to generation of NF-κB dimers in the first place.

Equilibrium States

The generation of the cell-type-specific NF-κB dimer repertoire is a function of cell-type-specific homeostatic regulation that may be subject to signals conditioning the cell in a particular microenvironment. Homeostatic regulation may be tuned at several different mechanistic levels: (1) stimulus-responsive and cell-type-specific RHD polypeptide synthesis; (2) dimerization specificities governing the association and dissociation of each of the 15 monomer pairs; (3) IκB-NF-κB interactions contributing to dimer stability (Fig. 2A,B). As IκBs may have differential affinities for NF-κB dimers, the relative abundances of various IκB proteins may affect the repertoire of latent dimers available for activation.

Different cell types have indeed been reported to contain different repertoires of NF-κB dimers. Murine embryonic fibroblasts (MEFs), just like many of the transformed cell lines commonly used (HeLa, HEK293), contain primarily latent RelA:p50 heterodimer, as well as RelA:RelA and p50:p50 homodimers. cRel and RelB expression can be detected in these cells, but appears functionally negligible. In contrast, B-cells are abundant with readily activatable cRel and p50 containing dimers

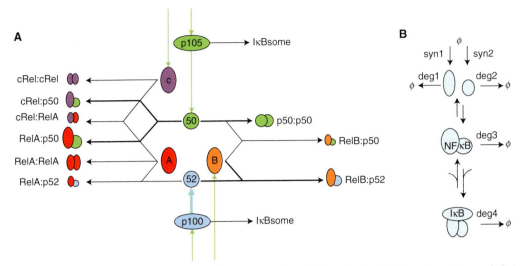

Figure 2. Mechanisms determining NF-κB dimer generation. (*A*) Synthesis of RHD polypeptides and their dimerization affinities control the generation of NF-κB dimers, whose relative abundances in MEFs are indicated by their relative size. Green arrows indicate regulation by canonical IKK and the blue arrow regulation by noncanonical IKK. The amount of processing of the p105 and p100 proteins alters the amount of p50 and p52 available for dimer interaction, and abundances of certain subunits is influenced by the amount of others. (*B*) IκB-NF-κB interactions may play a role in dimer generation. A theoretical model of NF-κB dimer metabolism indicates that dimerizaton may reduce monomer degradation (if deg3<deg2 and/or deg1), and further, that IκB interactions with the dimer may not only block dimer dissociation into monomers, but also dimer degradation (if deg4<deg3).

(Sen 2006). Dendritic cells (DCs) depend on both cRel and RelB for maturation (Ouaaz et al. 2002; Gerondakis et al. 2006), although the regulation of these cRel and RelB-containing dimer activities remain to be characterized.

Although cell-type-specific expression of RHD polypeptides and NF-κB dimers has been established, the molecular regulation that underlies the shifts in the NF-κB dimer repertoire during cell differentiation remains to be delineated. An early study examined the expression of NF-κB monomers and dimers in transformed cell lines that represent B-cell subtypes along the B-cell differentiation pathway (Liou et al. 1994). This work tracked differential NF-κB dimer availability, but what molecular mechanisms mediate these changes in the homeostatic NF-κB dimer repertoire has remained unclear. First insights about these molecular mechanisms come from the detailed analysis of NF-κB dimer misregulation in knockout cells deficient in one or more NF-κB or IκB genes (Basak et al. 2008). The mechanisms

by which compensation or interdependence of NF-κB dimer regulation is mediated can thereby be characterized and the insights may be applicable to normal differentiation processes.

There are several levels in which the generation of different NF-κB dimers is interdependent, allowing for cross talk and/or robustness through compensation. Cross talk may be mediated by the fact that synthesis of the four RHD polypeptides RelB, cRel, p50, and p52 is dependent on RelA-containing dimers at the level of gene expression (Fig. 2A). Hence, phenotypes resulting from RelA deficiency may (in part) be mediated by the consequently diminished availability of other RelA-dependent NF-κB dimers. Second, potential competition in dimerization interactions between limited pools of NF-κB monomers may mediate cross-regulation in dimer generation when specific RHD polypeptide synthesis rates change. However, a primary binding partner for the TAD-containing RelA, RelB, and cRel proteins is p50, which is generally produced in excess by abundant synthesis and processing of p105, leading to system robustness. When p50 is lacking in *nfkb1*$^{-/-}$, then the mature *nfkb2* gene product p52 compensates. In this scenario, however, the demand for p52 depletes the pool of p100, which is required for the noncanonical IκBδ activity, thereby attenuating noncanonical signal responsiveness. This demonstrates the capacity for one means of cross talk regulation between the canonical and noncanonical pathways (Basak et al. 2008).

One motivation for understanding the regulation of the cell-type-specific NF-κB dimer repertoire is that disease-associated cells may well exhibit an altered equilibrium state. Such altered states may be caused by chronic exposure to external signals within the disease microenvironment. Inflammatory signals such as TNF, IL-1, IL-6, or IFNγ are associated with inflammatory conditions or tumors that can skew the NF-κB dimer repertoire and NF-κB responsiveness of macrophages. Such changes may mitigate or potentiate the pathology. In other pathological conditions, altered NF-κB dimer repertoires may be a cause of the pathology. In some subsets of B-cell lymphomas, an enhanced pro-proliferative NF-κB-cRel gene expression signature can be traced back to cRel gene amplifications (Shaffer et al. 2002). Enhanced cRel expression may in other subsets be the result of cell intrinsic alterations that are not immediately obvious, such as an altered chromatin homeostasis caused by misregulated HAT (histone acetyltransferase) and HDAC (histone deacetylase) activities. The effects of disease-associated altered NF-κB dimer repertoires continue to pose important questions for research.

Stimulus-responsive Alterations

As four RHD polypeptides are encoded by known NF-κB-RelA response genes, the NF-κB dimer repertoire is generally thought to be readily alterable in response to inflammatory stimulation or other RelA-dimer-inducing stimuli (Fig. 3). However, it is unclear how dynamic or reversible the resulting changes are, whether they are occurring within hours or days, or whether they are transient or long

lasting. Most observations suggest that changes in the NF-κB dimer repertoire because of stimulus-induced transcription are occurring on long timescales and are long-lasting; we therefore suggest that transcriptional control of RHD polypeptide expression determines the homeostatic state of the NF-κB signaling system.

Stimulus-responsive alterations in the NF-κB dimer pool do, however, occur as a result of stimulus-responsive processing of p105 and p100 to p50 and p52. Both processing events occur cotranslationally or at least shortly following translation, before the precursors form higher order complexes (including IκBsomes) through homo- and heterotypic interactions that render them processing incompetent. Processing of p105 to p50 is mediated by long-term constitutive canonical IKK activity that ensures elevated expression of p50. The resulting p50:p50 homodimers have been observed in TLR-stimulated macrophages as well as other cell types, but their physiological role is unclear. Processing of p100 to p52 is mediated by noncanonical IKK activity and has been studied in detail. This processing event generates primarily the RelB:p52 dimer, which regulates expression of organogenic chemokines and is implicated in survival of certain differentiating and maturing cell types. Thus, noncanonical pathway activating stimuli, such as BAFF, LTβ, RANKL, and CD40L, can be thought of as altering the NF-κB dimer repertoire to mediate their functional effects (in addition to regulating the activities of NF-κB dimers via IκB regulation, as discussed later).

REGULATION OF NF-κB ACTIVITY VIA IκB DEGRADATION AND RESYNTHESIS

The three canonical IκB proteins, IκBα, IκBβ, and IκBε, and the noncanonical IκBδ activity bind and inhibit NF-κB dimers and thereby allow for their stimulus-responsive activation. The degradation and resynthesis of IκBs is coordinately regulated to produce dynamic NF-κB activities that are stimulus-specific. For example, the ubiquitous RelA:p50 dimer is bound and inhibited by the canonical IκB proteins (IκBα, IκBβ, and IκBε), as well as the

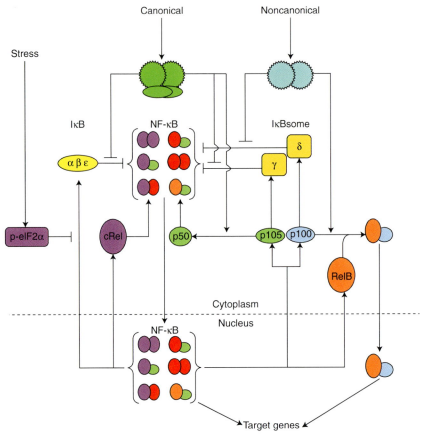

Figure 3. Mechanisms controlling stimulus-responsive NF-κB activities. Canonical signals activate NEMO-containing IKK complexes (green), which degrade the canonical IκB proteins (IκBα, β, and ε) and the IκBγ activity (composed of asymmetric p105 dimers) associated with NF-κB dimers. Released NF-κB dimers move to the nucleus to activate gene expression programs, including the expression of IκBα, IκBε, p105, p100, cRel, and RelB proteins. Noncanonical signals activate IKKα complexes, which degrade IκBδ complexes associated with NF-κB dimers. The resulting increase in synthesis of p100 and RelB, concomitant with IKKα activity, causes increased p100 processing to p52 and dimerization with RelB, to generate active RelB:p52 dimers to the nucleus. Stress signals can activate the eIF2α kinases, causing phosphorylation of eIF2α and resulting in inhibited translation. A block in IκB synthesis, in combination with constitutive IKK activity, results in the loss of IκB proteins and subsequent NF-κB dimer activation.

IκBδ activity. RelB-containing dimers, however, are known to have strong binding specificity only for IκBδ. In this manner, different IκB-NF-κB dimer associations allow for specific dimers to be activated in response to specific stimuli.

The regulated metabolism (synthesis and degradation) of the canonical IκB proteins, especially IκBα, has been studied in detail. Two degradation pathways control steady-state levels of canonical IκBs and NF-κB activity.

When associated with NF-κB, IκBα, -β, and -ε are degraded through a canonical IKK-dependent mechanism. IκB proteins that are not associated with NF-κB, however, are intrinsically unstable with a 5–10 minute half-life, being degraded in an IKK- and ubiquitin-independent manner (O'Dea et al. 2007; Mathes et al. 2008). Accumulating evidence indicates that the 20S proteasome can mediate degradation of free IκBα, whereby weakly folded regions in carboxy-terminal ankyrin repeats and other carboxy-terminal sequences mediate its instability (Alvarez-Castelao and Castano 2005; Truhlar et al. 2006). NF-κB association triggers complete IκB protein folding and removes IκB from the 20S proteasome pathway. In addition to determining the IκB degradation pathway, NF-κB activity also drives the transcription of IκBα and IκBε, thus providing another self-inhibitory mechanism to prevent constitutive NF-κB activity.

IκBδ activity is the result of the dimerization of two p100 molecules (through interaction between their RHDs), whereby the ARD of one p100 molecule folds back onto the RHD–RHD dimer interface to effect self-inhibition ("in *cis*"), leaving the ARD of the second p100 molecule capable of binding latent NF-κB dimers (primarily RelA:p50, but also RelB:p50) "in *trans*" (Basak and Hoffmann 2008). Subsequent studies with processing-defective mutants of p105 provided evidence for an analogous activity, termed IκBγ (Sriskantharajah et al. 2009). NF-κB also drives the synthesis of p105/IκBγ and p100/IκBδ, which may, like IκBα and IκBε, contribute to tight control over constitutive NF-κB activity. Recent studies revealed that these nonclassical IκB activities, IκBγ and IκBδ, reside within high molecular weight complexes, whose assembly is mediated by the helical dimerization domains present within p100 and p105 in addition to RHD dimerization and ARD interaction surfaces (Savinova et al. 2009). These complexes may be termed IκBsomes as they trap a variety of NF-κB dimers for slow homeostatic or stimulus-responsive release at later times.

IκB Signaling in Response to Inflammatory Signals

Endogenous inflammatory stimuli (e.g., cytokines-TNF, IL-1β) or pathogen-derived substances (e.g., LPS or CpG DNA) activate the canonical NF-κB pathway. Engagement of the TNF receptor (TNFR), IL-1β receptor (IL-1βR), or TLRs causes phosphorylation-dependent activation of a NEMO-containing kinase complex. Once activated, the canonical

IKK complex phosphorylates IκBα, -β, and -ε at two specific serine residues, as a signal for K48-linked ubiquitination at two specific lysine residues, leading to degradation by the 26S proteasome (Karin and Ben-Neriah 2000). Degradation of IκB releases NF-κB, allowing it to localize to the nucleus to bind DNA and activate gene expression. Interestingly, IκBα and IκBε proteins are among the large number of NF-κB response genes, thus functioning as negative-feedback regulators. Indeed, NF-κB activity induced by inflammatory stimuli shows complex and diverse temporal or dynamic profiles.

A mathematical model of the IκB–NF-κB signaling module recapitulates the signaling events triggered by TNF stimulation observed experimentally in MEFs (Kearns and Hoffmann 2009; Cheong et al. 2008). Combined experimental and computational studies showed that the three IκB proteins IκBα, -β, and -ε each have distinct roles in the dynamic control of NF-κB activation and termination. In response to TNF, IκBα is rapidly degraded and then rapidly resynthesized in an NF-κB-dependent manner. Continued TNF stimulation propagates a cycle of synthesis and degradation of IκBα that can result in oscillations of nuclear NF-κB activity (Hoffmann et al. 2002). IκBε expression is also strongly induced by NF-κB activity, but with a distinct 40-minute delay (Kearns et al. 2006). As IκBε protein accumulates at later time points, this antiphase negative-feedback loop acts to dampen the IκBα-driven oscillations.

Recent evidence indicates that the non-canonical IκB, IκBδ provides negative feedback on canonical NF-κB signaling when it is induced by pathogen-triggered TLR signals that produce longer lasting canonical IKK activities than inflammatory cytokines (e.g., TNF or IL-1). The induction of NF-κB by either LPS or TNF drives increased synthesis of p100, but on a delayed and longer timescale than that of IκBα. The slowly accumulating p100 can form into IκBsomes, within which IκBδ traps NF-κB dimers (Shih et al. 2009). Whereas TNF-induced signaling wanes within 8 hours, LPS-induced NF-κB remains elevated at late time points. This late NF-κB activity becomes subject to attenuation by IκBδ, which is un-responsive to and cannot be degraded by canonical signals. Cells deficient in IκBδ activity (*nfkb2* gene knockout), show similar NF-κB activation profiles in responses to TNF, but show significantly higher LPS-induced late NF-κB activity as compared to wild-type (IκBδ containing) cells (Shih et al. 2009). This study also emphasized that the prominent IκBα negative feedback is actually specific for transient cytokine signals. Because its inhibition is reversible by canonical IKK activity, it plays little role in shaping NF-κB temporal profiles in response to long lasting TLR-initiated signals. In fact, perinatal lethality of the *ikba*$^{-/-}$ mouse was rescued by compound deficiency with the *tnf* gene (Shih et al. 2009).

The ability of the IκB-NF-κB signaling module to mediate complex temporal control over NF-κB activity has led to research into the functionality of NF-κB dynamics. Different inflammatory stimuli elicit different IKK activation profiles, which induce distinct temporal profiles of NF-κB activity. For example, TNF stimulation induces a transient spike in IKK activity in which the second phase is rather small because of the short TNF half-life and the inhibitory functions of the deubiquitinase A20 (Werner et al. 2008). In contrast, LPS activates gene expression of cytokines that provide for positive autocrine feedback, amplifying late IKK activity (Werner et al. 2005). These studies have demonstrated that temporal control of NF-κB activity is important for stimulus-specific gene expression programs, yet it remains unclear how gene promoters decipher temporal profiles. If the temporal profile of NF-κB activity conveys information about the stimulus and therefore represents a code (i.e., "temporal code"), it remains unclear how this information is decoded. Also, given that temporal control depends on kinetic rate constants, it is likely that different cells encode stimulus information differently. Further experimental investigation and expansion of the current NF-κB mathematical model to include other cell types will be necessary to decipher how NF-κB dynamics plays a role in mammalian physiology.

IκB Signaling in Response to Developmental Signals

A group of noninflammatory stimuli have been shown to activate NF-κB through the noncanonical NF-κB signaling pathway. These developmental signals of the TNF-receptor super-family such as B-cell activation factor receptor (BAFFR), lymphotoxin β receptor (LTβR), and receptor activator of NF-κB (RANK), have been described to activate NF-κB activity at a low level for hours or days. The noncanonical pathway is not transduced by a canonical NEMO-containing kinase complex, but rather by an IKKα-containing complex, whose activity is also dependent on NIK. In addition to the noncanonical NIK/IKKα-dependent NF-κB activation, these stimuli may also, in certain cellular conditions and contexts, activate the canonical NEMO/IKKβ-dependent NF-κB activation pathway (Pomerantz and Baltimore 2002).

Initial studies of NF-κB activation by these developmental stimuli focused on the generation of the RelB:p52 dimer by cotranslational proteolytic processing of de novo synthesized p100 to p52. The p100 protein contains carboxy-terminal serines whose phosphorylation by IKKα is critical for stimulus-responsive processing. More recently, it was shown that the same pathway also activates latent RelA:p50 NF-κB complexes not through NEMO-dependent degradation of classical IκB, but rather via the degradation of IκBδ activity.

IκBδ activity and the IκBsomes, which mediate it, were first discovered in cells lacking all three classical IκB proteins, IκBα, -β, and -ε (Tergaonkar et al. 2005; Basak et al. 2007). In these cells, the ubiquitous RelA:p50 dimer was shown to be associated with cytoplasmic p100 proteins. Further biochemical analysis showed that all cells contain significant amounts of NF-κB associated with high molecular weight complexes of p100 and p105 that mediate IκBγ and IκBδ activities (Savinova et al. 2009). The functional importance of these complexes was revealed by the discovery that noncanonical signals, through NIK- and IKKα-dependent degradation of IκBδ, release associated RelA:p50 and RelB:p50 to the nucleus (Basak et al. 2007). Furthermore, conditions or specific stimuli that control the abundance of IκBδ are therefore able to tune the responsiveness of RelA:p50 activation through noncanonical signals. When IκBδ is highly abundant, usually weak developmental signals that engage the noncanonical pathway are able to activate inflammatory genes via strong RelA:p50 activation.

IκB Signaling in Response to Cellular Stresses

Recent work has uncovered homeostatic mechanisms of regulation of NF-κB in resting cells (i.e., in the absence of a stimulus) that predetermine the responsiveness of the NF-κB system to inducers, including stress stimuli.

Homeostatic regulation of IκB synthesis and degradation has emerged as an important characteristic that renders the NF-κB signaling system surprisingly insensitive to a variety of perturbations. First, the differential degradation rates of free and bound IκB allows for compensation between IκBs, evident by studies in IκB knockout cells (O'Dea et al. 2007). The loss of IκBα, for example, causes excess NF-κB to bind and thus stabilize IκBε, in addition to enhancing its synthesis. Second, the very short half-life of free IκB necessitates a high rate of constitutive IκB synthesis to maintain the small cellular free IκB pool (estimated at 15% of total [Rice and Ernst 1993]) that is critical for keeping basal NF-κB activity levels low. One consequence of the very high constitutive IκB synthesis/degradation flux is that the NF-κB system is remarkably resistant to transient alterations in translation rates that are a hallmark of metabolic stress agents. Indeed, UV, UPR (unfolded protein response), or other ribotoxic stress that cause partial inhibition of IκB synthesis rates were found to activate NF-κB only modestly (O'Dea et al. 2008).

Understanding homeostatic control of the IκB-NF-κB system also resolved some seemingly conflicting observations with regards to NF-κB activation by UV. Although UV does not induce IKK activity, basal IKK activity was found to be required (Huang et al. 2002;

O'Dea et al. 2008). The systems model revealed how basal IKK activity is critical for slowly degrading NF-κB-bound IκBα, allowing for accumulation of nuclear NF-κB in response to translational inhibition. Whereas stimuli that induce canonical IKK activity cause the degradation of NF-κB-bound IκB proteins, UV irradiation was shown to cause the depletion of the free IκB pool. Indeed, mutations that stabilize free IκB (but allow for IKK-dependent degradation of NF-κB-bound IκB) abolished activation of NF-κB by UV (O'Dea et al. 2008). UV-induced CK2 activity and phosphorylation of the carboxyl terminus of IκBα may further accelerate IκBα degradation but the molecular mechanism of this pathway remain unclear (Kato et al. 2003).

DNA damage, caused by irradiation or chemotherapeutic drugs, however, has been reported to induce IKK activation. Until recently, it was unclear how a nuclear signal could relay back to the cytoplasm. It was found that DNA damage not only initiates the activation of the nuclear kinase ataxia telangiectasia mutated (ATM), the primary regulator of the tumor suppressor and transcription factor p53, but it also initiates the sumoylation of NEMO by the sumo ligase PIASy, promoting its nuclear localization (Mabb et al. 2006). It was known that activated ATM was required for NF-κB activation by DNA damage, and recent work has uncovered the connection between ATM activity and IKK activation. Wu et al. (2006) showed that nuclear sumoylated NEMO associates with and is phosphorylated by the activated ATM, promoting mono-ubiquitination of NEMO that triggers its export to the cytoplasm. The ATM-NEMO complex associates with the protein ELKS, facilitating ATM-dependent activation of the canonical IKK complex, leading to IκBα degradation and NF-κB activation (Wu et al. 2006).

THE NF-κB SIGNALING SYSTEM AS AN INTEGRATOR OF SIGNALS

The NF-κB signaling system mediates the physiological effects of a variety of signals. It does so via a large number of biochemical reactions determining the abundance and kinetic regulation of two families of proteins (NF-κB and IκB) and their interactions with each other. Many NF-κB-inducing stimuli control the activity of the two IKK complexes, but others impact many other reactions to alter NF-κB activity, or modulate the responsiveness of the NF-κB signaling system. When stimuli affect different reaction rates within the same signaling system, there is potential for signaling cross talk. As such, the NF-κB signaling system has great potential for integrating diverse signals. In the following section, we summarize just a few of many possible examples.

Integrating Signals that Control IκB Degradation and Synthesis

Ribotoxic Stresses and Inflammatory Signals Cross Talk via IκBα

Whereas ribotoxic stress-inducing stimuli such as UV irradiation or the unfolded protein response (UPR) that act to inhibit the translation of IκBα can rapidly deplete the cellular pool of free, intrinsically unstable IκBα, NF-κB-bound IκBα remains associated with NF-κB until either constitutive or induced canonical IKK-mediated phosphorylation triggers its degradation. The concept of this differential degradation of IκBα, combined with computational simulations suggesting that the level of constitutive IKK activity predetermines the responsiveness of NF-κB activity to translational inhibition, led to the prediction that cells chronically exposed to low levels of inflammatory signals may have significantly enhanced responses to stress stimuli that inhibit protein synthesis (O'Dea et al. 2008).

Indeed, it was found that cells treated with low doses of inflammatory signals that only marginally enhance IKK and NF-κB activities, synergize with UV or ER stress agents that inhibit protein synthesis (via eIF2α phosphorylation) to produce a super-additive increase in NF-κB activity and NF-κB-dependent inflammatory gene expression (O'Dea et al. 2008). The combined effect of the metabolic stress reaction (eIF2α phosphorylation) and

IKK activation in response to inflammatory signals or immune receptors may be a more common mechanism to activating NF-κB than previously recognized. A potential function of super-activating NF-κB during metabolic stresses may be to override the concomitant decrease in translation rates for NF-κB controlled survival and immune response genes.

Inflammatory Signals and Developmental Signals Cross Talk via IκBδ

Inflammatory signaling can interact and coordinate responses with developmental cues. For example, lymph node development and homeostasis appears to require not only the developmental regulator lymphotoxin β (LTβ) but also the inflammatory cytokine TNF (Rennert et al. 1998). Conversely, inflammation can derail normal developmental or cellular homeostasis and be a major factor in cancer progression (Karin and Greten 2005). Over the last few years, it has become apparent that the NF-κB signaling system may be a primary integrator of these diverse signals to mediate cross talk in both physiological and pathological processes.

Following the identification of the fourth IκBδ activity that mediates noncanonical signals, its NF-κB-induced expression by canonical signals was explored as a cross talk mechanism (Basak et al. 2007). Inflammatory stimuli drive the expression of p100, which mediates inhibition of RelA:p50 complexes through its IκBδ activity. Because IκBδ is not degraded by canonical NEMO-mediated signals, the cellular pool of RelA:p50 bound to IκBδ increases, thereby amplifying RelA:p50 responsiveness to subsequent developmental signals (which degrade IκBδ) to the extent that developmental stimuli may cause inflammatory gene expression. Combined computational modeling and experimental studies demonstrated the potential for developmental signals such as LTβ to result in inflammatory responses. This cross talk mediated by IκBδ has been speculated to play a role in regulating cancer-associated inflammation (Basak and Hoffmann 2008).

Integrating Signals that Control NF-κB Synthesis and IκB Degradation

Several observations suggest that there may be synergistic cross talk between canonical and noncanonical IKK pathways via the NF-κB signaling system. RelA-containing dimers are known to increase the synthesis of RelB and p100/p52 polypeptides (www.nf-kb.org). Indeed, basal RelA NF-κB activity was shown to be required for LTβ-responsive RelB:p52 dimer generation (Basak et al. 2008). In B-cells, basal or tonic canonical IKK activity was shown to control the generation of the p100 substrate for noncanonical BAFFR signals (Stadanlick et al. 2008). By the same rationale and molecular mechanism, inflammatory signals such as TNF and LPS may potentially enhance RelB:p52 generation in the presence of developmental stimuli. However, at this time, no compelling physiological or molecular evidence has been presented, and the functional significance of the RelB versus RelA dependent gene expression remains to be elucidated.

CONCLUDING REMARKS

The regulatory logic of the NF-κB signaling system suggests regulation at two levels. The first pertains to the generation of the transcriptional activators, the NF-κB dimers, via monomer expression and dimerization reactions. The second is the regulation of the metabolism of the IκB inhibitors via synthesis and degradation control. Inflammatory signals largely involve the latter, and produce complex dynamic control of NF-κB activity. Homeostatic and cell differentiation associated mechanisms control NF-κB dimer generation. Interestingly, signals engaging the noncanonical IKK complex engage both mechanistic levels; such developmental signals are able to produce inflammatory responses (though generally weakly) and they affect the available NF-κB dimer repertoire. Recognizing the two levels also suggests that interesting, nonadditive or nonlinear signaling may result when the two interface. With multiple interdependencies apparent within the NF-κB signaling system,

these cross-regulatory connections must be further characterized mechanistically and quantified to allow for a better understanding of physiological regulation and pathological misregulation of the NF-κB signaling system. Biophysical descriptions at the molecular and cellular level are likely to result in further insights about a signaling system that has tremendous human health relevance.

ACKNOWLEDGMENTS

We thank the many researchers whose work informed the conceptual framework described in this article and apologize if it was not explicitly described or cited due to space restrictions. Research on NF-κB in our laboratory was funded by National Institutes of Health (NIH) grants GM071573 and GM071862. We thank S. Basak and B. Schroefelbauer for frequent discussions and critical reading.

REFERENCES

Alvarez-Castelao B, Castano JG. 2005. Mechanism of direct degradation of IκBα by 20S proteasome. *FEBS Lett* **579:** 4797–4802.

Basak S, Hoffmann A. 2008. Crosstalk via the NF-κB signaling system. *Cytokine Growth Factor Rev* **19:** 187–197.

Basak S, Shih VF, Hoffmann A. 2008. Generation and activation of multiple dimeric transcription factors within the NF-κB signaling system. *Mol Cell Biol* **28:** 3139–3150.

Basak S, Kim H, Kearns JD, Tergaonkar V, O'Dea E, Werner SL, Benedict CA, Ware CF, Ghosh G, Verma IM, et al. 2007. A fourth IκB protein within the NF-κB signaling module. *Cell* **128:** 369–381.

Cheong R, Hoffmann A, Levchenko A. 2008. Understanding NF-κB signaling via mathematical modeling. *Mol Syst Biol* **4:** 192.

Gerondakis S, Grumont R, Gugasyan R, Wong L, Isomura I, Ho W, Banerjee A. 2006. Unravelling the complexities of the NF-κB signalling pathway using mouse knockout and transgenic models. *Oncogene* **25:** 6781–6799.

Hayden MS, Ghosh S. 2004. Signaling to NF-κB. *Genes Dev* **18:** 2195–2224.

Hoffmann A, Baltimore D. 2006. Circuitry of nuclear factor κB signaling. *Immunol Rev* **210:** 171–186.

Hoffmann A, Natoli G, Ghosh G. 2006. Transcriptional regulation via the NF-κB signaling module. *Oncogene* **25:** 6706–6716.

Hoffmann A, Levchenko A, Scott ML, Baltimore D. 2002. The IκB-NF-κB signaling module: Temporal control and selective gene activation. *Science* **298:** 1241–1245.

Huang TT, Feinberg SL, Suryanarayanan S, Miyamoto S. 2002. The zinc finger domain of NEMO is selectively required for NF-κB activation by UV radiation and topoisomerase inhibitors. *Mol Cell Biol* **22:** 5813–5825.

Karin M, Greten FR. 2005. NF-κB: Linking inflammation and immunity to cancer development and progression. *Nat Rev Immunol* **5:** 749–759.

Karin M, Ben-Neriah Y. 2000. Phosphorylation meets ubiquitination: The control of NF-κB activity. *Annu Rev Immunol* **18:** 621–663.

Kato T Jr, Delhase M, Hoffmann A, Karin M. 2003. CK2 Is a C-Terminal IκB Kinase Responsible for NF-κB Activation during the UV Response. *Mol Cell* **12:** 829–839.

Kearns JD, Hoffmann A. 2009. Integrating computational and biochemical studies to explore mechanisms in NF-κB signaling. *J Biol Chem* **284:** 5439–5443.

Kearns JD, Basak S, Werner SL, Huang CS, Hoffmann A. 2006. IκBε provides negative feedback to control NF-κB oscillations, signaling dynamics, and inflammatory gene expression. *J Cell Biol* **173:** 659–664.

Liou HC, Sha WC, Scott ML, Baltimore D. 1994. Sequential induction of NF-κ B/Rel family proteins during B-cell terminal differentiation. *Mol Cell Biol* **14:** 5349–5359.

Mabb AM, Wuerzberger-Davis SM, Miyamoto S. 2006. PIASy mediates NEMO sumoylation and NF-κB activation in response to genotoxic stress. *Nat Cell Biol* **8:** 986–993.

Mathes E, O'Dea EL, Hoffmann A, Ghosh G. 2008. NF-κB dictates the degradation pathway of IκBα. *Embo J* **27:** 1357–1367.

O'Dea EL, Kearns JD, Hoffmann A. 2008. UV as an amplifier rather than inducer of NF-κB activity. *Mol Cell* **30:** 632–641.

O'Dea EL, Barken D, Peralta RQ, Tran KT, Werner SL, Kearns JD, Levchenko A, Hoffmann A. 2007. A homeostatic model of IκB metabolism to control constitutive NF-κB activity. *Mol Syst Biol* **3:** 111.

Ouaaz F, Arron J, Zheng Y, Choi Y, Beg AA. 2002. Dendritic cell development and survival require distinct NF-κB subunits. *Immunity* **16:** 257–270.

Pomerantz JL, Baltimore D. 2002. Two pathways to NF-κB. *Mol Cell* **10:** 693–695.

Rennert PD, James D, Mackay F, Browning JL, Hochman PS. 1998. Lymph node genesis is induced by signaling through the lymphotoxin β receptor. *Immunity* **9:** 71–79.

Rice NR, Ernst MK. 1993. In vivo control of NF-κB activation by I κ B α. *Embo J* **12:** 4685–4695.

Savinova OV, Hoffmann A, Ghosh G. 2009. The Nfkb1 and Nfkb2 proteins p105 and p100 function as the core of heterogeneous NF-κBsomes. *Mol Cell* **34:** 591–602.

Scheidereit C. 2006. IκB kinase complexes: gateways to NF-κB activation and transcription. *Oncogene* **25:** 6685–6705.

Sen R. 2006. Control of B lymphocyte apoptosis by the transcription factor NF-κB. *Immunity* **25:** 871–883.

Shaffer AL, Rosenwald A, Staudt LM. 2002. Lymphoid malignancies: The dark side of B-cell differentiation. *Nat Rev Immunol* **2:** 920–932.

Shih VF, Kearns JD, Basak S, Savinova OV, Ghosh G, Hoffmann A. 2009. Kinetic control of negative feedback regulators of NF-κB/RelA determines their pathogen- and cytokine-receptor signaling specificity. *Proc Natl Acad Sci* **106:** 9619–9624.

Sriskantharajah S, Belich MP, Papoutsopoulou S, Janzen J, Tybulewicz V, Seddon B, Ley SC. 2009. Proteolysis of NF-κB1 p105 is essential for T cell antigen receptor-induced proliferation. *Nat Immunol* **10:** 38–47.

Stadanlick JE, Kaileh M, Karnell FG, Scholz JL, Miller JP, Quinn WJ 3rd, Brezski RJ, Treml LS, Jordan KA, Monroe JG, et al. 2008. Tonic B cell antigen receptor signals supply an NF-κB substrate for prosurvival BLyS signaling. *Nat Immunol* **9:** 1379–1387.

Tergaonkar V, Correa RG, Ikawa M, Verma IM. 2005. Distinct roles of IκB proteins in regulating constitutive NF-κB activity. *Nat Cell Biol* **7:** 921–923.

Truhlar SM, Torpey JW, Komives EA. 2006. Regions of IκBα that are critical for its inhibition of NF-κB. DNA interaction fold upon binding to NF-κB. *Proc Natl Acad Sci* **103:** 18951–18956.

Werner SL, Barken D, Hoffmann A. 2005. Stimulus specificity of gene expression programs determined by temporal control of IKK activity. *Science* **309:** 1857–1861.

Werner SL, Kearns JD, Zadorozhnaya V, Lynch C, O'Dea E, Boldin MP, Ma A, Baltimore D, Hoffmann A. 2008. Encoding NF-κB temporal control in response to TNF: Distinct roles for the negative regulators IκBα and A20. *Genes Dev* **22:** 2093–2101.

Wu ZH, Shi Y, Tibbetts RS, Miyamoto S. 2006. Molecular linkage between the kinase ATM and NF-κB signaling in response to genotoxic stimuli. *Science* **311:** 1141–1146.

Cite this article as *Cold Spring Harb Perspect Biol* 2010;2:a000216

Oncogenic Activation of NF-κB

Louis M. Staudt

Metabolism Branch, Center for Cancer Research, National Cancer Institute, NIH, Bethesda, Maryland 20892–8322

Correspondence: lstaudt@mail.nih.gov

Recent genetic evidence has established a pathogenetic role for NF-κB signaling in cancer. NF-κB signaling is engaged transiently when normal B lymphocytes respond to antigens, but lymphomas derived from these cells accumulate genetic lesions that constitutively activate NF-κB signaling. Many genetic aberrations in lymphomas alter CARD11, MALT1, or BCL10, which constitute a signaling complex that is intermediate between the B-cell receptor and IκB kinase. The activated B-cell-like subtype of diffuse large B-cell lymphoma activates NF-κB by a variety of mechanisms including oncogenic mutations in CARD11 and a chronic active form of B-cell receptor signaling. Normal plasma cells activate NF-κB in response to ligands in the bone marrow microenvironment, but their malignant counterpart, multiple myeloma, sustains a variety of genetic hits that stabilize the kinase NIK, leading to constitutive activation of the classical and alternative NF-κB pathways. Various oncogenic abnormalities in epithelial cancers, including mutant K-ras, engage unconventional IκB kinases to activate NF-κB. Inhibition of constitutive NF-κB signaling in each of these cancer types induces apoptosis, providing a rationale for the development of NF-κB pathway inhibitors for the treatment of cancer.

Given the crucial role of NF-κB in signaling downstream of a multitude of surface receptors, cancer inevitably has found mechanisms to co-opt this pathway. NF-κB plays an important role in the initiation and promotion of cancer by fostering an inflammatory milieu in which various cytokines aid and abet malignant transformation (reviewed in Karin 2010; Karin et al. 2006). Some cancers are caused by viruses that encode activators of the NF-κB pathway, which block the cell death inherent in viral transformation (reviewed in Hiscott et al. 2006). In this article, I focus on mechanisms by which NF-κB is aberrantly and stably activated by genetic lesions in human cancer. The selective advantage imparted to a tumor cell on engagement of the NF-κB pathway derives in large measure from the ability of this pathway to block apoptosis. In a variety of lymphoid cancers, NF-κB is constitutively active owing to diverse somatic mutations, genomic amplifications and deletions, and chromosomal translocations. These abnormalities subvert the normal function of NF-κB in immune cell signaling. An oncogenic role for NF-κB has surfaced in epithelial cancers as well. This emerging genetic evidence shows that the NF-κB pathway is central to the pathogenesis of many cancer types, providing impetus for the development of therapeutics targeting this pathway.

NF-κB signaling can be dichotomized into a "classical" pathway in which IκB kinase β (IKKβ) phosphorylates IκBα and an "alternative" NF-κB pathway in which IKKα phosphorylates the p100 precursor of the NF-κB p52 subunit. The IKK complex in the classical pathway requires the regulatory IKKγ subunit, whereas the IKK complex in the alternative pathway does not. The result of these signaling events is the accumulation of the heterodimeric NF-κB transcription factors in the nucleus, with the classical pathway regulating mainly p50/p65 and p50/c-Rel dimers and the alternative pathway regulating p52/relB dimers. In addition, NF-κB can be activated by other kinases, including the unconventional IKK family members, IKKε and TBK1, although the exact mechanisms linking these kinases to NF-κB activation need clarification. Numerous signaling pathways converge on these NF-κB regulators, providing ample means by which cancers can aberrantly stimulate NF-κB.

NF-κB IN LYMPHOMA

As outlined in the following discussion, many subtypes of human lymphoma rely on constitutive activity of the NF-κB pathway for survival. This dependency likely has its roots in the pervasive role of the NF-κB pathway in normal B-cell maturation and activation. Genetic deletion of NF-κB subunits in B cells blocks B-cell differentiation at a variety of steps, depending on which subunit is ablated (reviewed in Vallabhapurapu et al. 2009). The alternative NF-κB pathway is activated in response to exposure of B cells to BAFF, a tumor necrosis factor (TNF) family member made by myeloid-derived cells in secondary lymphoid organs. Signals from BAFF are essential for development of mature follicular B cells from transitional B cells (Claudio et al. 2002). NF-κB is also required for the maintenance of all mature resting B cells because conditional deletion of the IKKβ or IKKγ subunits causes B cells to be lost from the follicular compartment (Pasparakis et al. 2002), and a small molecule inhibitor of IKKβ depletes

the mature B-cell pool (Nagashima et al. 2006). During antigenic challenge, the classical NF-κB pathway is strongly activated by B-cell receptor signaling, via formation of the "CBM" signaling complex consisting of CARD11, MALT1, and BCL10 (Thome 2004). The CBM pathway is pathologically altered in several lymphoma subtypes.

Roughly 90% of human lymphomas arise from B lymphocytes at various stages of differentiation, with the remainder derived from T lymphocytes. The most prevalent type of non-Hodgkin's lymphoma is diffuse large B-cell lymphoma, comprising ~40% of cases. This diagnostic category harbors several molecularly and clinically distinct diseases, as originally defined by gene expression profiling (Alizadeh et al. 2000). The three well-delineated DLBCL subtypes are termed germinal center B-cell-like (GCB) DLBCL, activated B-cell-like (ABC) DLBCL, and primary mediastinal B-cell lymphoma (PMBL). These subtypes apparently arise from different stages of normal B-cell differentiation and use distinct oncogenic pathways (Staudt et al. 2005). The DLBCL subtypes respond differentially to standard chemotherapeutic regimens (Lenz et al. 2008) and to regimens incorporating newer, more targeted agents (Dunleavy et al. 2009).

ABC DLBCL

A hallmark of ABC DLBCL is constitutive activation of the NF-κB pathway (Davis et al. 2001). Gene expression profiling of tumor biopsy samples revealed preferential expression of NF-κB target genes in ABC DLBCL compared with GCB DLBCL, and this was also true for cell line models of ABC DLBCL. ABC DLBCLs engage the classical NF-κB pathways because they have rapid phosphorylation and turnover of IκBα and prominent nuclear accumulation of p50/p65 heterodimers with lesser accumulation of p50/c-rel heterodimers (Davis et al. 2001). Interruption of NF-κB signaling with an IκB super repressor or with a small molecule inhibitor of IKKβ induces apoptosis in ABC DLBCL but not GCB DLBCL cell lines (Davis et al. 2001; Lam et al. 2005).

A biological consequence of NF-κB signaling in ABC DLBCL is to propel the malignant cell forward toward the plasma cell stage of differentiation. An NF-κB target in ABC DLBCL is IRF4 (Lam et al. 2005), a key transcription factor that drives plasmacytic differentiation (Klein et al. 2006; Sciammas et al. 2006; Shaffer et al. 2008; Shaffer et al. 2009). In normal lymphocytes, IRF4 drives terminal differentiation by transactivating *PRDM1*, which encodes Blimp-1 (Sciammas et al. 2006; Shaffer et al. 2008), another master regulator of plasmacytic differentiation (Shaffer et al. 2002; Shapiro-Shelef et al. 2005; Shapiro-Shelef et al. 2003; Shapiro-Shelef et al. 2005; Turner et al. 1994). Together, IRF4 and Blimp-1 directly or indirectly promote expression of XBP-1, a transcription factor that specifies the secretory phenotype (Iwakoshi et al. 2003; Reimold et al. 2001; Shaffer et al. 2004). Interestingly, ABC DLBCLs express IRF4 and XBP-1, but frequently have genetic lesions that inactivate Blimp-1 (Iqbal et al. 2007; Pasqualucci et al. 2006; Shaffer et al. 2000; Tam et al. 2006). These observations suggest a model in which NF-κB activation in ABC DLBCL transactivates IRF4 and initiates plasmacytic differentiation, but full plasmacytic differentiation is blocked by lesions that inactivate Blimp-1. In this way, the tumor escapes the cell cycle arrest that typifies normal plasma cells, which is due in part to repression of MYC by Blimp-1 (Lin et al. 1997).

Another important influence of NF-κB signaling on ABC DLBCL biology is the production of the cytokines IL-6 and IL-10 (Ding et al. 2008; Lam et al. 2008). Both cytokines are secreted by ABC DLBCL cells and signal in an autocrine fashion, activating the transcription factor STAT3. A signature of STAT3 target genes typifies a subset of ABC DLBCL tumors, and these tumors have high STAT3 protein levels and have phosphorylated STAT3 in the nucleus (Lam et al. 2008). In contrast, GCB DLBCL biopsies lack both STAT3 target gene expression and phosphorylated STAT3. ABC DLBCLs with STAT3 activation also have higher expression of NF-κB target genes, in keeping with the fact that STAT3 physically interacts with NF-κB factors, thereby increasing their ability

to transactivate their targets (Yang et al. 2007). As a consequence of this interaction, a JAK kinase inhibitor, which extinguished STAT3 phosphorylation, synergized with an IKKβ inhibitor in killing ABC DLBCL cells (Lam et al. 2008).

Engagement of the CARD11/BCL10/MALT1 Signaling Module in ABC DLBCL

The mechanisms responsible for constitutive NF-κB signaling in ABC DLBCL have been elucidated by RNA interference genetic screens (Ngo et al. 2006). In these so-called "Achilles heel" screens, shRNAs are screened for those that block cancer cell proliferation and survival. An initial Achilles heel screen revealed toxicity of shRNAs targeting CARD11, MALT1, and BCL10 for ABC but not GCB DLBCL cell lines (Ngo et al. 2006). In normal lymphocytes, these three proteins form a "CBM" signaling complex that is required for NF-κB signaling downstream of the antigen receptors (reviewed in Blonska et al. 2009; Rawlings et al. 2006; Thome 2004) (Fig. 1A). CARD11 is a multidomain signaling scaffold protein, consisting of an amino-terminal CARD domain, a coiled-coil domain, and a carboxy-terminal MAGUK domain. In a resting lymphocyte, CARD11 resides in a latent form in the cytoplasm. On antigen receptor engagement, CARD11 is phosphorylated in a "linker" region residing between its coiled-coil and MAGUK domains (Matsumoto et al. 2005; Sommer et al. 2005). In B cells, PKCβ carries out this function, whereas PKCθ is responsible in T cells. Phosphorylated CARD11 relocalizes to the plasma membrane, where it recruits a multiprotein complex consisting of MALT1, BCL10, TRAF6, TAK1, caspase 8, and c-Flip (McCully et al. 2008). A series of ubiquitin-dependent interactions leads to IKK activation (reviewed in Chen 2005). On recruitment to the CBM complex, TRAF6 becomes active as a ubiquitin ligase, attaching K63-linked polyubiquitin chains to itself and MALT1 (Oeckinghaus et al. 2007; Sun et al. 2004). The IKK complex is subsequently recruited, most likely using its ubiquitin binding domain to attach to ubiquitinated TRAF6 and MALT1. TRAF6 ubiquitinates IKKγ, which is a required step in its activation

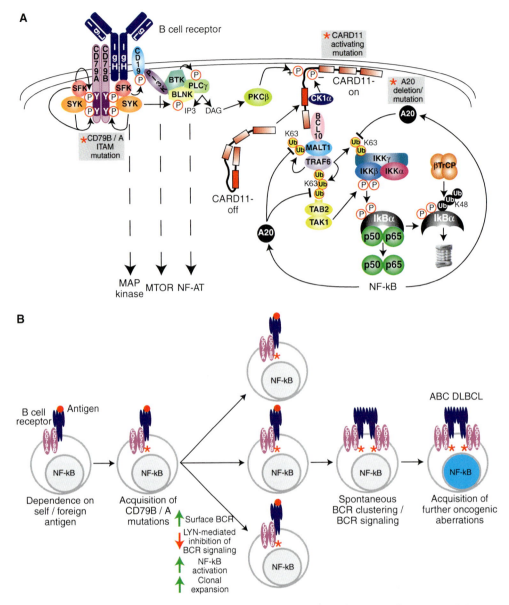

Figure 1. Role of BCR signaling to NF-κB in ABC DLBCL. (*A*) Schematic of BCR signaling to NF-κB. Recurrent genetic alterations in ABC DLBCL that result in constitutive NF-κB activation are indicated by the gray boxes. Proximal signaling by the BCR is initiated by SRC-family kinases (SFK; e.g., LYN, FYN, FGR, and BLK), which phosphorylate the ITAM motifs in the CD79A and CD79B components of the BCR receptor. SYK is recruited to the phosphorylated ITAMs and activated to phosphorylate many downstream proteins. The PI(3) kinase pathway is activated by SRC-family kinase phosphorylation of the BCR coreceptor CD19. The generation of PIP3 by PI(3) kinase recruits BTK and associated BLNK and phospholipase Cγ2 (PLCγ2) to the plasma membrane. PLCγ2 generates inositol triphosphate (IP3), which leads to opening of the capacitative calcium channel, thereby activating the NF-AT pathway. Diacylglycerol (DAG) is also generated, which activates protein kinase Cβ (PKCβ). PKCβ phosphorylates the latent form of CARD11 (CARD-off) in the cytoplasm, causing it to adopt an active conformation (CARD11-on), translocate to the plasma membrane, and recruit the signaling adapters BCL10 and MALT1. (*Figure legend continued on the following page.*)

 Cite this article as *Cold Spring Harb Perspect Biol* 2010;00:a000109

as a kinase. Ubiquitinated TRAF6 also interacts with TAB2, leading to activation of the associated kinase TAK1. TAK1 carries out the second required step in IKK activation, phosphorylation of IKKβ in its activation loop.

RNA interference screening in DLBCL also revealed casein kinase 1α (CK1α) as a new component of the CBM pathway (Bidere et al. 2009). shRNAs targeting CK1α were toxic for ABC but not GCB DLBCL cell lines and blocked the formation of the CBM complex in activated lymphocytes. CK1α performs two opposing functions in this pathway. First, CK1α interacts with CARD11 and promotes the formation of the CBM complex; this function does not require its kinase activity. Second, CK1α phosphorylates serine 608 in the CARD11 linker region, decreasing NF-κB signaling, perhaps by enhancing turnover of CARD11. Given these opposing activities, the only effective way in which to target CK1α therapeutically would be to inhibit its interaction with CARD11.

Oncogenic CARD11 Mutations

Whereas the function of the CBM complex in normal antigen receptor signaling is transient, being limited by several negative feedback loops, the CBM complex is continuously required for the survival of ABC DLBCL cells. An explanation of this paradox was revealed by the discovery of somatic CARD11 mutations

in DLBCL (Lenz et al. 2008). CARD11 mutations are present in ~10% of ABC DLBCLs, and all mutations are confined to the coiled-coil domain. At a lower frequency, CARD11 mutations are present in GCB DLBCLs, and these rare cases have high NF-κB target gene expression, unlike most GCB DLBCLs. The CARD11 mutants can constitutively activate NF-κB when introduced into heterologous cells. Moreover, these mutants potentiated the activation of NF-κB following antigen receptor stimulation.

The oncogenic activation of CARD11 by mutations DLBCL is likely explained by their effect on CARD11 subcellular localization (Lenz et al. 2008). Whereas wild-type CARD11 localizes diffusely in the cytoplasm, the CARD11 mutations form prominent cytoplasmic aggregates. These aggregates colocalize with other components of the CBM signalosome, including MALT1 and phosphorylated IKK, suggesting that they are sites of active signaling. The quantitative degree of aggregation by the various CARD11 mutations correlates directly with the degree of IKK kinase activity, again suggesting the aggregates are functional.

The coiled-coil domain mediates several essential functions in CARD11 (Tanner et al. 2007). Point mutations in the coiled-coil can destroy the function of the protein in lymphocyte activation (Jun et al. 2003). Certain mutations in the coiled-coil domain impair its self association, whereas others block membrane

---

Figure 1. (*Continued*) MALT1 binds TRAF6, causing TRAF6 to catalyze K63-linked polyubiquitination of the IKKγ subunit, MALT1 and itself. TAB2 recognizes the polyubiquitin chains on TRAF6, leading to phosphorylation of IKKβ in its activation loop by TAK1 kinase. Ubiquitination of IKKγ and phosphorylation of IKKβ activate IKKβ to phosphorylate IκBα. Phosphorylated IκBα is ubiquitinated by the ubiquitin ligase βTrCP, leading to its proteasomal degradation. Nuclear NF-κB heterodimers activate multiple target genes, including A20. A20 terminates NF-κB signaling by removing K63-linked ubiquitin chains from IKKγ, TRAF6, and MALT1, and attaching K48-linked ubiquitin chains. (*B*) The role of chronic active BCR signaling in the pathogenesis of ABC DLBCL. Genetic mutations in CD79B and CD79A ITAM regions in ABC DLBCLs suggest that BCR signaling is key to the pathogenesis of ABC DLBCL. CD79B and CD79A ITAM mutations in the mouse cause hyperactive BCR signaling, suggesting that CD79 ITAM mutations in ABC DLBCL may amplify antigen-stimulated BCR signaling. CD79 ITAM mutations increase surface BCR expression and decrease activation of LYN kinase, a negative regulator of BCR signaling, potentially resulting in increased signaling to NF-κB and greater clonal expansion. A separate step leading to chronic active BCR signaling in ABC DLBCL is the acquisition of spontaneous BCR clustering, which is not caused by the ITAM mutations. The BCR clustering phenotype could theoretically be acquired either before or after the CD79 mutations. Finally, ABC DLBCLs must acquire additional oncogenic hits to become fully malignant.

localization of CARD11 (Tanner et al. 2007). The coiled-coil domain appears to be under the negative influence of the linker domain. Deletion of the linker regions results in a constitutively active form of CARD11 (McCully et al. 2008; Sommer et al. 2005), suggesting that the linker region may fold back on the coiled-coil domain, inhibiting its function. Indeed, an isolated linker peptide can associate with a peptide consisting of the amino-terminal CARD and coiled-coil domains (McCully et al. 2008). Certain phosphorylations of the linker domain activate CARD11 (Matsumoto et al. 2005; Sommer et al. 2005), and these presumably interfere with the association of the linker regions with the coiled-coil domain, although this needs to be directly tested experimentally. Once relieved of the negative influence of the linker region, the coiled-coil domain contributes to binding of BCL10, Caspase-8, TRAF6, and IKKγ (McCully et al. 2008).

These results suggest a simple model in which the CARD11 mutations in DLBCL prevent association of the coiled-coil and linker domains, creating an isoform that binds several NF-κB pathway components constitutively. Less clear is why the CARD11 mutations cause the protein to form such prominent cytosoloic aggregates. Presumably, this represents uncontrolled lattice formation involving both the coiled-coil domain and the carboxy-terminal MAGUK domain. MAGUK proteins form large protein interaction lattices that contribute to the postsynaptic density in neurons and tight junctions (Funke et al. 2005). By analogy, the CARD11 aggregates may be signaling lattices that massively nucleate components of the NF-κB signaling cascade. Therapeutic targeting of CARD11 coiled-coil domain interactions should be considered because overexpression of an isolated coiled-coil domain abrogates NF-κB signaling and kills ABC DLBCL cell lines with either wild-type or mutant CARD11 (Lenz et al. 2008). A structural analysis of the CARD11 coiled-coil domain will be required to determine how a small molecule might interfere with the function of the mutant coiled-coil domains.

Chronic Active B-cell Receptor Signaling

In the majority of ABC DLBCLs, CARD11 is not mutated, yet the tumors have high expression of NF-κB target genes. Further, ABC DLBCL cell lines with wild-type CARD11 are killed by CARD11 knockdown. These observations led to the hypothesis that an upstream signaling pathway activates CARD11 and NF-κB in these cases (Davis et al. 2010). An RNA interference screen revealed that the kinase BTK is required for NF-κB signaling in ABC DLBCLs with wild-type CARD11. Mutations in BTK cause a failure of B-cell production in Bruton's X-linked agammaglobulinemia (Satterthwaite et al. 2000). BTK is also a key kinase required to connect B-cell receptor (BCR) signaling to NF-κB (Bajpai et al. 2000; Petro et al. 2000). Following BCR signaling, BTK forms a complex with the adapter BLNK and phospholipase Cγ2 (PLCγ2). PLCγ2 then produces the second messenger diacyl glycerol, which activates PKCβ, leading to CARD11 phosphorylation and NF-κB signaling. In keeping with this model, mouse B-cells deficient in BLNK, PLCγ2, and PKCβ are all defective in NF-κB activation (Leitges et al. 1996; Petro et al. 2001; Saijo et al. 2002; Tan et al. 2001).

ABC DLBCLs with wild-type CARD11 activate BTK as a consequence of signals emanating from the BCR itself. In these lymphoma cells, knockdown of the immunoglobulin heavy or light chains is lethal, as is knockdown of the essential signaling subunits of the BCR, CD79A (Ig-α) and CD79B (Ig-β) (Davis et al. 2010). BCR disruption blocks several downstream signaling pathways in these cells, including NF-κB, AKT/mTOR, ERK MAP kinase, and NF-AT. Given that BCR signaling engages these potent cell growth and survival pathways, it is perhaps not surprising that lymphomas have found ways to use BCR signaling pathologically.

Total internal reflection microscopy revealed that the BCRs in these lymphoma cells form prominent clusters in the plasma membrane that have low diffusion (Davis et al. 2010). These are likely to be the sites of active BCR signaling because the clusters colocalize with phosphotyrosine. In contrast, the BCRs in cell lines representing GCB DLBCL, Burkitt's lymphoma, and mantle cell lymphoma are distributed diffusely in the plasma membrane. In normal B cells, interaction with a membrane-bound antigen causes the BCRs to form immobile clusters within seconds of contact (Tolar et al. 2009). Hence, the clustered BCRs in ABC DLBCLs could represent the influence of an antigen. Alternatively, these lymphoma cells may have defects in the regulation of BCR assembly in the plasma membrane.

The constitutive BCR activity in these ABC DLBCLs has been termed "chronic active BCR signaling" to deliberately distinguish it from "tonic" BCR signaling. Tonic BCR signaling was initially defined based on experiments in which the BCR was conditionally ablated in mouse B cells (Kraus et al. 2004; Lam et al. 1997). In these mice, all mature B cells disappeared over the course of 1–2 weeks, demonstrating an ongoing requirement for BCR signaling in the survival of mature B cells. Tonic BCR signaling most likely does not require antigen engagement by the BCR because artificial BCR mimics that lack immunoglobulin components can rescue the development and survival of B cells that lack immunoglobulin (Monroe 2006). Recent experiments implicate PI(3) kinase signaling as an important survival signal that is delivered by tonic BCR signaling (Srinivasan et al. 2009). An activated allele of the p110 subunit of PI(3) kinase could rescue B cells following BCR ablation but an activated IKKβ allele could not. Nonetheless, NF-κB may still contribute to tonic signaling because conditional ablation of IKKβ or IKKγ in mouse B-cells impairs their survival (Pasparakis et al. 2002).

Chronic active BCR signaling can be distinguished from tonic BCR signaling by two main criteria. First, tonic BCR signaling does not appear to require the CBM complex because mice deficient in CBM components have relatively normal numbers of follicular B cells (Thome 2004), unlike mice in which tonic signaling has been blocked. Because the BCR signaling in ABC DLBCL acts through the CBM complex, it cannot be characterized as

"tonic." Second, tonic BCR signaling in the mouse is required to maintain resting mature B cells, which have unclustered BCRs. In contrast, the BCRs in ABC DLBCLs are clustered, more akin to antigen-stimulated B cells than to resting B cells. Although it remains possible that tonic BCR signaling is required for the survival of some lymphoma subtypes, it does not appear to play a role in the pathogenesis of ABC DLBCL.

Cancer gene resequencing has revealed recurrent somatic mutations in BCR components in ABC DLBCL, providing a genetic "smoking gun" implicating BCR signaling in the pathogenesis of this lymphoma subtype (Davis et al. 2010). In over one fifth of ABC DLBCL specimens, somatic mutations are present in CD79B and less frequently in CD79A, but such mutations are rare or absent in other lymphoma subtypes. All mutations affect the critical "ITAM" signaling motifs of CD79A and CD79B. The ITAM motif is an evolutionarily conserved signaling module in several immune receptors that includes two invariant tyrosine residues (Reth 1989). SRC-family tyrosine kinases phosphorylate the ITAM tyrosines and then become further active as kinases when they bind the phosphorylated ITAMs through their SH2 domains. SYK is also recruited via its tandem SH2 domains to phosphorylated ITAMs, becoming active as a kinase in the process. One sixth of ABC DLBCLs have mutations that change the first ITAM tyrosine of CD79B to a variety of other amino acids (Davis et al. 2010). One ABC DLBCL tumor was identified that has a surgical three-base-pair deletion that removed this tyrosine, further accentuating its importance. At a lower frequency, ABC DLBCL tumors have point mutations affecting other conserved CD79B ITAM residues or deletions that remove all or part of the ITAM regions of CD79B or CD79A.

The CD79 mutations in ABC DLBCL are not loss-of-function because they sustain survival of ABC DLBCL cells and promote signaling to NF-κB (Davis et al. 2010). Unlike CARD11 coiled-coil mutants, however, the CD79 mutants do not activate NF-κB when introduced into a heterologous cell type. Clues to the function of the CD79 mutations come from studies of knockin mice with loss or mutation of CD79A or CD79B ITAMs (Gazumyan et al. 2006; Kraus et al. 1999; Torres et al. 1999). Unexpectedly, mature B cells develop in these animals despite disruption of one or the other CD79 ITAM motif. In fact, B cells from these mice are hyperresponsive to certain antigenic stimuli. For example, mice with both CD79B ITAM residues changed to alanine have higher serum IgM levels and respond with 10-fold higher antibody titers to immunization with a T cell-independent antigen (Gazumyan et al. 2006). Further, B cells from these mice have higher expression of the BCR on the cell surface because of decreased receptor internalization. This latter phenotype is also a feature of the CD79 mutants from ABC DLBCL tumors: Reconstitution of the BCR with a CD79B mutant in the first ITAM tyrosine yields higher BCR expression than reconstitution with wild-type CD79B (Davis et al. 2010). In contrast, a CD79B mutant affecting the second ITAM tyrosine, which has not been observed in ABC DLBCL tumors, does not raise surface BCR expression, highlighting the functional specificity of the human CD79B mutants. Chronic active BCR signaling lowers surface expression of the BCRs containing wild-type CD79B but not those with mutant CD79B, suggesting that CD79B mutations may have been selected in part to prevent signaling-induced receptor internalization (Davis et al. 2010).

A second functional attribute of the CD79 mutations in ABC DLBCL is to block negative autoregulation by the kinase LYN. LYN performs a dual function in B cells (Gauld et al. 2004; Xu et al. 2005). Like other SRC-family kinases, LYN participates in the initial activation of BCR signaling by phosphorylating CD79 ITAMs. However, LYN is unique among SRC-family kinases in initiating a feedback loop that attenuates BCR signaling. LYN can phosphorylate the "ITIM" modules in CD22 and the Fc γ-receptor, allowing recruitment of the phosphatase SHP1 to the BCR, leading to ITAM tyrosine dephosphorylation. LYN can also phosphorylate SYK at a negative regulatory site, decreasing SYK kinase activity. Because of

these negative regulatory mechanisms, mice deficient in LYN have hyperactive BCR signaling that leads to a severe and sometimes fatal autoimmune disease (Chan et al. 1997). ABC DLBCL cells reconstituted with mutant CD79B have consistently less LYN kinase activity than those reconstituted with wild-type CD79B (Davis et al. 2010). Because LYN is an important regulator of BCR internalization (Ma et al. 2001; Niiro et al. 2004), the inhibition of LYN kinase activity might explain the higher surface BCR levels in ABC DLBCL cells with mutant CD79B.

These observations suggest a model in which the CD79 mutations are selected in ABC DLBCLs to avoid negative influences on BCR signaling (Fig. 1B). This could occur at an early stage of lymphomagenesis in which a self or foreign antigen-dependent B cell might acquire a CD79 ITAM mutation, allowing it to undergo greater clonal expansion, enhance activation of NF-κB, and avoid cell death. An independent step in the formation of ABC DLBCL appears to be the acquisition of BCR clustering, which is independent of the CD79 mutations (Davis et al. 2010), but is likely responsible for the strong constitutive BCR signaling that typifies these lymphomas. Finally, ABC DLBCLs accumulate a number of other oncogenic hits that are required to create a fully malignant clone (Lenz et al. 2008).

Other Genetic Alterations Deregulating NF-κB in ABC DLBCL

Taken together, the CARD11 and CD79 mutations could potentially account for the NF-κB activation in roughly one third of ABC DLBCLs, implying that many more genetic alterations will be discovered that affect NF-κB in this lymphoma subtype. One common target of genetic inactivation in ABC DLBCL is A20, a negative regulator of NF-κB signaling. ABC DLBCLs have a variety of nonsense mutations and genomic deletions that inactivate A20 in roughly one-quarter of cases, but these events have been observed in only 2% of GCB DLBCLs (Compagno et al. 2009). Reintroduction of wild-type A20 into an ABC DLBCL cell line with an A20 mutation extinguished NF-κB signaling and caused apoptosis. A20 serves as a negative regulator of NF-κB signaling by deubiquitinating TRAF2, TRAF6, MALT1, and the IKKγ subunit (Boone et al. 2004; Lin et al. 2008; Mauro et al. 2006; Wertz et al. 2004). An important point to emphasize is that A20 serves as a brake on NF-κB signaling, but its loss most likely cannot by itself cause NF-κB signaling. This concept is illustrated by A20-deficient mice, which die from a lethal inflammatory disease (Boone et al. 2004). The inflammatory disease in these mice depends on homeostatic MyD88 signaling in response to commensal gut flora (Turer et al. 2008). In ABC DLBCLs, it is likely that A20 mutations augment but do not cause constitutive NF-κB signaling. A20 mutations may cooperate with events that activate NF-κB signaling, such as CARD11 mutations or chronic active BCR signaling. For example, one cell line with an A20 mutation, RC-K8 (Compagno et al. 2009), also harbors inactivating mutations in IκBα (Kalaitzidis et al. 2002).

One TRAF2 mutant was isolated that can activate NF-κB when overexpressed (Compagno et al. 2009). However, the mutant protein appeared to be significantly less stable than wild-type TRAF2 protein, making the interpretation of this mutant complicated. Other NF-κB modulators reported to be mutated in DLBCLs include RANK, TRAF5, and TAK1, but the effect of these mutations on NF-κB signaling was not studied (Compagno et al. 2009).

GCB DLBCL

GCB DLBCL likely derives from the rapidly proliferating centroblasts within the germinal center. Following activation by antigen, B-cells acutely activate NF-κB via the CBM complex. Thereafter, by mechanisms that are poorly understood, the B cell adopts a new gene expression program that is strikingly distinct from that of an acutely activated B cell. In particular, the NF-κB pathway is essentially silent in germinal center B cells, as measured by expression of NF-κB target genes and by nuclear accumulation of NF-κB heterodimers (Basso et al. 2004;

Shaffer et al. 2001). However, it is likely that NF-κB signaling is periodically activated in germinal center B cells because CD40, a known activator of the classical and alternative NF-κB pathway, is continuously required to sustain the germinal center reaction (Han et al. 1995). A small fraction of germinal center B cells have nuclear NF-κB, indicating that they may be receiving stronger signals either from CD40 ligand on the surface of follicular helper T cells or from antigen (Basso et al. 2004). This stronger NF-κB signal can promote transcription of the NF-κB target gene *IRF4*, which encodes a key inducer of plasmacytic differentiation.

Because of their germinal center derivation, GCB DLBCLs have generally lower expression of NF-κB target genes than other DLBCL subtypes. A confounding issue in these analyses is that NF-κB target genes can be expressed by nonmalignant immune cells (macrophages and T cells) that infiltrate the lymphoma. However, GCB DLBCL cell lines have much lower expression of NF-κB target genes than ABC DLBCL cell lines (Davis et al. 2001). Moreover, in situ analysis of nuclear NF-κB abundance confirms that the majority of GCB DLBCLs have little NF-κB activation (Compagno et al. 2009), and thus inherit from their nonmalignant counterparts a low NF-κB set point. Accordingly, treatment of GCB DLBCL cell lines with inhibitors of IκB kinase has no effect (Lam et al. 2005).

A minority of GCB DLBCLs may activate NF-κB, as evidenced by the presence of oncogenic CARD11 mutations in 3.7% of cases (Lenz et al. 2008). These rare GCB DLBCLs retain expression of germinal center B cell signature genes, on which is layered the expression of NF-κB target genes (Lenz et al. 2008). An occasional GCB DLBCL may also use the chronic active BCR signaling mechanism because 3% of these cases had genetic alterations of the CD79B ITAM (Davis et al. 2010).

Curiously, a recurrent genetic abnormality in GCB DLBCL is amplification of the locus on chromosome 2 encoding c-rel (Lenz et al. 2008). This amplicon occurs in 27% of GCB DLBCLs but in only 5% of ABC DLBCLs. These GCB DLBCLs are not associated with nuclear c-Rel (Feuerhake et al. 2005), nor with higher expression of the NF-κB target gene signature (Davis et al. 2001). Therefore, the selective pressure for GCB DLBCLs to amplify this locus may relate to another gene in this genomic interval. Alternatively, the amplification of the c-rel locus could have been selected early in the evolution of the tumor to enhance responsiveness to extracellular stimuli that engage the NF-κB pathway.

PRIMARY MEDIASTINAL B-CELL LYMPHOMA

Primary mediastinal B-cell lymphoma (PMBL) is a subtype of DLBCL that is clinically distinct, occurring most commonly in young women under 40. As the name implies, this lymphoma subtype typically arises in the mediastinum and most likely originates from a rare thymic B-cell subpopulation (Copie-Bergman et al. 2002). Gene expression profiling revealed that PMBL has a distinctive gene expression signature that can be used to differentiate this entity from other DLBCL subtypes (Rosenwald et al. 2003; Savage et al. 2003).

Unexpectedly, PMBL shares an extensive gene expression signature with the malignant cells of Hodgkin lymphoma, known as Hodgkin Reed-Sternberg (HRS) cells (Rosenwald et al. 2003; Savage et al. 2003). Included in the shared gene expression signature are NF-κB target genes (Rosenwald et al. 2003; Savage et al. 2003). Inhibition of NF-κB signaling in a PMBL cell line, using either an IKKβ inhibitor or an IκBα super-repressor, extinguished NF-κB target gene expression and induced cell death (Feuerhake et al. 2005; Lam et al. 2005).

Clues regarding the molecular nature of constitutive NF-κB signaling in PMBL are beginning to emerge. Inactivating mutations and deletions of A20 are recurrent in PMBL, occurring in roughly one third of cases (Schmitz et al. 2009). As in ABC DLBCL, the removal of the negative influence of A20 may potentiate other mechanisms that activate NF-κB in these cells. Roughly one fifth of PMBL cases have amplification of the c-rel locus (Lenz et al. 2008), and this is associated with nuclear c-rel (Feuerhake et al. 2005). Overexpression of c-rel likely

augments the constitutive NF-κB activation in these cases. Despite these clues, the root cause of constitutive IKKβ activation in PMBL remains elusive.

HODGKIN LYMPHOMA

Hodgkin lymphoma is characterized by tumors in which the malignant HRS cells are imbedded in a sea of inflammatory cells from a variety of hematopoietic lineages. In roughly one half of Hodgkin lymphomas, the HRS cell harbors the Epstein-Barr virus (EBV), which encodes LMP1, a potent activator of NF-κB. Although LMP1 may account for NF-κB signaling in EBV-positive cases, analysis of primary HRS cells and cell lines from EBV-negative cases also revealed constitutive nuclear p50/p65 heterodimers, consistent with classical NF-κB signaling (Bargou et al. 1996). Inhibition of NF-κB using an IκBα super-repressor was toxic to Hodgkin cell lines (Bargou et al. 1997).

In ~20% of Hodgkin lymphoma cases, the HRS cells have inactivating mutations or deletions of NFKBIA, encoding IκBα (Cabannes et al. 1999; Emmerich et al. 1999; Jungnickel et al. 2000; Lake et al. 2009). Less commonly, genetic events inactivate NFKBIE, encoding IκBε (Emmerich et al. 2003). Notably, genetic inactivation of IκBα occurs preferentially in EBV-negative Hodgkin lymphomas, presumably because the LMP1-mediated NF-κB eliminates the selective pressure for IκBα abnormalities (Lake et al. 2009).

In some EBV-negative Hodgkin lymphoma cell lines, IKK constitutively phosphorylates and degrades wild-type IκBα (Krappmann et al. 1999). Although the molecular underpinnings of this constitutive IKK activity are not fully understood, HRS cells from 44% of Hodgkin lymphomas sustain genomic mutations and deletions that inactivate A20, the majority of which are in EBV-negative cases (Schmitz et al. 2009). Reconstitution of A20-deficient Hodgkin cell lines with A20 causes cell death by shutting down NF-κB signaling. However, as in ABC DLBCL, loss of the negative regulator A20 may be insufficient to activate NF-κB, and additional mechanisms may contribute in positively activating IKKβ in Hodgkin lymphoma in a cell autonomous fashion.

Within the inflammatory milieu of the Hodgkin lymphoma tumor, various TNF receptor signaling pathways could conceivably contribute to NF-κB activation of the HRS cell. For example, activated T cells in the microenvironment of the HRS cells often express CD40 ligand (Carbone et al. 1995; Gruss et al. 1994), which may active the classical and alternative pathways in HRS cells. Additionally, the TNF family receptor CD30 is expressed highly in Hodgkin lymphoma (Schwab et al. 1982) and its ligand is expressed on mast cells and eosinophils, which are present in Hodgkin lymphoma tumors (Molin et al. 2002; Pinto et al. 1996).

A less common form of Hodgkin lymphoma, termed lymphocyte-predominant Hodgkin lymphoma, has a cellular composition resembling a normal germinal center reaction, including the presence of follicular dendritic cells and T-follicular helper cells (Schmitz et al. 2009). This Hodgkin lymphoma subtype also has constitutive NF-κB activation as evidenced by NF-κB target gene expression and nuclear p65 (Brune et al. 2008). The mechanisms responsible are unknown but do not include A20 inactivation.

MALT LYMPHOMA

Mucosa-associated lymphoid tissue (MALT) lymphoma, a form of marginal zone lymphoma, is an often indolent disease that occurs most commonly in the stomach but can also arise at a variety of anatomical sites near mucosal surfaces (Isaacson et al. 2004). In most cases, gastric MALT lymphoma is associated with persistent infection with *Helicobacter pylori*. Other marginal zone lymphomas have been associated with *Chlamydia psittaci*, *Campylobacter jejuni*, *Borelia burgdorferi*, and hepatitis C virus infections (Ferreri et al. 2007). Ex vivo culture experiments suggest that *H. pylori* antigens trigger T-cell proliferation but do not affect the malignant MALT lymphoma clone (Hussell et al. 1996). Instead, the BCRs of the malignant B-cell clone often possess rheumatoid factor activity (Bende et al. 2005). This suggests a

scenario in which nonmalignant B cells in the inflammatory microenvironment produce polyclonal IgG antibodies specific for *H. pylori* that generate immune complexes, which then bind to the BCR of the malignant B-cell clone owing to its rheumatoid factor activity. These gastric MALT lymphomas presumably activate NF-κB via BCR signaling although this has not been directly tested. Remarkably, eradication of *H. pylori* by antibiotic treatment causes sustained complete remissions in many of these patients with gastric MALT lymphoma (Wotherspoon et al. 1993).

However, antibiotic therapy is ineffective in a substantial proportion of gastric MALT cases, suggesting a different pathogenesis. Approximately 25% of gastric MALT lymphoma harbor a t(11;18) translocation that creates a fusion protein between c-IAP2 and MALT1 (Akagi et al. 1999; Dierlamm et al. 1999; Morgan et al. 1999). These cases do not appear to have BCRs with rheumatoid factor activity (Bende et al. 2005) and are not responsive to antibiotic treatment (Liu et al. 2002). Less commonly, MALT1 can be overexpressed in MALT lymphoma as a consequence of a t(14;18) translocation linking it to the immunoglobulin heavy chain (IgH) locus or by chromosomal amplification of the MALT1 locus (Sanchez-Izquierdo et al. 2003). In rare cases, MALT lymphomas are associated with a t(1;14) translocation that juxtaposes BCL10 translocation with the IgH locus (Willis et al. 1999; Zhang et al. 1999).

Transgenic mice expressing either the c-IAP2-MALT1 fusion protein or BCL10 under control of immunoglobulin heavy chain enhancer elements develop splenic marginal zone hyperplasia (Baens et al. 2006; Li et al. 2009). These findings underscore the relationship between CARD11/BCL10/MALT1 signaling and marginal zone B-cell differentiation, which is also evident from the defective marginal zone development in mice deficient in MALT1 or BCL10 (Thome 2004). The low frequency and delayed onset of lymphoma in these mice suggests that uncontrolled NF-κB activation is insufficient for lymphomagenesis and that additional cooperating oncogenic events are necessary.

Although the chromosomal breakpoints in t(11;18) translocations are somewhat variable (Isaacson et al. 2004), in most cases the c-IAP2-MALT1 fusion protein includes 3 BIR domains of c-IAP2 and the paracaspase domain of MALT1 (Fig. 2A). Missing is the c-IAP2 RING finger domain, presumably because it mediates c-IAP2 autoubiquitination and destabilization. Also missing is the death domain of MALT1, which is required for BCL10 interaction. During normal antigen receptor signaling, BCL10 is required to oligimerize MALT1 to activate NF-κB (Lucas et al. 2001). The role of BCL10

in c-IAP2-MALT1 signaling to NF-κB is controversial: Two studies found that BCL10 is not required for NF-κB activation by c-IAP2-MALT1 (Noels et al. 2007; Ruland et al. 2003), whereas another report found that BCL10 knockdown abrogated NF-κB activation by the fusion protein in HeLa cells (Hu et al. 2006). This latter study reported that c-IAP2 binds BCL10 through its BIR domains (Hu et al. 2006), but this was not observed in another study (Noels et al. 2007). BCL10 protein is more abundant in MALT lymphomas with the t(11;18) translocation than in cases lacking

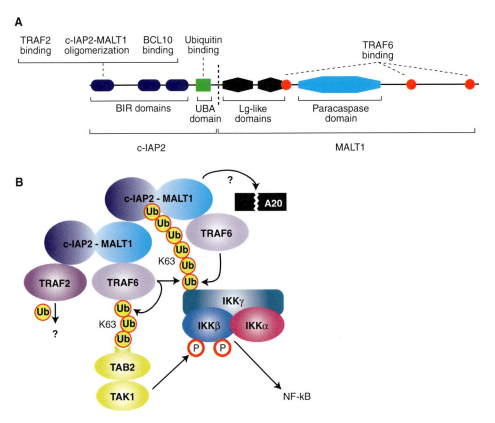

Figure 2. Pathogenesis of MALT lymphoma. (*A*) Schematic of the c-IAP2-MALT1 fusion oncoprotein produced by the t (11;18) translocation in MALT lymphomas. Shown is one common fusion breakpoint (dashed line), but other breakpoints occur less frequently (for review, see Isaacson et al. 2004). (*B*) Molecular mechanisms of NF-κB activation by the c-IAP2-MALT1 fusion protein. c-IAP2-MALT1 forms multimers by heterotypic interactions between its BIR and MALT1 regions. c-IAP2-MALT1 binds TRAF6, activating the K63-linked ubiquitin ligase activity of TRAF6 for IKKγ and itself. Polyubiquitated TRAF6 binds TAB2, thereby activating the associated TAK1 kinase to phosphorylate IKKβ. The ubiquitin-binding domain of c-IAP2-MALT1 stabilizes its interaction with ubiquitinated IKK. The proteolytic cleavage of a substrate protein, potentially A20, by the paracaspase domain of MALT1 is required for the function of c-IAP2-MALT1. TRAF2 is also required for c-IAP2-MALT1 activity, but its substrate is unknown.

this translocation (Hu et al. 2006), perhaps because of the ability of c-IAP2-MALT1 to dominantly interfere with the ability of c-IAP2 to ubiquitinate and degrade BCL10 (Hu et al. 2006). Some of the discrepancies in the aforementioned studies may be traced to different model systems and conditions, most of which use overexpression of proteins in heterologous cell types. Clearly lacking in these studies is an analysis of these biochemical interactions in MALT lymphoma cells.

The c-IAP2-MALT1 fusion protein activates NF-κB when overexpressed (Lucas et al. 2001; Ruland et al. 2003; Uren et al. 2000). The carboxy-terminal BIR1 domain appears to mediate two functions required for NF-κB activation. First, this domain binds TRAF2, which contributes to NF-κB activation by the c-IAP2-MALT1 fusion protein by an unknown mechanism (Garrison et al. 2009; Lucas et al. 2007; Samuel et al. 2006). Second, the BIR1 domain mediates oligomerization of the c-IAP2-MALT1 fusion protein by interacting heterotypically with the carboxy-terminal region from MALT1 (Lucas et al. 2007; Zhou et al. 2005). In the IAP family of proteins, the BIR domains interact with caspase domains, thereby inhibiting their activation and blocking apoptosis (Wright et al. 2005). Hence, it is conceivable that the BIR1 domain may bind the paracaspase domain in an analogous fashion, although this has not been tested experimentally. Because artificial dimerization of wild-type MALT1 activates NF-κB (Lucas et al. 2001; Sun et al. 2004; Zhou et al. 2004), BIR1-mediated oligomerization of c-IAP2-MALT1 may play an important role in the oncogenic activity of this fusion protein.

In normal lymphocytes, activation of IKK downstream of antigen receptor signaling requires lysine-63 polyubiquitination of the IKKγ subunit, which may be mediated by TRAF6 recruitment by the CBM complex (Sun et al. 2004; Zhou et al. 2004). c-IAP2-MALT1 activation of NF-κB also requires ubiquitination of IKKγ (Zhou et al. 2005). This activity may require TRAF6 (Noels et al. 2007) and/or the ability of the fusion protein to bind an E2 ubiquitin conjugating enzyme

containing Ubc13 (Zhou et al. 2005). TRAF2 binding to the BIR domains could also contribute to IKKγ ubiquitination, but this has not been explicitly shown (Garrison et al. 2009; Lucas et al. 2007; Samuel et al. 2006).

c-IAP2 also contains a ubiquitin binding domain (UBD) that is retained in most of the c-IAP1-MALT1 fusion proteins from MALT lymphomas (Gyrd-Hansen et al. 2008) (Fig. 2A). Mutations that prevent ubiquitin binding by this domain attenuate NF-κB activation by the c-IAP1-MALT1 fusion protein. As mentioned earlier, lysine-63 ubiquitination of IKKγ is required for IKK activation by c-IAP2-MALT1. Notably, the c-IAP2 UBD interacts with ubiquitinated IKKγ (Gyrd-Hansen et al. 2008).

The paracaspase domain of MALT1 has recently been shown to have proteolytic activity directly toward arginine residues of its substrates A20 and BCL10 (Coornaert et al. 2008; Rebeaud et al. 2008). A c-IAP2-MALT1 fusion protein that is proteolytically inactive is less efficient in stimulating NF-κB than the corresponding wild-type fusion protein (Coornaert et al. 2008; Uren et al. 2000). Activation of caspases is mediated by induced proximity (Yang et al. 1998), raising the possibility that the MALT1 paracaspase proteolytic activity is induced by c-IAP2-MALT1 oligomerization. Conceivably, the c-IAP2-MALT1 might cleave and inactivate A20, thereby preserving IKKγ ubiquitination and potentiating NF-κB signaling. It is possible that an inhibitor of MALT paracaspase activity (see the following) might promote apoptosis in MALT lymphoma cells, but this has not yet been directly tested.

Figure 2B presents a plausible model for NF-κB activation by c-IAP2-MALT1. Oligomerization of the fusion protein may provide a multivalent platform for the recruitment of the ubiquitin ligases TRAF6 and TRAF2. Although it is unclear how IKK is recruited initially to this complex, once it is ubiquitinated, the ubiquitin binding domain of the fusion protein can help retain IKK in the complex, possibly facilitating its phosphorylation by TAK1. The c-IAP2-MALT1 oligomers may have active proteolytic activity for A20, thereby augmenting the NF-κB signaling output.

MULTIPLE MYELOMA

Multiple myeloma is a malignancy derived from plasmacytic cells dwelling in the bone marrow. Its progression is often slow, but all patients eventually need treatment. Multimodality therapy, including chemotherapy and more targeted agents such as bortezomib, can delay disease progression for years but is not usually curative. The search for mechanisms of therapy resistance has focused on cell survival signals contributed by stromal cells in the bone marrow microenvironment (Dalton 2003; Hideshima et al. 2002). Eventually, myeloma cells may sustain genetic hits that confer independence from the bone marrow microenvironment.

In this scenario, the NF-κB pathway appears to play a pivotal role. Two complementary studies uncovered recurrent genetic alterations that cause constitutive NF-κB signaling in multiple myeloma (Annunziata et al. 2007; Keats et al. 2007) (Fig. 3A). In one study, a substantial fraction of multiple myeloma cell lines were found to be sensitive to a highly specific IKKβ inhibitor, suggesting that the classical NF-κB pathway was active (Annunziata et al. 2007). These cell lines had phosphorylated unstable IκBα, nuclear NF-κB complexes containing p65, and expression of NF-κB target genes, whereas the IKKβ-resistant cell lines did not. The NF-κB target gene signature was evident in over 80% of primary samples of multiple myeloma, and these malignant cells had nuclear p65. Thus, a substantial fraction of multiple myeloma cases engage the classical NF-κB pathway. In addition, myeloma cells with NF-κB pathway activation have nuclear p52 and RelB, which has been taken as evidence for alternative NF-κB pathway activation, although the detailed mechanism responsible has not been clarified (see later discussion).

In part, NF-κB activation in multiple myeloma may be caused by signals from the bone marrow microenvironment (Fig. 3B). In a survey of normal B-cell subpopulations, bone marrow-derived plasma cells had the highest expression of NF-κB target genes (Annunziata et al. 2007). Quite conceivably, NF-κB could be induced in these cells by two TNF family ligands in the bone marrow microenvironment, BAFF and APRIL (Hideshima et al. 2002; Marsters et al. 2000; Moreaux et al. 2005; O'Connor et al. 2004). Indeed, interference with BAFF signaling leads to the disappearance of plasma cells from the bone marrow (O'Connor et al. 2004). Plasma cells express two receptors for BAFF and APRIL highly—BCMA and TACI—and multiple myeloma cells variably retain these receptors (Moreaux et al. 2005). Multiple myeloma cases with high TACI expression have a gene expression signature of mature plasma cells and a more favorable prognosis, whereas those with low TACI expression have a more plasmablastic phenotype and an inferior prognosis (Moreaux et al. 2005). Potentially, the TACI-high myeloma subset may retain their dependence on BAFF and APRIL the bone marrow microenvironment whereas the TACI-low subset may be less dependent on the microenvironment for NF-κB survival signals.

In many myeloma cases, however, a diverse array of genetic aberrations target regulators of NF-κB (Annunziata et al. 2007; Keats et al. 2007) (Fig. 3A). At various frequencies, multiple myeloma cells harbor amplifications and/or translocations of NIK, CD40, lymphotoxin-β receptor, and TACI, as well as deletions, inactivating mutations, or transcriptional silencing of TRAF3, TRAF2, and CYLD. Occasionally, myeloma acquire abnormalities that alter NF-κB transcription factors directly, including massive overexpression of NF-κB p50 and mutations of *NFKB2* that produce truncated NF-κB p100 isoforms that do not need proteolytic processing to enter the nucleus (Annunziata et al. 2007; Keats et al. 2007; Migliazza et al. 1994). Generally, these abnormalities are non-overlapping, however much additional work needs to be performed to determine if additional genetic abnormalities affect NF-κB in this disease.

Many of the aberrations focus attention on the kinase NIK, a known regulator of both classical and alternative NF-κB signaling (Claudio et al. 2002; Coope et al. 2002; Ramakrishnan et al. 2004; Yin et al. 2001). In some cases, myelomas have high-level amplifications of the NIK genomic locus, resulting in overexpression. In

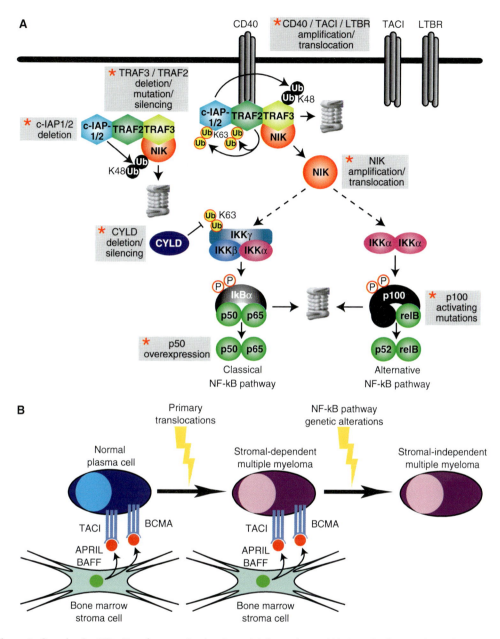

Figure 3. Constitutive NF-κB pathway activation in multiple myeloma. (*A*) Genetic abnormalities that activate NF-κB in multiple myeloma. Gray boxes highlight recurrent genetic aberrations in multiple myeloma involving NF-κB pathway components. The kinase NIK is tethered to the ubiquitin ligases c-IAP1 and/c-IAP2 (c-IAP1/2) by TRAF2 and TRAF3, leading to rapid turnover of NIK protein because of c-IAP1/2-catalyzed K46-linked polyubiquitination. On recruitment to a subset of TNF receptor-family proteins—notably CD40, TACI, and lymphotoxin-β receptor (LTBR)—TRAF2 ubiquitin ligase activity is induced, leading to K63-linked polyubiquitination of c-IAP1/2. The K46-linked ubiquitin ligase activity of c-IAP1/2 is directed toward TRAF3, leading to its proteasomal degradation. NIK is liberated and stabilized in the process. NIK overexpression stimulates both the classical and alternative NF-κB pathways. (*Figure legend continued on the following page.*)

others, NIK is deregulated by translocations, some of which lead to overexpression of NIK because of juxtaposition with strong immunoglobulin enhancer elements. One interesting translocation fuses the *EFTUD* and *NIK* genes, creating a fusion protein that lacks the amino-terminal 86 amino acids of NIK (Annunziata et al. 2007), which are required for TRAF3-mediated NIK degradation (Liao et al. 2004). The resultant fusion protein is highly stable, an observation that raised the possibility that TRAF3 might mediate NIK turnover in myeloma cells.

Indeed, roughly one eighth of myeloma cases sustain inactivating mutations or homozygous deletions of the TRAF3 gene (Annunziata et al. 2007; Keats et al. 2007). These genetic aberrations are scattered through the length of the protein, but include some carboxy-terminal events that affect the MATH domain, which is important for interaction of TRAF3 with NIK and TNF family receptors (Li et al. 2003; Liao et al. 2004; Ni et al. 2004; Ni et al. 2000). In cell lines that have inactivated TRAF3, NIK protein is stabilized. Less commonly, some myelomas have biallelic deletion of a genomic region housing *BIRC2* and *BIRC3*, encoding c-IAP1 and c-IAP2, respectively. Myeloma cell lines with this deletion also have high protein expression of NIK and TRAF3, which provided the first clue that c-IAP1 and c-IAP2 are linked in a biochemical pathway with NIK and TRAF3.

Myelomas also recurrently inactivate CYLD, a negative regulator of NF-κB signaling, either by deletion, mutation, or transcriptional silencing (Annunziata et al. 2007, Keats et al. 2007). CYLD is capable of deubiquitinating several NF-κB regulators, including the IKKγ subunit, TRAF2, TRAF6, and BCL3 (Brummelkamp

et al. 2003; Kovalenko et al. 2003; Massoumi et al. 2006; Regamey et al. 2003; Trompouki et al. 2003), but the functionally relevant target of CYLD in multiple myeloma is not known.

By a number of criteria, these diverse genetic abnormalities activate both the classical and alternative NF-κB pathways in multiple myeloma. Both NIK overexpression and TRAF3 deletion induce IκBα kinase activity in myeloma cells and the nuclear accumulation of p50/p65 (Annunziata et al. 2007). This finding is in keeping with the sensitivity of these cells to an IKKβ inhibitor and shows classical NF-κB activation. However, p52/RelB heterodimers also accumulate in the nucleus as a result of these genetic aberrations, consistent with alternative NF-κB activation (Annunziata et al. 2007; Keats et al. 2007). In normal cells, the alternative NF-κB pathway relies on activated IKKα to phosphorylate p100, leading to its processing by the proteasome into p52 (Senftleben et al. 2001). However, a strong knockdown of IKKα by RNA interference was not toxic to myeloma cells with NIK overexpression, yet knockdown of NIK was toxic. Hence, p100 processing in these cells may depend on other IKKα-independent mechanisms in these myeloma cells.

Recent elegant biochemical studies have integrated these genetic clues from multiple myeloma into a signaling pathway regulating NIK stability (Matsuzawa et al. 2008; Vallabhapurapu et al. 2008; Zarnegar et al. 2008). In most cells, NIK protein is highly unstable, because of proteasomal degradation. This proteasomal targeting requires c-IAP-1 or c-IAP-2 (c-IAP1/2) to add K48-liked polyubiquitin chains to NIK. c-IAP1/2 are brought into complex with NIK via a molecular bridge in which TRAF3 binds to the NIK N-terminus, recruiting

---

Figure 3. (*Continued*) The deubiquitinase CYLD negative regulates this signaling pathway by removing K63-linked polyubiquitin chains from IKKγ. (*B*) Model of NF-κB activation during the genesis of multiple myeloma. Normal plasma cells receive signals from BAFF and APRIL, two TNF family ligands in the bone marrow microenvironment. Their receptors, TACI and BCMA, are highly expressed in multiple myeloma and signal to NF-κB through both the classical and alternative NF-κB pathways. Initial transformation is often caused by oncogenic translocations, but the myeloma cell may still remain dependent on the bone marrow microenvironment to receive prosurvival NF-κB signals. Myeloma cells that acquire mutations that cause constitutive NF-κB pathway activation are selected because they allow the malignant cells to survive and proliferate without being limited by the bone marrow microenvironment.

TRAF2 and c-IAP1/2. On ligand engagement of certain TNF family receptors, notably CD40, the TRAF3/TRAF2/c-IAP1/2 complex is recruited to the receptor, which stimulates K48-linked ubiquitination of TRAF3 by c-IAP1/2 and rapid TRAF3 degradation. Subsequently, NIK is stabilized, presumably because TRAF3 depletion dissociates c-IAP1/2 from NIK. It is not completely clear why c-IAP1/2 ubiquitination of TRAF3 is stimulated by receptor engagement, but TRAF2-mediated K63-linked ubiquitination of c-IAP1/2 may stimulate its activity as an E3 ligase for TRAF3 (Vallabhapurapu et al. 2008).

The fact that the genetic aberrations in myeloma target a pathway regulating NIK stability has led to a suggestion that these aberrations are selected to activate the alternative NF-κB pathway in myeloma cells (Keats et al. 2007). This is a natural conclusion given that in normal cells, NIK stabilization primarily stimulates IKKα-mediated NFKB2 processing into p52, leading to the nuclear accumulation of p52/RelB heterodimers. However, TRAF3 knockout fibroblasts also have elevated classical NF-κB signaling, as indicated by higher IκBα kinase activity and nuclear accumulation of the p65 subunit, possibly as a result of autocrine stimulation of NF-κB by cytokines (Vallabhapurapu et al. 2008; Zarnegar et al. 2008). Moreover, in cells stimulated with lymphotoxin-β, classical NF-κB activation depends on both NIK and IKKα (Zarnegar et al. 2008).

These considerations lead to the view that pathological overexpression of NIK, as occurs in multiple myeloma, may directly activate IKKβ and the classical NF-κB pathway (Annunziata et al. 2007). Indeed, NIK can activate classical NF-κB signaling when overexpressed (Malinin et al. 1997; O'Mahony et al. 2000; Woronicz et al. 1997) and certain TNF family members (CD40 ligand and BAFF) can activate the classical NF-κB pathway through NIK (Ramakrishnan et al. 2004). Another possibility is that alternative NF-κB signaling in myeloma cells causes the synthesis of cytokines that stimulate the classical NF-κB pathway in an autocrine fashion. As mentioned earlier, the role of IKKα in this process in myeloma cells is not supported by knockdown studies (Annunziata et al. 2007), but cannot be ruled out given the role of IKKα in stimulating the classical NF-κB pathway in cells treated with LTβ (Zarnegar et al. 2008). Irrespective of the mechanism, it is important to appreciate that myeloma cells activate the classical and alternative NF-κB pathways in concert and that therapeutic targeting of the classical pathway with IKKβ inhibitors could be developed as a strategy to treat this disease (Annunziata et al. 2007; Hideshima et al. 2002; Hideshima et al. 2006).

NF-κB IN EPITHELIAL CANCERS

A rich literature shows the central role of NF-κB signaling in the inflammatory milieu that fosters many epithelial cancers (Karin 2010; Karin et al. 2006). In colon cancer, NF-κB expression in the premalignant epithelium prevents apoptosis and promotes transformation (Greten et al. 2004). However, it has been less clear that fully formed epithelial cancers depend on NF-κB. Unlike lymphomas, solid tumors often genetically alter receptor tyrosine kinases, which provide prosurvival signals through the PI(3) kinase pathway. Perhaps as a result, genetic lesions in the NF-κB pathway have not been uncovered in most instances.

CYLD as a Tumor Suppressor

Familial cylindromatosis is a malignancy of the hair follicle stem cell in which CYLD functions as a classic tumor suppressor (reviewed in Sun 2009). Individuals with this syndrome inherit one loss-of-function CYLD allele and the other CYLD allele is somatically inactivated in the cylindromas. A related disorder, known as familial skin appendage tumors, also involves CYLD inactivation. The CYLD gene was positionally cloned and had no known function until three groups showed that it negatively regulated the NF-κB pathway (Brummelkamp et al. 2003; Kovalenko et al. 2003; Trompouki et al. 2003). CYLD can deubiquitinate a wide range of cellular proteins, including the NF-κB regulators TRAF2, TRAF6, TRAF7, BCL3, TAK1, and the IKKγ subunit (Sun 2009). CYLD may function as a tumor suppressor by negatively regulating classical NF-κB pathway because ubiquitination of the IKKγ is required for IKK activation. The importance of NF-κB signaling in familial cylindromatosis is highlighted by a recent clinical trial in which salicylic acid, an NF-κB inhibitor, was used topically to treat the cylindromas (Kovalenko et al. 2003). Although the study was small and not blinded, several cylindromas in one patient regressed completely with treatment.

Unconventional IKKs and Epithelial Cancers

Recent experiments suggest that two unconventional members of the IKK family, IKKε and TBK1, activate NF-κB in epithelial cancers. In the innate immune response to viruses, Toll receptors and cytosolic receptors for nucleic acids activate these unconventional IKKs, thereby engaging NF-κB and type-I interferon pathways (Clement et al. 2008; Hacker et al. 2006; Pichlmair et al. 2007). The mechanism by which these kinases regulate the interferon pathway is by phosphorylating IRF3 and IRF7, causing their nuclear translocation, but their mode of action in the NF-κB pathway is less well understood. One mechanism may be induction of TNFα by IRF3, thereby activating the classical NF-κB pathway in a feed-forward loop (Covert et al. 2005; Werner et al. 2005). It is also possible that the unconventional IKKs may play a more direct role in IκBα turnover. IKKε is part of a multiprotein complex that can target IκBα for degradation (Peters et al. 2000; Shimada et al. 1999). Although IKKε and TBK1 can promote IκBα degradation, they can only phosphorylate serine 36 of IκBα (Peters et al. 2000; Pomerantz et al. 1999; Tojima et al. 2000). Hence, another kinase must cooperate with these unconventional IKKs to phosphorylate serine 32 of IκBα, allowing it to be targeted for degradation (Brown et al. 1995). The unconventional IKKs have several other substrates, including the NF-κB subunits p65 and c-rel, which might contribute to their ability to activate NF-κB target genes (reviewed in Clement et al. 2008).

Functional genomics experiments nominated IKKε as an oncogene in breast cancer (Boehm et al. 2007) (Fig. 4A). In a screen of activated kinase alleles, activated IKKε took the place of activated AKT in a transformation assay using human embryonic kidney cells. In a complementary approach, roughly one sixth of breast cancer cases were found to harbor a gain or high-level amplification of a region on chromosome 1 encoding IKKε. Generally, the region of copy number gain spans many megabases, suggesting that several genes in this interval may play a role in breast cancer pathogenesis. However, IKKε appears to be one important actor because knockdown of IKKε was toxic to breast cancer cell lines harboring the IKKε amplicon. This finding is in keeping with the observation that a dominant negative IKKε isoform was toxic to breast cancer cell lines (Eddy et al. 2005). In breast cancer biopsies, IKKε expression was correlated with an increase in nuclear c-rel, suggesting activation of the NF-κB pathway may participate in the pathogenesis of these malignancies (Boehm et al. 2007). This finding is in keeping with previous work showing overexpression of IKKε in both human and mouse breast cancer in association with NF-κB activation (Eddy et al. 2005).

In cell transformation experiments, the activated IKKε allele promoted IκBα degradation and the induction of NF-κB target genes (Boehm et al. 2007), but the exact mechanism by which IKKε achieves NF-κB activation has not been elucidated. A newly defined substrate of IKKε is the deubiquitinase CYLD (Peters et al. 2000; Shimada et al. 1999). Phosphorylation of serine 418 of CYLD diminishes its deubiquitinase action toward IKKγ and TRAF2. Thus, IKKε overexpression may relieve the negative influence of CYLD on the NF-κB pathway. Nonetheless, IKK requires phosphorylation of the IKKβ subunit in its activation loop in addition to IKKγ ubiquitination to become a functioning kinase, and it is currently unclear how IKKε overexpression promotes these posttranscriptional modifications.

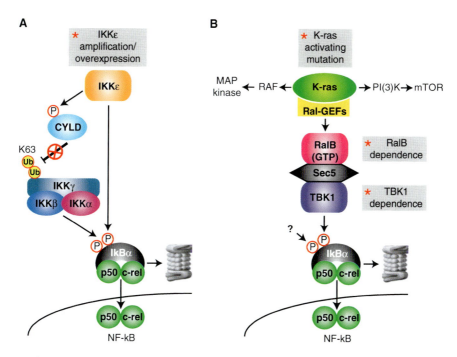

Figure 4. Solid tumors activate the NF-κB pathway to maintain cell survival. (*A*) IKKε is frequently overexpressed in breast cancer in association with a gain or amplification of its genomic locus. IKKε phosphorylates a serine residue in CYLD, thereby lowering its deubiquitinase activity toward the IKKγ subunit, leading to a more active IKK enzyme. In addition, IKKε can directly phosphorylate serine-46 of IκBα. (*B*) Regulation of NF-κB by TBK1 downstream of mutant K-ras in lung cancer. Mutant K-ras signals to several downstream pathways, including the MAP kinase, PI(3) kinase, and RAL GTPase pathways. K-ras-associated RAL guanine nucleotide exchange factors (Ral-GEFs) promote the active, GTP-bound state of the RAL proteins. RalB interacts with Sec5, which in turn recruits TBK1, causing kinase activation. TBK1 triggers classical NF-κB pathway activation, as judged by the accumulation of nuclear p50/c-rel heterodimers. Because TBK1 can only phosphorylate one serine in IκBα, an as yet unknown kinase must cooperate with TBK1 to achieve dual IκBα phosphorylation.

Two complementary experimental approaches have revealed a connection between oncogenic signaling by mutant K-ras and NF-κB (Barbie et al. 2009; Meylan et al. 2009) (Fig. 4B). In a mouse model of lung cancer caused by mutant K-ras overexpression and p53 deletion, the classical NF-κB pathway is activated (Meylan et al. 2009). Lung cancer cell lines from these mice were killed when NF-κB was inhibited, and in vivo delivery of an IκBα super-repressor slowed growth of established lung tumors.

Mutant K-ras engages several downstream effector pathways by stimulating RAF and PI(3) kinases. Additional effectors of mutant K-ras are the RAL GTPases (RALA and RALB), which are converted into the active, GTP-bound state by K-ras-associated RAL guanine nucleotide exchange factors (RalGDS, Rgl, and Rlf). Recent experiments have implicated the unconventional IKK TBK1 in a signaling pathway that links mutant K-ras and NF-κB. TBK1 is activated by RALB in concert with SEC5, a component of the exocyst, and is required for K-ras-mediated transformation of mouse embryo fibroblasts (Chien et al. 2006). A synthetic lethal RNA interference genetic screen revealed that TBK1 shRNAs kill mutant K-ras-transformed cells but not cancer cells with wild-type K-ras (Barbie et al. 2009). TBK1 knockdown decreases viability of human lung cancer cell lines with mutant K-ras as well as lung epithelial cells acutely transformed with mutant K-ras. Knockdown of RALB, but not c-RAF, B-RAF, or AKT, is also synthetically lethal for mutant K-ras-transformed cells, suggesting a link between RALB and TBK1. Overexpression of mutant K-ras activates a set of K-ras signature genes, which includes a number of known targets of NF-κB signaling. Knockdown of TBK1 blocks expression of these NF-κB target genes in mutant K-ras-transformed cells and blocks nuclear accumulation of c-rel. Knockdown of c-rel or expression of an IκBα super-repressor is lethal to lung cancer cells with mutant K-ras, directly demonstrating the requirement for NF-κB signaling to block cell death. Notably, roughly three quarters of human lung adenocarcinomas with mutant K-ras coexpress the K-ras and NF-κB target gene signatures. Among lung adenocarcinomas with wild-type K-ras, one quarter of cases coexpress the K-ras and NF-κB signatures, suggesting that these cancers may have upstream signals that activate wild-type K-ras and engage TBK1 to activate NF-κB. Alternatively, these tumors may be an independent mechanism to activate NF-κB.

THERAPY OF NF-κB-DRIVEN CANCERS

Many prospects and hurdles can be envisaged when considering pharmacological interference with NF-κB signaling (reviewed in Baud et al. 2009; Karin et al. 2004). Inhibitors of IKKβ could have activity against ABC DLBCL, PMBL, MALT lymphoma, and multiple myeloma. However, long-term, full inhibition of IKKβ would be expected to impair the adaptive and innate immune system (Vallabhapurapu et al. 2009). Because IKKβ counteracts IL-1β processing and secretion (Greten et al. 2007), IKKβ inhibition might release IL-1β, causing febrile reactions and possibly inflammation. During development, IKKβ and p65 deficiency provokes massive apoptosis of hepatocytes, but the role of NF-κB in adult hepatocytes has not been fully resolved (reviewed in Schwabe et al. 2007). Conditional deficiency of IKKβ in adult

hepatocytes causes little if any sensitivity to TNFα-dependent apoptosis (Luedde et al. 2005). However, deficiency in IKKγ or p65 makes adult hepatocytes markedly more sensitive to TNFα and other injuries to the liver (Beraza et al. 2007; Geisler et al. 2007). Moreover, conditional deficiency in IKKγ caused hepatic steatohepatitis and hepatocellular carcinoma (Luedde et al. 2007). Whether or not liver injury occurs may therefore depend on the extent and/or nature of NF-κB pathway inhibition as well as on the presence of concurrent infections that might elevate proapoptotic inflammatory stimuli such as TNFα.

Despite these cautions, IKKβ-directed therapy might find a role in cancer treatment if cancer cells are unusually addicted to NF-κB signaling and the right degree of NF-κB inhibition can be achieved. Not only might IKKβ inhibitors kill NF-κB-dependent cancer cells directly, but they could potentiate the apoptotic response to cytotoxic chemotherapy (Baldwin 2001; Wang et al. 1998). Many pharmaceutical companies have developed IKKβ inhibitors but none have entered into clinical trials in cancer, presumably because of the concerns expressed earlier. An indirect way to block IKK activity is by inhibiting HSP90. HSP90 is an intrinsic component of the IKK supramolecular complex, and its inhibition blocks the induction of IKK enzymatic activity by various stimuli as well as the biogenesis of IKKβ and IKKα (Broemer et al. 2004; Chen et al. 2002). HSP90 inhibitors are in clinical trials for a number of cancers (Workman et al. 2007) and might show particular activity for IKK-dependent tumors. In fact, the HSP90 inhibitor geldanimycin kills Hodgkin lymphoma lines with constitutive NF-κB activation (Broemer et al. 2004) as well as ABC DLBCL cells (unpubl. data).

Inhibitors of IKKα might potentiate the effects of IKKβ inhibition because IKKα can serve as an IκBα kinase, particularly under conditions of IKKβ inhibition (Lam et al. 2008). In an RNA interference genetic screen in ABC DLBCL, shRNAs targeting IKKα synergized with an IKKβ inhibitor in killing the lymphoma cells. However, knockdown of IKKα

Cite this article as *Cold Spring Harb Perspect Biol* 2010;00:a000109

without IKKβ inhibition had no effect. Similar "compensatory" IKKα activity was observed in cells in which NF-κB was induced by PMA/ionomycin treatment, CARD11 pathway activation, or TNFα treatment. Because IKKα is in a macromolecular complex with IKKβ, it is conceivable that IKKα may be activated by the same upstream kinases as IKKβ, such as TAK1. Recombinant IKKβ is more active than IKKα as an IκBα kinase (Huynh et al. 2000). Hence, in an IKK complex containing both subunits, IKKβ would mediate most of the IκBα phosphorylation. But when IKKβ is pharmacologically inhibited, IKKα may take over this function, presumably relying on the high local concentration of IκBα.

Given the role of the proteasome in degrading IκBα, the proteasome inhibitor bortezomib might be harnessed to inactivate NF-κB in cancer. All cells require proteasome function and full inhibition of the proteasome is lethal to many cell types. Nonetheless, bortezomib was successfully developed as a cancer agent using a novel approach in which the dose-escalation end-point was not systemic toxicity but rather the degree of proteasome inhibition in vivo (Adams 2004).

Based on the constitutive activation of the NF-κB pathway in ABC DLBCL (Davis et al. 2001), a clinical trial was initiated in relapsed and refractory DLBCL that combined bortezomib with EPOCH, a type of multiagent cytotoxic chemotherapy (Dunleavy et al. 2009). All patients had tumor biopsies to allow the molecular diagnosis of ABC versus GCB DLBCL to be made. Initially, bortezomib was given as a single agent, but no responses were seen. However, the combination of bortezomib with EPOCH chemotherapy yielded complete and partial responses, and the vast majority of these were in patients with ABC DLBCL. This response difference translated into a significantly superior overall survival rate for patients with ABC DLBCL. This was notable because in all previous trials with chemotherapy alone, patients with ABC DLBCL had a decidedly inferior cure rate (Alizadeh et al. 2000; Hummel et al. 2006; Lenz et al. 2008; Monti et al. 2005; Rosenwald et al. 2002).

Bortezomib has been approved for the treatment of multiple myeloma because it delays disease progression (Richardson et al. 2005), but it is unclear whether its efficacy in this disease relates to NF-κB inhibition. In a trial comparing bortezomib and dexamethasone, myeloma cases with low TRAF3 expression had an increased response to bortezomib than those with high TRAF3 (Keats et al. 2007). However, patients stratified according to high or low expression of NF-κB target genes did not differ in their response to bortezomib (Keats et al. 2007), suggesting that this issue needs to be readdressed in a prospective fashion.

An intriguing new way to attack the NF-κB pathway is by interfering with the βTrCP ubiquitin ligase that targets IκBα for destruction. This SCF-type E3 ubiquitin ligase consists of a cullin subunit that serves as a scaffold to assemble SKP1, the RING finger E3 ubiquitin ligase ROC1, and the Fbox protein βTrCP, which is the receptor that recognizes phosphorylated IκBα (Ben-Neriah 2002; Yaron et al. 1998). The cullin subunit requires modification by a NEDD8 moiety for activity of the complex as an E3 ubiquitin ligase (Chiba et al. 2004; Podust et al. 2000; Read et al. 2000). NEDD8 is added by the concerted action of a series of enzymes that functionally mirror the ubiquitin conjugation machinery. The E1 enzyme in this system, termed NEDD8 activating enzyme (NAE), can be inhibited by the small molecule MLN4924 (Soucy et al. 2009). MLN4924 potently inhibits the βTrCP ubiquitin ligase by interfering with its neddylation, thereby causing the protein levels of βTrCP substrates to rise. One substrate CDT1, causes DNA re-replication during S phase, which triggers the S-phase DNA replication checkpoint to initiate apoptosis in cancer cells. MLN4924 is also expected to inhibit NF-κB by preventing IκBα degradation, making it an attractive candidate for clinical trials in cancers characterized by constitutive NF-κB activation.

Given the inherent risks of inhibiting NF-κB in all cell types, strategies that target specific upstream pathways leading to NF-κB in cancer are attractive. The CBM complex is a compelling target in this regard because mice with deficiency in CARD11 or MALT1 only have defects in the adaptive immune system (Thome 2004). Recently, the proteolytic activity of the MALT1 paracaspase domain has been directly shown and two of its proteolytic targets have been identified, BCL10 and A20 (Coornaert et al. 2008; Rebeaud et al. 2008). Unlike conventional caspases that cleave after aspartic acid residues, the paracaspase domain of MALT1 cleaves after arginine. Overexpression of a proteolytically inactive form of MALT1 is less active than wild-type MALT1 in enhancing IL-2 secretion by activated T cells and in activating NF-κB (Lucas et al. 2001; Rebeaud et al. 2008). Likewise, inhibition of MALT1 proteolytic activity with a cell-permeable peptide mimetic reduces NF-κB activation and IL-2 secretion in T cells stimulated through the antigen receptor (Rebeaud et al. 2008). However, the effect of MALT1 proteolytic activity on NF-κB signaling is quantitative rather than complete, suggesting that the proteolytic activity of MALT1 modulates a regulatory circuit engaged during NF-κB activation. Indeed, MALT1 cleavage inactivates A20, a negative regulator of NF-κB signaling (Coornaert et al. 2008).

Encouragingly, two studies have shown that the peptide inhibitors of MALT1 block NF-κB activity and kills ABC DLBCL cell lines but are not toxic for the NF-κB-independent GCB DLBCL cell lines (Ferch et al. 2009; Hailfinger et al. 2009). ABC DLBCL cell lines with wild-type or mutant forms of CARD11 are equally sensitive to MALT1 protease inhibition, in keeping with the dependence of both types of ABC DLBCL on MALT1 (Ngo et al. 2006). These studies pave the way for the development of small molecule peptidomimetic inhibitors of MALT1 for the therapy of ABC DLBCL, and possibly also MALT lymphomas with the c-IAP2-MALT1 fusion oncoprotein.

Chronic active BCR signaling in ABC DLBCL offers a wealth of therapeutic targets upstream of IKK (Davis et al. 2010). BTK kinase activity couples chronic active BCR signaling to NF-κB via IKK activation. Accordingly, a small molecule inhibitor of BTK, PCI-32765 (Pan et al. 2007), kills ABC DBLCL cells with chronic

active BCR signaling but not other lymphoma lines (Davis et al. 2010). Dasatinib is a multi-kinase inhibitor that has especially high activity against SRC-family tyrosine kinases and BTK (Hantschel et al. 2007). This drug extinguished signaling to the NF-κB, AKT and ERK MAP kinase pathways and killed ABC DLBCL cells with chronic active BCR signaling (Davis et al. 2010). Clinical trials based on this observation can begin immediately because dasatinib is already approved as a second-line treatment of chronic myelogenous leukemia that is refractory to imatinib. Other kinases in the BCR pathway leading to NF-κB might also serve as therapeutic targets, such as PKCβ, which phosphorylates and activates wild-type CARD11 in ABC DLBCL. The tyrosine kinase SYK, although essential for normal BCR signaling, is only required by some ABC DLBCLs with chronic active BCR signaling, suggesting that drugs targeting this kinase may have a more restricted efficacy (Davis et al. 2010). By targeting BCR signaling in ABC DLBCL, it should be possible to achieve full inhibition of NF-κB in malignant (and normal) B cells while sparing other normal cells, providing an acceptable therapeutic window.

In multiple myeloma, an intriguing therapeutic target is the kinase NIK, because many genetic alterations in this cancer stabilize NIK protein expression, causing a critical dependence on this kinase (Annunziata et al. 2007; Keats et al. 2007). Although NIK-deficient mice have defects in the organization of secondary lymphoid organs (Miyawaki et al. 1994; Shinkura et al. 1999), this defect might not occur with short-term inhibition of NIK pharmacologically. In response to viral challenge, NIK-deficient B cells are impaired in isotype-switching, but NIK-deficient T cells are relatively unaffected (Karrer et al. 2000), suggesting that pharmacological inhibition of this kinase might only modestly impair adaptive immune responses. Some multiple myeloma cases may activate NIK and NF-κB in a quasi-physiological mechanism through BAFF and/or APRIL signaling in the bone marrow microenvonment. Disruption of this signaling by a TACI-Fc fusion protein should be tested for activity in this setting (Moreaux et al. 2005; Yaccoby et al. 2008).

As should be clear from this review, the wise deployment of NF-κB-based therapies in cancer will require molecular profiling to determine the subtype of cancer, the presence of genetic lesions in NF-κB regulators, and the activity of the NF-κB pathway. In the era of targeted cancer therapy, it is important to evaluate whether a targeted agent hits its mark. In trials of NF-κB inhibitors, this could be easily achieved by measuring the expression of NF-κB target genes before and during therapy, potentially by real-time RT-PCR or other quantitative methods. By this cautious and reasoned approach, I hope that NF-κB pathway inhibitors will find their way into the anticancer armamentarium.

ACKNOWLEDGMENTS

Supported by the Intramural Research Program of the National Institutes of Health, National Cancer Institute, Center for Cancer Research.

REFERENCES

Adams J. 2004. The proteasome: a suitable antineoplastic target. *Nat Rev Cancer* **4:** 349–360.

Akagi T, Motegi M, Tamura A, Suzuki R, Hosokawa Y, Suzuki H, Ota H, Nakamura S, Morishima Y, Taniwaki M, et al. 1999. A novel gene, MALT1 at 18q21, is involved in t(11;18) (q21;q21) found in low-grade B-cell lymphoma of mucosa-associated lymphoid tissue. *Oncogene* **18:** 5785–5794.

Alizadeh AA, Eisen MB, Davis RE, Ma C, Lossos IS, Rosenwald A, Boldrick JC, Sabet H, Tran T, Yu X, et al. 2000. Distinct types of diffuse large B-cell lymphoma identified by gene expression profiling. *Nature* **403:** 503–511.

Annunziata CM, Davis RE, Demchenko Y, Bellamy W, Gabrea A, Zhan F, Lenz G, Hanamura I, Wright G, Xiao W, et al. 2007. Frequent engagement of the classical and alternative NF-kappaB pathways by diverse genetic abnormalities in multiple myeloma. *Cancer Cell* **12:** 115–130.

Baens M, Fevery S, Sagaert X, Noels H, Hagens S, Broeckx V, Billiau AD, De Wolf-Peeters C, Marynen P. 2006. Selective expansion of marginal zone B cells in Emicro-API2-MALT1 mice is linked to enhanced IkappaB kinase gamma polyubiquitination. *Cancer Res* **66:** 5270–5277.

Bajpai UD, Zhang K, Teutsch M, Sen R, Wortis HH. 2000. Bruton's tyrosine kinase links the B cell receptor to nuclear factor kappaB activation. *J Exp Med* **191:** 1735–1744.

Baldwin AS. 2001. Control of oncogenesis and cancer therapy resistance by the transcription factor NF-kappaB. *J Clin Invest* **107:** 241–246.

Barbie DA, Tamayo P, Boehm JS, Kim SY, Moody SE, Dunn IF, Schinzel AC, Sandy P, Meylan E, Scholl C, et al. 2009. Systematic RNA interference reveals that oncogenic KRAS-driven cancers require TBK1. *Nature* **462:** 108–112.

Bargou RC, Emmerich F, Krappmann D, Bommert K, Mapara MY, Arnold W, Royer HD, Grinstein E, Greiner A, Scheidereit C, et al. 1997. Constitutive nuclear factor-kappaB-RelA activation is required for proliferation and survival of Hodgkin's disease tumor cells. *J Clin Invest* **100:** 2961–2969.

Bargou RC, Leng C, Krappmann D, Emmerich F, Mapara MY, Bommert K, Royer HD, Scheidereit C, Dorken B. 1996. High-level nuclear NF-kappa B and Oct-2 is a common feature of cultured Hodgkin/Reed-Sternberg cells. *Blood* **87:** 4340–4347.

Basso K, Klein U, Niu H, Stolovitzky GA, Tu Y, Califano A, Cattoretti G, Dalla-Favera R. 2004. Tracking CD40 signaling during germinal center development. *Blood* **104:** 4088–4096.

Baud V, Karin M. 2009. Is NF-kappaB a good target for cancer therapy? Hopes and pitfalls. *Nat Rev Drug Discov* **8:** 33–40.

Ben-Neriah Y. 2002. Regulatory functions of ubiquitination in the immune system. *Nat Immunol* **3:** 20–26.

Bende RJ, Aarts WM, Riedl RG, de Jong D, Pals ST, van Noesel CJ. 2005. Among B cell non-Hodgkin's lymphomas, MALT lymphomas express a unique antibody repertoire with frequent rheumatoid factor reactivity. *J Exp Med* **201:** 1229–1241.

Beraza N, Ludde T, Assmus U, Roskams T, Vander Borght S, Trautwein C. 2007. Hepatocyte-specific IKK gamma/NEMO expression determines the degree of liver injury. *Gastroenterology* **132:** 2504–2517.

Bidere N, Ngo VN, Lee J, Collins C, Zheng L, Wan F, Davis RE, Lenz G, Anderson DE, Arnoult D, et al. 2009. Casein kinase 1alpha governs antigen-receptor-induced NF-kappaB activation and human lymphoma cell survival. *Nature* **458:** 92–96.

Blonska M, Lin X. 2009. CARMA1-mediated NF-kappaB and JNK activation in lymphocytes. *Immunol Rev* **228:** 199–211.

Boehm JS, Zhao JJ, Yao J, Kim SY, Firestein R, Dunn IF, Sjostrom SK, Garraway LA, Weremowicz S, Richardson AL, et al. 2007. Integrative genomic approaches identify IKBKE as a breast cancer oncogene. *Cell* **129:** 1065–1079.

Boone DL, Turer EE, Lee EG, Ahmad RC, Wheeler MT, Tsui C, Hurley P, Chien M, Chai S, Hitotsumatsu O, et al. 2004. The ubiquitin-modifying enzyme A20 is required for termination of Toll-like receptor responses. *Nat Immunol* **5:** 1052–1060.

Broemer M, Krappmann D, Scheidereit C. 2004. Requirement of Hsp90 activity for IkappaB kinase (IKK) biosynthesis and for constitutive and inducible IKK and NF-kappaB activation. *Oncogene* **23:** 5378–5386.

Brown K, Gerstberger S, Carlson L, Franzoso G, Siebenlist U. 1995. Control of I kappa B-alpha proteolysis by site-specific, signal-induced phosphorylation. *Science* **267:** 1485–1488.

Brummelkamp TR, Nijman SM, Dirac AM, Bernards R. 2003. Loss of the cylindromatosis tumour suppressor inhibits apoptosis by activating NF-kappaB. *Nature* **424:** 797–801.

Brune V, Tiacci E, Pfeil I, Doring C, Eckerle S, van Noesel CJ, Klapper W, Falini B, von Heydebreck A, Metzler D, et al. 2008. Origin and pathogenesis of nodular lymphocyte-predominant Hodgkin lymphoma as revealed by global gene expression analysis. *J Exp Med* **205:** 2251–2268.

Cabannes E, Khan G, Aillet F, Jarrett RF, Hay RT. 1999. Mutations in the IkBa gene in Hodgkin's disease suggest a tumour suppressor role for IkappaBalpha. *Oncogene* **18:** 3063–3070.

Carbone A, Gloghini A, Gruss HJ, Pinto A. 1995. CD40 ligand is constitutively expressed in a subset of T cell lymphomas and on the microenvironmental reactive T cells of follicular lymphomas and Hodgkin's disease. *Am J Pathol* **147:** 912–922.

Chan VW, Meng F, Soriano P, DeFranco AL, Lowell CA. 1997. Characterization of the B lymphocyte populations in Lyn-deficient mice and the role of Lyn in signal initiation and down-regulation. *Immunity* **7:** 69–81.

Chen G, Cao P, Goeddel DV. 2002. TNF-induced recruitment and activation of the IKK complex require Cdc37 and Hsp90. *Mol Cell* **9:** 401–410.

Chen ZJ. 2005. Ubiquitin signalling in the NF-kappaB pathway. *Nat Cell Biol* **7:** 758–765.

Chiba T, Tanaka K. 2004. Cullin-based ubiquitin ligase and its control by NEDD8-conjugating system. *Curr Protein Pept Sci* **5:** 177–184.

Chien Y, Kim S, Bumeister R, Loo YM, Kwon SW, Johnson CL, Balakireva MG, Romeo Y, Kopelovich L, Gale M Jr., et al. 2006. RalB GTPase-mediated activation of the IkappaB family kinase TBK1 couples innate immune signaling to tumor cell survival. *Cell* **127:** 157–170.

Claudio E, Brown K, Park S, Wang H, Siebenlist U. 2002. BAFF-induced NEMO-independent processing of NF-kappa B2 in maturing B cells. *Nat Immunol* **3:** 958–965.

Clement JF, Meloche S, Servant MJ. 2008. The IKK-related kinases: from innate immunity to oncogenesis. *Cell Res* **18:** 889–899.

Compagno M, Lim WK, Grunn A, Nandula SV, Brahmachary M, Shen Q, Bertoni F, Ponzoni M, Scandurra M, Califano A, et al. 2009. Mutations of multiple genes cause deregulation of NF-kappaB in diffuse large B-cell lymphoma. *Nature* **459:** 717–721.

Coope HJ, Atkinson PG, Huhse B, Belich M, Janzen J, Holman MJ, Klaus GG, Johnston LH, Ley SC. 2002. CD40 regulates the processing of NF-kappaB2 pp100 to p52. *Embo J* **21:** 5375–5385.

Coornaert B, Baens M, Heyninck K, Bekaert T, Haegman M, Staal J, Sun L, Chen ZJ, Marynen P, Beyaert R. 2008. T cell antigen receptor stimulation induces MALT1 paracaspase-mediated cleavage of the NF-kappaB inhibitor A20. *Nat Immunol* **9:** 263–271.

Copie-Bergman C, Plonquet A, Alonso MA, Boulland ML, Marquet J, Divine M, Moller P, Leroy K, Gaulard P. 2002. MAL expression in lymphoid cells: further evidence for MAL as a distinct molecular marker of primary mediastinal large B-cell lymphomas. *Mod Pathol* **15:** 1172–1180.

Covert MW, Leung TH, Gaston JE, Baltimore D. 2005. Achieving stability of lipopolysaccharide-induced NF-kappaB activation. *Science* **309:** 1854–1857.

Dalton WS. 2003. The tumor microenvironment: focus on myeloma. *Cancer Treat Rev* **29:** 11–19.

Davis RE, Brown KD, Siebenlist U, Staudt LM. 2001. Constitutive nuclear factor kappa B activity is required for survival of activated B Cell-like diffuse large B cell lymphoma cells. *J Exp Med* **194:** 1861–1874.

Davis RE, Ngo VN, Lenz G, Tolar P, Young RM, Romesser PB, Kohlhammer H, Lamy L, Zhao H, Yang Y, et al. 2010. Chronic active B-cell-receptor signalling in diffuse large B-cell lymphoma. *Nature* **463:** 88–92.

Dierlamm J, Baens M, Wlodarska I, Stefanova-Ouzounova M, Hernandez JM, Hossfeld DK, De Wolf-Peeters C, Hagemeijer A, Van den Berghe H, Marynen P. 1999. The apoptosis inhibitor gene API2 and a novel 18q gene, MLT, are recurrently rearranged in the t(11;18) (q21;q21)p6ssociated with mucosa- associated lymphoid tissue lymphomas. *Blood* **93:** 3601–3609.

Ding BB, Yu JJ, Yu RY, Mendez LM, Shaknovich R, Zhang Y, Cattoretti G, Ye BH. 2008. Constitutively activated STAT3 promotes cell proliferation and survival in the activated B-cell subtype of diffuse large B-cell lymphomas. *Blood* **111:** 1515–1523.

Dunleavy K, Pittaluga S, Czuczman MS, Dave SS, Wright G, Grant N, Shovlin M, Jaffe ES, Janik JE, Staudt LM, et al. 2009. Differential efficacy of bortezomib plus chemotherapy within molecular subtypes of diffuse large B-cell lymphoma. *Blood* **113:** 6069–6076.

Eddy SF, Guo S, Demicco EG, Romieu-Mourez R, Landesman-Bollag E, Seldin DC, Sonenshein GE. 2005. Inducible IkappaB kinase/IkappaB kinase epsilon expression is induced by CK2 and promotes aberrant nuclear factor-kappaB activation in breast cancer cells. *Cancer Res* **65:** 11375–11383.

Emmerich F, Meiser M, Hummel M, Demel G, Foss HD, Jundt F, Mathas S, Krappmann D, Scheidereit C, Stein H, et al. 1999. Overexpression of I kappa B alpha without inhibition of NF-kappaB activity and mutations in the I kappa B alpha gene in reed-sternberg cells. *Blood* **94:** 3129–3134.

Emmerich F, Theurich S, Hummel M, Haeffker A, Vry MS, Dohner K, Bommert K, Stein H, Dorken B. 2003. Inactivating I kappa B epsilon mutations in Hodgkin/Reed-Sternberg cells. *J Pathol* **201:** 413–420.

Ferch U, Kloo B, Gewies A, Pfander V, Duwel M, Peschel C, Krappmann D, Ruland J. 2009. Inhibition of MALT1 protease activity is selectively toxic for activated B cell-like diffuse large B cell lymphoma cells. *J Exp Med* **206:** 2313–2320.

Ferreri AJ, Zucca E. Marginal-zone lymphoma. 2007. *Crit Rev Oncol Hematol* **63:** 245–256.

Feuerhake F, Kutok JL, Monti S, Chen W, LaCasce AS, Cattoretti G, Kurtin P, Pinkus GS, de Leval L, Harris NL, et al. 2005. NFkappaB activity, function, and target-gene signatures in primary mediastinal large B-cell lymphoma and diffuse large B-cell lymphoma subtypes. *Blood* **106:** 1392–1399.

Funke L, Dakoji S, Bredt DS. 2005. Membrane-associated guanylate kinases regulate adhesion and plasticity at cell junctions. *Annu Rev Biochem* **74:** 219–245.

Cite this article as *Cold Spring Harb Perspect Biol* 2010;00:a000109

Garrison JB, Samuel T, Reed JC. 2009. TRAF 2-binding BIR 1 domain of c-IAP 2/MALT 1 fusion protein is essential for activation of NF-kappaB. *Oncogene* **28:** 1584–1593.

Gauld SB, Cambier JC. 2004. Src-family kinases in B-cell development and signaling. *Oncogene* **23:** 8001–8006.

Gazumyan A, Reichlin A, Nussenzweig MC. 2006. Ig beta tyrosine residues contribute to the control of B cell receptor signaling by regulating receptor internalization. *J Exp Med* **203:** 1785–1794.

Geisler F, Algul H, Paxian S, Schmid RM. 2007. Genetic inactivation of RelA/p65 sensitizes adult mouse hepatocytes to TNF-induced apoptosis in vivo and in vitro. *Gastroenterology* **132:** 2489–2503.

Greten FR, Arkan MC, Bollrath J, Hsu LC, Goode J, Miething C, Goktuna SI, Neuenhahn M, Fierer J, Paxian S, et al. 2007. NF-kappaB is a negative regulator of IL-1beta secretion as revealed by genetic and pharmacological inhibition of IKKbeta. *Cell* **130:** 918–931.

Greten FR, Eckmann L, Greten TF, Park JM, Li ZW, Egan LJ, Kagnoff MF, Karin M. 2004. IKKbeta links inflammation and tumorigenesis in a mouse model of colitis-associated cancer. *Cell* **118:** 285–296.

Gruss HJ, Hirschstein D, Wright B, Ulrich D, Caligiuri MA, Barcos M, Strockbine L, Armitage RJ, Dower SK. 1994. Expression and function of CD40 on Hodgkin and Reed-Sternberg cells and the possible relevance for Hodgkin's disease. *Blood* **84:** 2305–2314.

Gyrd-Hansen M, Darding M, Miasari M, Santoro MM, Zender L, Xue W, Tenev T, da Fonseca PC, Zvelebil M, Bujnicki JM, et al. 2008. IAPs contain an evolutionarily conserved ubiquitin-binding domain that regulates NF-kappaB as well as cell survival and oncogenesis. *Nat Cell Biol* **10:** 1309–1317.

Hacker H, Karin M. 2006. Regulation and function of IKK and IKK-related kinases. *Sci STKE* **2006:** pre13.

Hailfinger S, Lenz G, Ngo V, Posvitz-Fejfar A, Rebeaud F, Guzzardi M, Penas EM, Dierlamm J, Chan WC, Staudt LM, et al. 2009. Essential role of MALT1 protease activity in activated B cell-like diffuse large B-cell lymphoma. *Proc Natl Acad Sci USA* **106:** 19946–19951.

Han S, Hathcock K, Zheng B, Kepler TB, Hodes R, Kelsoe G. 1995. Cellular interaction in germinal centers. Roles of CD40 ligand and B7-2 in established germinal centers. *J Immunol* **155:** 556–567.

Hantschel O, Rix U, Schmidt U, Burckstummer T, Kneidinger M, Schutze G, Colinge J, Bennett KL, Ellmeier W, Valent P, et al. 2007. The Btk tyrosine kinase is a major target of the Bcr-Abl inhibitor dasatinib. *Proc Natl Acad Sci USA* **104:** 13283–13288.

Hideshima T, Anderson KC. 2002. Molecular mechanisms of novel therapeutic approaches for multiple myeloma. *Nat Rev Cancer* **2:** 927–937.

Hideshima T, Chauhan D, Richardson P, Mitsiades C, Mitsiades N, Hayashi T, Munshi N, Dang L, Castro A, Palombella V, et al. 2002. NF-kappa B as a therapeutic target in multiple myeloma. *J Biol Chem* **277:** 16639–16647.

Hideshima T, Neri P, Tassone P, Yasui H, Ishitsuka K, Raje N, Chauhan D, Podar K, Mitsiades C, Dang L, et al. 2006. MLN120B, a novel IkappaB kinase beta inhibitor, blocks multiple myeloma cell growth in vitro and in vivo. *Clin Cancer Res* **12:** 5887–5894.

Hiscott J, Nguyen TL, Arguello M, Nakhaei P, Paz S. 2006. Manipulation of the nuclear factor-kappaB pathway and the innate immune response by viruses. *Oncogene* **25:** 6844–6867.

Hu S, Du MQ, Park SM, Alcivar A, Qu L, Gupta S, Tang J, Baens M, Ye H, Lee TH, et al. 2006. cIAP2 is a ubiquitin protein ligase for BCL10 and is dysregulated in mucosa-associated lymphoid tissue lymphomas. *J Clin Invest* **116:** 174–181.

Hummel M, Bentink S, Berger H, Klapper W, Wessendorf S, Barth TF, Bernd HW, Cogliatti SB, Dierlamm J, Feller AC, et al. 2006. A biologic definition of Burkitt's lymphoma from transcriptional and genomic profiling. *N Engl J Med* **354:** 2419–2430.

Hussell T, Isaacson PG, Crabtree JE, Spencer J. 1996. Helicobacter pylori-specific tumour-infiltrating T cells provide contact dependent help for the growth of malignant B cells in low-grade gastric lymphoma of mucosa-associated lymphoid tissue. *J Pathol* **178:** 122–127.

Huynh QK, Boddupalli H, Rouw SA, Koboldt CM, Hall T, Sommers C, Hauser SD, Pierce JL, Combs RG, Reitz BA, et al. 2000. Characterization of the recombinant IKK1/IKK2 heterodimer. Mechanisms regulating kinase activity. *J Biol Chem* **275:** 25883–25891.

Iqbal J, Greiner TC, Patel K, Dave BJ, Smith L, Ji J, Wright G, Sanger WG, Pickering DL, Jain S, et al. 2007. Distinctive patterns of BCL6 molecular alterations and their functional consequences in different subgroups of diffuse large B-cell lymphoma. *Leukemia* **21:** 2332–2343.

Isaacson PG, Du MQ. 2004. MALT lymphoma: from morphology to molecules. *Nat Rev Cancer* **4:** 644–653.

Iwakoshi NN, Lee AH, Glimcher LH. 2003. The X-box binding protein-1 transcription factor is required for plasma cell differentiation and the unfolded protein response. *Immunol Rev* **194:** 29–38.

Jun JE, Wilson LE, Vinuesa CG, Lesage S, Blery M, Miosge LA, Cook MC, Kucharska EM, Hara H, Penninger JM, et al. 2003. Identifying the MAGUK protein Carma-1 as a central regulator of humoral immune responses and atopy by genome-wide mouse mutagenesis. *Immunity* **18:** 751–762.

Jungnickel B, Staratschek-Jox A, Br]auninger A, Spieker T, Wolf J, Diehl V, Hansmann ML, Rajewsky K, R K. 2000. Clonal deleterious mutations in the IkappaBalpha gene in the malignant cells in Hodgkin's lymphoma. *J Exp Med* **191:** 395–402.

Kalaitzidis D, Davis RE, Rosenwald A, Staudt LM, Gilmore TD. 2002. The human B-cell lymphoma cell line RC-K8 has multiple genetic alterations that dysregulate the Rel/NF-kappaB signal transduction pathway. *Oncogene* **21:** 8759–8768.

Karin M. 2010. NF-κB as a critical link between inflammation and cancer. In: Karin M, Staudt LM, eds. NF-κB: A network hub controlling immunity, inflammation, and cancer: Cold Spring Harbor Laboratory Press.

Karin M, Lawrence T, Nizet V. 2006. Innate immunity gone awry: linking microbial infections to chronic inflammation and cancer. *Cell* **124:** 823–835.

Karin M, Yamamoto Y, Wang QM. 2004. The IKK NF-kappa B system: a treasure trove for drug development. *Nat Rev Drug Discov* **3:** 17–26.

Karrer U, Althage A, Odermatt B, Hengartner H, Zinkernagel RM. 2000. Immunodeficiency of alymphoplasia mice (aly/aly) in vivo: structural defect of secondary lymphoid organs and functional B cell defect. *Eur J Immunol* **30:** 2799–2807.

Keats JJ, Fonseca R, Chesi M, Schop R, Baker A, Chng WJ, Van Wier S, Tiedemann R, Shi CX, Sebag M, et al. 2007. Promiscuous mutations activate the noncanonical NF-kappaB pathway in multiple myeloma. *Cancer Cell* **12:** 131–144.

Klein U, Casola S, Cattoretti G, Shen Q, Lia M, Mo T, Ludwig T, Rajewsky K, Dalla-Favera R. 2006. Transcription factor IRF4 controls plasma cell differentiation and class-switch recombination. *Nat Immunol* **7:** 773–782.

Kovalenko A, Chable-Bessia C, Cantarella G, Israel A, Wallach D, Courtois G. 2003. The tumour suppressor CYLD negatively regulates NF-kappaB signalling by deubiquitination. *Nature* **424:** 801–805.

Krappmann D, Emmerich F, Kordes U, Scharschmidt E, Dorken B, Scheidereit C. 1999. Molecular mechanisms of constitutive NF-kappaB/Rel activation in Hodgkin/Reed-Sternberg cells. *Oncogene* **18:** 943–953.

Kraus M, Alimzhanov MB, Rajewsky N, Rajewsky K. 2004. Survival of resting mature B lymphocytes depends on BCR signaling via the Igalpha/beta heterodimer. *Cell* **117:** 787–800.

Kraus M, Saijo K, Torres RM, Rajewsky K. 1999. Ig-alpha cytoplasmic truncation renders immature B cells more sensitive to antigen contact. *Immunity* **11:** 537–545.

Lake A, Shield LA, Cordano P, Chui DT, Osborne J, Crae S, Wilson KS, Tosi S, Knight SJ, Gesk S, et al. 2009. Mutations of NFKBIA, encoding IkappaB alpha, are a recurrent finding in classical Hodgkin lymphoma but are not a unifying feature of non-EBV-associated cases. *Int J Cancer* **125:** 1334–1342.

Lam KP, Kuhn R, Rajewsky K. 1997. In vivo ablation of surface immunoglobulin on mature B cells by inducible gene targeting results in rapid cell death. *Cell* **90:** 1073–1083.

Lam LT, Davis RE, Ngo VN, Lenz G, Wright G, Xu W, Zhao H, Yu X, Dang L, Staudt LM. 2008. Compensatory IK-Kalpha activation of classical NF-kappaB signaling during IKKbeta inhibition identified by an RNA interference sensitization screen. *Proc Natl Acad Sci U S A* **105:** 20798–20803.

Lam LT, Davis RE, Pierce J, Hepperle M, Xu Y, Hottelet M, Nong Y, Wen D, Adams J, Dang L, et al. 2005. Small molecule inhibitors of IkappaB kinase are selectively toxic for subgroups of diffuse large B-cell lymphoma defined by gene expression profiling. *Clin Cancer Res* **11:** 28–40.

Lam LT, Wright G, Davis RE, Lenz G, Farinha P, Dang L, Chan JW, Rosenwald A, Gascoyne RD, Staudt LM. 2008. Cooperative signaling through the signal transducer and activator of transcription 3 and nuclear factor-{kappa}B pathways in subtypes of diffuse large B-cell lymphoma. *Blood* **111:** 3701–3713.

Leitges M, Schmedt C, Guinamard R, Davoust J, Schaal S, Stabel S, Tarakhovsky A. 1996. Immunodeficiency in protein kinase cbeta-deficient mice. *Science* **273:** 788–791.

Lenz G, Davis RE, Ngo VN, Lam L, George TC, Wright GW, Dave SS, Zhao H, Xu W, Rosenwald A, et al. 2008. Oncogenic CARD11 mutations in human diffuse large B cell lymphoma. *Science* **319:** 1676–1679.

Lenz G, Wright G, Dave SS, Xiao W, Powell J, Zhao H, Xu W, Tan B, Goldschmidt N, Iqbal J, et al. 2008. Stromal gene signatures in large-B-cell lymphomas. *N Engl J Med* **359:** 2313–2323.

Lenz G, Wright GW, Emre NC, Kohlhammer H, Dave SS, Davis RE, Carty S, Lam LT, Shaffer AL, Xiao W, et al. 2008. Molecular subtypes of diffuse large B-cell lymphoma arise by distinct genetic pathways. *Proc Natl Acad Sci USA* **105:** 13520–13525.

Li C, Norris PS, Ni CZ, Havert ML, Chiong EM, Tran BR, Cabezas E, Reed JC, Satterthwait AC, Ware CF, et al. 2003. Structurally distinct recognition motifs in lymphotoxin-beta receptor and CD40 for tumor necrosis factor receptor-associated factor (TRAF)-mediated signaling. *J Biol Chem* **278:** 50523–50529.

Li Z, Wang H, Xue L, Shin DM, Roopenian D, Xu W, Qi CF, Sangster MY, Orihuela CJ, Tuomanen E, et al. 2009. Emu-BCL10 mice exhibit constitutive activation of both canonical and noncanonical NF-kappaB pathways generating marginal zone (MZ) B-cell expansion as a precursor to splenic MZ lymphoma. *Blood* **114:** 4158–68.

Liao G, Zhang M, Harhaj EW, Sun SC. 2004. Regulation of the NF-kappaB-inducing kinase by tumor necrosis factor receptor-associated factor 3-induced degradation. *J Biol Chem* **279:** 26243–26250.

Lin SC, Chung JY, Lamothe B, Rajashankar K, Lu M, Lo YC, Lam AY, Darnay BG, Wu H. 2008. Molecular basis for the unique deubiquitinating activity of the NF-kappaB inhibitor A20. *J Mol Biol* **376:** 526–540.

Lin Y, Wong K, Calame K. 1997. Repression of c-myc transcription by Blimp-1, an inducer of terminal B cell differentiation. *Science* **276:** 596–599.

Liu H, Ye H, Ruskone-Fourmestraux A, De Jong D, Pileri S, Thiede C, Lavergne A, Boot H, Caletti G, Wundisch T, et al. 2002. T(11;18) is a marker for all stage gastric MALT lymphomas that will not respond to H. pylori eradication. *Gastroenterology* **122:** 1286–1294.

Lucas PC, Kuffa P, Gu S, Kohrt D, Kim DS, Siu K, Jin X, Swenson J, McAllister-Lucas LM. 2007. A dual role for the API2 moiety in API2-MALT1-dependent NF-kappaB activation: heterotypic oligomerization and TRAF2 recruitment. *Oncogene* **26:** 5643–5654.

Lucas PC, Yonezumi M, Inohara N, McAllister-Lucas LM, Abazeed ME, Chen FF, Yamaoka S, Seto M, Nunez G. 2001. Bcl10 and MALT1, independent targets of chromosomal translocation in malt lymphoma, cooperate in a novel NF-kappa B signaling pathway. *J Biol Chem* **276:** 19012–19019.

Luedde T, Assmus U, Wustefeld T, Meyer zu Vilsendorf A, Roskams T, Schmidt-Supprian M, Rajewsky K, Brenner DA, Manns MP, Pasparakis M, et al. 2005. Deletion of IKK2 in hepatocytes does not sensitize these cells to TNF-induced apoptosis but protects from ischemia/reperfusion injury. *J Clin Invest* **115:** 849–859.

Luedde T, Beraza N, Kotsikoris V, van Loo G, Nenci A, De Vos R, Roskams T, Trautwein C, Pasparakis M. 2007. Deletion of NEMO/IKKgamma in liver parenchymal cells causes steatohepatitis and hepatocellular carcinoma. *Cancer Cell* **11:** 119–132.

Ma H, Yankee TM, Hu J, Asai DJ, Harrison ML, Geahlen RL. 2001. Visualization of Syk-antigen receptor interactions

Cite this article as *Cold Spring Harb Perspect Biol* 2010;00:a000109

using green fluorescent protein: differential roles for Syk and Lyn in the regulation of receptor capping and internalization. *J Immunol* **166:** 1507–1516.

Malinin NL, Boldin MP, Kovalenko AV, Wallach D. 1997. MAP3K-related kinase involved in NF-kappaB induction by TNF, CD95 and IL-1. *Nature* **385:** 540–544.

Marsters SA, Yan M, Pitti RM, Haas PE, Dixit VM, Ashkenazi A. 2000. Interaction of the TNF homologues BLyS and APRIL with the TNF receptor homologues BCMA and TACI. *Curr Biol* **10:** 785–788.

Massoumi R, Chmielarska K, Hennecke K, Pfeifer A, Fassler R. 2006. Cyld inhibits tumor cell proliferation by blocking Bcl-3-dependent NF-kappaB signaling. *Cell* **125:** 665–677.

Matsumoto R, Wang D, Blonska M, Li H, Kobayashi M, Pappu B, Chen Y, Lin X. 2005. Phosphorylation of CARMA1 plays a critical role in T Cell receptor-mediated NF-kappaB activation. *Immunity* **23:** 575–585.

Matsuzawa A, Tseng PH, Vallabhapurapu S, Luo JL, Zhang W, Wang H, Vignali DA, Gallagher E, Karin M. 2008. Essential cytoplasmic translocation of a cytokine receptor-assembled signaling complex. *Science* **321:** 663–668.

Mauro C, Pacifico F, Lavorgna A, Mellone S, Iannetti A, Acquaviva R, Formisano S, Vito P, Leonardi A. 2006. ABIN-1 binds to NEMO/IKKgamma and co-operates with A20 in inhibiting NF-kappaB. *J Biol Chem* **281:** 18482–18488.

McCully RR, Pomerantz JL. 2008. The protein kinase C-responsive inhibitory domain of CARD11 functions in NF-kappaB activation to regulate the association of multiple signaling cofactors that differentially depend on Bcl10 and MALT1 for association. *Mol Cell Biol* **28:** 5668–5686.

Meylan E, Dooley AL, Feldser DM, Shen L, Turk E, Ouyang C, Jacks T. 2009. Requirement for NF-kappaB signalling in a mouse model of lung adenocarcinoma. *Nature* **462:** 104–107.

Migliazza A, Lombardi L, Rocchi M, Trecca D, Chang CC, Antonacci R, Fracchiolla NS, Ciana P, Maiolo AT, Neri A. 1994. Heterogeneous chromosomal aberrations generate 3′ truncations of the NFKB2/lyt-10 gene in lymphoid malignancies. *Blood* **84:** 3850–3860.

Miyawaki S, Nakamura Y, Suzuka H, Koba M, Yasumizu R, Ikehara S, Shibata Y. 1994. A new mutation, aly, that induces a generalized lack of lymph nodes accompanied by immunodeficiency in mice. *Eur J Immunol* **24:** 429–434.

Molin D, Edstrom A, Glimelius I, Glimelius B, Nilsson G, Sundstrom C, Enblad G. 2002. Mast cell infiltration correlates with poor prognosis in Hodgkin's lymphoma. *Br J Haematol* **119:** 122–124.

Monroe JG. 2006. ITAM-mediated tonic signalling through pre-BCR and BCR complexes. *Nat Rev Immunol* **6:** 283–294.

Monti S, Savage KJ, Kutok JL, Feuerhake F, Kurtin P, Mihm M, Wu B, Pasqualucci L, Neuberg D, Aguiar RC, et al. 2005. Molecular profiling of diffuse large B-cell lymphoma identifies robust subtypes including one characterized by host inflammatory response. *Blood* **105:** 1851–1861.

Moreaux J, Cremer FW, Reme T, Raab M, Mahtouk K, Kaukel P, Pantesco V, De Vos J, Jourdan E, Jauch A, et al. 2005.

The level of TACI gene expression in myeloma cells is associated with a signature of microenvironment dependence versus a plasmablastic signature. *Blood* **106:** 1021–1030.

Morgan JA, Yin Y, Borowsky AD, Kuo F, Nourmand N, Koontz JI, Reynolds C, Soreng L, Griffin CA, Graeme-Cook F, et al. 1999. Breakpoints of the t(11;18) (q21;q21) in mucosa-associated lymphoid tissue (MALT) lymphoma lie within or near the previously undescribed gene MALT1 in chromosome 18. *Cancer Res* **59:** 6205–6213.

Nagashima K, Sasseville VG, Wen D, Bielecki A, Yang H, Simpson C, Grant E, Hepperle M, Harriman G, Jaffee B, et al. 2006. Rapid TNFR1-dependent lymphocyte depletion in vivo with a selective chemical inhibitor of IKKbeta. *Blood* **107:** 4266–4273.

Ngo VN, Davis RE, Lamy L, Yu X, Zhao H, Lenz G, Lam LT, Dave S, Yang L, Powell J, et al. 2006. A loss-of-function RNA interference screen for molecular targets in cancer. *Nature* **441:** 106–110.

Ni CZ, Oganesyan G, Welsh K, Zhu X, Reed JC, Satterthwait AC, Cheng G, Ely KR. 2004. Key molecular contacts promote recognition of the BAFF receptor by TNF receptor-associated factor 3: implications for intracellular signaling regulation. *J Immunol* **173:** 7394–7400.

Ni CZ, Welsh K, Leo E, Chiou CK, Wu H, Reed JC, Ely KR. 2000. Molecular basis for CD40 signaling mediated by TRAF3. *Proc Natl Acad Sci U S A* **97:** 10395–10399.

Niiro H, Allam A, Stoddart A, Brodsky FM, Marshall AJ, Clark EA. 2004. The B lymphocyte adaptor molecule of 32 kilodaltons (Bam32) regulates B cell antigen receptor internalization. *J Immunol* **173:** 5601–5609.

Noels H, van Loo G, Hagens S, Broeckx V, Beyaert R, Marynen P, Baens M. 2007. A Novel TRAF6 binding site in MALT1 defines distinct mechanisms of NF-kappaB activation by API2middle dotMALT1 fusions. *J Biol Chem* **282:** 10180–10189.

O'Connor BP, Raman VS, Erickson LD, Cook WJ, Weaver LK, Ahonen C, Lin LL, Mantchev GT, Bram RJ, Noelle RJ. 2004. BCMA is essential for the survival of long-lived bone marrow plasma cells. *J Exp Med* **199:** 91–98.

O'Mahony A, Lin X, Geleziunas R, Greene WC. 2000. Activation of the heterodimeric IkappaB kinase alpha (IKKalpha)-IKKbeta complex is directional: IKKalpha regulates IKKbeta under both basal and stimulated conditions. *Mol Cell Biol* **20:** 1170–1178.

Oeckinghaus A, Wegener E, Welteke V, Ferch U, Arslan SC, Ruland J, Scheidereit C, Krappmann D. 2007. Malt1 ubiquitination triggers NF-kappaB signaling upon T-cell activation. *Embo J* **26:** 4634–4645.

Pan Z, Scheerens H, Li SJ, Schultz BE, Sprengeler PA, Burrill LC, Mendonca RV, Sweeney MD, Scott KC, Grothaus PG, et al. 2007. Discovery of selective irreversible inhibitors for Bruton's tyrosine kinase. *ChemMedChem* **2:** 58–61.

Pasparakis M, Schmidt-Supprian M, Rajewsky K. 2002. IkappaB kinase signaling is essential for maintenance of mature B cells. *J Exp Med* **196:** 743–752.

Pasqualucci L, Compagno M, Houldsworth J, Monti S, Grunn A, Nandula SV, Aster JC, Murty VV, Shipp MA, Dalla-Favera R. 2006. Inactivation of the PRDM1/BLIMP1 gene in diffuse large B cell lymphoma. *J Exp Med* **203:** 311–317.

Peters RT, Liao SM, Maniatis T. 2000. IKKepsilon is part of a novel PMA-inducible IkappaB kinase complex. *Mol Cell* **5:** 513–522.

Petro JB, Khan WN. 2001. Phospholipase C-gamma 2 couples Bruton's tyrosine kinase to the NF-kappaB signaling pathway in B lymphocytes. *J Biol Chem* **276:** 1715–1719.

Petro JB, Rahman SM, Ballard DW, Khan WN. 2000. Bruton's tyrosine kinase is required for activation of IkappaB kinase and nuclear factor kappaB in response to B cell receptor engagement. *J Exp Med* **191:** 1745–1754.

Pichlmair A, Reis e Sousa C. 2007. Innate recognition of viruses. *Immunity* **27:** 370–383.

Pinto A, Aldinucci D, Gloghini A, Zagonel V, Degan M, Improta S, Juzbasic S, Todesco M, Perin V, Gattei V, et al. 1996. Human eosinophils express functional CD30 ligand and stimulate proliferation of a Hodgkin's disease cell line. *Blood* **88:** 3299–3305.

Podust VN, Brownell JE, Gladysheva TB, Luo RS, Wang C, Coggins MB, Pierce JW, Lightcap ES, Chau V. 2000. A Nedd 8 conjugation pathway is essential for proteolytic targeting of p27Kip1 by ubiquitination. *Proc Natl Acad Sci U S A* **97:** 4579–4584.

Pomerantz JL, Baltimore D. 1999. NF-kappaB activation by a signaling complex containing TRAF2, TANK and TBK1, a novel IKK-related kinase. *EMBO J* **18:** 6694–704.

Ramakrishnan P, Wang W, Wallach D. 2004. Receptor-specific signaling for both the alternative and the canonical NF-kappaB activation pathways by NF-kappaB-inducing kinase. *Immunity* **21:** 477–89.

Rawlings DJ, Sommer K, Moreno-Garcia ME. 2006. The CARMA 1 signalosome links the signalling machinery of adaptive and innate immunity in lymphocytes. *Nat Rev Immunol* **6:** 799–812.

Read MA, Brownell JE, Gladysheva TB, Hottelet M, Parent LA, Coggins MB, Pierce JW, Podust VN, Luo RS, Chau V, et al. 2000. Nedd8 modification of cul-1 activates SCF(beta(TrCP))-dependent ubiquitination of Ikappa-Balpha. *Mol Cell Biol* **20:** 2326–2333.

Rebeaud F, Hailfinger S, Posevitz-Fejfar A, Tapernoux M, Moser R, Rueda D, Gaide O, Guzzardi M, Iancu EM, Rufer N, et al. 2008. The proteolytic activity of the paracaspase MALT1 is key in T cell activation. *Nat Immunol* **9:** 272–281.

Regamey A, Hohl D, Liu JW, Roger T, Kogerman P, Toftgard R, Huber M. 2003. The tumor suppressor CYLD interacts with TRIP and regulates negatively nuclear factor kappaB activation by tumor necrosis factor. *J Exp Med* **198:** 1959–1964.

Reimold AM, Iwakoshi NN, Manis J, Vallabhajosyula P, Szomolanyi-Tsuda E, Gravallese EM, Friend D, Grusby MJ, Alt F, Glimcher LH. 2001. Plasma cell differentiation requires the transcription factor XBP-1. *Nature* **412:** 300–307.

Reth M. 1989. Antigen receptor tail clue. *Nature* **338:** 383–384.

Richardson PG, Sonneveld P, Schuster MW, Irwin D, Stadtmauer EA, Facon T, Harousseau JL, Ben-Yehuda D, Lonial S, Goldschmidt H, et al. 2005. Bortezomib or high-dose dexamethasone for relapsed multiple myeloma. *N Engl J Med* **352:** 2487–2498.

Rosenwald A, Wright G, Chan WC, Connors JM, Campo E, Fisher RI, Gascoyne RD, Muller-Hermelink HK, Smeland EB, Giltnane JM, et al. 2002. The use of molecular profiling to predict survival after chemotherapy for diffuse large-B-cell lymphoma. *N Engl J Med* **346:** 1937–1947.

Rosenwald A, Wright G, Leroy K, Yu X, Gaulard P, Gascoyne RD, Chan WC, Zhao T, Haioun C, Greiner TC, et al. 2003. Molecular diagnosis of primary mediastinal B cell lymphoma identifies a clinically favorable subgroup of diffuse large B cell lymphoma related to Hodgkin lymphoma. *J Exp Med* **198:** 851–862.

Ruland J, Duncan GS, Wakeham A, Mak TW. 2003. Differential requirement for Malt1 in T and B cell antigen receptor signaling. *Immunity* **19:** 749–758.

Saijo K, Mecklenbrauker I, Santana A, Leitger M, Schmedt C, Tarakhovsky A. 2002. Protein kinase C beta controls nuclear factor kappaB activation in B cells through selective regulation of the IkappaB kinase alpha. *J Exp Med* **195:** 1647–1652.

Samuel T, Welsh K, Lober T, Togo SH, Zapata JM, Reed JC. 2006. Distinct BIR domains of cIAP1 mediate binding to and ubiquitination of tumor necrosis factor receptor-associated factor 2 and second mitochondrial activator of caspases. *J Biol Chem* **281:** 1080–1090.

Sanchez-Izquierdo D, Buchonnet G, Siebert R, Gascoyne RD, Climent J, Karran L, Marin M, Blesa D, Horsman D, Rosenwald A, et al. 2003. MALT1 is deregulated by both chromosomal translocation and amplification in B-cell non-Hodgkin lymphoma. *Blood* **101:** 4539–4546.

Satterthwaite AB, Witte ON. 2000. The role of Bruton's tyrosine kinase in B-cell development and function: a genetic perspective. *Immunol Rev* **175:** 120–127.

Savage KJ, Monti S, Kutok JL, Cattoretti G, Neuberg D, De Leval L, Kurtin P, Dal Cin P, Ladd C, Feuerhake F, et al. 2003. The molecular signature of mediastinal large B-cell lymphoma differs from that of other diffuse large B-cell lymphomas and shares features with classical Hodgkin lymphoma. *Blood* **102:** 3871–3879.

Schmitz R, Hansmann ML, Bohle V, Martin-Subero JI, Hartmann S, Mechtersheimer G, Klapper W, Vater I, Giefing M, Gesk S, et al. 2009. TNFAIP3 (A20) is a tumor suppressor gene in Hodgkin lymphoma and primary mediastinal B cell lymphoma. *J Exp Med* **206:** 981–989.

Schmitz R, Stanelle J, Hansmann ML, Kuppers R. 2009. Pathogenesis of classical and lymphocyte-predominant Hodgkin lymphoma. *Annual review of pathology* **4:** 151–174.

Schwab U, Stein H, Gerdes J, Lemke H, Kirchner H, Schaadt M, Diehl V. 1982. Production of a monoclonal antibody specific for Hodgkin and Sternberg-Reed cells of Hodgkin's disease and a subset of normal lymphoid cells. *Nature* **299:** 65–67.

Schwabe RF, Brenner DA. Nuclear factor-kappaB in the liver: friend or foe? 2007. *Gastroenterology* **132:** 2601–2604.

Sciammas R, Shaffer AL, Schatz JH, Zhao H, Staudt LM, Singh H. 2006. Graded expression of interferon regulatory factor-4 coordinates isotype switching with plasma cell differentiation. *Immunity* **25:** 225–236.

Senftleben U, Cao Y, Xiao G, Greten FR, Krahn G, Bonizzi G, Chen Y, Hu Y, Fong A, Sun SC, et al. 2001. Activation by

IKKalpha of a second, evolutionary conserved, NF-kappa B signaling pathway. *Science* **293**: 1495–1499.

Shaffer AL, Emre NC, Lamy L, Ngo VN, Wright G, Xiao W, Powell J, Dave S, Yu X, Zhao H, et al. 2008. IRF4 addiction in multiple myeloma. *Nature* **454**: 226–231.

Shaffer AL, Emre NC, Romesser PB, Staudt LM. 2009. IRF4: Immunity. Malignancy! Therapy? *Clin Cancer Res* **15**: 2954–2961.

Shaffer AL, Lin KI, Kuo TC, Yu X, Hurt EM, Rosenwald A, Giltnane JM, Yang L, Zhao H, Calame K, et al. 2002. Blimp-1 orchestrates plasma cell differentiation by extinguishing the mature B cell gene expression program. *Immunity* **17**: 51–62.

Shaffer AL, Rosenwald A, Hurt EM, Giltnane JM, Lam LT, Pickeral OK, Staudt LM. 2001. Signatures of the immune response. *Immunity* **15**: 375–385.

Shaffer AL, Shapiro-Shelef M, Iwakoshi NN, Lee A-H, Qian S-B, Zhao H, Yu X, Yang L, Tan BK, Rosenwald A, et al. 2004. XBP1, Downstream of Blimp-1, Expands the Secretory Apparatus and Other Organelles, and Increases Protein Synthesis in Plasma Cell Differentiation. *Immunity* **21**: 81–93.

Shaffer AL, Yu X, He Y, Boldrick J, Chan EP, Staudt LM. 2000. BCL-6 represses genes that function in lymphocyte differentiation, inflammation, and cell cycle control. *Immunity* **13**: 199–212.

Shapiro-Shelef M, Calame K. 2005. Regulation of plasma-cell development. *Nat Rev Immunol* **5**: 230–242.

Shapiro-Shelef M, Lin KI, McHeyzer-Williams LJ, Liao J, McHeyzer-Williams MG, Calame K. 2003. Blimp-1 is required for the formation of immunoglobulin secreting plasma cells and pre-plasma memory B cells. *Immunity* **19**: 607–620.

Shapiro-Shelef M, Lin KI, Savitsky D, Liao J, Calame K. 2005. Blimp-1 is required for maintenance of long-lived plasma cells in the bone marrow. *J Exp Med* **202**: 1471–1476.

Shimada T, Kawai T, Takeda K, Matsumoto M, Inoue J, Tatsumi Y, Kanamaru A, Akira S. 1999. IKK-i, a novel lipopolysaccharide-inducible kinase that is related to IkappaB kinases. *Int Immunol* **11**: 1357–1362.

Shinkura R, Kitada K, Matsuda F, Tashiro K, Ikuta K, Suzuki M, Kogishi K, Serikawa T, Honjo T. 1999. Alymphoplasia is caused by a point mutation in the mouse gene encoding Nf-kappa b-inducing kinase. *Nat Genet* **22**: 74–77.

Sommer K, Guo B, Pomerantz JL, Bandaranayake AD, Moreno-Garcia ME, Ovechkina YL, Rawlings DJ. 2005. Phosphorylation of the CARMA1 linker controls NF-kappaB activation. *Immunity* **23**: 561–574.

Soucy TA, Smith PG, Milhollen MA, Berger AJ, Gavin JM, Adhikari S, Brownell JE, Burke KE, Cardin DP, Critchley S, et al. 2009. An inhibitor of NEDD8-activating enzyme as a new approach to treat cancer. *Nature* **458**: 732–736.

Srinivasan L, Sasaki Y, Calado DP, Zhang B, Paik JH, DePinho RA, Kutok JL, Kearney JF, Otipoby KL, Rajewsky K. 2009. PI3 kinase signals BCR-dependent mature B cell survival. *Cell* **139**: 573–586.

Staudt LM, Dave S. 2005. The biology of human lymphoid malignancies revealed by gene expression profiling. *Adv Immunol* **87**: 163–208.

Sun L, Deng L, Ea CK, Xia ZP, Chen ZJ. 2004. The TRAF 6 ubiquitin ligase and TAK 1 kinase mediate IKK activation by BCL 10 and MALT 1 in T lymphocytes. *Mol Cell* **14**: 289–301.

Sun SC. 2009. CYLD: a tumor suppressor deubiquitinase regulating NF-kappaB activation and diverse biological processes. *Cell death and differentiation* **17**: 25–34.

Tam W, Gomez M, Chadburn A, Lee JW, Chan WC, Knowles DM. 2006. Mutational analysis of PRDM1 indicates a tumor-suppressor role in diffuse large B-cell lymphomas. *Blood* **107**: 4090–4100.

Tan JE, Wong SC, Gan SK, Xu S, Lam KP. 2001. The adaptor protein BLNK is required for B cell antigen receptor-induced activation of NFk-B and cell cycle entry and Survival of B lymphocytes. *J Biol Chem* **23**: 23.

Tanner MJ, Hanel W, Gaffen SL, Lin X. 2007. CARMA1 coiled-coil domain is involved in the oligomerization and subcellular localization of CARMA1 and is required for T cell receptor-induced NF-kappaB activation. *J Biol Chem* **282**: 17141–17147.

Thome M. 2004. CARMA1, BCL-10 and MALT 1 in lymphocyte development and activation. *Nat Rev Immunol* **4**: 348–359.

Tojima Y, Fujimoto A, Delhase M, Chen Y, Hatakeyama S, Nakayama K, Kaneko Y, Nimura Y, Motoyama N, Ikeda K, et al. 2000. NAK is an IkappaB kinase-activating kinase. *Nature* **404**: 778–782.

Tolar P, Hanna J, Krueger PD, Pierce SK. 2009. The constant region of the membrane immunoglobulin mediates B cell-receptor clustering and signaling in response to membrane antigens. *Immunity* **30**: 44–55.

Torres RM, Hafen K. 1999. A negative regulatory role for Ig-alpha during B cell development. *Immunity* **11**: 527–536.

Trompouki E, Hatzivassiliou E, Tsichritzis T, Farmer H, Ashworth A, Mosialos G. 2003. CYLD is a deubiquitinating enzyme that negatively regulates NF-kappaB activation by TNFR family members. *Nature* **424**: 793–796.

Turer EE, Tavares RM, Mortier E, Hitotsumatsu O, Advincula R, Lee B, Shifrin N, Malynn BA, Ma A. 2008. Homeostatic MyD 8 8-dependent signals cause lethal inflamMation in the absence of A 20. *J Exp Med* **205**: 451–464.

Turner CA Jr., Mack DH, Davis MM. 1994. Blimp-1, a novel zinc finger-containing protein that can drive the maturation of B lymphocytes into immunoglobulin-secreting cells. *Cell* **77**: 297–306.

Uren AG, O'Rourke K, Aravind LA, Pisabarro MT, Seshagiri S, Koonin EV, Dixit VM. 2000. Identification of paracaspases and metacaspases: two ancient families of caspase-like proteins, one of which plays a key role in MALT lymphoma. *Mol Cell* **6**: 961–967.

Vallabhapurapu S, Karin M. 2009. Regulation and function of NF-kappaB transcription factors in the immune system. *Annu Rev Immunol* **27**: 693–733.

Vallabhapurapu S, Matsuzawa A, Zhang W, Tseng PH, Keats JJ, Wang H, Vignali DA, Bergsagel PL, Karin M. 2008. Nonredundant and complementary functions of TRAF2 and TRAF3 in a ubiquitination cascade that activates NIK-dependent alternative NF-kappaB signaling. *Nat Immunol* **9**: 1364–1370.

Wang CY, Mayo MW, Korneluk RG, Goeddel DV, Baldwin AS. 1998. NF-kappaB antiapoptosis: induction of TRAF1 and TRAF2 and c-IAP1 and c-IAP2 to suppress caspase-8 activation. *Science* **281**: 1680–1683.

Werner SL, Barken D, Hoffmann A. 2005. Stimulus specificity of gene expression programs determined by temporal control of IKK activity. *Science* **309**: 1857–1861.

Wertz IE, O'Rourke KM, Zhou H, Eby M, Aravind L, Seshagiri S, Wu P, Wiesmann C, Baker R, Boone DL, et al. 2004. De-ubiquitination and ubiquitin ligase domains of A20 downregulate NF-kappaB signalling. *Nature* **430**: 694–699.

Willis TG, Jadayel DM, Du MQ, Peng H, Perry AR, Abdul-Rauf M, Price H, Karran L, Majekodunmi O, Wlodarska I, et al. 1999. Bcl10 is involved in t(1;14)(p22;q32) of MALT B cell lymphoma and mutated in multiple tumor types. *Cell* **96**: 35–45.

Workman P, Burrows F, Neckers L, Rosen N. 2007. Drugging the cancer chaperone HSP90: combinatorial therapeutic exploitation of oncogene addiction and tumor stress. *Ann N Y Acad Sci* **1113**: 202–216.

Woronicz JD, Gao X, Cao Z, Rothe M, Goeddel DV. 1997. IkappaB kinase-beta: NF-kappaB activation and complex formation with IkappaB kinase-alpha and NIK. *Science* **278**: 866–869.

Wotherspoon AC, Doglioni C, Diss TC, Pan L, Moschini A, de Boni M, Isaacson PG. 1993. Regression of primary low-grade B-cell gastric lymphoma of mucosa- associated lymphoid tissue type after eradication of Helicobacter pylori. *Lancet* **342**: 575–577.

Wright CW, Duckett CS. 2005. Reawakening the cellular death program in neoplasia through the therapeutic blockade of IAP function. *J Clin Invest* **115**: 2673–2678.

Xu Y, Harder KW, Huntington ND, Hibbs ML, Tarlinton DM. 2005. Lyn tyrosine kinase: accentuating the positive and the negative. *Immunity* **22**: 9–18.

Yaccoby S, Pennisi A, Li X, Dillon SR, Zhan F, Barlogie B, Shaughnessy JD Jr. 2008. Atacicept (TACI-Ig) inhibits growth of TACI (high) primary myeloma cells in SCID-hu mice and in coculture with osteoclasts. *Leukemia* **22**: 406–413.

Yang J, Liao X, Agarwal MK, Barnes L, Auron PE, Stark GR. 2007. Unphosphorylated STAT 3 accumulates in response to IL- 6 and activates transcription by binding to NFkappaB. *Genes Dev* **21**: 1396–1408. Epub 2007 May 17.

Yang X, Chang HY, Baltimore D. 1998. Autoproteolytic activation of pro-caspases by oligomerization. *Mol Cell* **1**: 319–325.

Yaron A, Hatzubai A, Davis M, Lavon I, Amit S, Manning AM, Andersen JS, Mann M, Mercurio F, Ben-Neriah Y. 1998. Identification of the receptor component of the IkappaBalpha-ubiquitin ligase. *Nature* **396**: 590–594.

Yin L, Wu L, Wesche H, Arthur CD, White JM, Goeddel DV, Schreiber RD. 2001. Defective lymphotoxin-beta receptor-induced NF-kappaB transcriptional activity in NIK-deficient mice. *Science* **291**: 2162–2165.

Zarnegar B, Yamazaki S, He JQ, Cheng G. 2008. Control of canonical NF-kappaB activation through the NIK-IKK complex pathway. *Proc Natl Acad Sci U S A* **105**: 3503–3508.

Zarnegar BJ, Wang Y, Mahoney DJ, Dempsey PW, Cheung HH, He J, Shiba T, Yang X, Yeh WC, Mak TW, et al. 2008. Noncanonical NF-kappaB activation requires coordinated assembly of a regulatory complex of the adaptors cIAP1, cIAP2, TRAF2 and TRAF3 and the kinase NIK. *Nat Immunol* **9**: 1371–1378.

Zhang Q, Siebert R, Yan M, Hinzmann B, Cui X, Xue L, Rakestraw KM, Naeve CW, Beckmann G, Weisenburger DD, et al. 1999. Inactivating mutations and overexpression of BCL10, a caspase recruitment domain-containing gene, in MALT lymphoma with t(1;14)(p22;q32). *Nat Genet* **22**: 63–68.

Zhou H, Du MQ, Dixit VM. 2005. Constitutive NF-kappaB activation by the t(11;18)(q21;q21) product in MALT lymphoma is linked to deregulated ubiquitin ligase activity. *Cancer Cell* **7**: 425–431.

Zhou H, Wertz I, O'Rourke K, Ultsch M, Seshagiri S, Eby M, Xiao W, Dixit VM. 2004. Bcl10 activates the NF-kappaB pathway through ubiquitination of NEMO. *Nature* **427**: 167–171.

NF-κB as a Critical Link Between Inflammation and Cancer

Michael Karin

Laboratory of Gene Regulation and Signal Transduction, Department of Pharmacology and Pathology, Moores Cancer Canter, UCSD School of Medicine, La Jolla, California 92093-0723

Correspondence: mkarin@ucsd.edu

NF-κB transcription factors have been suspected to be involved in cancer development since their discovery because of their kinship with the v-Rel oncogene product. Subsequent work led to identification of oncogenic mutations that result in NF-κB activation in lymphoid malignancies, but most of these mutations affect upstream components of NF-κB signaling pathways, rather than NF-κB family members themselves. NF-κB activation has also been observed in many solid tumors, but so far no oncogenic mutations responsible for NF-κB activation in carcinomas have been identified. In such cancers, NF-κB activation is a result of underlying inflammation or the consequence of formation of an inflammatory microenvironment during malignant progression. Most importantly, through its ability to up-regulate the expression of tumor promoting cytokines, such as IL-6 or TNF-α, and survival genes, such as Bcl-X_L, NF-κB provides a critical link between inflammation and cancer.

An important chapter in the long saga of NF-κB is the one dealing with its role as a pivotal link between inflammation and cancer. A possible association between NF-κB and cancer has emerged during the early days of RelA/p65 cloning and sequencing, which instantaneously revealed its kinship to c-Rel and its oncogenic derivative v-Rel (Gilmore 2003). However, oncogenic mutations that endow RelA, c-Rel, or other NF-κB proteins with transforming activity were found to be rare and mainly limited to lymphoid malignancies (Gilmore 2003). Yet, not only lymphoid cancers, but most solid tumors as well, exhibit activated NF-κB (Karin et al. 2002). As in most of these cases, no loss-of-function IκB

mutations or gain-of-function IKK mutations have been detected. We have suggested that NF-κB activation in cancer may be the result of either exposure to proinflammatory stimuli in the tumor microenvironment or mutational activation of upstream components in IKK–NF-κB signaling pathways (Karin et al. 2002). Further bolstering our belief in the oncogenic potential of "normal" NF-κB activated by stimuli that are extrinsic to the cancer cell were the findings that NF-κB can inhibit apoptosis (Beg and Baltimore 1996; Liu et al. 1996; Van Antwerp et al. 1996; Wang et al. 1996), stimulate cell proliferation (Joyce et al. 2001), as well as promote a migratory and invasive phenotype that is associated with tumor progression

M. Karin

(Huang et al. 2001). Concurrently, we became cognizant of a large body of epidemiological and experimental data providing new support for a causal link between inflammation and cancer, an association that was first proposed by Virchow during the 19th century (Balkwill and Mantovani 2001). Considering these findings, together with sightings of activated NF-κB in a large number of cancers, most of which are not associated with genetic alterations in NF-κB, IKK, or upstream components of this signaling system, we proposed that NF-κB may provide a critical mechanistic link between inflammation and cancer (Karin et al. 2002). During the past seven years, this proposal has been subjected to intense scrutiny by a number of labs, in a variety of experimental systems, and although complex and occasionally unpredictable, the role of the NF-κB signaling system in bridging inflammation and cancer is currently well appreciated (Karin 2006). It was also found that some IKK subunits (IKKα) and closely related protein kinases (e.g., IKKε) can play NF-κB independent roles in a variety of cancers (Boehm et al. 2007; Luo et al. 2007). In addition, new work has resulted in the identification of cancer-associated mutations in upstream components of the IKK-NF-κB signaling system that can lead to cell autonomous activation of NF-κB in multiple myeloma (Annunziata et al. 2007; Keats et al. 2007). The goal of this article is to review the experimental evidence for the pathogenic function of NF-κB in cancer and discuss whether and how IKK-NF-κB targeted interventions can be used in cancer prevention and/or therapy.

NF-κB IN LYMPHOID MALIGNANCIES: FROM CELL AUTONOMY TO PARACRINE EFFECTS

As mentioned above, the first hint to a link between NF-κB and cancer had emerged with the cloning of RelA and the realization of its close kinship to the viral oncoprotein v-Rel and its cellular homolog c-Rel (Gilmore 2003). Soon thereafter, the Bcl-3 oncoprotein, a product of a gene activated by chromosomal translocation in B-cell chronic lymphocytic leukemia, was identified as a member of the IκB family (Franzoso et al. 1992; Bours et al. 1993). Later, the *NF-κB2* gene was also found to be rearranged in B- and T-cell lymphomas, giving rise to a truncated NF-κB2/p100 protein devoid of the IκB-like activity that is exhibited by native p100 (Neri et al. 1991). These early findings led to an extensive search for mutations affecting the IκB-NF-κB system in other lymphoid malignancies. This effort, however, has netted few new results other than those described previously. For instance, IκBα gene mutations were detected in Hodgkin's lymphoma (Cabannes et al. 1999), but their contributions to pathogenesis is still not clear. Eventually, this has led to a broader view of the role played by NF-κB in tumorigenesis, according to which, mutations that cause NF-κB activation in malignant cells may occur in genes coding for signaling proteins that feed into the IKK–NF-κB module. Indeed, translocations that lead to Bcl-10 overexpression and activation of IKK–NF-κB signaling were identified in MALT lymphomas (Willis et al. 1999). Another product of a chromosomal translocation in MALT lymphoma is MALT1, a protein with paracaspase homology that interacts with Bcl-10 and Carma-1 to yield IKK activation (Uren et al. 2000). Given its well established antiapoptotic function, especially in B cells (Grossmann et al. 2000; Gugasyan et al. 2000; Pasparakis et al. 2002), activation of NF-κB through MALT1 or Bcl-10 is thought to be one of the hallmarks and a key pathogenic event in MALT lymphoma.

Another B-cell malignancy in which the CARD11:MALT1:Bcl-10 complex plays an important pathogenic role is diffuse large B-cell lymphoma (DLBCL). Staudt and coworkers made extensive use of DNA microarray technology to identify genes that are misregulated in different types of DLBCL and arrived at the conclusion that NF-κB is constitutively active in activated B-cell-like (ABC)-DLBCL, but not in germinal center B-cell-like (GCB)-DLBCL (Davis et al. 2001). Importantly, constitutively active NF-κB is required for the survival of ABC-DLBCL (Davis et al. 2001). An shRNA-based screen for genes, whose expression is required for the survival of DLBCL cells, identified CARD11 as the driver of constitutive NF-κB activity in ABC-DLBCL (Ngo et al. 2006). Furthermore, the *CARD11* gene was found to be mutated in about 10% of ABC-DLBCL (Lenz et al. 2008). The mutations all affect residues within the coiled-coil domain of CARD11 and generate a protein that is a constitutive activator of IKK-NF-κB signaling. These results indicate that in a subpopulation of ABC-DLBCL, *CARD11* acts as a *bona fide* oncogene and that its coiled-coil domain serves a negative regulatory function. However, we still do not know what causes the CARD11-dependent activation of NF-κB in the remaining 90% of DLBCL. Nonetheless, it is clear that the CARD11:MALT1:Bcl-10 complex is an important driver of malignant B-cell survival in more than one type of lymphoma.

Another lymphoid malignancy associated with NF-κB activation is multiple myeloma. Although activated NF-κB is a common feature of multiple myeloma, no mutations in NF-κB or IκB encoding genes have been discovered in this disease either. However, extensive genetic analysis of primary tumors and multiple myeloma cell lines have revealed a number of mutations in genes encoding upstream signaling molecules that lead to stabilization and accumulation of NF-κB inducing kinase (NIK), a member of the MAPK kinase kinase (MAP3K) family (Annunziata et al. 2007; Keats et al. 2007). Normally, NIK is a very unstable protein whose activity is kept at a low level because of its rapid turnover (Vallabhapurapu et al. 2008). However, mutations in genes encoding components of an ubiquitin ligase complex responsible for NIK turnover or in the *NIK* gene itself result in accumulation and self-activation of NIK. These mutations include alterations in either NIK or in TRAF3 that disrupt the interactions between the two proteins (Annunziata et al. 2007; Keats et al. 2007). Normally, the binding of TRAF3 to NIK in nonstimulated cells results in the recruitment to NIK of a protein complex composed of the ubiquitin ligases cIAP1 or cIAP2 and TRAF2 and this complex leads to degradative NIK ubiquitination (Vallabhapurapu et al. 2008).

Other multiple myeloma-linked mutations include large deletions affecting the closely linked *cIAP1* and *cIAP2* loci, resulting in the complete absence of their protein products, thereby preventing degradative polyubiquitination of NIK (Annunziata et al. 2007; Keats et al. 2007). More rare mutations abolish the expression of TRAF2 (Keats et al. 2007). Although related in structure to TRAF3, TRAF2 does not directly interact with NIK and instead serves as an activating ubiquitin ligase for cIAP1 and cIAP2, enhancing their ability to polyubiquitinate NIK (Vallabhapurapu et al. 2008). Based on its known ability to activate IKKα and thereby induce processing of NF-κB2/p100 to NF-κB2/p52, it was expected that the elevated and activated NIK in multiple myeloma exerts its oncogenic activity via IKKα (Senftleben et al. 2001). It was, therefore, much of a surprise that only IKKβ inhibition and not IKKα depletion affected the proliferation and survival of multiple myeloma cells (Annunziata et al. 2007).

NF-κB can also be activated in several other lymphoid malignancies as a result of infection with either DNA or RNA tumor viruses. For instance, Epstein-Barr virus (EBV) activates NF-κB through expression of latent membrane protein 1 (LMP1), a protein that can induce lymphomas when expressed in transgenic mice (Eliopoulos and Young 2001; Thornburg et al. 2006). Curiously, LMP1 can induce NIK-dependent NF-κB2/p100 processing (Luftig et al. 2004), but the specific contribution of NF-κB2/p52 formation to lymphomagenesis is not entirely clear, unless NF-κB2/p100 acts as a general NF-κB inhibitor in nontransformed lymphocytes. Kaposi Sarcoma-associated herpesvirus (KSHV) can induce primary effusion lymphoma (PEL) through expression of vFLIP, a viral version of the cFLIP protein that can lead to IKK activation (Liu et al. 2002). Inhibition of NF-κB induces the apoptotic death of PEL cells (Keller et al. 2000). Human T-cell lymphoma virus (HTLV) leads to NF-κB activation through expression of the Tax oncoprotein, which binds to IKKγ/NEMO and induces IKK activation (Carter et al. 2001). Tax can also activate the alternative,

IKKα-dependent, NF-κB pathway (Xiao et al. 2001), but in light of what has been discussed previously, the contribution of alternative NF-κB signaling to lymphomagenesis is not entirely clear.

NF-κB IN COLITIS-ASSOCIATED CANCER: ONCOGENIC COOPERATION BETWEEN NEIGHBORS

Whereas the involvement of NF-κB and its activators in lymphomagenesis was somewhat anticipated, identifying a role for NF-κB in solid malignancies required a conviction in this possibility and the use of specialized mouse models, in which tumor induction depends on inflammation, thus mimicking inflammation-driven cancers in humans. The first such model was a mouse model for colitis-associated cancer (CAC), a type of colon cancer that appears in patients suffering from ulcerative colitis, a chronic inflammatory bowel disease. In this particular model, mice are given azoxymethane (AOM), a procarcinogen that undergoes metabolic activation in intestinal epithelial cells (IEC) and can give use to oncogenic mutations, such as those that lend to activation of β catenin (Greten et al. 2004). Although β-catenin is the most commonly activated oncogene in colon cancer (Morin et al. 1997), AOM alone gives rise to only a small number of large bowel adenomas, which can be strongly augmented through concomitant induction of colonic inflammation that in this model is elicited by repeated administration of the irritant dextrane sulfate sodium (DSS). Using the AOM and DSS model for CAC induction (Okayasu et al. 1996) and conditional disruption of the *Ikk*β gene in mice, we found that IKKβ-driven NF-κB activation in IEC is essential for the development of colonic adenomas (Greten et al. 2004). The oncogenic role of NF-κB in IEC appears to be mediated through its antiapoptotic function (Lin and Karin 2003), mainly through induction of Bcl-X_L, which prevents the apoptotic elimination of premalignant cells (Greten et al. 2004). In addition to its cell-autonomous function in premalignant IEC, IKKβ-driven NF-κB contributes to CAC

development by acting within myeloid cells, most likely within lamina propria macrophages. Activation of NF-κB in these cells was found to stimulate the proliferation of premalignant IEC, through the secretion of growth factors (Greten et al. 2004). No effect of myeloid IKKβ on the survival of IEC was found.

We have searched for NF-κB-dependent factors produced by lamina propria macrophages that stimulate CAC growth. As earlier experiments suggested that IL-6 produced by T cells at late stages of CAC progression enhances adenoma growth (Becker et al. 2004), we first examined the involvement of this cytokine in early tumor promotion. We confirmed that during CAC development, IL-6 is mainly produced by lamina propria macrophages and dendritic cells as initially suspected (Grivennikov et al. 2009). Most importantly, ablation of IL-6 reduced both the multiplicity and size of colonic adenomas in AOM plus DSS-treated mice (Grivennikov et al. 2009). However, unlike the ablation of IKKβ in myeloid cells, which had no effect on the survival of IEC and their premalignant derivatives (Greten et al. 2004), the IL-6 deficiency compromised IEC survival (Grivennikov et al. 2009). Both the proliferative and the survival effects of IL-6 are mediated through activation of the STAT3 transcription factor and the ablation of STAT3 in IEC dramatically compromised IEC survival and greatly reduced CAC growth (Grivennikov et al. 2009). These results suggest that some of the protumorigenic effects of NF-κB activation in myeloid cells could be caused by paracrine signaling to STAT3 in epithelial cells (Fig. 1). In addition, these results suggest that the proproliferative and prosurvival effect of myeloid cell NF-κB on premalignant IEC is predominantly mediated via other cytokines. Likely candidates are IL-22, IL-11, and EGF family members (Bollrath et al. 2009; Pickert et al. 2009). IL-12 family members also play an important role in CAC development and growth, as ablation of the gene encoding the p40 subunit, which is common to both IL-12 and IL-23, greatly diminishes CAC induction and growth (Gri-vennikov and Karin, unpubl.). However, the effect of these cytokines on IEC

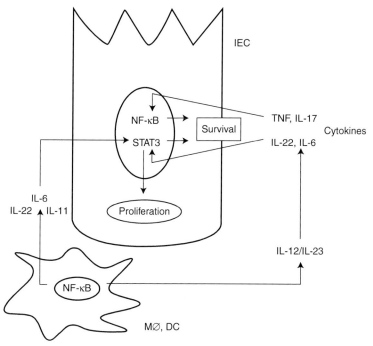

Figure 1. NF-κB-dependent interactions between myeloid (MØ, DC) and intestinal epithelial cells (IEC) drive the development of colitis-associated cancer.

appears to be indirect, because IEC do not express IL-12/IL-23 receptors.

Activation of NF-κB in IEC results in induction of antiapoptotic genes that increase the survival of premalignant cells. In MØ, however, the activation of NF-κB results in production of cytokines, particularly IL-6, IL-11, and IL-22, which drive the proliferation of premalignant IEC. IL-6 and IL-11 exert their proliferative effect via STAT3, which further synergizes with NF-κB to increase the expression of survival genes. NF-κB also drives the production of IL-12/IL-23 cytokines, which amplify the production of prosurvival cytokines.

An alternative mode of action for myeloid cell NF-κB was suggested by Hagemann and coworkers (Hagemann et al. 2008), who found that inhibition of NF-κB activity in tumor-associated macrophages (TAM) through the conditional deletion of IKKβ re-educates these immunosuppressive and protumorigenic cells to acquire a cytotoxic, antitumorigenic phenotype. Importantly, the adoptive transfer of TAMs infected with a dominant–negative

IKKβ adenovirus into mice bearing transplanted ovarian carcinomas resulted in inhibition of tumor growth, which was associated with enhanced tumoricidal activity (Hagemann et al. 2008). Thus, according to this work, another tumor promoting function exerted by IKKβ and NF-κB in TAM is the maintenance of a tumor suppressive phenotype characterized by low levels of inducible NO synthase (iNOS) and IL-12 expression, and high levels of IL-10, TNF-α, and arginase-1. Exactly how NF-κB maintains this immunosuppressive phenotype is not clear, but inhibition of NF-κB in TAMs was seen to result in up-regulation of iNOS and IL-12, leading to elevated production of tumoricidal NO and enhanced activation of NK-cell-mediated antitumor immunity, respectively (Hagemann et al. 2008). It remains to be seen whether this outcome of IKKβ inhibition in myeloid cells also contributes to the results obtained in the CAC model that were described previously, as well as those found in the liver cancer model described later, where IKKβ activation in liver myeloid cells also promotes tumor development. So far, however, inhibition of IKKβ-driven NF-κB in lamina propria myeloid cells was not found to result in increased IL-12 production (Greten et al. 2004).

THE COMPLEX ROLE OF NF-κB IN HEPATOCELLULAR CARCINOMA: LOCATION, LOCATION, LOCATION

Another inflammation-linked cancer is hepatocellular carcinoma (HCC), the most common form of liver cancer. HCC most commonly develops in the context of chronic viral hepatitis caused by either HBV or HCV infection. However, as neither virus infects mice, mouse models of HCC are not based on viral hepatitis. Nonetheless, one mouse model in which spontaneous HCC development is dependent on chronic liver inflammation is the $Mdr2^{-/-}$ knockout mouse, which develops hepatosteatosis caused by defective phospholipid and bile acid export (Mauad et al. 1994). Hepatosteatosis in these mice leads to low grade hepatitis, which eventually results in the development of HCC. In this model, Pikarsky and colleagues have examined the role of hepatocyte NF-κB by expressing a nondegradable form of IκBα from a doxycycline-regulated liver-specific promoter (Pikarsky et al. 2004). Inhibition of NF-κB activation in hepatocytes of $Mdr2^{-/-}$ mice retarded and reduced HCC development. Although the initial stimulus leading to NF-κB activation in $Mdr2^{-/-}$ mice has not been fully identified, it appears to be associated with a chronic inflammatory response that is propagated via paracrine TNF-α production, as treatment of these mice with a neutralizing anti-TNF-α antibody inhibits NF-κB activation in hepatocytes and decreases expression of NF-κB-dependent antiapoptotic genes. The major mechanism by which NF-κB was suggested to exert its tumor promoting function in $Mdr2^{-/-}$ mice is the suppression of apoptosis (Pikarsky et al. 2004). However, the published results are also consistent with a role for hepatocyte NF-κB in the maintenance of chronic inflammation in $Mdr2^{-/-}$ mice that is critical for tumor development.

An entirely different scenario applies to the role of NF-κB in HCC development in mice injected with the procarcinogen diethylnitrosamine (DEN). DEN undergoes metabolic activation in zone 3 hepatocytes and if injected into 2-week-old mice, it acts as a "complete" carcinogen that, unlike AOM, does not require assistance from concurrent inflammation. Nonetheless, DEN-induced HCC requires NF-κB activation in myeloid cells, in this case Kupffer cells, the resident liver macrophages (Maeda et al. 2005). As found in CAC, DEN-induced HCC requires the NF-κB-dependent production of IL-6 by Kupffer cells (Naugler et al. 2007) and the activation of STAT3 by IL-6 in hepatocytes (Yu and Karin, unpubl.) (Fig. 2). However, in a striking difference from CAC and HCC in $Mdr2^{-/-}$ mice, development of DEN-induced HCC is strongly enhanced by inhibition of NF-κB activation in hepatocytes through the targeted deletion of IKKβ (Maeda et al. 2005). An even more striking effect on HCC development is seen upon the conditional deletion of hepatocyte IKKγ/NEMO (Luedde et al. 2007). In this case, the "deleted" mice exhibit spontaneous liver damage and sequentially develop hepatosteatosis, hepatitis, liver fibrosis, and HCC even without any injection of a carcinogen. Enhanced chemical hepatocarcinogenesis was also observed in hepatocyte-specific $p38\alpha$ knockout mice (Hui et al. 2007; Sakurai et al. 2008). Mice lacking either IKKβ ($Ikk\beta^{\Delta hep}$) or p38α ($p38\alpha^{\Delta hep}$) in their hepatocytes exhibit greatly enhanced accumulation of reactive oxygen species (ROS) in zone-3 hepatocytes after DEN exposure (Maeda et al. 2005; Sakurai et al. 2008). As a result of elevated ROS accumulation, which can be prevented by oral administration of antioxidant butylated hydroxyanisol (BHA), both $Ikk\beta^{\Delta hep}$ and $p38\alpha^{\Delta hep}$ mice show increased hepatocyte death. However, in the liver, an organ with unusually high regenerative capacity, cell death triggers compensatory proliferation. We proposed that compensatory proliferation acts as a tumor promoter in situations in which liver tumorigenesis is driven by circles of injury and regeneration, rather than low-grade chronic inflammation, and is therefore the major cause

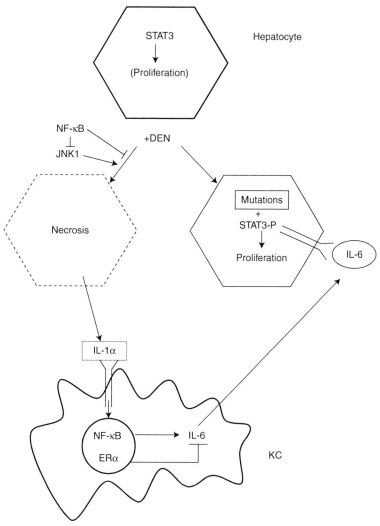

Figure 2. NF-κB in Kupffer cells (KC) and STAT3 in hepatocytes drive the development of DEN-induced hepatocellular carcinoma (HCC).

of enhanced hepatocarcinogenesis in $Ikk\beta^{\Delta hep}$, $p38\alpha^{\Delta hep}$, and $Ikk\gamma^{\Delta hep}$ mice (Sakurai et al. 2006). Indeed, in all of these mutant-mouse strains, administration of BHA prevents liver damage and inhibits compensatory proliferation and, where tested, it fully blocks the increase in hepatocellular carcinogenesis (Maeda et al. 2005; Luedde et al. 2007; Sakurai et al. 2008). Reduced hepatocyte death, compensatory proliferation, and hepatocarcinogenesis were also seen upon crossing of $Ikk\beta^{\Delta hep}$ mice with JNK1-deficient $Jnk1^{-/-}$ mice (Sakurai

et al. 2006). Contrary to IKKβ, JNK1 promotes the death of DEN-exposed hepatocytes and at the same time stimulates compensatory proliferation. Furthermore, ablation of hepatocyte IKKβ results in increased JNK activity (Maeda et al. 2005) because of increased ROS accumulation (Kamata et al. 2005). Collectively, these results indicate that the major function of hepatocyte NF-κB in DEN-administered mice or even in unchallenged mice is to maintain hepatocyte survival and liver homeostasis, in part by suppressing cytotoxic ROS accumulation.

In the mouse, after 2 weeks of age, most hepatocytes withdraw from the cell cycle and arrest in G_0. The same applies to human liver, although in this case cell-cycle withdrawal occurs at a later time point. Carcinogen exposure in a tissue that does not undergo active proliferation, such as the uninjured liver, can not easily give rise to cancer. Therefore, any injury or an alteration that augments hepatocyte death and gives rise to compensatory proliferation will enhance HCC development. However, in the colon, IEC undergo continuous renewal and the absence of NF-κB in such cells does not further enhance cell proliferation, resulting in a net increase in cell death. Under these circumstances, increased elimination of premalignant cells is the dominant outcome of NF-κB inhibition, resulting in reduced tumorigenesis. Thus, by acting in different cells subject to different tissue kinetics, NK-κB can either enhance or suppress tumorigenesis.

Administration of DEN results in induction of oncogenic mutations in some hepatocytes and the necrotic death of others. Necrotic hepatocytes release IL-1α that leads to activation of NF-κB in Kupffer cells. This results in induction of IL-6, which is negatively regulated by estrogen receptor (ER)α. The IL-6 produced by Kupffer cells acts on neighboring hepatocytes to activate STAT3 and induce the expression of proliferation-promoting genes. If these cells harbor oncogenic mutations, their proliferation would eventually give rise to HCC.

Importantly, in all of the models discussed so far, inhibition of NF-κB in myeloid cells reduces tumor development. Furthermore, as found in CAC, the major protumorigenic effect of NF-κB in Kupffer cells is mediated through the induction of IL-6, which is inhibited by activation of estrogen receptor (ER)α (Naugler et al. 2007). We found that DEN administration, especially in $Ikk\beta^{\Delta hep}$ and $p38\alpha^{\Delta hep}$ mice, gives rise to NF-κB activation in Kupffer cells in a manner that depends on induction of hepatocyte necrosis (Sakurai et al. 2008). In this case, the primary mediator of NF-κB activation in Kupffer cells is IL-1α, which is released in large amounts by necrotic hepatocytes (Fig. 2). Importantly, mice that are deficient in IL-1 receptor or its adaptor protein MyD88 are quite refractory to DEN-induced hepatocarcinogenesis, demonstrating the importance of the IL-1α-mediated cross talk between dying hepatocytes and Kupffer cells (Sakurai et al. 2008).

Interestingly, the incidence of HCC is three to five times higher in men than in women (Bosch et al. 2004) and the same applies to DEN-induced HCC in mice (Naugler et al. 2007). As mentioned previously, production of IL-6 by Kupffer cells exposed to IL-1α or other NF-κB activators is negatively regulated by ERα. Thus, DEN-treated female mice produce less IL-6 than similarly treated male mice and contain less activated STAT3 in their hepatocytes (Naugler et al. 2007). Ablation of the *Il6* gene abolishes the gender difference in HCC induction, whereas ovariectomy enhances IL-6 production and augments HCC induction in female mice (Naugler et al. 2007). It is likely that gender-specific differences in IL-6 expression also affect the incidence of human HCC, as serum IL-6 is higher after menopause (Jilka et al. 1992; Ershler and Keller 2000) and postmenopausal women display higher HCC incidence than premenopausal women (Bosch et al. 2004). Recently, elevated serum IL-6 was found to be associated with rapid progression from chronic viral hepatitis to frank HCC in a large cohort of HBV-positive patients in Hong Kong (Wong et al. 2009).

AN IKK ACTIVATION CASCADE IN PROSTATE CANCER: AN IKKβ-IKKα RELAY

Although HCC and CAC are clearly inflammation-linked cancers, there are many other cancers that rarely arise in the context of underlying inflammation or infection and yet are dependent on inflammatory processes, most of which occur as a consequence of tumor progression. One such cancer is prostate cancer (CaP), which is the most common malignancy in older men. We have used the TRAMP mouse in which CaP development and progression are driven by expression of SV40 T antigen in prostate epithelial cells (Greenberg et al. 1995; Gingrich et al. 1996) to study the role of IKK signaling in prostate tumorigenesis. We first examined whether deletion of IKKβ in prostate epithelial cells has any effect on CaP development and found no effect whatsoever, neither on tumor development and progression, nor on the development of androgen-independent (AI) cancer after castration (Ammirante et al., in prep.). The latter results were surprising, as AI CaP usually exhibits activated NF-κB (Gasparian et al. 2002). These findings led us to examine whether IKKβ in hematopoietic-derived cells has a role in CaP development. Although deletion of IKKβ in the hematopoietic compartment had no effect on development and progression of primary CaP in TRAMP mice, it slowed down the development of AI CaP after castration and inhibited the appearance of metastases (Ammirante et al., in prep.). Similar results were obtained in a different model based on subcutaneous implantation of the androgen-dependent (AD) mouse CaP cell line, Myc-CaP. In this case, the tumors were allowed to grow to a size of 1000 mm^3 before castration of the hosts, which subjects the tumor to androgen deprivation, causing near complete regression because of necrotic and apoptotic death of CaP cells, which depend on androgens (testosterone) for survival. However, through mechanisms that are not entirely understood, almost as soon as the original AD CaP disappear, an AI tumor starts growing, as is often the case in prostate cancer patients undergoing androgen ablation therapy. Silencing of endogenous IKKβ in Myc-CaP cells had no effect on their primary tumorigenic growth, regression upon castration, and regrowth as AI CaP. However, deletion of IKKβ in bone-marrow-derived cells (BMDC) of the host substantially slowed down the regrowth of AI CaP in castrated tumor-bearing mice (Ammirante et al., in prep.). A similar delay in the regrowth of AI CaP was seen upon treatment of tumor-bearing hosts with IKKβ inhibitors. As found in both the CAC and HCC models described above, IKKβ ablation in BMDC-inhibited STAT3 activation in CaP cells and a STAT3 inhibitor slowed down the emergence of AI CaP (Ammirante et al., in prep.).

Curiously, the development of AI CaP is associated with the accumulation of activated IKKα in nuclei of CaP cells (Ammirante et al., in prep.). Accumulation of nuclear IKKα was previously found to be linked with and necessary for metastatic progression of CaP in TRAMP mice (Luo et al. 2007). Furthermore, accumulation of nuclear IKKα correlated with progression and clinical grade in human CaP (Luo et al. 2007). Importantly, IKKβ in BMCD is required for activation of nuclear IKKα in CaP cells through the production of IKKα-activating cytokines. The silencing of IKKα in Myc-CaP cells delays the emergence of AI CaP as effectively as the inhibition of IKKβ does.

The mechanism by which nuclear IKKα contributes to the growth of AI CaP remains to be determined, but previous studies on metastatic progression in TRAMP mice revealed that nuclear IKKα enhances metastatic progression by repressing transcription of the metastasis inhibitor maspin (Luo et al. 2007). Although the details of maspin repression by IKKα are not fully known, it is clear that it does not involve activation of either canonical or noncanonical NF-κB signaling (Luo et al. 2007). Repression of maspin requires the kinase activity of IKKα, suggesting that it is exerted through the phosphorylation of another protein, possibly a component of chromatin, which is involved in the regulation of maspin transcription. It remains to be seen whether repression of maspin contributes to the emergence of AI CaP, but maspin was shown to have antiproliferative and proapoptotic activities (Lockett et al. 2006).

Concurrent with the death of AD CaP, androgen withdrawal results in massive inflammatory infiltration of the tumor remnant. Most immune and inflammatory cell types are present within the regressing tumor transiently and only B cells are the ones that remain within the newly emerging AI CaP for at least 3 weeks after castration (Ammirante et al., in prep.). Curiously, we found that about 90% of human prostate tumors, but not normal, hyperplastic, or malignant tissue, contain B cells as well. Most importantly, B cells, but not T cells, were found to be required for the rapid emergence of AI tumors as well as for IKKα and STAT3 activation in CaP cells of such tumors (Ammirante et al., in prep.). Most likely, B cells are required for production of cytokines that lead to IKKα and STAT3 activation within newly emerging AI CaP cells (Fig. 3). The activation of IKKα and STAT3 is likely to be required for the survival and proliferation of these cells.

In androgen-dependent (AD) CaP, IKKα, and STAT3 are not activated and are located in the cytosol. Androgen ablation results in the death of most AD CaP cells. In response to the release of inflammatory mediators by the dying AD CaP cells, inflammatory and immune cells, including B cells, are recruited into the tumor remnant. In these tumor-infiltrating B cells, IKKβ is activated, resulting in the production of NF-κB-dependent cytokines that lead to activation of STAT3 and nuclear translocation of activated IKKα in remaining CaP cells. STAT3 and IKKα promote the survival of cells that have become independent of androgens,

leading to the development of AI CaP. Because of the presence of nuclear IKKα, AI CaP is often metastatic.

In summary, the findings described previously indicate that even a cancer whose development is not associated with an underlying inflammatory condition depends on an NF-κB-regulated inflammatory response. In the case of prostate cancer, tumor-associated inflammation is part of normal progression, but can also be elicited and accelerated as a result of therapy (in this case, androgen ablation)-induced death of the primary tumor. It is possible that in both cases, the localized tumor-associated inflammatory response is triggered by the necrotic death of malignant cells, either as the result of hypoxia during normal progression or as the consequence of therapeutic intervention. In the prostate cancer models described previously, the inflammatory response elicited by androgen deprivation is a major contributor to the emergence of AI CaP. The dependence of this response on IKKβ, B cells, STAT3, and IKKα (Fig. 3) suggests that

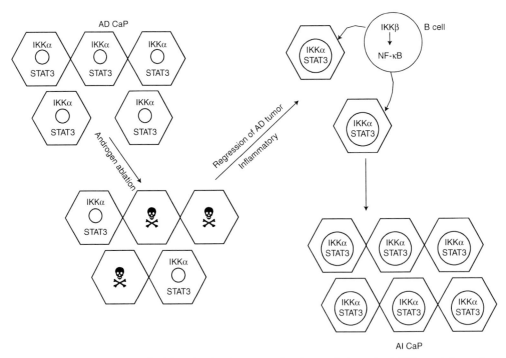

Figure 3. IKKβ–IKKα cross talk drives the development of androgen-independent (AI) prostate carcinoma (CaP).

therapeutic interventions targeting any of these four elements may be used to improve the outcome of androgen ablation therapy and delay the appearance of AI CaP.

CONCLUSIONS AND TRANSLATIONAL IMPLICATIONS

In the three types of epithelial cancers described previously, inactivation of IKKβ in premalignant tumor progenitors or in the neoplastic cell itself can either inhibit tumor development (CAC), enhance tumor development (HCC), or have no discernable effect (CaP). However, in all three cancers, the inactivation or inhibition of IKKβ in cells of the hematopoietic compartment (lamina propria macrophages, Kupffer cells, or B cells) inhibits tumor growth and progression. We assume that the tumor-promoting effect of IKKβ in such cells is mediated via NF-κB and at least in two cases (CAC and HCC), we know that it depends, at least in part, on the production of the NF-κB regulated cytokine IL-6. In all three cases, NF-κB signaling in cells of the hematopoietic compartment results in activation of STAT3, a transcription factor that controls the expression of proliferative and survival genes in premalignant cells and their fully neoplastic derivatives. In one case (CaP), IKKβ signaling in tumor-infiltrating lymphocytes results in the accumulation of activated IKKα in cancer cell nuclei. Thus, it can be generalized that IKKβ-dependent signaling to NF-κB in tumor-associated inflammatory and immune cells results in the production of cytokines that activate signaling pathways that stimulate the proliferation and enhance the survival of malignant carcinoma cells.

These findings suggest that regardless of the direct effect of NF-κB on the survival of neoplastic cells, IKKβ inhibition can be used to slow down tumor growth and enhance susceptibility to cytocidal therapeutics (e.g., genotoxic chemicals, microtubule disruptors, ionizing radiation, and apoptosis-inducing cytokines). In this regard, it is important to realize that the mere disruption of NF-κB or STAT3 signaling does not lead to cell death.

Hence, the requirement for classical cytocidal therapy for which IKKβ inhibitors can serve as adjuvants.

It should also be recognized that unless the cytocidal agent being used in conjunction with an IKKβ or another NF-κB inhibitor has a wide safety margin, the systemic inhibition of NF-κB function may result in enhanced toxicity to normal cells and tissues. Two ways to circumvent this problem are: (a) direct the IKKβ or NF-κB inhibitor to the relevant inflammatory or immune cell type; and (b) direct the cytocidal therapy to the neoplastic cell. Another potential complication of IKKβ or NF-κB inhibition is long-lasting immune suppression and unpredictable effects on the inflammatory response. For instance, we have found that instead of reducing sepsis-induced inflammation, systemic administration of a specific IKKβ inhibitor or the ablation of IKKβ in myeloid cells resulted in greatly enhanced inflammation and mortality driven by elevated IL-1β production in mice infected with bacteria or challenged with endotoxin (Greten et al. 2007). Thus, a great deal of caution needs to be exerted during clinical development of IKKβ inhibitors as adjuvants to chemotherapy or radiation. Limiting the duration of treatment with an IKKβ inhibitor to a short period preceding or concurrent with cytocidal therapy may provide a solution to this potential complication.

Given the complexities and unpredictable nature of anti-IKKβ therapy, it is worthwhile considering a more specific approach that targets the IKKβ-dependent cytokine responsible for tumor growth and survival. For instance, in the case of HCC and CAC, it may make sense to target IL-6 by the use of anti-IL-6 receptor antibodies. Although the cytokines that are critical for the growth of AI CaP and its metastatic spread remain to be identified, our initial analysis suggests that they may be RANK ligand (RANKL) or lymphotoxin (LT) (Luo et al. 2007). As both anti-RANKL and anti-LT therapeutics have already been developed for other indications, it is worthwhile to assess their effect on the emergence of AI CaP and its progression into metastatic disease, first in animal models and if positive, in prostate cancer patients receiving androgen ablation therapy.

In summary, much has been learned about the role of inflammation and inflammatory processes in the pathogenesis of cancer by studying the oncogenic functions of NF-κB. It is our hope that the next stage in this long endeavor would see this basic knowledge arriving at the bedside to result in new and improved cancer therapies and preventive agents.

REFERENCES

Annunziata CM, Davis RE, Demchenko Y, Bellamy W, Gabrea A, Zhan F, Lenz G, Hanamura I, Wright G, Xiao W, et al. 2007. Frequent engagement of the classical and alternative NF-κB pathways by diverse genetic abnormalities in multiple myeloma. *Cancer Cell* **12:** 115–130.

Balkwill F, Mantovani A. 2001. Inflammation and cancer: Back to Virchow? *Lancet* **357:** 539–545.

Becker C, Fantini MC, Schramm C, Lehr HA, Wirtz S, Nikolaev A, Burg J, Strand S, Kiesslich R, Huber S, et al. 2004. TGF-β suppresses tumor progression in colon cancer by inhibition of IL-6 trans-signaling. *Immunity* **21:** 491–501.

Beg AA, Baltimore D. 1996. An essential role for NF-κB in preventing TNF-α-induced cell death. *Science* **274:** 782–784.

Boehm JS, Zhao JJ, Yao J, Kim SY, Firestein R, Dunn IF, Sjostrom SK, Garraway LA, Weremowicz S, Richardson AL, et al. 2007. Integrative genomic approaches identify IKBKE as a breast cancer oncogene. *Cell* **129:** 1065–1079.

Bollrath J, Phesse TJ, von Burstin VA, Putoczki T, Bennecke M, Bateman T, Nebelsiek T, Lundgren-May T, Canli O, Schwitalla S, et al. 2009. gp130-mediated Stat3 activation in enterocytes regulates cell survival and cell-cycle progression during colitis-associated tumorigenesis. *Cancer Cell* **15:** 91–102.

Bosch FX, Ribes J, Diaz M, Cleries R. 2004. Primary liver cancer: Worldwide incidence and trends. *Gastroenterology* **127:** S5–S16.

Bours V, Franzoso G, Azarenko V, Park S, Kanno T, Brown K, Siebenlist U. 1993. The oncoprotein Bcl-3 directly transactivates through κ B motifs via association with DNA-binding p50B homodimers. *Cell* **72:** 729–739.

Cabannes E, Khan G, Aillet F, Jarrett RF, Hay RT. 1999. Mutations in the IkBa gene in Hodgkin's disease suggest a tumour suppressor role for IκBα. *Oncogene* **18:** 3063–3070.

Carter RS, Geyer BC, Xie M, Acevedo-Suarez CA, Ballard DW. 2001. Persistent activation of NF-κ B by the tax transforming protein involves chronic phosphorylation of IκB kinase subunits IKKβ and IKKγ. *J Biol Chem* **276:** 24445–24448.

Davis RE, Brown KD, Siebenlist U, Staudt LM. 2001. Constitutive nuclear factor κB activity is required for survival of activated B cell-like diffuse large B cell lymphoma cells. *J Exp Med* **194:** 1861–1874.

Eliopoulos AG, Young LS. 2001. LMP1 structure and signal transduction. *Semin Cancer Biol* **11:** 435–444.

Ershler WB, Keller ET. 2000. Age-associated increased interleukin-6 gene expression, late-life diseases, and frailty. *Annu Rev Med* **51:** 245–270.

Franzoso G, Bours V, Park S, Tomita-Yamaguchi M, Kelly K, Siebenlist U. 1992. The candidate oncoprotein Bcl-3 is an antagonist of p50/NF-κB-mediated inhibition. *Nature* **359:** 339–342.

Gasparian AV, Yao YJ, Kowalczyk D, Lyakh LA, Karseladze A, Slaga TJ, Budunova IV. 2002. The role of IKK in constitutive activation of NF-κB transcription factor in prostate carcinoma cells. *J Cell Sci* **115:** 141–151.

Gilmore TD. 2003. The Re1/NF-κ B/I κ B signal transduction pathway and cancer. *Cancer Treat Res* **115:** 241–265.

Gingrich JR, Barrios RJ, Morton RA, Boyce BF, DeMayo FJ, Finegold MJ, Angelopoulou R, Rosen JM, Greenberg NM. 1996. Metastatic prostate cancer in a transgenic mouse. *Cancer Res* **56:** 4096–4102.

Greenberg NM, DeMayo F, Finegold MJ, Medina D, Tilley WD, Aspinall JO, Cunha GR, Donjacour AA, Matusik RJ, Rosen JM. 1995. Prostate cancer in a transgenic mouse. *Proc Natl Acad Sci* **92:** 3439–3443.

Greten FR, Arkan MC, Bollrath J, Hsu LC, Goode J, Miething C, Goktuna SI, Neuenhahn M, Fierer J, Paxian S, et al. 2007. NF-κB is a negative regulator of IL-1β secretion as revealed by genetic and pharmacological inhibition of IKKβ. *Cell* **130:** 918–931.

Greten FR, Eckmann L, Greten TF, Park JM, Li ZW, Egan LJ, Kagnoff MF, Karin M. 2004. IKKβ links inflammation and tumorigenesis in a mouse model of colitis-associated cancer. *Cell* **118:** 285–296.

Grivennikov S, Karin E, Terzic J, Mucida D, Yu GY, Vallabhapurapu S, Scheller J, Rose-John S, Cheroutre H, Eckmann L, et al. 2009. IL-6 and stat3 are required for survival of intestinal epithelial cells and development of colitis-associated cancer. *Cancer Cell* **15:** 103–113.

Grossmann M, O'Reilly LA, Gugasyan R, Strasser A, Adams JM, Gerondakis S. 2000. The anti-apoptotic activities of Rel and RelA required during B-cell maturation involve the regulation of Bcl-2 expression. *EMBO J* **19:** 6351–6360.

Gugasyan R, Grumont R, Grossmann M, Nakamura Y, Pohl T, Nesic D, Gerondakis S. 2000. Rel/NF-κB transcription factors: Key mediators of B-cell activation. *Immunol Rev* **176:** 134–140.

Hagemann T, Lawrence T, McNeish I, Charles KA, Kulbe H, Thompson RG, Robinson SC, Balkwill FR. 2008. "Re-educating" tumor-associated macrophages by targeting NF-κB. *J Exp Med* **205:** 1261–1268.

Huang S, Pettaway CA, Uehara H, Bucana CD, Fidler IJ. 2001. Blockade of NF-κB activity in human prostate cancer cells is associated with suppression of angiogenesis, invasion, and metastasis. *Oncogene* **20:** 4188–4197.

Hui L, Bakiri L, Mairhorfer A, Schweifer N, Haslinger C, Kenner L, Komnenovic V, Scheuch H, Beug H, Wagner

EF. 2007. p38α suppresses normal and cancer cell proliferation by antagonizing the JNK-c-Jun pathway. *Nat Genet* **39:** 741–749.

Jilka RL, Hangoc G, Girasole G, Passeri G, Williams DC, Abrams JS, Boyce B, Broxmeyer H, Manolagas SC. 1992. Increased osteoclast development after estrogen loss: Mediation by interleukin-6. *Science* **257:** 88–91.

Joyce D, Albanese C, Steer J, Fu M, Bouzahzah B, Pestell RG. 2001. NF-κB and cell-cycle regulation: The cyclin connection. *Cytokine Growth Factor Rev* **12:** 73–90.

Kamata H, Honda S, Maeda S, Chang L, Hirata H, Karin M. 2005. Reactive oxygen species promote TNFα-induced death and sustained JNK activation by inhibiting MAP kinase phosphatases. *Cell* **120:** 649–661.

Karin M. 2006. Nuclear factor-κB in cancer development and progression. *Nature* **441:** 431–436.

Karin M, Cao Y, Greten FR, Li ZW. 2002. NF-κB in cancer: From innocent bystander to major culprit. *Nat Rev Cancer* **2:** 301–310.

Keats JJ, Fonseca R, Chesi M, Schop R, Baker A, Chng WJ, Van Wier S, Tiedemann R, Shi CX, Sebag M, et al. 2007. Promiscuous mutations activate the noncanonical NF-κB pathway in multiple myeloma. *Cancer Cell* **12:** 131–144.

Keller SA, Schattner EJ, Cesarman E. 2000. Inhibition of NF-κB induces apoptosis of KSHV-infected primary effusion lymphoma cells. *Blood* **96:** 2537–2542.

Lenz G, Davis RE, Ngo VN, Lam L, George TC, Wright GW, Dave SS, Zhao H, Xu W, Rosenwald A, et al. 2008. Oncogenic CARD11 mutations in human diffuse large B cell lymphoma. *Science* **319:** 1676–1679.

Lin A, Karin M. 2003. NF-κB in cancer: A marked target. *Semin Cancer Biol* **13:** 107–114.

Liu L, Eby MT, Rathore N, Sinha SK, Kumar A, Chaudhary PM. 2002. The human herpes virus 8-encoded viral FLICE inhibitory protein physically associates with and persistently activates the Iκ B kinase complex. *J Biol Chem* **277:** 13745–13751.

Liu ZG, Hsu H, Goeddel DV, Karin M. 1996. Dissection of TNF receptor 1 effector functions: JNK activation is not linked to apoptosis while NF-κB activation prevents cell death. *Cell* **87:** 565–576.

Lockett J, Yin S, Li X, Meng Y, Sheng S. 2006. Tumor suppressive maspin and epithelial homeostasis. *J Cell Biochem* **97:** 651–660.

Luedde T, Beraza N, Kotsikoris V, van Loo G, Nenci A, De Vos R, Roskams T, Trautwein C, Pasparakis M. 2007. Deletion of NEMO/IKKγ in liver parenchymal cells causes steatohepatitis and hepatocellular carcinoma. *Cancer Cell* **11:** 119–132.

Luftig M, Yasui T, Soni V, Kang MS, Jacobson N, Cahir-McFarland E, Seed B, Kieff E. 2004. Epstein-Barr virus latent infection membrane protein 1 TRAF-binding site induces NIK/IKK α-dependent noncanonical NF-κB activation. *Proc Natl Acad Sci* **101:** 141–146.

Luo JL, Tan W, Ricono JM, Korchynskyi O, Zhang M, Gonias SL, Cheresh DA, Karin M. 2007. Nuclear cytokine-activated IKKα controls prostate cancer metastasis by repressing Maspin. *Nature* **446:** 690–694.

Maeda S, Kamata H, Luo JL, Leffert H, Karin M. 2005. IKKβ couples hepatocyte death to cytokine-driven compensatory proliferation that promotes chemical hepatocarcinogenesis. *Cell* **121:** 977–990.

Mauad TH, van Nieuwkerk CM, Dingemans KP, Smit JJ, Schinkel AH, Notenboom RG, van den Bergh Weerman MA, Verkruisen RP, Groen AK, Oude Elferink RP, et al. 1994. Mice with homozygous disruption of the mdr2 P-glycoprotein gene. A novel animal model for studies of nonsuppurative inflammatory cholangitis and hepatocarcinogenesis. *Am J Pathol* **145:** 1237–1245.

Morin PJ, Sparks AB, Korinek V, Barker N, Clevers H, Vogelstein B, Kinzler KW. 1997. Activation of β-catenin-Tcf signaling in colon cancer by mutations in β-catenin or APC. *Science* **275:** 1787–1790.

Naugler WE, Sakurai T, Kim S, Maeda S, Kim K, Elsharkawy AM, Karin M. 2007. Gender disparity in liver cancer due to sex differences in MyD88-dependent IL-6 production. *Science* **317:** 121–124.

Neri A, Chang CC, Lombardi L, Salina M, Corradini P, Maiolo AT, Chaganti RS, Dalla-Favera R. 1991. B cell lymphoma-associated chromosomal translocation involves candidate oncogene lyt-10, homologous to NF-κ B p50. *Cell* **67:** 1075–1087.

Ngo VN, Davis RE, Lamy L, Yu X, Zhao H, Lenz G, Lam LT, Dave S, Yang L, Powell J, et al. 2006. A loss-of-function RNA interference screen for molecular targets in cancer. *Nature* **441:** 106–110.

Okayasu I, Ohkusa T, Kajiura K, Kanno J, Sakamoto S. 1996. Promotion of colorectal neoplasia in experimental murine ulcerative colitis. *Gut* **39:** 87–92.

Pasparakis M, Schmidt-Supprian M, Rajewsky K. 2002. IκB kinase signaling is essential for maintenance of mature B cells. *J Exp Med* **196:** 743–752.

Pickert G, Neufert C, Leppkes M, Zheng Y, Wittkopf N, Warntjen M, Lehr H-A, Hirth S, Weigmann B, Wirtz S, et al. 2009. STAT3 links IL-22 signaling in intestinal epithelial cells to mucosal wound healing. *J Exp Med* **206:** 1465–1472.

Pikarsky E, Porat RM, Stein I, Abramovitch R, Amit S, Kasem S, Gutkovich-Pyest E, Urieli-Shoval S, Galun E, Ben-Neriah Y. 2004. NF-κB functions as a tumour promoter in inflammation-associated cancer. *Nature* **431:** 461–466.

Sakurai T, He G, Matsuzawa A, Yu GY, Maeda S, Hardiman G, Karin M. 2008. Hepatocyte necrosis induced by oxidative stress and IL-1 α release mediate carcinogen-induced compensatory proliferation and liver tumorigenesis. *Cancer Cell* **14:** 156–165.

Sakurai T, Maeda S, Chang L, Karin M. 2006. Loss of hepatic NF-κ B activity enhances chemical hepatocarcinogenesis through sustained c-Jun N-terminal kinase 1 activation. *Proc Natl Acad Sci* **103:** 10544–10551.

Senftleben U, Cao Y, Xiao G, Greten FR, Krahn G, Bonizzi G, Chen Y, Hu Y, Fong A, Sun SC, et al. 2001. Activation by IKKα of a second, evolutionary conserved, NF-κ B signaling pathway. *Science* **293:** 1495–1499.

Thornburg NJ, Kulwichit W, Edwards RH, Shair KH, Bendt KM, Raab-Traub N. 2006. LMP1 signaling and activation of NF-κB in LMP1 transgenic mice. *Oncogene* **25:** 288–297.

Uren AG, O'Rourke K, Aravind LA, Pisabarro MT, Seshagiri S, Koonin EV, Dixit VM. 2000. Identification of paracaspases and metacaspases: Two ancient families of caspase-like proteins, one of which plays a key role in MALT lymphoma. *Mol Cell* **6:** 961–967.

Vallabhapurapu S, Matsuzawa A, Zhang W, Tseng PH, Keats JJ, Wang H, Vignali DA, Bergsagel PL, Karin M. 2008. Nonredundant and complementary functions of TRAF2 and TRAF3 in a ubiquitination cascade that activates NIK-dependent alternative NF-κB signaling. *Nat Immunol* **9:** 1364–1370.

Van Antwerp DJ, Martin SJ, Kafri T, Green DR, Verma IM. 1996. Suppression of TNF-α-induced apoptosis by NF-κB. *Science* **274:** 787–789.

Wang CY, Mayo MW, Baldwin AS Jr. 1996. TNF- and cancer therapy-induced apoptosis: Potentiation by inhibition of NF-κB. *Science* **274:** 784–787.

Willis TG, Jadayel DM, Du MQ, Peng H, Perry AR, Abdul-Rauf M, Price H, Karran L, Majekodunmi O, Wlodarska I, et al. 1999. Bcl10 is involved in t(1;14)(p22;q32) of MALT B cell lymphoma and mutated in multiple tumor types. *Cell* **96:** 35–45.

Wong VW, Yu J, Cheng AS, Wong GL, Chan HY, Chu ES, Ng EK, Chan FK, Sung JJ, Chan HL. 2009. High serum interleukin-6 level predicts future hepatocellular carcinoma development in patients with chronic hepatitis B. *Int J Cancer* **124:** 2766–2770.

Xiao G, Cvijic ME, Fong A, Harhaj EW, Uhlik MT, Waterfield M, Sun SC. 2001. Retroviral oncoprotein Tax induces processing of NF-κB2/p100 in T cells: Evidence for the involvement of IKKα. *EMBO J* **20:** 6805–6815.

The Nuclear Factor NF-κB Pathway in Inflammation

Toby Lawrence

Inflammation Biology Group, Centre d'Immunologie Marseille-Luminy, Parc Scientifique de Luminy, Case 906, 13288 Marseille, France

Correspondence: lawrence@ciml.univ-mrs.fr

The nuclear factor NF-κB pathway has long been considered a prototypical proinflammatory signaling pathway, largely based on the role of NF-κB in the expression of proinflammatory genes including cytokines, chemokines, and adhesion molecules. In this article, we describe how genetic evidence in mice has revealed complex roles for the NF-κB in inflammation that suggest both pro- and anti-inflammatory roles for this pathway. NF-κB has long been considered the "holy grail" as a target for new anti-inflammatory drugs; however, these recent studies suggest this pathway may prove a difficult target in the treatment of chronic disease. In this article, we discuss the role of NF-κB in inflammation in light of these recent studies.

NF-κB has long been considered a prototypical proinflammatory signaling pathway, largely based on the activation of NF-κB by proinflammatory cytokines such as interleukin 1 (IL-1) and tumor necrosis factor α (TNFα), and the role of NF-κB in the expression of other proinflammatory genes including cytokines, chemokines, and adhesion molecules, which has been extensively reviewed elsewhere. But inflammation is a complex physiological process and the role of NF-κB in the inflammatory response cannot be extrapolated from in vitro studies. In this article, we describe how genetic evidence in mice has revealed complex roles for the NF-κB pathway in inflammation.

ACTIVATION OF NF-κB IN INFLAMMATION

The inflammatory response is characterized by coordinate activation of various signaling pathways that regulate expression of both pro- and anti-inflammatory mediators in resident tissue cells and leukocytes recruited from the blood. Currently, most of our knowledge of signaling in inflammation is gained from studying members of the IL-1 and TNF receptor families and the Toll-like microbial pattern recognition receptors (TLRs), which belong to the IL-1R family. IL-1 and TNFα represent the archetypal proinflammatory cytokines that are rapidly released on tissue injury or infection. TLRs recognize microbial molecular patterns, hence the term pattern recognition receptor (PRR). TLRs represent a germline encoded nonself recognition system that is hardwired to trigger inflammation (Akira et al. 2006). However, there is some suggestion that endogenous ligands may trigger TLRs during tissue injury and certain disease states, which may act to promote inflammation in the absence of infection (Karin et al. 2006). Although structurally different, these receptors use similar signal transduction mechanisms that include activation of IκB kinase (IKK) and NF-κB (Ghosh and Karin 2002). In recent years, it has become clear that there are at least two separate pathways for NF-κB activation (Fig. 1). The "canonical" pathway is triggered by microbial products and proinflammatory cytokines such as TNFα and IL-1 as described previously, usually leading to activation of RelA- or cRel-containing complexes (Karin and Ben-Neriah 2000). An "alternative" NF-κB pathway is activated by TNF-family cytokines—lymphotoxin β (TNFSF3) (Senftleben et al. 2001a; Dejardin et al. 2002), CD40 ligand (CD40L and TNFSF5) (Senftleben et al. 2001a), B cell activating factor (BAFF and TNFSF13B) (Bonizzi et al. 2004), and receptor activator of NF-κB ligand (RANKL and TNFSF11) (Novack et al. 2003)—but not TNFα (Matsushima et al. 2001; Dejardin et al. 2002; Bonizzi et al. 2004),

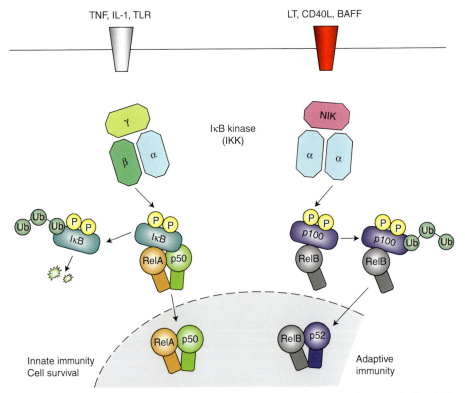

Figure 1. Canonical and alternative NF-κB pathways. This diagram illustrates the canonical and alternative pathways for NF-κB activation. The canonical pathway is triggered by TLRs and proinflammatory cytokines such as TNFα and IL-1, leading to activation of RelA that regulates expression of proinflammatory and cell survival genes. The alternative NF-κB pathway is activated by LT β, CD40L, BAFF, and RANKL, but not TNFα, and results in activation of RelB/p52 complexes. Activation of the alternative pathway regulates genes required for lymph-organogenesis and B-cell activation. These pathways are characterized by the differential requirement for IKK subunits. IKKβ regulates activation of the canonical pathway through phosphorylation of IκBs and requires the IKKγ subunit but not IKKα, whereas IKKα is required for activation of the alternative pathway through the phosphorylation and processing of p100, the precursor for p52, but this is independent of both IKKβ and IKKγ.

Cite this article as *Cold Spring Harb Perspect Biol* 2009;1:a001651

resulting in activation of RelB/p52 complexes (Bonizzi and Karin 2004). These pathways are characterized by the differential requirement for IKK subunits. The IKK complex consists of two kinase subunits, IKKα (IKK1) and IKKβ (IKK2), and a regulatory subunit IKKγ (NEMO). IKKβ regulates activation of the canonical pathway through phosphorylation of IκBs and requires the IKKγ subunit but not IKKα (Zandi et al. 1997). IKKα is required for activation of the alternative pathway through the phosphorylation and processing of p100, the precursor for p52 (Senftleben et al. 2001a), and this is independent of both IKKβ and IKKγ (Ghosh and Karin 2002).

THE CANONICAL NF-κB PATHWAY

The canonical NF-κB pathway has been defined primarily in response to TNFα and IL-1 signaling, prototypical proinflammatory cytokines that have important roles in the pathogenesis of chronic inflammatory diseases such as rheumatoid arthrtitis (RA), inflammatory bowel disease (IBD), asthma, and chronic obstructive pulmonary disease (COPD) (Holgate 2004; Chung 2006; Williams et al. 2007). NF-κB activation is also widely implicated in inflammatory diseases (Table 1) (Tak and Firestein 2001) and much attention has focused on the development of anti-inflammatory drugs targeting NF-κB (Karin et al. 2004).

Invariably, NF-κB activity at sites of inflammation is associated with activation of the canonical pathway and RelA- or cRel-containing complexes. There have been several studies to show proinflammatory cytokine and chemokine

Table 1. Chronic inflammatory diseases associated with NF-κB activation

NF-κB activation in human inflammatory diseases
Rheumatoid arthritis
Atherosclerosis
Chronic obstructive pulmonary disease (COPD)
Asthma
Multiple sclerosis
Inflammatory bowel disease (IBD)
Ulcerative colitis

Adapted from Tak and Firestein 2001.

production by disease tissue is NF-κB-dependent; for example, using fibroblast-like synoviocytes from RA patients (Aupperle et al. 1999; Aupperle et al. 2001). Similar studies have shown that proinflammatory cytokine production in human atherosclerotic plaques is also NF-κB-dependent (Monaco et al. 2004). However, these studies relied on ex vivo tissue culture systems and expression of dominant–negative inhibitors or overexpression of IκBα that may not reflect the role of NF-κB in the disease context. There has also been correlation of NF-κB activation with inflammatory disease in animal models of arthritis (Miagkov et al. 1998) and allergic airway disease (Poynter et al. 2002). But the association of NF-κB activity and inflammatory disease is not easy to interpret because both pro- and anti-inflammatory mediators are produced during inflammation and the balance between these factors is likely to dictate disease progression (Lawrence and Gilroy 2007). It is clear from genetic experiments in mice that NF-κB activation is not necessarily proinflammatory and has complex roles in the inflammatory response. The role of RelA as a critical effector of the canonical pathway has been demonstrated with the development of RelA and IKKβ knockout mice (Beg and Baltimore 1996; Li et al. 1999). Using radiation chimeras, Alcamo et al. showed that RelA expression in radiation-resistant tissue cells is required for the leukocyte recruitment in the lung after challenge with the bacterial product lipopolysaccharide (LPS), but RelA was not required in hematopoietic cells for inflammation (Alcamo et al. 2001). This was quite surprising considering the strong activation of NF-κB in lung macrophages in response to LPS and suggested a different role for NF-κB in cells of the immune system.

Cre/lox gene targeting technology (Sauer 1998) has made it possible to specifically target NF-κB activation in different cell lineages, an approach that has shown that NF-κB plays a tissue-specific role in the inflammatory response. The deletion of IKKβ or IKKγ in intestine epithelial cells clearly revealed a cytoprotective role for NF-κB. The resulting breakdown in epithelial barrier integrity leads to increased inflammation because of commensal bacteria activating tissue macrophages (Chen et al. 2003; Nenci et al. 2007; Eckmann et al. 2008). Interestingly, macrophage-driven inflammation in response to a loss of barrier function was also suggested to be NF-κB-dependent (Eckmann et al. 2008). In contrast, the specific targeting of NF-κB in lung epithelial cells did not apparently affect the integrity of the epithelium but impaired inflammation by inhibiting the expression of proinflammatory cytokines and chemokines (Poynter et al. 2003; Poynter et al. 2004; Broide et al. 2005).

In 2001, we showed the involvement of NF-κB in both the onset and resolution of acute inflammation in a single model system using pharmacological inhibitors (Lawrence et al. 2001). These studies confirmed the expected role of NF-κB in proinflammatory gene induction during the onset of inflammation but also showed a role for NF-κB in expression of anti-inflammatory genes and induction of leukocyte apoptosis during the resolution of inflammation. Inhibition of NF-κB during the resolution of inflammation prolonged inflammatory response and inhibited apoptosis, in conflict with the generally accepted view that NF-κB was antiapoptotic in inflammatory cells. More recently, Greten et al. also showed an anti-inflammatory role for IKKβ in sepsis (Greten et al. 2007). Specific deletion of IKKβ in myeloid cells increased sensitivity of mice to endotoxin (LPS)-induced shock caused by elevated plasma IL-1β levels resulting from increased pro-IL-1β processing in macrophages and neutrophils. In addition, Greten et al. confirmed a proapoptotic role for NF-κB in neutrophils, which may also contribute to an anti-inflammatory role of NF-κB as previously described (Lawrence et al. 2001). More recent studies by our group have shown both pro- and anti-inflammatory roles for IKKβ during bacterial infection (Fong et al. 2008). In a model of Streptococcal pneumonia, IKKβ was deleted in either macrophages or lung epithelial cells, and neutrophil recruitment and bacterial clearance were inhibited in mice lacking IKKβ in lung epithelial cells but were enhanced in mice with IKKβ deletion in

macrophages. In addition, IKKβ-deficient macrophages showed increased MHC II, iNOS, and IL-12 expression, which are hallmarks of "classical" or M1 macrophage activation (Gordon and Taylor 2005). CD124 (IL-4 receptor) expression was absent on IKKβ-deficient macrophages, suggesting that these cells have lost the ability to respond to IL-4 and develop an anti-inflammatory M2 phenotype (Gordon 2003). These data suggest that IKKβ suppresses the proinflammatory M1 phenotype and favors the development of anti-inflammatory M2 macrophages. M2 macrophages are also thought to be important in promoting inflammation-associated cancer (Mantovani et al. 2008). Hagemann et al. showed that inhibiting IKKβ in tumor-associated macrophages (TAM) switched the phenotype from M2 to M1, characterized by increased IL-12, iNOS, and MHC II (Hagemann et al. 2008). Interestingly, Saccani et al. have also shown that NF-κB inhibits the proinflammatory phenotype of TAM (Saccani et al. 2006). These studies suggest an anti-inflammatory role for NF-κB that limits the bactericidal and tumoricidal function of macrophages.

Gene knockout studies have also shown that NF-κB proteins can have both pro- and anti-inflammatory roles. Homodimers of the p50 subunit of NF-κB, which lack transactivation domains, have been shown to repress expression of NF-κB target genes and inhibit inflammation (Bohuslav et al. 1998). A homodimeric complex of p50 was found in resting T cells and reduced p50 expression was observed after T-cell activation. Furthermore, overexpression of p50 was shown to repress IL-2 expression in T cells (Kang 1992). Although increased p50 expression was reported to suppress TNFα production in LPS tolerance (Bohuslav et al. 1998; Kastenbauer and Ziegler-Heitbrock 1999), Gadjeva et al. showed that p50-deficient mice that are heterozygous for RelA ($p50^{-/-}$ $p65^{+/-}$) were extremely sensitive to LPS-induced shock (Gadjeva et al. 2004). These studies suggest anti-inflammatory roles of p50 homodimer and p50/p65 heterodimers in septic shock in keeping with the studies of Greten et al. targeting the canonical pathway

Cite this article as *Cold Spring Harb Perspect Biol* 2009;1:a001651

through IKKβ (Greten et al. 2007). Apart from sepsis, an anti-inflammatory role of NF-κB was also reported in inflammatory bowel disease in which p50$^{-/-}$p65$^{+/-}$ mice were more susceptible to *Helicobacter hepaticus* induced colitis (Erdman et al. 2001). Later studies have shown that colitis was associated with increased IL-12p40 expression in the colon (Tomczak et al. 2003), and a further study has shown administration of IL-10 fusion protein inhibited IL-12p40 production and *H. hepaticus* induced colitis, which was dependent on p50/p105 expression in macrophages (Tomczak et al. 2006). These studies suggest that NF-κB can have anti-inflammatory roles by directly inhibiting expression of proinflammatory genes and by manipulating the expression or activity of anti-inflammatory cytokines such as IL-10.

Apoptosis is an essential mechanism that prevents prolonged inflammation: Neutrophil apoptosis during acute inflammation and activation induced cell death (AICD) of antigen-specific T cells are important mechanisms that limit inflammatory and immune responses (Lawrence and Gilroy 2007). As described previously, NF-κB has a proapoptotic role in neutrophils during inflammation (Lawrence et al. 2001; Greten et al. 2007), which may represent an important anti-inflammatory mechanism for NF-κB during acute inflammation. However, NF-κB has also been shown to be an important inhibitor of pathogen-induced apoptosis in macrophages, at least in vitro (Park et al. 2005). In this context, NF-κB may have a proinflammatory role by enabling prolonged macrophage activation. This would increase innate resistance to infection and therefore block pathogen-induced inflammation during infection. Studies from Teixeiro et al. and Kasibhatla et al. have shown that inhibition of NF-κB activation decreases Fas (CD95) ligand expression on T cells, which is required for AICD (Ju et al. 1995; Emma Teixeiro 1999; Kasibhatla et al. 1999). Overexpression of the endogenous NF-κB inhibitor IκBα, specifically in T cells, also suggests a proapoptotic role for NF-κB in double-positive thymocytes (Hettmann et al. 1999). These studies contradict the antiapoptotic role of NF-κB in inducing

expression Bcl-x$_L$, TRAF1, TRAF2, c-IAP1, and cIAP2 (Martin SJ 1995; Wang et al. 1998). IKKβ was also shown to inhibit T-cell apoptosis in radiation chimera experiments using fetal embryonic liver cells from IKKβ knockout embryos (Senftleben et al. 2001b). Studies from Lin et al. (1999) have shown the involvement of NF-κB in both pro- and antiapoptotic function in T cells. Inhibiting NF-κB reduced FasL induction and apoptosis in T cells but increased glucocorticoid-mediated apoptosis. Glucocorticoids are produced in the thymus and function to induce thymocyte apoptosis during positive selection. However, Fas and FasL interaction is important in AICD and peripheral T-cell deletion. These data suggest that NF-κB inhibits glucocorticoid-mediated apoptosis and survival during positive selection. On the other hand, NF-κB has the opposite role in mature peripheral T cells, promoting apoptosis by increasing FasL expression, which may be linked to termination of T-cell responses (Lin et al. 1999). FasL knockout mice provide a well characterized model of autoimmune disease because of hyperactivation of autoreactive lymphocytes, demonstrating the importance of this pathway in eliminating potentially pathological cells (Roths et al. 1984). These studies suggest that NF-κB activation can also have contrasting roles in same-cell lineage, depending on the physiological context.

THE ALTERNATIVE NF-κB PATHWAY

The alternative NF-κB pathway is characterized by the inducible phosphorylation of p100 by IKKα, leading to activation of RelB/p52 heterodimers. The upstream kinase that activates IKKα in this pathway has been identified as an NIK (NF-κB inducing kinase) (Senftleben et al. 2001a). Genetic studies in mice have showed the important role for this pathway in lymphoid organogenesis and B-lymphocyte function (Senftleben et al. 2001a; Bonizzi et al. 2004), but the role this pathway plays in inflammation is still unclear (Bonizzi and Karin 2004; Lawrence and Bebien 2007). Gene disruption studies have shown that IKKγ and IKKβ subunits are required for IκBα phosphorylation and canonical NF-κB activation, whereas the alternative pathway is independent of both IKKγ and IKKβ (Ghosh and Karin 2002). This raised the question as to why the IKK complex invariably contains IKKα. We addressed this using transgenic mice that express a mutant form of IKKα in which two serine residues in the activation loop of the kinase were mutated to alanine (IKKα^{AA}) (Cao et al. 2001). Cells from these mice express a native IKK complex but lack the NIK-inducible activity of IKKα. Using cells from these mice IKKα was shown to regulate the stability and promoter recruitment of RelA and c-Rel-containing NF-κB through carboxy-terminal phosphorylation and proteosomal degradation (Lawrence et al. 2005). IKKα activation was shown to limit the inflammatory response during bacterial infection and inhibit canonical NF-κB activation. Subsequent studies also showed that IKKα negatively regulates canonical NF-κB activation, using macrophages derived from fetal liver cells of IKKα knockout embryos (Li et al. 2005) or zebrafish with a targeted mutation in the mammalian IKKα ortholog (Correa et al. 2005). IKKα-deficient macrophages showed increased expression of proinflammatory cytokines and an enhanced ability to stimulate T-cell proliferation (Li et al. 2005). However, interpretation of these studies may be clouded by the use of IKKα knockout cells: These experiments showed elevated IKKβ activity toward IκBα, which is not seen in cells from IKKα^{AA} mice (Cao et al. 2001; Lawrence et al. 2005). One would presume that the absence of IKKα protein generates IKKβ homodimers with increased activity toward IκBα and therefore the context of these experiments is less physiological than those performed with IKKα^{AA} cells. IKKα has also been shown to have an anti-inflammatory role through regulation of the SUMO ligase activity of PIAS (protein inhibitor of activated STAT) 1 (Liu et al. 2007). PIAS proteins were originally described as inhibitors of STAT transcription factor activation but have also been shown to regulate NF-κB activity (Liu et al. 1998; Tahk et al. 2007). IKKα-mediated phosphorylation of PIAS1 was shown to block binding of both

STAT-1 and NF-κB to proinflammatory gene promoters (Liu et al. 2007), but the significance of this pathway in the inflammatory response in vivo was not tested. It is yet to be determined how regulation of the canonical NF-κB pathway by IKKα affects the cell-specific roles of NF-κB in inflammation described previously. One assumes the anti-inflammatory roles of IKKα would only be present in the context of proinflammatory NF-κB activation.

It is interesting that studies with RelB deficient mice have also revealed an anti-inflammatory role for RelB (Weih et al. 1995; Xia et al. 1997), although this has not been connected with IKKα activity, suggesting other components of the alternative NF-κB pathway may have anti-inflammatory functions. RelB-deficient mice die of multiorgan inflammation (Weih et al. 1995), a phenotype that has been attributed to the breakdown of immunological tolerance caused by abnormal development of the thymus. Indeed, the pathology in *Relb*−/− mice is driven by autoreactive T cells (Burkly et al. 1995; DeKoning et al. 1997). However, *Relb*−/− fibroblasts show increased expression of proinflammatory cytokines and chemokines on stimulation with LPS in vitro (Xia et al. 1997). A more recent study has also shown that RelB has a role in endotoxin tolerance (Yoza et al. 2006), again suggesting that components of the alternative pathway have an anti-inflammatory role. The mechanism by which RelB confers this anti-inflammatory effect is not clear. Work from David Lo and colleagues suggests that RelB regulates IκBα stability and therefore limits canonical NF-κB activation (Xia et al. 1999). More recent work suggests that RelB may interfere with NF-κB activity in the nucleus through protein–protein interactions with RelA (Jacque et al. 2005). Other work has described the reciprocal recruitment of RelA and RelB to NF-κB target gene promoters and showed that the replacement of RelA-containing dimers with RelB complexes results in the down-regulation of certain NF-κB target genes (Saccani et al. 2003). The physiological significance of these putative mechanisms has not yet been established in vivo.

Genetic "knockout" of several components of the alternative pathway, including RelB and p52, have established an important role in lymphoid organogenesis (Bonizzi and Karin 2004). Analysis of IKKα[AA] mice (Senftleben et al. 2001a; Bonizzi et al. 2004) and adoptive transfer of IKKα-deficient hematopoietic cells to lethally irradiated mice (Kaisho et al. 2001) revealed an important role for IKKα in the organization of the splenic marginal zone and germinal center reaction in response to antigenic challenge, implicating the alternative pathway in humoral immunity. The role of IKKα in lymphoid organogenesis is attributed to its role in lymphotoxin β receptor (LTBR)-signaling in spleen stromal cells (Bonizzi et al. 2004; Bonizzi and Karin 2004). LTBR-mediated induction of organogenic chemokines— CCL19, CCL21, CCL22—is dependent on IKKα-mediated activation of RelB/p52 complexes (Bonizzi et al. 2004). IKKα has also been described to have a role in B-cell maturation (Senftleben et al. 2001a), and recent studies have shown that this may contribute to the pathogenesis of B-cell mediated autoimmunity (Enzler et al. 2006). Our studies have also established that IKKα is required for the generation of cell-mediated immune responses, independent of humoral immunity, such as the delayed-type hypersensitivity reaction (DTH) in mice (unpublished observations). This suggests that IKKα regulates both humoral and cell-mediated adaptive immune responses. Studies of RelB- and p52-deficient mice have established an important role for these proteins in dendritic cell (DC) function and the generation of cell-mediated immunity (Caamano et al. 1998; Franzoso et al. 1998; Wu et al. 1998; Weih et al. 2001; Speirs et al. 2004). The role of IKKα in DC function and maturation has not been examined, although recent studies have shown that LTBR signaling is important to maintain DC populations in vivo (Kabashima et al. 2005). The function of IKKα in organogenic chemokine production may also be important in the homing of antigen-loaded DCs to secondary lymphoid tissues where they can prime naïve T cells. Alternatively, the homing of antigen-specific T cells could be disregulated in the absence of these chemokines. The role of IKKα in adaptive immunity may well stretch beyond its role in stromal cells and the regulation of lymphoid-organogenesis.

These recent studies suggest that IKKα has evolved distinct, but possibly complementary, roles in inflammation and adaptive immunity. IKKα functions to promote the resolution of inflammation by switching off the canonical NF-κB pathway, but regulates the development of adaptive immunity through the alternative pathway. Although inflammation is classically considered to prime the adaptive response, for example through promoting DC maturation, the resolution of inflammation is required to avoid tissue injury while supporting the development of immunological memory. Cross talk between the alternative and canonical NF-κB pathways may regulate the transition from acute inflammation to antigen-specific immune responses that drive autoimmune diseases such as RA and multiple sclerosis. Ultimately, inhibition of IKKα may represent a therapeutic target to prevent autoimmune inflammation while maintaining innate immunity.

SUMMARY

NF-κB has long been considered the holy grail as a target for new anti-inflammatory drugs; however, data from elegant genetic studies in mice suggest that NF-κB could equally be a difficult therapeutic target in inflammatory diseases. The NF-κB pathway does indeed regulate proinflammatory cytokine production, leukocyte recruitment, or cell survival, which are important contributors to the inflammatory response. But, the antiapoptotic functions of NF-κB can both protect against inflammation, in the case of epithelial cell survival and mucosal barrier integrity, and also maintain the inflammatory response through persistent leukocyte activation. In contrast, NF-κB can promote leukocyte apoptosis in certain contexts and contribute to the resolution of inflammation. It is also clear that NF-κB contributes to the feedback control of inflammation by various mechanisms to affect the magnitude and duration of the inflammatory response. Future studies to evaluate the status of these varied roles for the NF-κB pathway in inflammatory disease are required to determine if this pathway could be a therapeutic target and in which context.

REFERENCES

Akira S, Uematsu S, Takeuchi O. 2006. Pathogen recognition and innate immunity. *Cell* **124:** 783–801.

Alcamo E, Mizgerd JP, Horwitz BH, Bronson R, Beg AA, Scott M, Doerschuk CM, Hynes RO, Baltimore D. 2001. Targeted mutation of TNF receptor I rescues the RelA-deficient mouse and reveals a critical role for NF-κ B in leukocyte recruitment. *J Immunol* **167:** 1592–1600.

Aupperle KR, Bennett BL, Boyle DL, Tak PP, Manning AM, Firestein GS. 1999. NF-κ B regulation by I κ B kinase in primary fibroblast-like synoviocytes. *J Immunol* **163:** 427–433.

Aupperle K, Bennett B, Han Z, Boyle D, Manning A, Firestein G. 2001. NF-κ B regulation by I κ B kinase-2 in rheumatoid arthritis synoviocytes. *J Immunol* **166:** 2705–2711.

Beg AA, Baltimore D. 1996. An essential role for NF-κB in preventing TNF-α-induced cell death. *Science* **274:** 782–784.

Bohuslav J, Kravchenko VV, Parry GC, Erlich JH, Gerondakis S, Mackman N, Ulevitch RJ. 1998. Regulation of an essential innate immune response by the p50 subunit of NF-κB. *J Clin Invest* **102:** 1645–1652.

Bonizzi G, Karin M. 2004. The two NF-κB activation pathways and their role in innate and adaptive immunity. *Trends Immunol* **25:** 280–288.

Bonizzi G, Bebien M, Otero DC, Johnson-Vroom KE, Cao Y, Vu D, Jegga AG, Aronow BJ, Ghosh G, Rickert RC, Karin M. 2004. Activation of IKKα target genes depends on recognition of specific κB binding sites by RelB:p52 dimers. *Embo J* **23:** 4202–4210.

Broide DH, Lawrence T, Doherty T, Cho JY, Miller M, McElwain K, McElwain S, Karin M. 2005. Allergen-induced peribronchial fibrosis and mucus production mediated by IκB kinase β-dependent genes in airway epithelium. *Proc Natl Acad Sci* **102:** 17723–17728.

Burkly L, Hession C, Ogata L, Reilly C, Marconi LA, Olson D, Tizard R, Cate R, Lo D. 1995. Expression of relB is required for the development of thymic medulla and dendritic cells. *Nature* **373:** 531–536.

Caamano JH, Rizzo CA, Durham SK, Barton DS, Raventos-Suarez C, Snapper CM, Bravo R. 1998. Nuclear factor NF-κ B2 (p100/p52) is required for normal splenic microarchitecture and B cell-mediated immune responses. *J Exp Med* **187:** 185–196.

Cao Y, Bonizzi G, Seagroves TN, Greten FR, Johnson R, Schmidt EV, Karin M. 2001. IKKα provides an essential link between RANK signaling and cyclin D1 expression during mammary gland development. *Cell* **107:** 763–775.

Chen LW, Egan L, Li ZW, Greten FR, Kagnoff MF, Karin M. 2003. The two faces of IKK and NF-κB inhibition: Prevention of systemic inflammation but increased local injury following intestinal ischemia-reperfusion. *Nat Med* **9:** 575–581.

Chung KF. 2006. Cytokines as targets in chronic obstructive pulmonary disease. *Curr Drug Targets* **7:** 675–681.

Correa RG, Matsui T, Tergaonkar V, Rodriguez-Esteban C, Izpisua-Belmonte JC, Verma IM. 2005. Zebrafish IκB kinase 1 negatively regulates NF-κB activity. *Curr Biol* **15:** 1291–1295.

Dejardin E, Droin NM, Delhase M, Haas E, Cao Y, Makris C, Li ZW, Karin M, Ware CF, Green DR. 2002. The lymphotoxin-β receptor induces different patterns of gene expression via two NF-κB pathways. *Immunity* **17:** 525–535.

DeKoning J, DiMolfetto L, Reilly C, Wei Q, Havran WL, Lo D. 1997. Thymic cortical epithelium is sufficient for the development of mature T cells in relB-deficient mice. *J Immunol* **158:** 2558–2566.

Eckmann L, Nebelsiek T, Fingerle AA, Dann SM, Mages J, Lang R, Robine S, Kagnoff MF, Schmid RM, Karin M, et al. 2008. Opposing functions of IKKβ during acute and chronic intestinal inflammation. *Proc Natl Acad Sci* **105:** 15058–15063.

Emma Teixeiro AG-SBARB. 1999. Apoptosis-resistant T cells have a deficiency in NF-κB-mediated induction of Fas ligand transcription. *Eur J Immunol* **29:** 745–754.

Enzler T, Bonizzi G, Silverman GJ, Otero DC, Widhopf GF, Anzelon-Mills A, Rickert RC, Karin M. 2006. Alternative and classical NF-κB signaling retain autoreactive B cells in the splenic marginal zone and result in lupus-like disease. *Immunity* **25:** 403–415.

Erdman SE, Fox JG, Dangler CA, Feldman D, Horwitz BH. 2001. Cutting edge: Typhlocolitis in NF-κB-deficient mice. *J Immunol* **166:** 1443–1447.

Fong CHY, Bebien M, Didierlaurent A, Nebauer R, Hussell T, Broide D, Karin M, Lawrence T. 2008. An antiinflammatory role for IKKβ through the inhibition of "classical" macrophage activation. *J Exp Med* **205:** 1269–1276.

Franzoso G, Carlson L, Poljak L, Shores EW, Epstein S, Leonardi A, Grinberg A, Tran T, Scharton-Kersten T, Anver M, et al. 1998. Mice deficient in nuclear factor NF-κ B/p52 present with defects in humoral responses, germinal center reactions, and splenic microarchitecture. *J Exp Med* **187:** 147–159.

Gadjeva M, Tomczak MF, Zhang M, Wang YY, Dull K, Rogers AB, Erdman SE, Fox JG, Carroll M, Horwitz BH. 2004. A role for NF-κB subunits p50 and p65 in the inhibition of lipopolysaccharide-induced shock. *J Immunol* **173:** 5786–5793.

Ghosh S, Karin M. 2002. Missing pieces in the NF-κB puzzle. *Cell* **109 Suppl:** S81–96.

Gordon S. 2003. Alternative activation of macrophages. *Nat Rev Immunol* **3:** 23–35.

Gordon S, Taylor PR. 2005. Monocyte and macrophage heterogeneity. *Nat Rev Immunol* **5:** 953–964.

Greten FR, Arkan MC, Bollrath J, Hsu LC, Goode J, Miething C, Goktuna SI, Neuenhahn M, Fierer J, Paxian S, et al. 2007. NF-κB is a negative regulator of

IL-1β secretion as revealed by genetic and pharmacological inhibition of IKKβ. *Cell* 130: 918–931.

Hagemann T, Lawrence T, McNeish I, Charles KA, Kulbe H, Thompson RG, Robinson SC, Balkwill FR. 2008. "Re-educating" tumor-associated macrophages by targeting NF-κB. *J Exp Med* 205: 1261–1268.

Hettmann T, DiDonato J, Karin M, Leiden JM. 1999. An essential role for nuclear factor κ B in promoting double positive thymocyte apoptosis. *J Exp Med* 189: 145–158.

Holgate ST. 2004. Cytokine and anti-cytokine therapy for the treatment of asthma and allergic disease. *Cytokine* 28: 152–157.

Jacque E, Tchenio T, Piton G, Romeo PH, Baud V. 2005. RelA repression of RelB activity induces selective gene activation downstream of TNF receptors. *Proc Natl Acad Sci* 102: 14635–14640.

Ju S-T, Panka DJ, Cui H, Ettinger R, Ei-Khatib M, Sherr DH, Stanger BZ, Marshak-Rothstein A. 1995. Fas(CD95)/FasL interactions required for programmed cell death after T-cell activation. *Nature* 373: 444–448.

Kabashima K, Banks TA, Ansel KM, Lu TT, Ware CF, Cyster JG. 2005. Intrinsic lymphotoxin-β receptor requirement for homeostasis of lymphoid tissue dendritic cells. *Immunity* 22: 439–450.

Kaisho T, Takeda K, Tsujimura T, Kawai T, Nomura F, Terada N, Akira S. 2001. IκB kinase α is essential for mature B cell development and function. *J Exp Med* 193: 417–426.

Kang SM TA, Grilli M, Lenardo MJ. 1992. NF-κ B subunit regulation in nontransformed CD4+ T lymphocytes. *Science* 5: 1452–1456.

Karin M, Ben-Neriah Y. 2000. Phosphorylation meets ubiquitination: The control of NF-κB activity. *Annu Rev Immunol* 18: 621–663.

Karin M, Lawrence T, Nizet V. 2006. Innate immunity gone awry: Linking microbial infections to chronic inflammation and cancer. *Cell* 124: 823–835.

Karin M, Yamamoto Y, Wang QM. 2004. The IKK NF-κB system: A treasure trove for drug development. *Nature Reviews Drug Discovery* 3: 17–26.

Kasibhatla S, Genestier L, Green DR. 1999. Regulation of Fas-Ligand Expression during Activation-induced Cell Death in T Lymphocytes via Nuclear Factor κ B. *J Biol Chem* 274: 987–992.

Kastenbauer S, Ziegler-Heitbrock HWL. 1999. NF-κ B1 (p50) is upregulated in lipopolysaccharide tolerance and can block tumor necrosis factor gene expression. *Infect Immun* 67: 1553–1559.

Lawrence T, Bebien M. 2007. IKKα in the regulation of inflammation and adaptive immunity. *Biochem Soc Trans* 35: 270–272.

Lawrence T, Gilroy DW. 2007. Chronic inflammation: A failure of resolution? *Int J Exp Pathol* 88: 85–94.

Lawrence T, Bebien M, Liu GY, Nizet V, Karin M. 2005. IKKα limits macrophage NF-κB activation and contributes to the resolution of inflammation. *Nature* 434: 1138–1143.

Lawrence T, Gilroy DW, Colville-Nash PR, Willoughby DA. 2001. Possible new role for NF-κB in the resolution of inflammation. *Nat Med* 7: 1291–1297.

Li Q, Lu Q, Bottero V, Estepa G, Morrison L, Mercurio F, Verma IM. 2005. Enhanced NF-κB activation and cellular function in macrophages lacking IκB kinase 1 (IKK1). *Proc Natl Acad Sci* 102: 12425–12430.

Li Q, Van Antwerp D, Mercurio F, Lee KF, Verma IM. 1999. Severe liver degeneration in mice lacking the IκB kinase 2 gene. *Science* 284: 321–325.

Lin B, Williams-Skipp C, Tao Y, Schleicher MS, Cano LL, Duke RC, Scheinman RI. 1999. NF-κB functions as both a proapoptotic and antiapoptotic regulatory factor within a single cell type. *Cell Death Differ* 6: 570–582.

Liu B, Liao J, Rao X, Kushner SA, Chung CD, Chang DD, Shuai K. 1998. Inhibition of Stat1-mediated gene activation by PIAS1. *Proc Natl Acad Sci* 95: 10626–10631.

Liu B, Yang Y, Chernishof V, Loo RR, Jang H, Tahk S, Yang R, Mink S, Shultz D, Bellone CJ, et al. 2007. Proinflammatory stimuli induce IKKα-mediated phosphorylation of PIAS1 to restrict inflammation and immunity. *Cell* 129: 903–914.

Mantovani A, Allavena P, Sica A, Balkwill F. 2008. Cancer-related inflammation. *Nature* 454: 436–444.

Martin SJ RC, McGahon AJ, Rader JA, van Schie RC, LaFace DM, Green DR. 1995. Early redistribution of plasma membrane phosphatidylserine is a general feature of apoptosis regardless of the initiating stimulus: Inhibition by overexpression of Bcl-2 and Abl. *J Exp Med* 182: 1545–1556.

Matsushima A, Kaisho T, Rennert PD, Nakano H, Kurosawa K, Uchida D, Takeda K, Akira S, Matsumoto M. 2001. Essential role of nuclear factor NF-κB-inducing kinase and inhibitor of κB (IκB) kinase α in NF-κB activation through lymphotoxin β receptor, but not through tumor necrosis factor receptor I. *J Exp Med* 193: 631–636.

Miagkov AV, Kovalenko DV, Brown CE, Didsbury JR, Cogswell JP, Stimpson SA, Baldwin AS, Makarov SS. 1998. NF-κB activation provides the potential link between inflammation and hyperplasia in the arthritic joint. *Proc Natl Acad Sci U S A* 95: 13859–13864.

Monaco C, Andreakos E, Kiriakidis S, Mauri C, Bicknell C, Foxwell B, Cheshire N, Paleolog E, Feldmann M. 2004. Canonical pathway of nuclear factor κ B activation selectively regulates proinflammatory and prothrombotic responses in human atherosclerosis. *Proc Natl Acad Sci U S A* 101: 5634–5639.

Nenci A, Becker C, Wullaert A, Gareus R, van Loo G, Danese S, Huth M, Nikolaev A, Neufert C, Madison B, et al. 2007. Epithelial NEMO links innate immunity to chronic intestinal inflammation. *Nature* 446: 557–561.

Novack DV, Yin L, Hagen-Stapleton A, Schreiber RD, Goeddel DV, Ross FP, Teitelbaum SL. 2003. The IκB function of nuclear factor κB2 p100 controls stimulated osteoclastogenesis. *J Exp Med* 198: 771–781.

Park JM, Greten FR, Wong A, Westrick RJ, Arthur JS, Otsu K, Hoffmann A, Montminy M, Karin M. 2005. Signaling pathways and genes that inhibit pathogen-induced macrophage apoptosis—CREB and NF-κB as key regulators. *Immunity* 23: 319–329.

Poynter ME, Irvin CG, Janssen-Heininger YM. 2002. Rapid activation of nuclear factor-κB in airway epithelium in a murine model of allergic airway inflammation. *Am J Pathol* 160: 1325–1334.

Poynter ME, Irvin CG, Janssen-Heininger YM. 2003. A prominent role for airway epithelial NF-κ B activation in lipopolysaccharide-induced airway inflammation. *J Immunol* **170**: 6257–6265.

Poynter ME, Cloots R, van Woerkom T, Butnor KJ, Vacek P, Taatjes DJ, Irvin CG, Janssen-Heininger YM. 2004. NF-κB activation in airways modulates allergic inflammation but not hyperresponsiveness. *J Immunol* **173**: 7003–7009.

Roths JB, Murphy ED, Eicher EM. 1984. A new mutation, gld, that produces lymphoproliferation and autoimmunity in C3H/HeJ mice. *J Exp Med* **159**: 1–20.

Saccani S, Pantano S, Natoli G. 2003. Modulation of NF-κB activity by exchange of dimers. *Mol Cell* **11**: 1563–1574.

Saccani A, Schioppa T, Porta C, Biswas SK, Nebuloni M, Vago L, Bottazzi B, Colombo MP, Mantovani A, Sica A. 2006. p50 nuclear factor-κB overexpression in tumor-associated macrophages inhibits M1 inflammatory responses and antitumor resistance. *Cancer Res* **66**: 11432–11440.

Sauer B. 1998. Inducible gene targeting in mice using the Cre/lox system. *Methods* **14**: 381–392.

Senftleben U, Cao Y, Xiao G, Greten FR, Krahn G, Bonizzi G, Chen Y, Hu Y, Fong A, Sun SC, Karin M. 2001a. Activation by IKKα of a second, evolutionary conserved, NF-κB signaling pathway. *Science* **293**: 1495–1499.

Senftleben U, Li ZW, Baud V, Karin M. 2001b. IKKβ is essential for protecting T cells from TNFα-induced apoptosis. *Immunity* **14**: 217–230.

Speirs K, Lieberman L, Caamano J, Hunter CA, Scott P. 2004. Cutting edge: NF-κB2 is a negative regulator of dendritic cell function. *J Immunol* **172**: 752–756.

Tahk S, Liu B, Chernishof V, Wong KA, Wu H, Shuai K. 2007. Control of specificity and magnitude of NF-κB and STAT1-mediated gene activation through PIASy and PIAS1 cooperation. *Proc Natl Acad Sci U S A* **104**: 11643–11648.

Tak PP, Firestein GS. 2001. NF-κB: A key role in inflammatory diseases. *J Clin Invest* **107**: 7–11.

Tomczak MF, Erdman SE, Davidson A, Wang YY, Nambiar PR, Rogers AB, Rickman B, Luchetti D, Fox JG, Horwitz BH. 2006. Inhibition of *Helicobacter hepaticus*-induced colitis by IL-10 requires the p50/p105 subunit of NF-κB. *J Immunol* **177**: 7332–7339.

Tomczak MF, Erdman SE, Poutahidis T, Rogers AB, Holcombe H, Plank B, Fox JG, Horwitz BH. 2003. NF-κB is required within the innate immune system to inhibit microflora-induced colitis and expression of IL-12 p40. *J Immunol* **171**: 1484–1492.

Wang C-Y, Mayo MW, Korneluk RG, Goeddel DV, Baldwin ASJr. 1998. NF-B antiapoptosis: Induction of TRAF1 and TRAF2 and c-IAP1 and c-IAP2 to suppress caspase-8 activation. *Science* **281**: 1680–1683.

Weih DS, Yilmaz ZB, Weih F. 2001. Essential role of RelB in germinal center and marginal zone formation and proper expression of homing chemokines. *J Immunol* **167**: 1909–1919.

Weih F, Carrasco D, Durham SK, Barton DS, Rizzo CA, Ryseck RP, Lira SA, Bravo R. 1995. Multiorgan inflammation and hematopoietic abnormalities in mice with a targeted disruption of RelB, a member of the NF-κB/Rel family. *Cell* **80**: 331–340.

Williams RO, Paleolog E, Feldmann M. 2007. Cytokine inhibitors in rheumatoid arthritis and other autoimmune diseases. *Curr Opin Pharmacol* **7**: 412–417.

Wu L, D'Amico A, Winkel KD, Suter M, Lo D, Shortman K. 1998. RelB is essential for the development of myeloid-related CD8α- dendritic cells but not of lymphoid-related CD8α+ dendritic cells. *Immunity* **9**: 839–847.

Xia Y, Chen S, Wang Y, Mackman N, Ku G, Lo D, Feng L. 1999. RelB modulation of IκBα stability as a mechanism of transcription suppression of interleukin-1α (IL-1α), IL-1β, and tumor necrosis factor α in fibroblasts. *Mol Cell Biol* **19**: 7688–7696.

Xia Y, Pauza ME, Feng L, Lo D. 1997. RelB regulation of chemokine expression modulates local inflammation. *Am J Pathol* **151**: 375–387.

Yoza BK, Hu JY, Cousart SL, Forrest LM, McCall CE. 2006. Induction of RelB participates in endotoxin tolerance. *J Immunol* **177**: 4080–4085.

Zandi E, Rothwarf DM, Delhase M, Hayakawa M, Karin M. 1997. The IκB kinase complex (IKK) contains two kinase subunits, IKKα and IKKβ, necessary for IκB phosphorylation and NF-κB activation. *Cell* **91**: 243–252.

Roles of the NF-κB Pathway in Lymphocyte Development and Function

Steve Gerondakis[1] and Ulrich Siebenlist[2]

[1]The Burnet Institute, 85 Commercial Rd, Prahran, Victoria 3004, Australia

[2]Laboratory of Immunoregulation, National Institutes of Allergy and Infectious Diseases, National Institutes of Health, Bethesda, Maryland 20892

Correspondence: gerondakis@burnet.edu.au, usiebenlist@niaid.nih.gov

This article focuses on the functions of NF-κB that vitally impact lymphocytes and thus adaptive immunity. NF-κB has long been known to be essential for many of the responses of mature lymphocytes to invading pathogens. In addition, NF-κB has important functions in shaping the immune system so it is able to generate adaptive responses to pathogens. In both contexts, NF-κB executes critical cell-autonomous functions within lymphocytes as well as within supportive cells, such as antigen-presenting cells or epithelial cells. It is these aspects of NF-κB's physiologic impact that we address in this article.

CELL-AUTONOMOUS ROLES OF NF-κB IN LYMPHOCYTE DEVELOPMENT

NF-κB makes numerous cell-autonomous contributions to the development of mature T and B lymphocytes. Given the importance of NF-κB in adaptive immune responses mediated by mature lymphocytes, it seems prudent for developing lymphocytes to have adopted a strategy in which their maturation hinges on a properly functioning NF-κB system. As discussed later, the primary, though not exclusive, contribution of NF-κB to lymphocyte development is to assure cell survival. These antiapoptotic functions of NF-κB remain crucial for the health of lymphocytes even after they mature. Unfortunately, these functions also aid tumorigenesis when NF-κB is dysregulated (Vallabhapurapu and Karin 2009). Most of the insights about the role of NF-κB in development of lymphocytes have come from analyses of genetically manipulated mice in which NF-κB components are missing or in which NF-κB activation has been compromised or is constitutively induced. B and T lymphocytes will be discussed in parallel to highlight similarities at related stages of their development. Figures 1 and 2 summarize some of the findings described later.

Early Lymphocyte Progenitors

Despite a clear role for the NF-κB homolog Dorsal in early *Drosophila* development (Hong et al. 2008), in mammalian development no such role for NF-κB has emerged, including development of early lymphocyte precursors. NF-κB can, however, play a protective role in precursors to

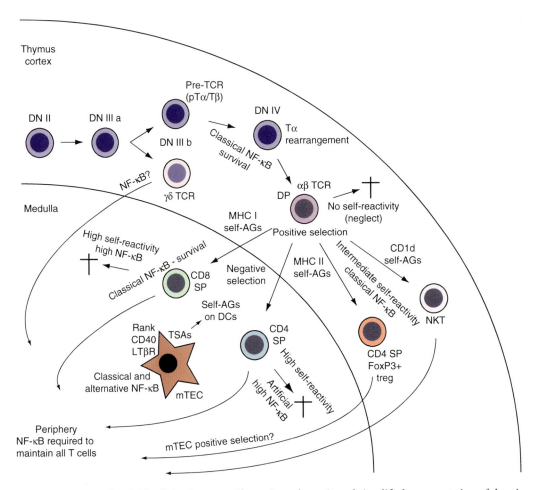

Figure 1. NF-κB in thymic T-cell development. Shown is a schematic and simplified representation of thymic T-cell development, highlighting stages at which NF-κB contributes in a cell-autonomous fashion. Also highlighted is the requirement of NF-κB for generation of medullary thymic epithelial cells (mTECs). γδ T cells and Tβ-expressing thymocytes can be distinguished at the (CD4, CD8) double negative (DN) stage III b. The pre-TCR (pTα/Tβ) drives development of thymocytes into DN IV cells, which in turn give rise to double positive (DP) cells (αβTCR). Positive selection of DP thymocytes to become CD4 or CD8 single-positive (SP) thymocytes is driven by weak recognition of self-AGs presented on cortical thymic epithelial cells in the context of MHC class II or class I, respectively. T-regulatory cells (Tregs, FoxP3+) and NKT cells may develop from DP thymocytes by recognition of self-AGs with intermediate strength (lipids presented on CD1d in the case of NKT cells). Failure to recognize self-AGs leads to elimination of thymocytes (death by neglect); strong recognition of self-AGs also leads to elimination (negative selection). Negative selection begins in the cortex but may occur predominantly in the medulla, where self-AGs are presented on dendritic cells (DCs) and on mTECs. mTECs produce tissue-specific (self)-AGs (TSAs) and can cross-prime DCs with these antigens. See text for further details.

protect them from TNFα-induced apoptosis. Artificially high levels of TNFα arise during adoptive transfers of hematopoietic stem cells into lethally irradiated hosts, so when donor cells were compromised in their ability to activate NF-κB, reconstitution of lymphocytes failed (Grossmann et al. 2000; Senftleben et al. 2001b; Claudio et al. 2006; Gerondakis et al. 2006; Igarashi et al. 2006). It is possible that a minimum of NF-κB activity may yet be necessary even during normal development (normal levels of TNFα). Female mice heterozygous for

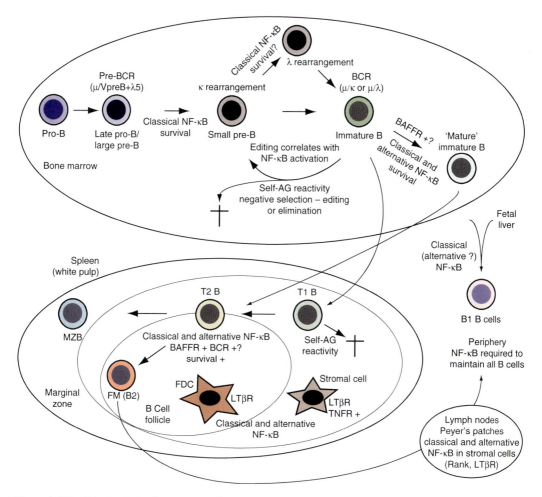

Figure 2. NF-κB in B cell development. A schematic and simplified representation of bone marrow and splenic B-cell development, highlighting stages at which NF-κB contributes in a cell-autonomous fashion to formation of marginal zone (MZ B) and follicular mature (FM) B cells; the latter are also known as B2 B cells and enter the peripheral circulation. Also highlighted is the requirement for NF-κB in B1 B-cell development, a peripherally self-renewing population with precursors in fetal liver and possibly bone marrow. Also highlighted is the importance of NF-κB in stromal cells/follicular dendritic cells (FDCs) in forming a proper splenic architecture (B-cell follicles, marginal zone) and in forming Peyer's patches and lymph nodes. B-cell development commences in the bone marrow, where the pre-BCR on large pre-B cells (a.k.a. late pro-B) drives development into small (late) pre-B cells, which in turn give rise to immature B cells (first to express a full BCR [IgM]). Self-antigen (AG)-reactive immature B cells edit their receptors by further light chain gene rearrangements or they are eliminated (negative selection). Surviving immature and "more mature" immature B cells (T2-like) then migrate to the spleen (white pulp), where they progress through the transitional 1 (T1) and T2 stages to become FM (located in B cell follicles) and MZB cells (located in marginal zones). Early transitional-staged cells continue to be subject to negative selection. Generation of FMs and MZBs from early transitional stages is driven by signals from the BCR, BAFFR, and other receptors ("+") to assure survival, but also to regulate cell differentiation ("+") in part via NF-κB. (Additional minor pathways and populations have been postulated, but are not shown here.) See text for further details.

loss of X-chromosome-encoded NEMO only generated NEMO sufficient, but not NEMO-deficient lymphocytes, even though random lyonization should have generated equal numbers (Makris et al. 2000; Schmidt-Supprian et al. 2000). NEMO (IKKγ) is an essential component of the classical pathway for NF-κB activation, and absolutely required for NF-κB activation by TNFα (reviewed in Hayden and Ghosh 2008; Vallabhapurapu and Karin 2009). It remains to be shown, however, if NEMO-deficient lymphocyte precursors were indeed eliminated by "normal" levels TNFα

Pre-antigen Receptor Expressing Large Pre-B Cells and Double Negative (DN) Thymocytes

The appearance of pre-TCRs on DN thymocytes (stage III) and pre-BCRs on developing bone marrow large pre-B cells provides important, ligand-independent signals for expansion and progression to the DN stage IV/DP stage and to small pre-B cells, respectively (T- and B-cell development reviewed in Bommhardt et al. 2004; Hardy et al. 2007; Allman and Pillai 2008; Northrup and Allman 2008; Taghon and Rothenberg 2008). These cells contain significant levels of nuclear NF-κB activity, presumably because of activation by the pre-antigen receptors (Voll et al. 2000; Jimi et al. 2005; Derudder et al. 2009). Failure to assemble a pre-TCR receptor (pTα paired with rearranged TRCβ) or pre-BCR (VpreB and γ5 surrogate light chains paired with rearranged μ heavy chain) eliminates these cells.

IκB super-repressor-mediated interference with NF-κB activation in pre-TCR expressing thymocytes led to their apoptosis (Voll et al. 2000), because of interference with NF-κB-mediated induction of the antiapoptotic Bcl-2 family member A1 (Mandal et al. 2005; Aifantis et al. 2006). Furthermore, exogenous expression of a constitutively active IKKβ allowed some DN thymocytes to progress to the double positive (DP) stage, even in the absence of a pre-TCR in RAG-deficient mice (Voll et al. 2000). Similarly, in pre-BCR⁺ large pre-B cells, suppression of NF-κB by the IκB super-repressor also resulted

in apoptosis, which could be overcome by ectopic expression of Bcl-xL, another NF-κB-induced antiapoptotic member of the Bcl-2 family (Feng et al. 2004a; Jimi et al. 2005). At a minimum, these data imply a survival role for NF-κB in both pre-BCR and pre-TCR dependent development of B and T lymphocytes (this has previously been reviewed [Denk et al. 2000; Siebenlist et al. 2005; Claudio et al. 2006; Vallabhapurapu and Karin 2009]).

Small (Late) Pre-B Cells

Following cellular activation via pre-BCRs, cells no longer express the receptor and enter the small pre-B cell stage. On successful rearrangement of their light chains, they begin to express a full BCR (IgM) on their surface and are termed immature B cells (Hardy et al. 2007; Northrup and Allman 2008). Circumstantial evidence has implicated NF-κB in demethylation of the κ light chain (LC) locus, an obligate step during κ gene rearrangement (Goldmit et al. 2005). However, loss of the intronic κ enhancer containing the first known κB binding site had no effect on this process (Inlay et al. 2004; Sen 2004), and a recent report implies no or only a minor role for NF-κB in κ locus demethylation (Derudder et al. 2009). mb1 promoter-driven Cre-induced conditional ablation of NEMO (eliminating the classical pathway) or of both IKK kinase subunits (IKK1 and IKK2; eliminating classical and alternative NF-κB pathways) at this stage of development did not significantly reduce κ LC expressing (immature) B cells. Surprisingly though, λ LC expressing immature B cells were reduced. Rearrangement of the λ locus is temporally delayed relative to that of κ, so it is possible that these B cells require NF-κB activity to allow them to survive longer. In support, the antiapoptotic protein kinase Pim2 was down-modulated in NEMO/IKK deficient small B cells and ectopic expression of a Bcl-2 transgene resurrected λ LC rearrangements.

It is not known what signals activate NF-κB at this stage. CD40, Bcl-10, MyD88, and ATM kinase were not required (Derudder et al. 2009); the latter was thought to be potentially involved in NF-κB activation and developmental progression in response to double-stranded DNA breaks associated with DNA rearrangements (Bredemeyer et al. 2008). Because Bcl-10 was not required, one might conclude that antigen-receptor signaling is not involved, but it is possible that this receptor can activate NF-κB by an unconventional pathway (see further discussion later). Loss of TRAF6 did reduce the number of λ-expressing B cells (Derudder et al. 2009), but many signaling receptors in addition to the antigen receptor avail themselves of this adaptor.

Immature (Bone Marrow) B Cells and Negative Selection

Autoreactive immature (bone marrow) B cells are subject to negative selection, but rather than undergo immediate apoptosis, a portion of these cells reactivate the RAG recombinase to further rearrange (edit) their light chain in hope of generating a nonautoreactive BCR (Nemazee 2006). Strong engagement of self-antigens was reported to down-regulate BCR cell surface expression and de-differentiate immature B cells back to the small pre-B cell stage (Schram et al. 2008). Contrary to an earlier report that concluded a critical role for NF-κB in RAG regulation during negative selection (Verkoczy et al. 2005), a recent report instead attached an at best only minor role to NF-κB: receptor editing for κ light chains and RAG expression levels were not significantly affected in B cells conditionally ablated for NEMO or for both IKK1 and IKK2, although as noted previously, λ light chain rearrangement was affected (Derudder et al. 2009). On the other hand, another recent study found a strong correlation between elevated NF-κB activity (induced by autoreactive BCRs) and various indicators of receptor editing, suggesting a causal relationship, even though these authors also failed to find a link between NF-κB and RAG expression (Cadera et al. 2009). These divergent conclusions might yet be reconciled if for example some NF-κB could be activated by BCRs via a NEMO/IKK independent mechanism.

NF-κB may contribute to the survival of immature cells during the negative selection/

editing phase and thereafter. Initial in vitro studies with immature B-like WEHI321 cells suggested as much; α-IgM induced apoptis of these cells could be prevented if cells were also stimulated with α-CD40, which led to sustained NF-κB activity, and in turn, higher c-myc levels (Schauer et al. 1998). CD40 activates both the classical and the alternative (nonclassical, noncanonical) pathways for NF-κB (Mineva et al. 2007; Hayden and Ghosh 2008; Vallabhapurapu and Karin 2009). We recently documented a partial reduction in numbers of immature bone marrow B cells in mice lacking both NF-κB1 and NF-κB2 (this partially eliminates classically and alternatively activated complexes) (Claudio et al. 2009). This loss correlated with impaired survival of mutant immature B cells. Activation of the alternative pathway in wild-type immature B cells is likely to occur via BAFF receptor signaling (Claudio et al. 2009), which was also suggested by another study that implicated BAFF receptor signaling in the de novo generation of a small subset of "more mature" immature B cells in bone marrow (Lindsley et al. 2007). The previously cited report on conditional ablation of NEMO or of IKK1/IKK2 in B cells also showed a reduction in the more mature immature B cells (Derudder et al. 2009).

DP and SP "Conventional" Thymocytes: Positive and Negative Selection

After "β" selection of pre-TCR+ DN thymocytes (stage III), these cells rearrange their TCRα gene (DN stage IV) and become DP (CD4+ CD8+) TCRαβ+ thymocytes (Bommhardt et al. 2004; Taghon and Rothenberg 2008). At this time, the thymocytes are subjected first to positive and then, in an overlapping fashion, to negative selection, while they physically move from the thymic cortex to the medulla (Boehm 2008). TCRs that fail to recognize self-antigens presented by MHCs are deleted (neglect), whereas weakly self-reactive TCRs are positively selected and strongly self-reactive TCRs are negatively selected (death by apoptosis). Surviving thymocytes become CD4 or CD8 single positive (SP) (depending on recognition of MHC class II or I, respectively), undergo further maturation, and eventually exit the thymus to enter the peripheral circulation as mature naïve T cells (Bommhardt et al. 2004; Taghon and Rothenberg 2008).

The role of NF-κB in selection remains somewhat controversial; NF-κB appears to have both positive and negative effects, probably because TCR signal strengths can be sensed within cells via the level of NF-κB activation; too much or too little activity may be selected against. Early studies with mice expressing IκB super-repressor transgenes have suggested a role for NF-κB in positive selection of CD8, and, to a lesser degree, of CD4 SP thymocytes, including studies with mouse models expressing TCR transgenes that weakly recognize self-peptides presented on class I or class II MHCs, respectively (positive selection) (Boothby et al. 1997; Hettmann and Leiden 2000; Mora et al. 2001). On the other hand, NF-κB has also been implicated in promoting apoptosis of DP thymocytes mimicked by α-CD3 treatment in vivo (Hettmann et al. 1999; Ren et al. 2002) and in mouse models with TCRs that strongly recognize self-peptides presented on class I or class II (negative selection) (Mora et al. 2001), although one study failed to see a role for NF-κB in class I negative selection (Hettmann and Leiden 2000).

In these studies, it is difficult to differentiate roles of NF-κB during the initial positive/negative selection from roles during the subsequent, SP stage, although a more recent study implicates NF-κB in the latter stage, especially in CD8 SP thymocytes. Mice conditionally ablated in thymocytes for NEMO or that conditionally express a kinase deficient (potentially dominant negative-acting) form of IKKβ in thymocytes generated significantly fewer CD8 SP thymocytes, and had no peripheral T cells, including CD4 T cells (Schmidt-Supprian et al. 2003). However, straight loss of IKKβ was well tolerated, suggesting compensation by IKKα. This study only informed on IKK/NF-κB in SP but not DP thymocytes, because the lck- or CD4-driven genetic changes only became fully penetrant after thymocytes had traversed the DP stage. Those SP thymocytes that remained in this NF-κB impaired model exhibited increased expression of apoptotic markers,

consistent with an antiapoptotic role for NF-κB. This study was further able to conclude that NF-κB was continuously required in T cells even after they emerged from the thymus, i.e., the maintenance of the peripheral population depended on NF-κB. A very recent study has added further insight, strongly implicating NF-κB in positive and negative selection of CD8 (DP and SP stage), but surprisingly, not of CD4 thymocytes. This study relied on mouse models with transgenes featuring positively or negatively selectable TCRs as well as an IκBα super-repressor or a constitutively active mutant IKK2 kinase (Jimi et al. 2008). IκBα super-repressor mediated inhibition of NF-κB eliminated cells expressing TCRs that weakly recognized MHC class I-presented self-peptides (CD8 positive selection), while also preventing the elimination of cells expressing TCRs that strongly recognized class I-presented self-peptides (CD8 negative selection). Confirming a role for strong NF-κB activity in negative selection, high NF-κB activity induced by constitutively active IKK2 eliminated some cells during positive selection for CD8, presumably mimicking a strong TCR signal that pushed cells into a negative selection mode. Thus, the level of NF-κB activity appeared to reflect TCR signal strength during CD8 selection, thereby setting thresholds for positive and negative selection. A low level of NF-κB activity during CD8 positive selection allowed cells to avoid death by neglect, whereas a high level was needed during CD8 negative selection to induce cell death. Interestingly, CD8 but not CD4 thymocytes expressed significant levels of NF-κB, which may explain why the IκBα super-repressor had little to no effect on CD4 selection processes. However, constitutively active IKK2 also eliminated cells during positive selection for CD4, presumably pushing these cells into a "pseudo"-negative selection mode. CD8 T cells may depend on a minimum of NF-κB for survival because they have very low levels of Bcl-2, whereas CD4 T cells express high levels of this antiapoptotic regulator, rendering them independent of NF-κB's antiapoptotic functions. It is surprising that high levels of NF-κB activity promoted cell death during negative selection, given the almost universal linkage of NF-κB with survival functions; how this is accomplished remains an interesting, yet open question. It will also be of interest to determine how DP thymocytes differentially regulate for NF-κB depending on whether their TCRs recognize self-peptides in the context of class I or class II MHCs; this may involve other signals coming from antigen-presenting cells. The IκBα super-repressor transgene models cited previously do have one inherent shortcoming; NF-κB inhibition is likely to be only partial, with the degree of inhibition dependent on the level of expression of the transgene, which may vary between cells and models.

How NF-κB is activated in thymocytes during positive and negative selection is not fully understood. If TCR signaling is responsible, then it must activate NF-κB by an "unconventional" pathway, given obligate components of the "conventional" pathway (as defined in mature T cells) could be largely dispensed with for generation of more mature SP thymocytes/ peripheral T cells. Regardless of this finding, the conventional pathway is likely to contribute to NF-κB activation during pre-TCR signaling (Sun et al. 2000; Lin and Wang 2004; Thome 2004; Felli et al. 2005; Jost et al. 2007). Obligate components of the conventional pathways include PKCθ and "CBM" complex components Carma1, Bcl-10, and Malt1 (PKCθ may have an NF-κB-independent role though [Morley et al. 2008]). Interestingly, a recent study reports the existence of an unconventional signaling path for TCR-mediated activation of NF-κB in mature CD8, but not CD4 T cells (Kingeter and Schaefer 2008); if such a pathway was also functional during selection of CD8 thymocytes, it would be consistent with the role for NF-κB in the development of these cells as discussed previously. Alternatively, NF-κB may also be activated by other signals, independent of the TCR. In either case, Tak1 is most likely involved; Tak 1 functions just upstream of the IKK complex in response to many NF-κB activating signals, and loss of this kinase reduced the number of SP thymocytes, and especially peripheral T cells, because of increased apoptosis (Liu et al. 2006; Sato et al. 2006; Wan et al. 2006).

Nonconventional T Cells

T-regulatory (Treg) cells recognize self-antigens, yet escape negative selection, possibly because they recognize self-antigens with intermediate strength (Caton et al. 2004; Lu and Rudensky 2009). Invariant natural killer T cells (iNKT; a.k.a. NKT) also recognize self-antigens (along with foreign ones), although in this case the antigens are lipids presented to developing NKT cells by CD1d on thymocytes (MacDonald and Mycko 2007; Burrows et al. 2009). γδ T cells are also thought to recognize certain self-antigens, but not in the context of MHC or CD1d (Thedrez et al. 2007; Xiong and Raulet 2007; Taghon and Rothenberg 2008). Auto-antigen-mediated TCR signaling may thus promote the development of all of these cell types, and it may do so in an NF-κB dependent manner.

The conditional loss of IKKβ in thymocytes prevented the emergence of Treg and NKT cells (Schmidt-Supprian et al. 2003, 2004a); therefore, IKKα was unable to compensate, even though it did during development of conventional T cells. Furthermore, a compound deficiency of c-Rel and NF-κB1 abrogated Treg generation (Zheng et al. 2003), while loss of RelA or the presence of the IκBα super-repressor blocked NKT development (Sivakumar et al. 2003; Vallabhapurapu et al. 2008).

The commitment of early thymocytes to the Treg lineage (as determined by FoxP3 expression) was fully dependent on TCR-induced conventional signaling for activation of NF-κB (PKCθ, CBM, and Tak1 were all required) (Schmidt-Supprian et al. 2004a; Sato et al. 2006; Wan et al. 2006; Barnes et al. 2009; Medoff et al. 2009; Molinero et al. 2009). For NKT cells, PKCθ and the CBM adaptor complex were not absolutely required, although PKCθ contributed to thymic and Bcl-10 to peripheral NKT-cell numbers, respectively (Schmidt-Supprian et al. 2004a; Medoff et al. 2009). On the other hand, stimulation of thymocytes via NKT-specific Vα14i TCRs was reported to activate RelA/ NF-κB complexes and thereby induce expression of IL-15 receptor α and the common γchain (γc), which in turn allowed for IL-15 and IL-7 mediated expansion of NK1.1⁻ NKT precursors and maturation to NK1.1$^+$ NKT cells (Vallabhapurapu et al. 2008). It is possible that TCRs on NKT activated NF-κB by a non-conventional pathway; alternatively, TCR independent signals might also have contributed to NF-κB activation.

Like NKTs, γδ T cells have innate and adaptive functions; they are able to rapidly produce γIFN or IL-17 in response to a variety of stimuli (Taghon and Rothenberg 2008; Jensen and Chien 2009). Relatively little is known about their development, but recently their differentiation into γIFN producing cells was reported to be imprinted early in developing γδ thymocytes, correlating with expression of TNF receptor members CD27 and LTβRs (Ribot et al. 2009). The latter is thought to be engaged by LTαβ-expressing DP thymocytes. The presence of CD27 and LTβR suggests roles for the classical and alternative pathways in generating γIFN producing γδ T cells, although this remains to be shown.

Transitional and Follicular Mature B Cells and Nonconventional Marginal Zone and B1 B Cells

Immature B cells eventually leave the bone marrow and migrate to the spleen to complete their maturation process. There they undergo several phenotypic and functional changes to become follicular mature (FM) B cells (a.k.a. B2 B cells), which recirculate in the periphery, and marginal zone B cells (MZB), which are largely sessile, although they may shuttle between the marginal zone and follicles to transport and present antigens (Thomas et al. 2006; Hardy et al. 2007; Allman and Pillai 2008). In this way, MZBs contribute to T-dependent antigen responses, but MZBs can also rapidly respond to pathogens in a T-independent, innate-like fashion. MZBs are a first line of defense to blood-borne pathogens, as splenic marginal zones filter these pathogens out of the blood stream (Lopes-Carvalho et al. 2005). B1 B cells represent a distinct lineage from B2 B cells; they populate the peritoneal and pleural cavities and are thought to originate from fetal liver precursors that self-renew in the periphery, although some precursors

may reside in the bone marrow (Allman and Pillai 2008). Like MZBs, B1 B cells participate in rapid T-independent innate-like antibody production and appear to recognize some self-antigens.

Newly arrived immature B cells in the spleen are referred to as transitional-1 B cells (T1); these cells continue to be subject to negative selection pressures, as antigen-receptor stimulation induces apoptosis. At some time during the T2 stage, B cells instead begin to respond positively to antigen-receptor stimulation, whereupon they become FM and MZB cells. The mechanisms underlying the divergence of FM and MZB remain poorly understood. In some studies, an additional third transitional B cell stage has been distinguished phenotypically, but this population may also include anergic B cells (Thomas et al. 2006; Hardy et al. 2007; Allman and Pillai 2008).

NF-κB is absolutely essential for the survival of developing B cells in the spleen. Compound deficiencies in NF-κB1/NF-κB2 and c-Rel/RelA profoundly arrested B-cell development at or shortly after the transition from T1 to T2, and resulted in a near-complete absence of FM and MZB cells (Franzoso et al. 1997b; Grossmann et al. 2000; Claudio et al. 2002, 2006; Gerondakis et al. 2006); a milder reduction of FM B cells occurred with compound deficiencies of NF-κB1/c-Rel (Pohl et al. 2002; Gerondakis et al. 2006) and NF-κB2/Bcl-3 or NIK/Bcl-3 (U.S., unpubl. results). These results suggest involvement of the classical and alternative pathways. Loss of alternative pathway components only (NF-κB2, NIK, or IKKα) resulted in partial reductions of more mature B cells (Franzoso et al. 1998; Yamada et al. 2000; Kaisho et al. 2001; Senftleben et al. 2001a; Siebenlist et al. 2005; Claudio et al. 2006). Bcl-3 is an atypical member of the IκB family and its involvement was unexpected. Bcl-3 may be induced by classical NF-κB activation and it is likely to modulate gene transcription via association with DNA-binding p50 homodimers in the nucleus, although biologic targets during B-cell development are unknown (Palmer and Chen 2008; Yamamoto and Takeda 2008). The severe block in B-cell development in the compound deficient NF-κB models cited previously was because of loss of survival; early transitional B cells (T1) from these mutant mice expressed lower levels of the antiapoptotic proteins Bcl-2 and A1, and exhibited increased spontaneous apoptosis in culture. Ectopic expression of Bcl-2 allowed the transitional B cells to survive and progress further; however, these cells did not fully mature phenotypically or functionally, remaining unable to produce antibodies. Therefore, NF-κB is critical not only for survival of transitional B cells, but also for the complete development of functions that accompanies differentiation.

During the transitional phases, BAFF (a.k.a. BLyS, TNFSF13B) induces the alternative pathway in B cells via the BAFF receptor (BAFFR, a.k.a. BR3, TNFRSF13C) (Claudio et al. 2002; Stadanlick and Cancro 2008; Mackay and Schneider 2009; Rauch et al. 2009). Because loss of BAFFR blocks B-cell maturation more profoundly than loss of the alternative pathway, BAFFR must contribute additional signals for cell survival. Although the alternative pathway does promote Bcl-2 expression, and, indirectly, the retention in the cytoplasm of otherwise apoptotic nuclear PKCδ (Sasaki et al. 2006), BAFFR also activates AKT, independently of NF-κB (Otipoby et al. 2008). Interestingly, a mutant form of the alternative pathway component p100/NF-κB2 that cannot be processed to p52 caused a substantial block in T1 to T2/M maturation (Tucker et al. 2007); the mutant p100 eliminates the alternative NF-κB activation, but, in addition, it also inhibits p65 (RelA) complexes, which are normally activated by the classical pathway (Kanno et al. 1994).

Although the alternative pathway contributes to B-cell maturation, the classical pathway is absolutely required. This conclusion was already apparent from the results with c-Rel, RelA compound deficiency (see previous discussion) and was confirmed with NEMO (IKKγ) deficient B cells. Conditional loss of NEMO in B cells with CD19- or mb1-Cre caused significant arrest at or near the T1 to T2 transition with a near complete absence of mature B cells; similarly induced conditional expression of a kinase deficient form of IKKβ

(see previous discussion) caused a milder block, whereas outright (conditional) loss of IKKβ caused no discernable block at the T1 to T2 transition (Pasparakis et al. 2002; Sasaki et al. 2007; Derudder et al. 2009). However, all three (B-cell specific) genetic changes nearly eliminated or at least severely reduced numbers of mature recirculating B cells. The most straightforward interpretation of these results is that during the transitional phases, IKKα may partially compensate for the outright loss of IKKβ to activate the classical pathway (a similar situation was seen in conventional thymocyte development discussed previously). On the other hand, once B cells have matured, they absolutely require IKK2-mediated classical activation for long-term survival. The most severe block at the T1 to T2 transition occurred in mice conditionally ablated with mb1-Cre for both IKK1 and IKK2; the block was more complete than that observed in the absence of NEMO (Derudder et al. 2009). This suggests that IKK1 also contributes in a NEMO-independent fashion, presumably via the alternative pathway. Of note, a constitutively active form of IKK2 was able to completely overcome the absence of BAFFR, indicating that the classical pathway is both necessary and sufficient to allow full maturation of B cells in the absence of BAFFR/alternative activation (Sasaki et al. 2006). However, because the alternative pathway was required for optimal generation of B cells under normal conditions, it must be inferred that the classical pathway is not normally sufficiently activated.

It has been suggested that weak tonic BCR signaling activates the classical pathway, which in turn is required to set up the alternative pathway by inducing expression of BAFFR and NF-κB2 via c-Rel (Stadanlick et al. 2008; Castro et al. 2009). However, BCR-independent signals must also be considered, because loss of CBM complex components (Carma1, Bcl-10, or Malt1) only mildly reduced FM B-cell numbers (CBM is part of the conventional signaling path from the BCR to NF-κB; see TCR signaling discussed previously) (Thome 2004; Ferch et al. 2007). This situation is somewhat reminiscent of what was noted for development of naïve T cells, and again raises the possibility that BCRs and TCRs may activate NF-κB in developing lymphocytes by an unconventional pathway. Whatever the pathway or signal, TAK1 was reported to be required for B-cell maturation (Schuman et al. 2009), although an earlier report appears to be at odds with this finding (Sato et al. 2005), while TRAF6 was reported to be at least partially required (Kobayashi et al. 2009). However, these proteins can be part of many NF-κB signaling pathways, so their involvement does not inform on the nature of the signal.

The generation of MZB cells is particularly sensitive to perturbations in NF-κB activity. Single deficiency in NF-κB1, NF-κB2 or RelB and to a lesser extent c-Rel or RelA in B cells already reduced numbers of MZBs (Caamano et al. 1998; Franzoso et al. 1998; Cariappa et al. 2000; Weih et al. 2001; Guo et al. 2007), although there may be an unexplained rebound with advancing age in NF-κB1 knockouts (Ferguson and Corley 2005). Furthermore, loss of IKK2 essentially eliminated these cells (Pasparakis et al. 2002). NF-κB1 reportedly synergizes with Notch2 receptor signaling during generation of MZBs, but mechanisms remain to be elucidated (Moran et al. 2007; Allman and Pillai 2008). The exquisite dependence of MZBs on NF-κB activity is also highlighted by mouse models in which the numbers of these cells were augmented. CYLD is deubiquitinase for K63-linked ubiquitin chains, which down-modulates the classical pathway by targeting proteins such as NEMO and RIP (Courtois 2008). The loss of CYLD in B cells led to a marked increase in constitutive NF-κB activity and a significant rise in MZBs (Jin et al. 2007). Overactivation of the alternative pathway also led to a selective increase of MZBs, as is the case in mice with a BAFF transgene (Mackay and Schneider 2009) or in mice deficient in the IκB-like part of p100/NF-κB2, leaving p52 to form complexes with RelB and enter nuclei (Guo et al. 2007). Finally, loss of Bcl-3 unexpectedly increased MZB cell numbers; this implicates Bcl-3 as a gatekeeper during MZB formation (U.S., unpubl. observations). As discussed previously, Bcl-3 is an unusual IκB family member; how it normally suppresses

MZB development remains to be determined, although it might do so by dampening NF-κB target gene expression.

MZB cells have been suggested to be selected by expression of weakly self-reactive BCRs that avoid negative selection but are possibly strong enough to activate some NF-κB to aid in their survival and/or expansion (Allman and Pillai 2008). Such a scenario would be consistent with the fact that components of the conventional BCR signaling path to NF-κB were required for MZB development (e.g., Carma1, Malt1, Bcl-10, TRAF6, and Tak1) (Thome 2004; Pappu and Lin 2006; Kobayashi et al. 2009; Schuman et al. 2009).

Formation of B1 B cells is highly dependent on the function of the conventional BCR signaling pathway to NF-κB and on NF-κB activity, possibly because these cells are thought to be selected for fairly strong recognition of self-antigens (Allman and Pillai 2008). Loss of component parts of this pathway (Thome 2004; Sato et al. 2005; Kobayashi et al. 2009; Schuman et al. 2009) or interference with NF-κB activity blocked development of B1 B cells; for example, deficiency in NF-κB1 already reduced their numbers, whereas a deficiency in both c-Rel and NF-κB1 nearly eliminated them (FM cell numbers were only mildly affected) (Pohl et al. 2002; Gerondakis et al. 2006). c-Rel is particularly important for cell-cycle entry and proliferation in response to BCR stimulation (Gilmore et al. 2004), suggesting the possibility that B1 B cells may expand in response to BCR signaling induced by self-antigens. As expected, B1 B cells also failed to develop in the absence of IKK2 (Pasparakis et al. 2002). Interestingly, loss of NF-κB1 and NF-κB2 severely diminished B1 B-cell numbers (Claudio et al. 2002), implying a possible role for the alternative pathway of activation, even though BAFFR was not required (Mackay and Schneider 2009; Rauch et al. 2009).

NF-κB IN THYMIC EPITHELIAL CELLS

NF-κB activity in epithelial cells is critical for central tolerance, i.e., tolerance of self-antigens enforced in the thymus via elimination (negative selection) of autoreactive TCRs (see previous discussion). Negative selection of conventional thymocytes with strongly autoreactive TCRs begins during interaction with self-antigen presented on cortical thymic epithelial cells (cTECs; also responsible for positive selection), but the bulk of negative selection is thought to take place in the medulla, where self-antigens are presented by thymic dendritic cells (DCs) and medullary thymic epithelial cells (mTECs) (Kyewski and Klein 2006; Boehm 2008; Nitta et al. 2008). mTECs promiscuously express many tissue specific self-antigens (TSAs), such as insulin and salivary protein 1, driven in part via the AIRE regulator; mTECs also efficiently cross-prime thymic DCs with these self-antigens (Mathis and Benoist 2009). mTECs are likely required for accumulation of thymic DCs in the medulla (Tykocinski et al. 2008). Finally, mTECs may help promote Treg development, at least those with specificities to TSAs, although this remains to be shown (Zhang et al. 2007; Koble and Kyewski 2009).

Development/differentiation of mTECs is under control of NF-κB. In the prenatal stage, generation of mTECs is induced by interaction with hematopoietic lymphoid tissue inducer cells (LTi) (Kim et al. 2009). These cells express ligands for RANK and LTβ receptors and thereby activate NF-κB in mTEC progenitors via matching receptors (Akiyama et al. 2008). Thereafter, continued generation and maintenance of the mTEC network is controlled by positively selected DP thymocytes, which express ligands for RANK, LTβ, and CD40 to stimulate mTECs (Akiyama et al. 2008; Hikosaka et al. 2008). Autoreactive thymocytes in particular have been suggested to promote the generation of mTECs (Irla et al. 2008). mTECs turn over rapidly, express self-antigens optimally on terminal differentiation, and cross-prime thymic DCs, most likely in a process linked to their turnover (Yano et al. 2008; Koble and Kyewski 2009). The nuclear protein AIRE facilitates expression of a significant portion of the promiscuously produced TSAs and it may also have a role in terminal differentiation and antigen presentation (Yano et al. 2008; Koble and Kyewski 2009; Mathis and Benoist

2009). Expression of some TSAs is independent of AIRE and instead may be produced by a distinct subclass of mTECs, dependent on LTβR signaling (Martins et al. 2008; Seach et al. 2008).

RANK, CD40, and LTβR are TNF receptor family members that activate both classically and alternatively activated NF-κB complexes (Ghosh and Hayden 2008; Hayden and Ghosh 2008; Vallabhapurapu and Karin 2009). Consistent with a critical role for the alternative pathway, loss of NIK, IKKα, RelB (Weih et al. 1995; Kajiura et al. 2004; Kinoshita et al. 2006), or compound loss of NF-κB2 and Bcl-3 abrogated the appearance of differentiated mTECs (Zhang et al. 2007). This in turn engendered multiorgan inflammation and early death, directly mediated by autoreactive T cells emerging from the thymus (Zhang et al. 2007) (U.S., unpubl. observations). Loss of NF-κB2 alone caused only mild impairment of mTECs, presumably because of p50/RelB complexes that remain active in the absence of p100/NF-κB2, the primary inhibitor of RelB (Zhu et al. 2006; Zhang et al. 2007). As discussed previously, Bcl-3 is not part of the alternative pathway and is most likely induced by the classical pathway. It is not known how the combined loss of both NF-κB2 and Bcl-3 leads to such a profound defect. Bcl-3 most likely functions by interaction with NF-κB1; this is consistent with a complete block in mTEC development in mice lacking NF-κB1 and NF-κB2 (Zhang et al. 2007).

The classical NF-κB pathway is also indispensable. TRAF6 knockout mutants failed to develop mTECs and exhibited multiorgan inflammation (Inoue et al. 2007; Akiyama et al. 2008). TRAF6 is engaged by numerous signals and functions just upstream of the IKK complex in the classical pathway (Ghosh and Hayden 2008; Hayden and Ghosh 2008; Vallabhapurapu and Karin 2009). RelB and NF-κB2 are likely critical targets of the classical pathway, because continuous production of these components of the alternative pathway would be required for long-term activation of this pathway (Inoue et al. 2007; Akiyama et al. 2008). Chemokines, such as CCR7 ligands, are important targets of NF-κB in epithelial cells that permit the attraction of positively selected DP thymocytes (Förster et al. 2008; Nitta et al. 2008). But how the alternative and/or classical pathways control the generation of mTECs themselves is an open question.

The alternative pathway and RelB may also have a specific role in nonhematopoietic cells required for thymic iNKT formation (Elewaut et al. 2003; Sivakumar et al. 2003; Franki et al. 2005), although recent reports suggest that LTβR-mediated activation of this pathway may be primarily required for thymic export/peripheral colonization of these cells and not their development (Franki et al. 2006; Vallabhapurapu et al. 2008).

SECONDARY LYMPHOID ORGANS: NF-κB IN STROMAL CELLS

The classical and alternative activation of NF-κB in stromal cells is instrumental in establishing secondary lymphoid organs and/or lymphoid organ architecture. The lymphoid organs provide the anatomical niches required for optimal initiation and expansion of adaptive immune responses (Drayton et al. 2006; Mueller and Ahmed 2008; Randall et al. 2008). A number of mouse mutants in which the alternative pathway was fully or partially blocked failed, to variable degrees, in forming splenic B-cell follicles, proper marginal zones, lymph nodes, Peyer's patches, and germinal centers. These mutants include those deficient in RelB or NIK, or with compound deficiency in NF-κB2 and Bcl-3, or harboring inactivating mutations of IKKα or NIK (Matsushima et al. 2001; Senftleben et al. 2001a; Weih et al. 2001; Paxian et al. 2002; Yilmaz et al. 2003; Weih and Caamano 2003; Zhang et al. 2007). Loss of NF-κB2 alone had milder defects, consistent with only partial abrogation of RelB complexes (see previous discussion) (Franzoso et al. 1998; Weih and Caamano 2003; Guo et al. 2007). Interestingly, loss of Bcl-3 also displayed some mild defects in lymphoid architecture (Franzoso et al. 1997b; Poljak et al. 1999).

The absence of splenic B-cell follicles in mice impaired in alternative NF-κB activation in the models cited previously correlated with a lack of follicular dendritic cell (FDC) networks (Weih and Caamano 2003; Zhang et al. 2007). FDCs present antigen–antibody immune complexes to facilitate clonal selection/differentiation of antigen-activated B cells in germinal center reactions that generate plasma cells and memory B cells (Allen and Cyster 2008; Mueller and Ahmed 2008; Batista and Harwood 2009). Apart from the ability of both FDCs and thymic mTECs to present antigens, these cells have many other parallels. They attract their lymphocyte partners by expressing chemoattractants—in the case of FDCs, CXCL13 (a.k.a. BLC) to attract B and organize them into follicles. These cells also require continuous stimulation by lymphocytes to maintain their differentiated network (Allen and Cyster 2008; Mueller and Ahmed 2008; Batista and Harwood 2009). B cells express LTαβ (and Light) and signal FDCs via LTβRs to activate RelB complexes via the alternative pathway, in turn inducing expression of chemokines, e.g., CXCL13 (Suto et al. 2009). LTβR signaling also activates RelA complexes, either by direct activation of the classical pathway or by liberating RelA complexes inhibited by p100 (Muller and Siebenlist 2003; Basak et al. 2007). Mice blocked in activation of the alternative pathway share many of the same defects apparent in mice deficient in LTβR signaling (Matsushima et al. 2001; Hehlgans and Pfeffer 2005; Drayton et al. 2006; Ruddle and Akirav 2009). The alternative pathway is also required within stromal cells to form proper marginal zones. Impaired signaling via this pathway decreased the numbers of MZB cells, even if the pathway was intact within B cells (where it also plays a role; see previous discussion). RelB and the alternative pathway appear to control integrin-ligand interactions and chemokine expression required for MZB retention (Weih et al. 2001; Weih and Caamano 2003; Guo et al. 2007).

Activation of the alternative pathway via LTβR in stromal (mesenchymal organizer) cells is essential for lymph node and Peyer's patch organogenesis (Drayton et al. 2006; Elewaut and Ware 2007; Vondenhoff et al. 2007). Similar to the situation in prenatal mTEC development, hematopoietic LTi cells initiate the process in part via expression of ligands for LTβR and RANK (Evans and Kim 2009); subsequently, B cells provide the necessary ligands (Elewaut and Ware 2007). Interestingly, only organogenesis of lymph nodes, but not of Peyer's patches, requires RANK activation on stromal cells, a potent activator of both the alternative and classical pathway for NF-κB (see previous discussion) (Yoshida et al. 2002).

In addition to RANK and LTβR, the TNF receptor I may also play a role in lymphoid organogenesis and maintenance, although it only stimulates the classical pathway (Ruddle and Akirav 2009). The importance of the classical pathway is also underscored by the role of the adaptor protein TRAF6, which mediates classical activation by many receptors, including RANK, though not TNF receptor I (Ghosh and Hayden 2008; Hayden and Ghosh 2008; Vallabhapurapu and Karin 2009); TRAF6 was required within stromal cells for proper splenic architecture (e.g., B-cell follicles, marginal zones) as well as for lymph node and Peyer's patch organogenesis (Qin et al. 2007). Similar to the situation with mTECs, the classical pathway is likely required to induce expression of RelB and NF-κB2 to provide the components needed for continuous signaling via the alternative pathway (see previous discussion); in addition, the classical pathway may induce expression of Bcl-3. Little is known about how NF-κB controls the development/differentiation of stromal cells, such as FDCs, but as in mTECs, NF-κB does induce expression of chemokines essential for cellular interactions and compartmentalization in secondary lymphoid organs (Vondenhoff et al. 2007; Mueller and Ahmed 2008; Randall et al. 2008; Evans and Kim 2009; Suto et al. 2009).

NF-κB AND B-CELL ACTIVATION

Mature B-cell populations respond to two broad categories of antigen: so-called thymus-independent (TI) and thymus (T)-dependent antigens. Polyvalent antigens that typically possess multiple repeat determinants found on macromolecules such as polysaccharides can activate B cells without T-cell help. These T-independent humoral responses, dominated by CD5[+] and marginal zone B cells, typically involve the secretion of relatively low affinity IgM antibodies (McHeyzer-Williams 2003). Follicular B cells on the other hand, which serve as precursors of T-dependent B-cell responses to protein antigens are typically triggered by activated helper T cells through CD40 and receptors for cytokines such as IL-4 (McHeyzer-Williams and McHeyzer-Williams 2005). This leads to isotype switching and germinal center formation in secondary lymphoid organs, during which somatic hyper-mutation and antigen-driven selection produce long-lived affinity-matured plasma cells, plus memory B cells that can respond rapidly to antigen rechallenge (McHeyzer-Williams 2003; McHeyzer-Williams and McHeyzer-Williams 2005). A dichotomy between T-dependent and T-independent B-cell responses is unlikely to be as clear-cut in the context of a microbial infection, in which multivalent antigens are encountered in conjunction with protein antigens. Figure 3 outlines those stages during follicular B-cell activation and differentiation that involve NF-κB signaling.

Mature B cells express different forms of NF-κB. In resting cells, NF-κB1/c-Rel heterodimers and NF-κB1 homodimers predominate (Grumont and Gerondakis 1994), with RelA (NF-κB1/RelA, c-Rel/RelA) and NF-κB2 (NF-κB2/RelB, NF-κB2 homodimers) dimers also present, albeit at lower levels. Constitutive, low-grade nuclear expression of NF-κB factors downstream of both the classical and alternate pathway is a feature of nonactivated B cells that is thought to mainly serve a survival role (Gerondakis and Strasser 2003; Claudio et al. 2006). Low intensity, nonmitogenic BCR signals sustain the production of p100 NF-κB2, which in turn serves to transmit survival signals through the BLyS receptor, BR3 (a.k.a. BAFFR) (Stadanlick et al. 2008).

Functional and developmental outcomes for B cells, including proliferation, differentiation, and survival that arise from specific types of activation signals, are reflected in the qualitative and quantitative nature of NF-κB responses. BCR, CD40, and TLR receptors all individually engage NF-κB during B-cell activation

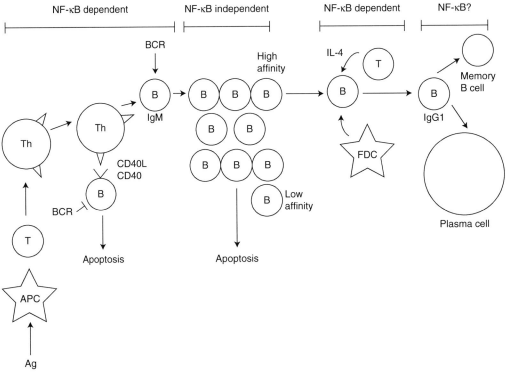

Figure 3. NF-κB in B-cell activation. The activation of follicular B cells by CD4 T-cell-dependent CD40 and BCR signals promotes a rapid antigen-driven expansion of B cells in the germinal centers (GC) of secondary lymphoid organs that is accompanied by isotype switching and affinity maturation. These antigen and cytokine driven events lead to the development of long-lived plasma cells and memory B cells. Shown are the requirements for NF-κB during the various phases of B-cell differentiation and proliferation.

(Vallabhapurapu and Karin 2009). Strong BCR signals rapidly lead to the IKKβ-dependent nuclear recruitment of NF-κB1/c-Rel, NF-κB1/RelA, and NF-κB1 homodimers from cytoplasmic reserves (Kontgen et al. 1995; Grumont et al. 1998). Lipopolysaccharide and CpG double stranded DNA, respective ligands for TLR4 and TLR9, also rapidly activate these same NF-κB complexes (Grumont and Gerondakis 1994; Krieg 2002). In contrast, CD40 signals activate both IKKα and IKKβ (Hayden and Ghosh 2008; Vallabhapurapu and Karin 2009), resulting in the nuclear expression of both RelB and c-Rel complexes. Although both NF-κB pathways participate in CD40-dependent humoral immunity (Gerondakis et al. 2006; Hayden and Ghosh 2008; Vallabhapurapu and Karin 2009), impaired antibody responses that arise from a block in the alternate NF-κB pathway are not caused by B-cell intrinsic defects (Franzoso

et al. 1998), highlighting the distinct roles served by the two arms of the NF-κB pathway. Cross talk among these receptors is crucial in determining how NF-κB influences functional outcomes during B-cell activation. BCR signals alone, although not able to efficiently trigger B-cell differentiation, nevertheless provide essential costimulatory signals that influence survival and differentiation. BCR induction of NF-κB protects CD40-stimulated B cells from activation-induced cell death (Rothstein et al. 2000) (see later discussion) and prevent B-cell hypo-responsiveness that can accompany continuous TLR signaling (Poovassery et al. 2009). Finally, the duration of B-cell stimulation also influences the temporal regulation of NF-κB activity. In addition to the initial rapid nuclear induction of NF-κB, sustained TLR4 (Grumont and Gerondakis 1994) and BCR (S.G., unpubl. results) signals lead to multiple, transient waves of nuclear NF-κB expression that are dependent on the "de novo" synthesis of these transcription factors (Grumont and Gerondakis 1994). It remains to be determined which functions during B-cell activation are linked to the specific phases of NF-κB expression.

NF-κB and B-cell Division

Both the IKKα and IKKβ-dependent NF-κB pathways contribute to mitogen-induced B-cell proliferation by regulating multiple B-cell intrinsic mechanisms. BCR, CD40, and TLR4 or TLR9 signals each use the classical NF-κB pathway to promote B-cell proliferation (Ghosh and Hayden 2008; Hayden and Ghosh 2008; Vallabhapurapu and Karin 2009); only CD40 appears to require the alternate pathway for this function (Vallabhapurapu and Karin 2009). All IKKβ-dependent B-cell mitogenic responses require c-Rel (Gerondakis et al 1998), albeit to varying degrees, with BCR activation being most dependent on c-Rel (Grumont et al. 1998). NF-κB1 on the other hand is only crucial for

TLR4-dependent B-cell proliferation (Sha et al. 1995; Grumont et al. 1998).

The classical NF-κB pathway serves several distinct roles in the cell cycle during BCR-induced proliferation (Fig. 4). c-Rel is required for progression from G1 to S-phase (Grumont et al. 1998). Although c-Rel regulates cell survival during BCR activation (Grumont et al. 1998, 1999; Cheng et al. 2003), the failure of c-Rel-deficient B cells to proliferate is a direct consequence of a cell cycle block and not merely a secondary consequence of cell death (Grumont et al. 1998; Cheng et al. 2003). The expression of E2F3, an E2F transcription factor involved in G1/S phase progression (Humbert et al. 2000) was shown to be impaired in BCR activated c-Rel-deficient B cells (Cheng et al. 2003). With E2F3 known to contribute to the suppression of p21Cip1 function (Wu et al. 2001), c-Rel may promote cell cycle progression by serving to inactivate Rb-dependent repression of S-phase entry. Whether E2F3 is the only direct c-Rel target important in cell

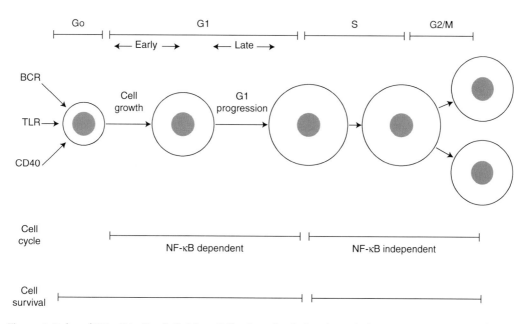

Figure 4. Roles of NF-κB in B-cell division. Following stimulation through the BCR, CD40, or TLR4/TLR9, mature quiescent B cells in G0 enter G1 and undergo an NF-κB dependent phase of growth. B-cell growth continues until late G1, at which point NF-κB is required for entry into S-phase. Subsequent steps in the B-cell cycle appear to be NF-κB independent. NF-κB also regulates survival signals associated with B-cell activation and division.

cycle progression remains to be determined. The expression of IRF4, a transcription factor shown to influence normal B cell (Mittrucker et al. 1997) and malignant B cell (Shaffer et al. 2008) proliferation is also impaired in mitogen activated c-Rel-deficient B cells (Grumont and Gerondakis 2000). Despite c-Rel being necessary for BCR or TLR4 entry into S phase (Grumont et al. 1998), these combined signals permit a majority of B cells to progress through the cell cycle in the absence of c-Rel. These findings are consistent with c-Rel setting a threshold for cell cycle progression by modulating the stimulus-dependent transcription of cell cycle regulators. Transcriptional targets of NF-κB1 involved in TLR4-dependent G1 to S phase progression remain to be identified.

c-Rel and NF-κB1 function redundantly to promote cell growth during G0 to G1 progression (Pohl et al. 2002; Grumont et al. 2002). This step in the cell cycle, associated with ribosome biogenesis (Stocker and Hafen 2000), is essential for assembling the cellular machinery necessary for DNA replication. This NF-κB-dependent control of B-cell growth, which is required for all activation signals, involves c-Rel and NF-κB1 inducing the expression of c-Myc (Lee et al. 1995; Grumont et al. 2002), which serves a key role in the growth of eukaryotic cells (Levens 2002). Despite c-Myc transgene expression promoting B-cell growth in the absence of c-Rel and NF-κB1, it is insufficient to drive the mitogen-dependent proliferation of B cells lacking these NF-κB transcription factors (Grumont et al. 2002), further highlighting the multiple role(s) c-Rel and possibly NF-κB1 serve in B-cell division.

In vivo, B cells undergo extensive antigen-dependent proliferation in germinal centers linked with BCR affinity maturation that occurs during the development of memory B cells and plasma cells (McHeyzer-Williams 2003). Remarkably, germinal center (GC) B cells fail to express most NF-κB target genes and NF-κB pathway components (Shaffer et al. 2001), indicating that NF-κB signaling appears to be dispensable for GC B-cell proliferation. A strong bias of the GC B-cell gene expression profile toward G2/M regulators rather than cell growth

regulators such as ribosomal components (Shaffer et al. 2001) is consistent with a shortened G1 that is characteristic of rapidly dividing GC B cells (McHeyzer-Williams 2003). The findings that the NF-κB pathway appears to be important for initiating the proliferation of naïve B cells, but not the division of GC B cells during differentiation, reinforces the link between NF-κB and G0/G1 control of the cell cycle. It also highlights how selective engagement of the NF-κB pathway during different stages of B-cell activation and differentiation can be used to control the rate of B-cell division.

NF-κB Control of Activated B-cell Survival

Elevated levels of apoptosis typically accompany B-cell activation. The prevailing models favor apoptosis in this instance as a mechanism used to eliminate B cells that are not programmed to undergo normal cell division. Such models are consistent with c-Rel-controlling BCR and TLR4-induced division and survival through the regulation of distinct target genes (Grumont et al. 1998, 1999; Cheng et al. 2003). The viability of BCR-activated B cells is dependent on c-Rel directly inducing the transcription of genes encoding the Bcl-2 prosurvival homologs, A1/Blf1 (Grumont et al. 1999) and Bcl-xL (Cheng et al. 2003). Although the coexpression of two related antiapoptotic genes has been viewed as a redundant, fail-safe survival mechanism (Lee et al. 1999), neither overexpression of A1 (Grumont e al. 1999) or Bcl-xL (S.G., unpubl. results) alone is sufficient to confer complete protection to BCR-activated B cells lacking c-Rel. However, with enforced Bcl-2 transgene expression able to block the BCR induced death of c-Rel deficient B cells (Grumont et al. 1998), thereby ruling out any involvement of the death receptor pathway, this reinforces the notion that A1 and Bcl-xL serve distinct survival roles. With the c-Rel induction of A1 preceding Bcl-xL in activated B cells (Grumont et al. 1999; Cheng et al. 2003), it is conceivable that A1-mediated B-cell survival is linked to early events in the cell cycle, such as cell growth, whereas Bcl-xL prevents death associated with the G1 to S phase transition.

 Cite this article as *Cold Spring Harb Perspect Biol* 2009;2:a000182

Unlike cell death associated with BCR activation, which is limited by c-Rel alone, both NF-κB1 and c-Rel cooperate to protect TLR4-stimulated B cells from apoptosis (Gerondakis et al. 2007). This cell death, driven by activation of the BH3-only proapoptotic protein Bim is blocked by c-Rel and NF-κB1 regulating distinct survival pathways (Gerondakis et al. 2007; Banerjee et al. 2008). c-Rel induction of A1 and Bcl-xL coincide with both survival proteins binding to Bim, thereby preventing it from engaging the cell intrinsic death pathway. NF-κB1 indirectly inhibits TLR4-induced death by controlling ERK phosphorylation of Bim (Banerjee et al. 2008), a posttranslational event that targets Bim for degradation (Ewings et al. 2007). NF-κB1 and ERK activation are linked through the NF-κB1 precursor, p105 serving as a scaffold for the ERK pathway specific MAP3 kinase, Tpl2 (Belich et al. 1999). Following TLR4 activation, IKKβ-induced phosphorylation and degradation of p105 serves as an essential step in Tpl2 activation (Waterfield et al. 2003; Banerjee et al. 2005). It remains to be determined if these distinct NF-κB-regulated survival mechanisms neutralize Bim in a spatial or temporal manner.

CD40 serves opposing roles in B-cell survival. In isolation, CD40 signals are able to promote B-cell survival through the NF-κB induction of A1 and Bcl-xL (Lee et al. 1999). However, under physiological conditions in which antigen-stimulated helper T cells express high levels of CD40 ligand, CD40 signals sensitize B cells to Fas-induced apoptosis (Rothstein et al. 2000). This death, which is prevented by costimulating B cells through the BCR, ensures that only appropriate antigen-specific B cells and not bystander cells are activated by CD40 (Rothstein et al. 2000). This BCR-induced protection of CD40-sensitized B cells is independent of c-Rel (Owyang et al. 2001) and requires the PI3K/AKT-dependent induction of c-FLIP (Moriyama and Yonehara 2007), a caspase-8 inhibitor essential for death-receptor-induced apoptosis (Strasser et al. 2009). Although the NF-κB pathway activated through CD40 has been shown in certain cell line models to prevent Fas-induced B-cell death (Lee et al. 1999; Zazzeroni et al. 2003), these seemingly

paradoxical findings may be reconciled with CD40 alone serving to protect B cells from transient, but not sustained Fas signals (Lee et al. 1999). With death-receptor-induced apoptosis largely resistant to prosurvival signals emanating from the cell intrinsic survival pathway (Strasser et al. 2009), CD40 induction of Bcl-xL and A1 most likely serves to prevent death linked to CD40-induced cell division.

Isotype Switching

NF-κB signaling is essential for isotype switching (Hayden et al. 2006), a mechanism used to fine-tune humoral immune responses that involves the transposition of the assembled VDJ gene located upstream of Cμ to a downstream heavy chain (C$_H$) gene. T-independent antigens and T-cell-dependent CD40 signals, in combination with specific cytokines, direct switching to specific classes of immunoglobulin (Manis et al. 2002). In the mouse, for example, LPS promotes switching to IgG3, whereas CD40 signals and IL-4 produced by Th2 T cells promote switching to IgG1 and IgE (Manis et al. 2002). Switch recombination occurs between regions of repeat sequence (S-regions) located upstream of each C$_H$ gene. S-region transcription activated by mitogen and cytokine-induced binding of proteins to cryptic promoters (I regions) flanking S-regions facilitates S-region access to the DNA recombination machinery (Manis et al. 2002). NF-κB transcription factors downstream of the classical and alternate pathways participate in switching through a variety of mechanisms (Hayden et al. 2006). NF-κB factors induced by mitogens and cytokines promote S-region transcription by binding to specific I regions (Delphin and Stavnezer 1995; Agresti and Vercelli 2002; Bhattacharya et al. 2002; Wang et al. 2006). Switching is also dependent on NF-κB factors binding to the 3′ IgH enhancer, a locus control region located downstream of the Cα gene (Zelazowski et al. 2000; Laurencikiene et al. 2001). Although NF-κB does not regulate the expression of the switch recombination enzymes (Chaudhuri et al. 2007), NF-κB1 has been implicated in the DNA rearrangement process (Kenter et al. 2004), and given DNA synthesis is essential for S-region rearrangement (Chaudhari et al. 2007), NF-κB signaling also probably indirectly influences switching through the promotion of B-cell proliferation.

NF-κB AND T-CELL ACTIVATION

The successful activation of naïve T cells requires two distinct signals: an antigen-specific signal arising from TCR engaging MHC bound peptides expressed on APC, plus a costimulatory signal in the form of the B7 ligand upregulated on APC that binds to CD28. These signals, in addition to inducing the rapid expression of IL-2 and activation-associated cell surface molecules, promote T-cell division and effector T-cell differentiation along different lineages, the later of which is dictated by cytokine signals. The careful regulation of T-cell numbers before, during, and following immune responses by controlling the competing mechanisms of cell survival and apoptosis, serve to prevent immune-related pathology associated with inappropriate T-cell expansion and function. Those stages during T-cell activation and differentiation that involve NF-κB function are summarized in Figure 5.

Before antigen activation, naïve CD4 and CD8 T cells express very low levels of NF-κB1 in the nucleus. In contrast to B cells, the absence of constitutive NF-κB activity in these T cells is consistent with this pathway being dispensable for the homeostatic survival of naïve T cells. Following TCR and costimulatory signals, different NF-κB proteins enter the nucleus in a temporally regulated fashion (Molitor et al. 1990). Initially, pre-exisiting NF-κB1/RelA dimers bound to IκBα are rapidly transported to the nucleus (Molitor et al. 1990; Venkataraman et al. 1996). Although c-Rel is present in the cytoplasm of naïve T cells, it is not rapidly mobilized by TCR signals (Venkataraman et al. 1995; Rao et al. 2003) due to its association with IκBβ (Banerjee et al. 2005), which is relatively resistant to degradation triggered by the initial TCR stimulus. Significant nuclear levels of c-Rel, which follow NF-κB1/RelA expression, depend on the TCR-dependent induction of *c-rel* transcription (Grumont and Gerondakis 1990; Venkataraman

et al. 1996; Rao et al. 2003). TCR-induced c-Rel expression is NFAT dependent (Venkataraman et al. 1995), consistent with c-Rel being a direct transcriptional target of NFAT (Grumont et al. 2004). Although c-Rel is necessary for the induction of IL-2 (see later discussion) following T-cell activation, the role served by the initial wave of NF-κB1/RelA expression remains unclear, although it may be related to cell-cycle control as reflected by the impaired proliferation of TCR-activated RelA-deficient T cells (Doi et al. 1997).

NF-κB and T-cell Proliferation

Both the classical and alternate NF-κB pathways regulate T-cell proliferation in culture and "in vivo" via T-cell autonomous and nonautonomous mechanisms (Gerondakis et al. 2006; Vallabhapurapu and Karin 2009). NF-κB signaling is pivotal in controlling the proliferation of naïve T cells (Gerondakis et al. 2006). NF-κB1, c-Rel, and RelA serves unique (Kontgen et al. 1995; Gerondakis et al. 2006; Doi et al. 1997; Sriskantharajah et al. 2009) and overlapping roles (Gerondakis et al. 2006) in T-cell proliferation during different stages of the cell cycle. In contrast to B cells, which require NF-κB1 or c-Rel to induce mitogen-dependent T-cell growth (Grumont et al. 2002), it is RelA and c-Rel that function redundantly to promote c-Myc-induced T-cell growth following TCR activation (Grumont et al. 2004). Although c-Myc expression is both necessary and sufficient to promote T-cell growth in the absence of NF-κB function, it is not to rescue the TCR-induced proliferation of NF-κB-deficient T cells (Grumont et al. 2004). Therefore, the combined activity of these transcription factors, as well NF-κB1 and c-Rel (Zheng et al. 2003), serve functions in addition to growth, which are required for T-cell proliferation. The finding that IKK2-deficient CD4 T cells proliferate in culture as effectively as normal cells in response to a range of strong polyclonal stimuli (Schmidt-Supprian et al. 2003) is seemingly at odds with the findings for T-cell proliferation that emerge from the analysis of the different NF-κB compound mutant mice. This can be reconciled by the ability of the residual

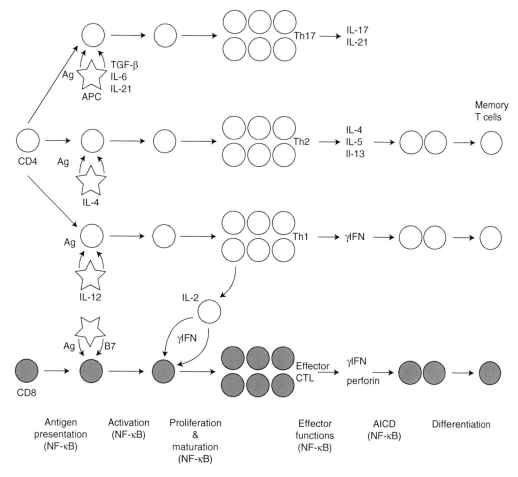

Figure 5. NF-κB in conventional T-cell activation. TCR-dependent activation of naïve CD4 and CD8 T cells by antigen presenting cells (APC) leads to T-cell activation, division, and effector T-cell differentiation. The various CD4 T helper cell subsets and CD8 cytotoxic T lymphocytes that develop in response to different cytokine and costimulatory signals delivered by APC and T cells undergo rapid antigen-driven expansion during the course of an infection. Following pathogen elimination, most effector T cells undergo activation-induced cell death, with the few surviving antigen-specific T cells differentiating into long-lived memory cells. Highlighted are roles for NF-κB in APC function, T-cell activation, CD4 Th differentiation, and T-cell survival and proliferation.

NEMO/IKKα complexes in IKKβ-deficient CD4 T cells to phosphorylate IκBα and induce a normal pattern of NF-κB complexes following T-cell activation (Schmidt-Supprian et al. 2003). However, the inability of OVA or KLH antigen-specific IKKβ-deficient T cells to proliferate in culture following restimulation with these same antigens that had previously been used to immunize mice in which IKKβ was selectively deleted in CD4 T lineage cells (Schmidt-Supprian et al. 2004b) suggests that IKKα can only

activate NF-κB complexes normally induced by IKKβ in response to strong stimuli.

IL-2, which is involved in autocrine-mediated T-cell proliferation, is dependent on c-Rel for its induction following TCR and costimulatory signaling (Kontgen et al. 1995). c-Rel appears to induce *il2* transcription in response to TCR and CD28 stimulation by remodeling chromatin in the vicinity of the CD28 response region of the IL-2 promoter, so that it is now conducive to transcription (Rao et al. 2003;

Chen et al. 2005). Despite c-Rel being essential for IL-2 dependent CD4 T-cell proliferation in culture (Kontgen et al. 1995), this cytokine is dispensable for conventional T-cell proliferation in vivo (Schimpl et al. 2002). Instead, IL-2-deficient mice develop lymphoproliferative and autoimmune disease caused by impaired CD4 regulatory T-cell development (Liston and Rudensky 2007). Although c-Rel-deficient mice have reduced numbers of CD4 regulatory T cells, this is caused by IL-2-independent mechanisms (S.G., unpubl. results). The failure of c-Rel-deficient mice to develop autoimmune disease indicates that IL-2 expression can be controlled in a c-Rel-independent fashion in vivo.

Although the mechanisms by which NF-κB controls T-cell proliferation in culture are reasonably well established, NF-κB regulation of T-cell proliferation in vivo is complex, with an in depth overview beyond the scope of this article. Nevertheless, it is clear that the extent to which the NF-κB pathway contributes to T-cell proliferation in vivo is dictated by the type of T cell, its state of differentiation, plus the nature and strength of the T-cell stimulatory signals. For example, the homeostatic expansion of naïve CD4 and CD8 T cells in a lymphopenic environment is absolutely dependent on IKKβ (Schmidt-Supprian et al. 2004b). Although IKKβ is not essential for polyclonal T-cell expansion in vivo that is driven by strong mitogenic stimuli such as super antigens (Schmidt-Supprian et al. 2004b), it does serve a nonredundant T-cell-intrinsic role in the antigen-dependent proliferation that accompanies a recall response (Schmidt-Supprian et al. 2004b). These results, supported by in vitro findings, indicate that IKKβ function is only dispensable for T-cell proliferation in vivo that is induced by strong TCR signals. Despite some confusion about the roles of NF-κB in TCR-induced T-cell proliferation that arise from the overlap of IKKα and IKKβ-dependent functions, individual NF-κB transcription factors do serve select roles in the T-cell proliferative responses triggered by infectious agents. In the case of c-Rel, it serves a T-cell-intrinsic role in the clonal expansion of Th1 effector cells responding to *Toxoplasma gondii* (Mason et al.

2004), whereas it is dispensable for the expansion of antigen-specific CD8 cytotoxic effector T cells during the primary response to influenza virus (Harling-McNabb et al. 1999). Although NF-κB is involved in the proliferative response of effector T cells, it remains to be determined whether this pathway features in the rapid clonal expansion of CD4 and CD8 memory T cells following antigen recall.

NF-κB and T-cell Survival

NF-κB regulates T-cell survival during antigen-dependent activation and proliferation by controlling both the cell intrinsic and death receptor pathways. Controversy that surrounds how some aspects of how T-cell survival is controlled appears to reflect the complexity of a process that is intimately linked to the strength and duration of immune signals.

TCR and costimulatory signals exploit the NF-κB pathway to promote T-cell survival following antigen-dependent activation. NF-κB activation triggered through antigen-dependent TCR recruitment of the IKK complex to a signaling hub at the immunological synapse that requires PKCθ and the signaling components CARM1, Bcl-10, and MALT1 (Weil and Israel 2006) appears to afford different degrees of protection to CD4 and CD8 T cells. Following TCR activation, PKCθ is critical for CD8 T-cell survival, which is mediated via c-Rel (Saibil et al. 2007). Although c-Rel is necessary for the induction of Bcl-xL, expression of this anti-apoptotic molecule alone appears to be insufficient to confer protection from TCR-induced cell death (Saibil et al. 2007). In contrast, PKCθ only serves a modest role in CD4 T-cell survival, which surprisingly is not dependent on either NF-κB1 or c-Rel (Saibil et al. 2007). However, the finding that a combined deficiency of c-Rel and NF-κB1 has a profound impact on TCR-induced CD4 T-cell survival (Zheng et al. 2003) points to these transcription factors serving overlapping survival roles, presumably mediated through PCKθ dependent and independent pathways. This difference serves to highlight that the NF-κB survival signals delivered through the TCR are programmed differently in CD4 and CD8 T cells. The NF-κB pathway also exploits or is used by the T-cell costimulatory molecules OX40 (CD134) and glucocorticoid induced TNF receptor (GITR) to promote the survival of activated T cells (Song et al. 2008; Zhan et al. 2008). In the case of GITR, NF-κB-dependent expression of this coreceptor on TCR-stimulated CD4 and CD8 T cells, appears to serve as a mechanism for promoting T-cell survival in response to weak, rather than strong TCR signals (Zhan et al. 2008).

NF-κB also serves as a lynch pin in coordinating cell survival during T-cell proliferation. Following TCR-dependent antigen stimulation, successful S-phase entry associated with the upregulation of E2F target gene expression and cyclin E-dependent kinase activity requires NF-κB to neutralize p73-induced apoptosis (Wan and DeGregori 2003). In the absence of appropriate signals that activate NF-κB during G1, although E2F activation and the capacity to enter S phase remains intact, cell death is triggered by CDK2-dependent p73 expression. To restrict the number of activated T cells during immune responses, ongoing T-cell proliferation associated with persistent TCR stimulation promotes activation-induced T-cell death (AICD) by engaging the death receptor Fas (Strasser et al. 2009). TCR-dependent activation of the NF-κB pathway through its inhibition of caspase-8 activation helps counter Fas-induced T-cell death (Jones et al. 2005). It remains to be determined how in the face of continuous TCR signals, T cells sense a need to prevent or succumb to AICD by regulating NF-κB activation. Once pathogenic agents have been eliminated, a small number of antigen-specific CD4 or CD8 effector T cells survive and differentiate into long-lived memory T cells capable of mounting rapid immune responses on recountering the pathogen. It remains to be determined if NF-κB signaling is required for memory T-cell survival.

NF-κB and Effector T-cell Differentiation

Naïve T can adopt a number of developmental options that are dictated by the types of signals these cells encounter during antigen-dependent activation. In the case of CD4 T helper cells (Th), these include Th1, Th2, Th17, and T regulatory cells, all of which are distinguished by the types of cytokines these cells produce (Fig. 5). The NF-κB pathway regulates CD4 Th differentiation through APC-dependent and T-cell intrinsic mechanisms. Aside from the role NF-κB plays in antigen-dependent and independent differentiation of certain dendritic cell (DC) subsets (Ouaaz et al. 2002; O'Keeffe et al. 2005), the NF-κB-regulated expression of specific cytokines in these key APC, influences Th cell differentiation and function. c-Rel, for example, which is required for Th1 immune responses (Gerondakis et al. 2006), is crucial for DC expression of p35, a subunit of the cytokine IL-12 that is essential in Th1 development (Grumont et al. 2001). The finding that in activated DC, c-Rel controls the expression of IL-12 (Grumont et al. 2001), whereas RelA is more important in regulating the production of inflammatory cytokines such as IL-1, IL-6, and TNF (Wang et al. 2007) that are involved in shaping Th17 responses (Laurence and O'Shea 2007), may indicate that the antigen-dependent activation of specific NF-κB complexes in DC, in part influences the outcome of Th cell development. However, a flexible rather than strict bifurcation of inflammatory versus regulatory cytokine production controlled by different NF-κB factors is more likely, with c-Rel in certain instances implicated in the expression of inflammatory cytokines such as TNF (Grigoriadis et al. 1996) and IL-23 (Mise-Omata et al. 2007). In contrast to the role of c-Rel in Th1 but not Th2 differentiation, NF-κB1 selectively regulates Th2 cell differentiation (Das et al. 2001). T cells lacking NF-κB1 fail to induce GATA3 expression, a transcription factor crucial in the differentiation of Th2 cells. The role of NF-κB1 in Th2 cells appears to be restricted to the development rather than the function of committed Th2 cells, with NF-κB1-deficient mice failing to mount airway inflammatory responses, a defect that coincides with the inability to produce the Th2 signature cytokines IL-4, IL-5, and IL-13 (Das et al. 2001).

In addition to helping polarize T-cell responses linked with the production of specific cytokines, NF-κB transcription factors also regulate cytokine gene expression in activated T cells. Aside from NF-κB having a direct role in the TCR and costimulatory signal dependent transcription of cytokine genes that include IL-2, GM-CSF, IL-3, and IFN-γ (Gerondakis et al. 2006), cytokine production by naïve T cells can be enhanced by the TLR-dependent induction in APC of T-cell costimulatory molecules and cytokines that alter c-Rel responsiveness during TCR activation (Banerjee et al. 2005). In naïve T cells, which are notoriously refractory for cytokine production, c-Rel is primarily bound to IκB-β, which is relatively resistant to degradation triggered by TCR signals. T cells exposed to the pro-inflammatory cytokines TNF-α and IL-1, shift c-Rel to IκB-α-associated complexes that are readily targeted by the TCR. As a consequence, in cytokine-primed T cells, IL-2 and IFN-γ mRNA are now produced rapidly, and at higher levels.

Finally, little is currently known about the genetic programs that regulate the development of memory cells from the pool of effector T cells. Despite an essential role for NF-κB function in memory T-cell development (Schmidt-Supprian et al. 2003), it remains unclear whether this dependence on the NF-κB pathway reflects a direct role in memory cell differentiation or a maintenance role through the control of memory cell survival.

CONCLUDING REMARKS

Despite remarkable progress in understanding the roles of the NF-κB pathway in normal lymphocyte development and function, much still remains to be learned. Areas in need of greater attention include gaining a better understanding of how NF-κB controls the development of the innate T-cell lineages, determining what roles the NF-κB pathway serves in memory T-cell differentiation, and cataloging the many and varied roles the NF-κB pathway serves in lymphocytes during different types of infectious diseases. With NF-κB involvement being documented in an increasing number of lymphoid-associated pathologies, answers to these and a range of other outstanding questions

will serve as a basis for preventing and treating these diseases.

REFERENCES

Agresti A, Vercelli D. 2002. c-Rel is a selective activator of a novel IL-4/CD40 responsive element in the human Ig γ4 germline promoter. *Mol Immunol* 38: 849–859.

Aifantis I, Mandal M, Sawai K, Ferrando A, Vilimas T. 2006. Regulation of T-cell progenitor survival and cell-cycle entry by the pre-T-cell receptor. *Immunol Rev* 209: 159–169.

Akiyama T, Shimo Y, Yanai H, Qin J, Ohshima D, Maruyama Y, Asaumi Y, Kitazawa J, Takayanagi H, Penninger JM, et al. 2008. The tumor necrosis factor family receptors RANK and CD40 cooperatively establish the thymic medullary microenvironment and self-tolerance. *Immunity* 29: 423–437.

Allen CD, Cyster JG. 2008. Follicular dendritic cell networks of primary follicles and germinal centers: phenotype and function. *Semin Immunol* 20: 14–25.

Allman D, Pillai S. 2008. Peripheral B cell subsets. *Curr Opin Immunol* 20: 149–157.

Banerjee D, Liou HC, Sen R. 2005. c-Rel-dependent priming of naive T cells by inflammatory cytokines. *Immunity* 23: 445–458.

Banerjee A, Grumont R, Gugasyan R, White C, Strasser A, Gerondakis S. 2008. NF-κB1 and c-Rel cooperate to promote the survival of TLR4-activated B cells by neutralizing Bim via distinct mechanisms. *Blood* 112: 5063–5073.

Banerjee A, Gugasyan R, McMahon M, Gerondakis S. 2006. Diverse Toll-like receptors utilize Tpl2 to activate extracellular signal-regulated kinase (ERK) in hemopoietic cells. *Proc Natl Acad Sci* 103: 3274–3279.

Barnes MJ, Krebs P, Harris N, Eidenschenk C, Gonzalez-Quintial R, Arnold CN, Crozat K, Sovath S, Moresco EM, Theofilopoulos AN, et al. 2009. Commitment to the regulatory T cell lineage requires CARMA1 in the thymus but not in the periphery. *PLoS Biol* 7: e51.

Basak S, Kim H, Kearns JD, Tergaonkar V, O'Dea E, Werner SL, Benedict CA, Ware CF, Ghosh G, Verma IM, et al. 2007. A fourth IκB protein within the NF-κB signaling module. *Cell* 128: 369–381.

Batista FD, Harwood NE. 2009. The who, how and where of antigen presentation to B cells. *Nat Rev Immunol* 9: 15–27.

Belich MP, Salmerón A, Johnston LH, Ley SC. 1999. TPL-2 kinase regulates the proteolysis of the NF-κB-inhibitory protein NF-κB1 p105. *Nature* 397: 363–368.

Bhattacharya D, Lee DU, Sha WC. 2002. Regulation of Ig class switch recombination by NF-κB: Retroviral expression of RelB in activated B cells inhibits switching to IgG1, but not to IgE. *Int Immunol* 14: 983–991.

Boehm T. 2008. Thymus development and function. *Curr Opin Immunol* 20: 178–184.

Bommhardt U, Beyer M, Hünig T, Reichardt HM. 2004. Molecular and cellular mechanisms of T cell development. *Cell Mol Life Sci* 61: 263–280.

Boothby MR, Mora AL, Scherer DC, Brockman JA, Ballard DW. 1997. Perturbation of the T lymphocyte lineage in transgenic mice expressing a constitutive repressor of nuclear factor (NF)-κB. *J Exp Med* **185:** 1897–1907.

Bredemeyer AL, Helmink BA, Innes CL, Calderon B, McGinnis LM, Mahowald GK, Gapud EJ, Walker LM, Collins JB, Weaver BK, et al. 2008. DNA double-strand breaks activate a multi-functional genetic program in developing lymphocytes. *Nature* **456:** 819–823.

Brown KD, Claudio E, Siebenlist U. 2008. The roles of the classical and alternative nuclear factor-κB pathways: Potential implications for autoimmunity and rheumatoid arthritis. *Arthritis Res Ther* **10:** 212.

Burrows PD, Kronenberg M, Taniguchi M. 2009. NKT cells turn ten. *Nat Immunol* **10:** 669–671.

Caamaño JH, Rizzo CA, Durham SK, Barton DS, Raventós-Suárez C, Snapper CM, Bravo R. 1998. Nuclear factor (NF)-κ B2 (p100/p52) is required for normal splenic microarchitecture and B cell-mediated immune responses. *J Exp Med* **187:** 185–96.

Cadera EJ, Wan F, Amin RH, Nolla H, Lenardo MJ, Schlissel MS. 2009. NF-κB activity marks cells engaged in receptor editing. *J Exp Med* **206:** 1803–1816.

Cariappa A, Liou HC, Horwitz BH, Pillai S. 2000. Nuclear factor κ B is required for the development of marginal zone B lymphocytes. *J Exp Med* **192:** 1175–1182.

Castro I, Wright JA, Damdinsuren B, Hoek KL, Carlesso G, Shinners NP, Gerstein RM, Woodland RT, Sen R, Khan WN. 2009. B cell receptor-mediated sustained c-Rel activation facilitates late transitional B cell survival through control of B cell activating factor receptor and NF-κB2. *J Immunol* **182:** 7729–7737.

Caton AJ, Cozzo C, Larkin J, Lerman MA 3rd, Boesteanu A, Jordan MS. 2004. CD4(+) CD25(+) regulatory T cell selection. *Ann N Y Acad Sci* **1029:** 101–114.

Chaudhuri J, Basu U, Zarrin A, Yan C, Franco S, Perlot T, Vuong B, Wang J, Phan RT, Datta A, et al. 2007. Evolution of the immunoglobulin heavy chain class switch recombination mechanism. *Adv Immunol* **94:** 157–214.

Chen X, Wang J, Woltring D, Gerondakis S, Shannon MF. 2005. Histone dynamics on the interleukin-2 gene in response to T-cell activation. *Mol Cell Biol* **25:** 3209–3219.

Cheng S, Hsia CY, Feng B, Liou ML, Fang X, Pandolfi PP, Liou HC. 2009. BCR-mediated apoptosis associated with negative selection of immature B cells is selectively dependent on Pten. *Cell Res* **19:** 196–207.

Cheng S, Hsia CY, Leone G, Liou HC. 2003. Cyclin E and Bcl-xL cooperatively induce cell cycle progression in c-Rel-/- B cells. *Oncogene* **22:** 8472–8486.

Claudio E, Brown K, Park S, Wang H, Siebenlist U. 2002. BAFF-induced NEMO-independent processing of NF-κ B2 in maturing B cells. *Nat Immunol* **3:** 958–965.

Claudio E, Brown K, Siebenlist U. 2006. NF-κB guides the survival and differentiation of developing lymphocytes. *Cell Death Differ* **13:** 697–701.

Claudio E, Saret S, Wang H, Siebenlist U. 2009. Cell-autonomous role for NF-κ B in immature bone marrow B cells. *J Immunol* **182:** 3406–3413.

Courtois G. 2008. Tumor suppressor CYLD: Negative regulation of NF-κB signaling and more. *Cell Mol Life Sci* **65:** 1123–1132.

Das J, Chen CH, Yang L, Cohn L, Ray P, Ray A. A critical role for NF-κ B in GATA3 expression and TH2 differentiation in allergic airway inflammation. *Nat Immunol* **2:** 45–50.

Davis RE, Brown KD, Siebenlist U, Staudt LM. 2001. Constitutive nuclear factor κB activity is required for survival of activated B cell-like diffuse large B cell lymphoma cells. *J Exp Med* **194:** 1861–1874.

Delphin S, Stavnezer J. 1995. Characterization of an interleukin 4 (IL-4) responsive region in the immunoglobulin heavy chain germline ε promoter: Regulation by NF-IL-4, a C/EBP family member and NF-κ B/p50. *J Exp Med* **181:** 181–192.

Denk A, Wirth T, Baumann B. 2000. NF-κB transcription factors: Critical regulators of hematopoiesis and neuronal survival. *Cytokine Growth Factor Rev* **11:** 303–320.

Derudder E, Cadera EJ, Vahl JC, Wang J, Fox CJ, Zha S, van Loo G, Pasparakis M, Schlissel MS, Schmidt-Supprian M, et al. 2009. Development of immunoglobulin lambda-chain-positive B cells, but not editing of immunoglobulin κ-chain, depends on NF-κB signals. *Nat Immunol* **10:** 647–654.

Doi TS, Takahashi T, Taguchi O, Azuma T, Obata Y. 1997. NF-κ B RelA-deficient lymphocytes: Normal development of T cells and B cells, impaired production of IgA and IgG1 and reduced proliferative responses. *J Exp Med* **185:** 953–61.

Drayton DL, Liao S, Mounzer RH, Ruddle NH. 2006. Lymphoid organ development: From ontogeny to neogenesis. *Nat Immunol* **7:** 344–353.

Elewaut D, Ware CF. 2007. The unconventional role of LT α β in T cell differentiation. *Trends Immunol* **28:** 169–175.

Elewaut D, Shaikh RB, Hammond KJ, De Winter H, Leishman AJ, Sidobre S, Turovskaya O, Prigozy TI, Ma L, Banks TA, et al. 2003. NIK-dependent RelB activation defines a unique signaling pathway for the development of V α 14i NKT cells. *J Exp Med* **197:** 1623–1633.

Evans I, Kim MY. 2009. Involvement of lymphoid inducer cells in the development of secondary and tertiary lymphoid structure. *BMB Rep* **42:** 189–193.

Ewings KE, Wiggins CM, Cook SJ. 2007. Bim and the pro-survival Bcl-2 proteins: Opposites attract, ERK repels. *Cell Cycle* **6:** 2236–2240.

Felli MP, Vacca A, Calce A, Bellavia D, Campese AF, Grillo R, Di Giovine M, Checquolo S, Talora C, Palermo R, et al. 2005. PKC θ mediates pre-TCR signaling and contributes to Notch3-induced T-cell leukemia. *Oncogene* **24:** 992–1000.

Feng B, Cheng S, Hsia CY, King LB, Monroe JG, Liou HC. 2004. NF-κB inducible genes BCL-X and cyclin E promote immature B-cell proliferation and survival. *Cell Immunol* **232:** 9–20.

Feng B, Cheng S, Pear WS, Liou HC. 2004. NF-kB inhibitor blocks B cell development at two checkpoints. *Med Immunol* **3:** 1.

Ferch U, zum Büschenfelde CM, Gewies A, Wegener E, Rauser S, Peschel C, Krappmann D, Ruland J. 2007. MALT1 directs B cell receptor-induced canonical nuclear factor-κB signaling selectively to the c-Rel subunit. *Nat Immunol* **8:** 984–991.

Ferguson AR, Corley RB. 2005. Accumulation of marginal zone B cells and accelerated loss of follicular dendritic cells in NF-κB p50-deficient mice. *BMC Immunol* **18:** 8.

Förster R, Davalos-Misslitz AC, Rot A. 2008. CCR7 and its ligands: Balancing immunity and tolerance. *Nat Rev Immunol* **8:** 362–371.

Franki AS, Van Beneden K, Dewint P, Hammond KJ, Lambrecht S, Leclercq G, Kronenberg M, Deforce D, Elewaut D. 2006. A unique lymphotoxin {α}β-dependent pathway regulates thymic emigration of V{α}14 invariant natural killer T cells. *Proc Natl Acad Sci* **103:** 9160–9165.

Franki AS, Van Beneden K, Dewint P, Meeus I, Veys E, Deforce D, Elewaut D. 2005. Lymphotoxin α 1 β 2: A critical mediator in V α 14i NKT cell differentiation. *Mol Immunol* **42:** 413–417.

Franzoso G, Carlson L, Poljak L, Shores EW, Epstein S, Leonardi A, Grinberg A, Tran T, Scharton-Kersten T, Anver M, et al. 1998. Mice deficient in nuclear factor (NF)-κB/p52 present with defects in humoral responses, germinal center reactions, and splenic microarchitecture. *J Exp Med* **187:** 147–159.

Franzoso G, Carlson L, Scharton-Kersten T, Shores EW, Epstein S, Grinberg A, Tran T, Shacter E, Leonardi A, Anver M, et al. 1997a. Critical roles for the Bcl-3 oncoprotein in T cell-mediated immunity, splenic microarchitecture, and germinal center reactions. *Immunity* **6:** 479–490.

Franzoso G, Carlson L, Xing L, Poljak L, Shores EW, Brown KD, Leonardi A, Tran T, Boyce BF, Siebenlist U. 1997b. Requirement for NF-κB in osteoclast and B-cell development. *Genes Dev* **11:** 3482–3496.

Gerondakis S, Strasser A. 2003. The role of Rel/NF-κB transcription factors in B lymphocyte survival. *Semin Immunol* **15:** 159–166.

Gerondakis S, Grumont RJ, Banerjee A. 2007. Regulating B-cell activation and survival in response to TLR signals. *Immunol Cell Biol* **85:** 471–475.

Gerondakis S, Grumont R, Rourke I, Grossmann M. 1998. The regulation and roles of Rel/NF-κ B transcription factors during lymphocyte activation. *Curr Opin Immunol* **10:** 353–359.

Gerondakis S, Grumont R, Gugasyan R, Wong L, Isomura I, Ho W, Banerjee A. 2006. Unravelling the complexities of the NF-κB signalling pathway using mouse knockout and transgenic models. *Oncogene* **25:** 6781–6799.

Ghosh S, Hayden MS. 2008. New regulators of NF-κB in inflammation. *Nat Rev Immunol* **8:** 837–848.

Gilmore TD, Kalaitzidis D, Liang MC, Starczynowski DT. 2004. The c-Rel transcription factor and B-cell proliferation: A deal with the devil. *Oncogene* **23:** 2275–2286.

Goldmit M, Ji Y, Skok J, Roldan E, Jung S, Cedar H, Bergman Y. 2005. Epigenetic ontogeny of the Igk locus during B cell development. *Nat Immunol* **6:** 198–203.

Greve B, Weissert R, Hamdi N, Bettelli E, Sobel RA, Coyle A, Kuchroo VK, Rajewsky K, Schmidt-Supprian M. 2007. I κ B kinase 2/β deficiency controls expansion of autoreactive T cells and suppresses experimental autoimmune encephalomyelitis. *J Immunol* **179:** 179–185.

Grigoriadis G, Zhan Y, Grumont RJ, Metcalf D, Handman E, Cheers C, Gerondakis S. 1996. The Rel subunit of NF-κB-like transcription factors is a positive and negative regulator of macrophage gene expression: Distinct roles for Rel in different macrophage populations. *EMBO J* **15:** 7099–7107.

Grossmann M, O'Reilly LA, Gugasyan R, Strasser A, Adams JM, Gerondakis S. 2000. The anti-apoptotic activities of Rel and RelA required during B-cell maturation involve the regulation of Bcl-2 expression. *EMBO J* **19:** 6351–6360.

Grumont RJ, Gerondakis S. 1990. Murine c-rel transcription is rapidly induced in T-cells and fibroblasts by mitogenic agents and the phorbol ester 12-O-tetradecanoylphorbol-13-acetate. *Cell Growth Differ* **1:** 345–350.

Grumont RJ, Gerondakis S. 1994. The subunit composition of NF-κ B complexes changes during B-cell development. *Cell Growth Differ* **5:** 1321–1331.

Grumont RJ, Gerondakis S. 2000. Rel induces interferon regulatory factor 4 (IRF-4) expression in lymphocytes: Modulation of interferon-regulated gene expression by rel/nuclear factor κB. *J Exp Med* **191:** 1281–1292.

Grumont RJ, Rourke IJ, Gerondakis S. 1999. Rel-dependent induction of A1 transcription is required to protect B cells from antigen receptor ligation-induced apoptosis. *Genes Dev* **13:** 400–411.

Grumont R, Hochrein H, O'Keeffe M, Gugasyan R, White C, Caminschi I, Cook W, Gerondakis S. 2001. c-Rel regulates interleukin 12 p70 expression in CD8(+) dendritic cells by specifically inducing p35 gene transcription. *J Exp Med* **194:** 1021–1032.

Grumont R, Lock P, Mollinari M, Shannon FM, Moore A, Gerondakis S. 2004. The mitogen-induced increase in T cell size involves PKC and NFAT activation of Rel/NF-κB-dependent c-myc expression. *Immunity* **21:** 19–30.

Grumont RJ, Rourke IJ, O'Reilly LA, Strasser A, Miyake K, Sha W, Gerondakis S. 1998. B lymphocytes differentially use the Rel and nuclear factor κB1 (NF-κB1) transcription factors to regulate cell cycle progression and apoptosis in quiescent and mitogen-activated cells. *J Exp Med* **187:** 663–674.

Grumont RJ, Strasser A, Gerondakis S. 2002. B cell growth is controlled by phosphatidylinositol 3-kinase-dependent induction of Rel/NF-κB regulated c-myc transcription. *Mol Cell* **10:** 1283–1294.

Guo F, Weih D, Meier E, Weih F. 2007. Constitutive alternative NF-κB signaling promotes marginal zone B-cell development but disrupts the marginal sinus and induces HEV-like structures in the spleen. *Blood* **110:** 2381–2389.

Hardy RR, Kincade PW, Dorshkind K. 2007. The protean nature of cells in the B lymphocyte lineage. *Immunity* **26:** 703–714.

Harling-McNabb L, Deliyannis G, Jackson DC, Gerondakis S, Grigoriadis G, Brown LE. 2009. Mice lacking the transcription factor subunit Rel can clear an influenza infection and have functional anti-viral cytotoxic T cells but do not develop an optimal antibody response. *Int Immunol* **11:** 1431–1439.

Hayden MS, Ghosh S. 2008. Shared principles in NF-κB signaling. *Cell* **132:** 344–362.

Hayden MS, West AP, Ghosh S. 2006. NF-κB and the immune response. *Oncogene* **25:** 6758–6780.

Hehlgans T, Pfeffer K. 2005. The intriguing biology of the tumour necrosis factor/tumour necrosis factor receptor superfamily: Players, rules and the games. *Immunology* **115:** 1–20.

Hettmann T, Leiden JM. 2000. NF-κ B is required for the positive selection of CD8+ thymocytes. *J Immunol* **165:** 5004–5010.

Hettmann T, DiDonato J, Karin M, Leiden JM. 1999. An essential role for nuclear factor κB in promoting double positive thymocyte apoptosis. *J Exp Med* **189:** 145–158.

Hikosaka Y, Nitta T, Ohigashi I, Yano K, Ishimaru N, Hayashi Y, Matsumoto M, Matsuo K, Penninger JM, Takayanagi H, et al. 2008. The cytokine RANKL produced by positively selected thymocytes fosters medullary thymic epithelial cells that express autoimmune regulator. *Immunity* **29:** 438–450.

Hong JW, Hendrix DA, Papatsenko D, Levine MS. 2008. How the Dorsal gradient works: Insights from postgenome technologies. *Proc Natl Acad Sci* **105:** 20072–20076.

Humbert PO, Verona R, Trimarchi JM, Rogers C, Dandapani S, Lees JA. 2000. E2f3 is critical for normal cellular proliferation. *Genes Dev* **14:** 690–703.

Igarashi H, Baba Y, Nagai Y, Jimi E, Ghosh S, Kincade PW. 2006. NF-κB is dispensable for normal lymphocyte development in bone marrow but required for protection of progenitors from TNFα. *Int Immunol* **18:** 653–659.

Inlay MA, Tian H, Lin T, Xu Y. 2004. Important roles for E protein binding sites within the immunoglobulin κ chain intronic enhancer in activating Vκ Jκ rearrangement. *J Exp Med* **200:** 1205–2011.

Inoue J, Gohda J, Akiyama T. 2007. Characteristics and biological functions of TRAF6. *Adv Exp Med Biol* **597:** 72–79.

Irla M, Hugues S, Gill J, Nitta T, Hikosaka Y, Williams IR, Hubert FX, Scott HS, Takahama Y, Holländer GA, et al. 2008. Autoantigen-specific interactions with CD4+ thymocytes control mature medullary thymic epithelial cell cellularity. *Immunity* **29:** 451–463.

Jensen KD, Chien YH. 2009. Thymic maturation determines γδ T cell function, but not their antigen specificities. *Curr Opin Immunol* **21:** 140–145.

Jimi E, Phillips RJ, Rincon M, Voll R, Karasuyama H, Flavell R, Ghosh S. 2005. Activation of NF-κB promotes the transition of large, CD43+ pre-B cells to small, CD43- pre-B cells. *Int Immunol* **17:** 815–825.

Jimi E, Strickland I, Voll RE, Long M, Ghosh S. 2008. Differential role of the transcription factor NF-κB in selection and survival of CD4+ and CD8+ thymocytes. *Immunity* **29:** 523–537.

Jin W, Reiley WR, Lee AJ, Wright A, Wu X, Zhang M, Sun SC. 2007. Deubiquitinating enzyme CYLD regulates the peripheral development and naive phenotype maintenance of B cells. *J Biol Chem* **282:** 15884–15893.

Jones RG, Saibil SD, Pun JM, Elford AR, Bonnard M, Pellegrini M, Arya S, Parsons ME, Krawczyk CM, Gerondakis S, et al. 2005. NF-κB couples protein kinase B/Akt signaling to distinct survival pathways and the regulation of lymphocyte homeostasis in vivo. *J Immunol* **175:** 3790–3799.

Jost PJ, Weiss S, Ferch U, Gross O, Mak TW, Peschel C, Ruland J. 2007. Bcl10/Malt1 signaling is essential for TCR-induced NF-κB activation in thymocytes but dispensable for positive or negative selection. *J Immunol* **178:** 953–960.

Kaisho T, Takeda K, Tsujimura T, Kawai T, Nomura F, Terada N, Akira S. 2001. IκB kinase α is essential for mature B cell development and function. *J Exp Med* **193:** 417–426.

Kajiura F, Sun S, Nomura T, Izumi K, Ueno T, Bando Y, Kuroda N, Han H, Li Y, Matsushima A, et al. 2004. NF-κ B-inducing kinase establishes self-tolerance in a thymic stroma-dependent manner. *J Immunol* **172:** 2067–2075.

Kanno T, Franzoso G, Siebenlist U. 1994. Human T-cell leukemia virus type I Tax-protein-mediated activation of NF-κ B from p100 (NF-κ B2)-inhibited cytoplasmic reservoirs. *Proc Natl Acad Sci* **91:** 12634–12638.

Kenter AL, Wuerffel R, Dominguez C, Shanmugam A, Zhang H. 2004. Mapping of a functional recombination motif that defines isotype specificity for μ→γ3 switch recombination implicates NF-κB p50 as the isotype-specific switching factor. *J Exp Med* **199:** 617–627.

Kim MY, Kim KS, McConnell F, Lane P. 2009. Lymphoid tissue inducer cells: Architects of CD4 immune responses in mice and men. *Clin Exp Immunol* **157:** 20–26.

Kingeter LM, Schaefer BC. 2008. Loss of protein kinase C θ, Bcl10, or Malt1 selectively impairs proliferation and NF-κ B activation in the CD4+ T cell subset. *J Immunol* **181:** 6244–6254.

Kinoshita D, Hirota F, Kaisho T, Kasai M, Izumi K, Bando Y, Mouri Y, Matsushima A, Niki S, Han H, et al. 2006. Essential role of IκB kinase α in thymic organogenesis required for the establishment of self-tolerance. *J Immunol* **176:** 3995–4002.

Kobayashi T, Kim TS, Jacob A, Walsh MC, Kadono Y, Fuentes-Pananá E, Yoshioka T, Yoshimura A, Yamamoto M, Kaisho T, et al. 2009. TRAF6 is required for generation of the B-1a B cell compartment as well as T cell-dependent and -independent humoral immune responses. *PLoS One* **4:** e4736.

Koble C, Kyewski B. 2009. The thymic medulla: A unique microenvironment for intercellular self-antigen transfer. *J Exp Med* **206:** 1505–1513.

Köntgen F, Grumont RJ, Strasser A, Metcalf D, Li R, Tarlinton D, Gerondakis S. 1995. Mice lacking the c-rel proto-oncogene exhibit defects in lymphocyte proliferation, humoral immunity, and interleukin-2 expression. *Genes Dev* **9:** 1965–1977.

Krieg AM. 2002. CpG motifs in bacterial DNA and their immune effects. *Annu Rev Immunol* **20:** 709–760.

Kyewski B, Klein L. 2006. A central role for central tolerance. *Annu Rev Immunol* **24:** 571–606.

Laurence A, O'Shea JJ. 2007. T(H)-17 differentiation: of mice and men. *Nat Immunol* **8:** 903–905.

Laurencikiene J, Deveikaite V, Severinson E. 2001. HS1,2 enhancer regulation of germline ε and γ2b promoters in murine B lymphocytes: Evidence for specific promoter-enhancer interactions. *J Immunol* **167:** 3257–3265.

Lee H, Arsura M, Wu M, Duyao M, Buckler AJ, Sonenshein GE. 1995. Role of Rel-related factors in control of c-myc gene transcription in receptor-mediated apoptosis of the murine B cell WEHI 231 line. *J Exp Med* **181:** 1169–1177.

Lee HH, Dadgostar H, Cheng Q, Shu J, Cheng G. 1999. NF-κB-mediated up-regulation of Bcl-x and Bfl-1/A1 is required for CD40 survival signaling in B lymphocytes. *Proc Natl Acad Sci* **96:** 9136–9141.

Levens D. 2002. Disentangling the MYC web. *Proc Natl Acad Sci* **99:** 5757–5759.

Lin X, Wang D. 2004. The roles of CARMA1, Bcl10, and MALT1 in antigen receptor signaling. *Semin Immunol* **16:** 429–435.

Lindsley RC, Thomas M, Srivastava B, Allman D. 2007. Generation of peripheral B cells occurs via two spatially and temporally distinct pathways. *Blood* **109:** 2521–2528.

Liston A, Rudensky AY. 2007. Thymic development and peripheral homeostasis of regulatory T cells. *Curr Opin Immunol* **19:** 176–185.

Liu HH, Xie M, Schneider MD, Chen ZJ. 2006. Essential role of TAK1 in thymocyte development and activation. *Proc Natl Acad Sci* **103:** 11677–11682.

Lopes-Carvalho T, Foote J, Kearney JF. 2005. Marginal zone B cells in lymphocyte activation and regulation. *Curr Opin Immunol* **17:** 244–250.

Lu LF, Rudensky A. 2009. Molecular orchestration of differentiation and function of regulatory T cells. *Genes Dev* **23:** 1270–1282.

MacDonald HR, Mycko MP. 2007. Development and selection of Vα 14i NKT cells. *Curr Top Microbiol Immunol* **314:** 195–212.

Mackay F, Schneider P. 2009. Cracking the BAFF code. *Nat Rev Immunol* **9:** 491–502.

Makris C, Godfrey VL, Krähn-Senftleben G, Takahashi T, Roberts JL, Schwarz T, Feng L, Johnson RS, Karin M. 2000. Female mice heterozygous for IKK γ/NEMO deficiencies develop a dermatopathy similar to the human X-linked disorder incontinentia pigmenti. *Mol Cell* **5:** 969–679.

Mandal M, Borowski C, Palomero T, Ferrando AA, Oberdoerffer P, Meng F, Ruiz-Vela A, Ciofani M, Zuniga-Pflucker JC, Screpanti I, et al. 2005. The BCL2A1 gene as a pre-T cell receptor-induced regulator of thymocyte survival. *J Exp Med* **201:** 603–614.

Manis JP, Tian M, Alt FW. 2002. Mechanism and control of class-switch recombination. *Trends Immunol* **23:** 31–39.

Martins VC, Boehm T, Bleul CC. 2008. Ltβr signaling does not regulate Aire-dependent transcripts in medullary thymic epithelial cells. *J Immunol* **181:** 400–407

Mason NJ, Liou HC, Hunter CA. 2004. T cell-intrinsic expression of c-Rel regulates Th1 cell responses essential for resistance to Toxoplasma gondii. *J Immunol* **172:** 3704–3711.

Mathis D, Benoist C. 2009. Aire. *Annu Rev Immunol* **27:** 287–312.

Matsushima A, Kaisho T, Rennert PD, Nakano H, Kurosawa K, Uchida D, Takeda K, Akira S, Matsumoto M. 2001. Essential role of nuclear factor (NF)-κB-inducing kinase and inhibitor of κB (IκB) kinase α in NF-κB activation through lymphotoxin β receptor, but not through tumor necrosis factor receptor I. *J Exp Med* **193:** 631–636.

McHeyzer-Williams MG. 2003. B cells as effectors. *Curr Opin Immunol* **15:** 354–361.

McHeyzer-Williams LJ, McHeyzer-Williams MG. 2005. Antigen-specific memory B cell development. *Annu Rev Immunol* **23:** 487–513.

Medoff BD, Sandall BP, Landry A, Nagahama K, Mizoguchi A, Luster AD, Xavier RJ. 2009. Differential requirement for CARMA1 in agonist-selected T-cell development. *Eur J Immunol* **39:** 78–84.

Mineva ND, Rothstein TL, Meyers JA, Lerner A, Sonenshein GE. 2007. CD40 ligand-mediated activation of the de novo RelB NF-κB synthesis pathway in transformed B cells promotes rescue from apoptosis. *J Biol Chem* **282:** 17475–17485.

Mise-Omata S, Kuroda E, Niikura J, Yamashita U, Obata Y, Doi TS. 2007. A proximal κB site in the IL-23 p19 promoter is responsible for RelA- and c-Rel-dependent transcription. *J Immunol* **179:** 6596–6603.

Mittrücker HW, Matsuyama T, Grossman A, Kündig TM, Potter J, Shahinian A, Wakeham A, Patterson B, Ohashi PS, Mak TW. 1997. Requirement for the transcription factor LSIRF/IRF4 for mature B and T lymphocyte function. *Science* **275:** 540–543.

Molinero LL, Yang J, Gajewski T, Abraham C, Farrar MA, Alegre ML. 2009. CARMA1 controls an early checkpoint in the thymic development of FoxP3+ regulatory T cells. *J Immunol* **182:** 6736–6743.

Molitor JA, Walker WH, Doerre S, Ballard DW, Greene WC. 1990. NF-κ B: A family of inducible and differentially expressed enhancer-binding proteins in human T cells. *Proc Natl Acad Sci* **87:** 10028–10032.

Mora AL, Stanley S, Armistead W, Chan AC, Boothby M. 2001. Inefficient ZAP-70 phosphorylation and decreased thymic selection in vivo result from inhibition of NF-κB/Rel. *J Immunol* **167:** 5628–5635.

Moran ST, Cariappa A, Liu H, Muir B, Sgroi D, Boboila C, Pillai S. 2007. Synergism between NF-κ B1/p50 and Notch2 during the development of marginal zone B lymphocytes. *J Immunol* **179:** 195–200.

Moriyama H, Yonehara S. 2007. Rapid up-regulation of c-FLIP expression by BCR signaling through the PI3K/Akt pathway inhibits simultaneously induced Fas-mediated apoptosis in murine B lymphocytes. *Immunol Lett* **109:** 36–46.

Morley SC, Weber KS, Kao H, Allen PM. 2008. Protein kinase C-θ is required for efficient positive selection. *J Immunol* **181:** 4696–4708.

Mueller SN, Ahmed R. 2008. Lymphoid stroma in the initiation and control of immune responses. *Immunol Rev* 2008 **224:** 284–294.

Müller JR, Siebenlist U. 2003. Lymphotoxin β receptor induces sequential activation of distinct NF-κ B factors via separate signaling pathways. *J Biol Chem* **278:** 12006–12012.

Nemazee D. 2006. Receptor editing in lymphocyte development and central tolerance. *Nat Rev Immunol* **6:** 728–740.

Nitta T, Murata S, Ueno T, Tanaka K, Takahama Y. 2008. Thymic microenvironments for T-cell repertoire formation. *Adv Immunol* **99:** 59–94.

Northrup DL, Allman D. 2008. Transcriptional regulation of early B cell development. *Immunol Res* **42:** 106–117.

O'Keeffe M, Grumont RJ, Hochrein H, Fuchsberger M, Gugasyan R, Vremec D, Shortman K, Gerondakis S. 2005. Distinct roles for the NF-κB1 and c-Rel transcription factors in the differentiation and survival of plasmacytoid and conventional dendritic cells activated by TLR-9 signals. *Blood* **106:** 3457–3464.

Otipoby KL, Sasaki Y, Schmidt-Supprian M, Patke A, Gareus R, Pasparakis M, Tarakhovsky A, Rajewsky K. 2008. BAFF activates Akt and Erk through BAFF-R in an IKK1-dependent manner in primary mouse B cells. *Proc Natl Acad Sci* **105:** 12435–12438.

Ouaaz F, Arron J, Zheng Y, Choi Y, Beg AA. 2002. Dendritic cell development and survival require distinct NF-κB subunits. *Immunity* **16:** 257–270.

Owyang AM, Tumang JR, Schram BR, Hsia CY, Behrens TW, Rothstein TL, Liou HC. 2001. c-Rel is required for the protection of B cells from antigen receptor-mediated, but not Fas-mediated, apoptosis. *J Immunol* **167:** 4948–4956.

Palmer S, Chen YH. 2008. Bcl-3, a multifaceted modulator of NF-κB-mediated gene transcription. *Immunol Res* **42:** 210–218.

Pappu BP, Lin X. 2006. Potential role of CARMA1 in CD40-induced splenic B cell proliferation and marginal zone B cell maturation. *Eur J Immunol* **36:** 3033–3043.

Pasparakis M, Schmidt-Supprian M, Rajewsky K. 2002. IκB kinase signaling is essential for maintenance of mature B cells. *J Exp Med* **196:** 743–752.

Paxian S, Merkle H, Riemann M, Wilda M, Adler G, Hameister H, Liptay S, Pfeffer K, Schmid RM. 2002. Abnormal organogenesis of Peyer's patches in mice deficient for NF-κB1, NF-κB2, and Bcl-3. *Gastroenterology* **122:** 1853–1868.

Pohl T, Gugasyan R, Grumont RJ, Strasser A, Metcalf D, Tarlinton D, Sha W, Baltimore D, Gerondakis S. 2002. The combined absence of NF-κB1 and c-Rel reveals that overlapping roles for these transcription factors in the B cell lineage are restricted to the activation and function of mature cells. *Proc Natl Acad Sci* **99:** 4514–4519.

Poljak L, Carlson L, Cunningham K, Kosco-Vilbois MH, Siebenlist U. 1999. Distinct activities of p52/NF-κB required for proper secondary lymphoid organ microarchitecture: Functions enhanced by Bcl-3. *J Immunol* **163:** 6581–6588.

Poovassery JS, Vanden Bush TJ, Bishop GA. 2009. Antigen receptor signals rescue B cells from TLR tolerance. *J Immunol* **183:** 2974–2983.

Qin J, Konno H, Ohshima D, Yanai H, Motegi H, Shimo Y, Hirota F, Matsumoto M, Takaki S, Inoue J, et al. 2007. Developmental stage-dependent collaboration between the TNF receptor-associated factor 6 and lymphotoxin pathways for B cell follicle organization in secondary lymphoid organs. *J Immunol* **179:** 6799–6807.

Randall TD, Carragher DM, Rangel-Moreno J. 2008. Development of secondary lymphoid organs. *Annu Rev Immunol* **26:** 627–650.

Rao S, Gerondakis S, Woltring D, Shannon MF. 2003. c-Rel is required for chromatin remodeling across the IL-2 gene promoter. *J Immunol* **170:** 3724–3731.

Rauch M, Tussiwand R, Bosco N, Rolink AG. 2009. Crucial role for BAFF-BAFF-R signaling in the survival and maintenance of mature B cells. *PLoS One* **4:** e5456.

Ren H, Schmalstieg A, van Oers NS, Gaynor RB. 2002. I-κ B kinases α and β have distinct roles in regulating murine T cell function. *J Immunol* **168:** 3721–3731.

Ribot JC, deBarros A, Pang DJ, Neves JF, Peperzak V, Roberts SJ, Girardi M, Borst J, Hayday AC, Pennington DJ, et al. 2009. CD27 is a thymic determinant of the balance between interferon-γ- and interleukin 17-producing γδ T cell subsets. *Nat Immunol* **10:** 427–436.

Rothstein TL, Zhong X, Schram BR, Negm RS, Donohoe TJ, Cabral DS, Foote LC, Schneider JL. 2000. Receptor-specific regulation of B-cell susceptibility to Fas-mediated apoptosis and a novel Fas apoptosis inhibitory molecule. *Immunol Rev* **176:** 116–133.

Ruddle NH, Akirav EM. 2009. Secondary lymphoid organs: Responding to genetic and environmental cues in ontogeny and the immune response. *J Immunol* **183:** 2205–2212.

Saibil SD, Jones RG, Deenick EK, Liadis N, Elford AR, Vainberg MG, Baerg H, Woodgett JR, Gerondakis S, Ohashi PS. 2007. CD4+ and CD8+ T cell survival is regulated differentially by protein kinase Cθ, c-Rel, and protein kinase B. *J Immunol* **178:** 2932–2939.

Sasaki Y, Calado DP, Derudder E, Zhang B, Shimizu Y, Mackay F, Nishikawa S, Rajewsky K, Schmidt-Supprian M. 2008. NIK overexpression amplifies, whereas ablation of its TRAF3-binding domain replaces BAFF:BAFF-R-mediated survival signals in B cells. *Proc Natl Acad Sci* **105:** 10883–10888.

Sasaki Y, Casola S, Kutok JL, Rajewsky K, Schmidt-Supprian M. 2004. TNF family member B cell-activating factor (BAFF) receptor-dependent and -independent roles for BAFF in B cell physiology. *J Immunol* **173:** 2245–2252.

Sasaki Y, Derudder E, Hobeika E, Pelanda R, Reth M, Rajewsky K, Schmidt-Supprian M. 2006. Canonical NF-κB activity, dispensable for B cell development, replaces BAFF-receptor signals and promotes B cell proliferation upon activation. *Immunity* **24:** 729–739.

Sasaki Y, Schmidt-Supprian M, Derudder E, Rajewsky K. 2007. Role of NFκB signaling in normal and malignant B cell development. *Adv Exp Med Biol* **596:** 149–154.

Sato S, Sanjo H, Takeda K, Ninomiya-Tsuji J, Yamamoto M, Kawai T, Matsumoto K, Takeuchi O, Akira S. 2005. Essential function for the kinase TAK1 in innate and adaptive immune responses. *Nat Immunol* **6:** 1087–1095.

Sato S, Sanjo H, Tsujimura T, Ninomiya-Tsuji J, Yamamoto M, Kawai T, Takeuchi O, Akira S. 2006. TAK1 is indispensable for development of T cells and prevention of colitis by the generation of regulatory T cells. *Int Immunol* **18:** 1405–1411.

Schauer SL, Bellas RE, Sonenshein GE. 1998. Dominant signals leading to inhibitor κB protein degradation mediate CD40 ligand rescue of WEHI 231 immature B cells from receptor-mediated apoptosis. *J Immunol* **160:** 4398–4405.

Schimpl A, Berberich I, Kneitz B, Krämer S, Santner-Nanan B, Wagner S, Wolf M, Hünig T. 2002. IL-2 and autoimmune disease. *Cytokine Growth Factor Rev* **13:** 369–378.

Schmidt-Supprian M, Bloch W, Courtois G, Addicks K, Israël A, Rajewsky K, Pasparakis M. 2000. NEMO/IKK γ-deficient mice model incontinentia pigmenti. *Mol Cell* **5:** 981–992.

Schmidt-Supprian M, Courtois G, Tian J, Coyle AJ, Israël A, Rajewsky K, Pasparakis M. 2003. Mature T cells depend on signaling through the IKK complex. *Immunity* **19:** 377–389.

Schmidt-Supprian M, Tian J, Grant EP, Pasparakis M, Maehr R, Ovaa H, Ploegh HL, Coyle AJ, Rajewsky K. 2004a. Differential dependence of CD4+CD25+ regulatory and natural killer-like T cells on signals leading to NF-κB activation. *Proc Natl Acad Sci* **101:** 4566–4571.

Schmidt-Supprian M, Tian J, Ji H, Terhorst C, Bhan AK, Grant EP, Pasparakis M, Casola S, Coyle AJ, Rajewsky K. 2004b. I κ B kinase 2 deficiency in T cells leads to defects in priming, B cell help, germinal center reactions, and homeostatic expansion. *J Immunol* **173:** 1612–1619.

Schram BR, Tze LE, Ramsey LB, Liu J, Najera L, Vegoe AL, Hardy RR, Hippen KL, Farrar MA, Behrens TW. 2008. B cell receptor basal signaling regulates antigen-induced Ig light chain rearrangements. *J Immunol* **180:** 4728–4741.

Schuman J, Chen Y, Podd A, Yu M, Liu HH, Wen R, Chen ZJ, Wang D. 2009. A critical role of TAK1 in B-cell receptor-mediated nuclear factor κB activation. *Blood* **113:** 4566–4574.

Seach N, Ueno T, Fletcher AL, Lowen T, Mattesich M, Engwerda CR, Scott HS, Ware CF, Chidgey AP, Gray DH, Boyd RL. 2008. The lymphotoxin pathway regulates Aire-independent expression of ectopic genes and chemokines in thymic stromal cells. *J Immunol* **180:** 5384–5392.

Sen R. 2004. NF-κB and the immunoglobulin κgene enhancer. *J Exp Med* **200:** 1099–1102.

Senftleben U, Cao Y, Xiao G, Greten FR, Krähn G, Bonizzi G, Chen Y, Hu Y, Fong A, Sun SC, Karin M. 2001a. Activation by IKKα of a second, evolutionary conserved, NF-κ B signaling pathway. *Science* **293:** 1495–1499.

Senftleben U, Li ZW, Baud V, Karin M. 2001b. IKKβ is essential for protecting T cells from TNFα-induced apoptosis. *Immunity* **14:** 217–230.

Sha WC, Liou HC, Tuomanen EI, Baltimore D. 1995. Targeted disruption of the p50 subunit of NF-κ B leads to multifocal defects in immune responses. *Cell* **80:** 321–330.

Shaffer AL, Emre NC, Lamy L, Ngo VN, Wright G, Xiao W, Powell J, Dave S, Yu X, Zhao H, et al. 2008. IRF4 addiction in multiple myeloma. *Nature* **454:** 226–231.

Shaffer AL, Rosenwald A, Hurt EM, Giltnane JM, Lam LT, Pickeral OK, Staudt LM. 2001. Signatures of the immune response. *Immunity* **15:** 375–385.

Siebenlist U, Brown K, Claudio E. 2005. Control of lymphocyte development by nuclear factor-κB. *Nat Rev Immunol* **5:** 435–445.

Sivakumar V, Hammond KJ, Howells N, Pfeffer K, Weih F. 2003. Differential requirement for Rel/nuclear factor κ B family members in natural killer T cell development. *J Exp Med* **197:** 1613–1621.

Song J, So T, Croft M. 2008. Activation of NF-κB1 by OX40 contributes to antigen-driven T cell expansion and survival. *J Immunol* **180:** 7240–7248.

Sriskantharajah S, Belich MP, Papoutsopoulou S, Janzen J, Tybulewicz V, Seddon B, Ley SC. 2009. Proteolysis of NF-κB1 p105 is essential for T cell antigen receptor-induced proliferation. *Nat Immunol* **10:** 38–47.

Stadanlick JE, Cancro MP. 2008. BAFF and the plasticity of peripheral B cell tolerance. *Curr Opin Immunol* **20:** 158–161.

Stadanlick JE, Kaileh M, Karnell FG, Scholz JL, Miller JP, Quinn WJ, Brezski RJ 3rd, Treml LS, Jordan KA, Monroe JG, et al. 2008. Tonic B cell antigen receptor signals supply an NF-κB substrate for prosurvival BLyS signaling. *Nat Immunol* **9:** 1379–1387.

Stanic AK, Bezbradica JS, Park JJ, Matsuki N, Mora AL, Van Kaer L, Boothby MR, Joyce S. 2004. NF-κ B controls cell fate specification, survival, and molecular differentiation of immunoregulatory natural T lymphocytes. *J Immunol* **172:** 2265–2273.

Stanic AK, Bezbradica JS, Park JJ, Van Kaer L, Boothby MR, Joyce S. 2004. Cutting edge: The ontogeny and function of Va14Ja18 natural T lymphocytes require signal processing by protein kinase C θ and NF-κ B. *J Immunol* **172:** 4667–4671.

Stocker H, Hafen E. 2000. Genetic control of cell size. *Curr Opin Genet Dev* **10:** 529–535.

Strasser A, Jost PJ, Nagata S. 2009. The many roles of FAS receptor signaling in the immune system. *Immunity* **30:** 180–192.

Sun Z, Arendt CW, Ellmeier W, Schaeffer EM, Sunshine MJ, Gandhi L, Annes J, Petrzilka D, Kupfer A, Schwartzberg PL, et al. 2000. PKC-θ is required for TCR-induced NF-κB activation in mature but not immature T lymphocytes. *Nature* **404:** 402–407.

Suto H, Katakai T, Sugai M, Kinashi T, Shimizu A. 2009. CXCL13 production by an established lymph node stromal cell line via lymphotoxin-β receptor engagement involves the cooperation of multiple signaling pathways. *Int Immunol* **21:** 467–476.

Taghon T, Rothenberg EV. 2008. Molecular mechanisms that control mouse and human TCR-αβ and TCR-γδ T cell development. *Semin Immunopathol* **30:** 383–398.

Thedrez A, Sabourin C, Gertner J, Devilder MC, Allain-Maillet S, Fournié JJ, Scotet E, Bonneville M. 2007. Self/non-self discrimination by human γδ T cells: Simple solutions for a complex issue? *Immunol Rev* **215:** 123–135.

Thomas MD, Srivastava B, Allman D. 2006. Regulation of peripheral B cell maturation. *Cell Immunol* **239:** 92–102.

Thome M. 2004. CARMA1, BCL-10 and MALT1 in lymphocyte development and activation. *Nat Rev Immunol* **4:** 348–359.

Tucker E, O'Donnell K, Fuchsberger M, Hilton AA, Metcalf D, Greig K, Sims NA, Quinn JM, Alexander WS, Hilton DJ, et al. 2007. A novel mutation in the Nfkb2 gene generates an NF-κ B2 "super repressor". *J Immunol* **179:** 7514–7522.

Tykocinski LO, Sinemus A, Kyewski B. 2008. The thymus medulla slowly yields its secrets. *Ann N Y Acad Sci* **1143:** 105–122.

Vallabhapurapu S, Karin M. 2009. Regulation and function of NF-κB transcription factors in the immune system. *Annu Rev Immunol* **27:** 693–733.

Venkataraman L, Burakoff SJ, Sen R. 1995. FK506 inhibits antigen receptor-mediated induction of c-rel in B and T lymphoid cells. *J Exp Med* 181: 1091–1099.

Venkataraman L, Wang W, Sen R. 1996. Differential regulation of c-Rel translocation in activated B and T cells. *J Immunol* 157: 1149–1155.

Vallabhapurapu S, Powolny-Budnicka I, Riemann M, Schmid RM, Paxian S, Pfeffer K, Körner H, Weih F. 2008. Rel/NF-κB family member RelA regulates NK1.1- to NK1.1+ transition as well as IL-15-induced expansion of NKT cells. *Eur J Immunol* 38: 3508–3519.

Verkoczy L, Aït-Azzouzene D, Skog P, Märtensson A, Lang J, Duong B, Nemazee D. 2005. A role for nuclear factor κ B/rel transcription factors in the regulation of the recombinase activator genes. *Immunity* 22: 519–531.

Voll RE, Jimi E, Phillips RJ, Barber DF, Rincon M, Hayday AC, Flavell RA, Ghosh S. 2000. NF-κ B activation by the pre-T cell receptor serves as a selective survival signal in T lymphocyte development. *Immunity* 13: 677–689.

Vondenhoff MF, Kraal G, Mebius RE. 2007. Lymphoid organogenesis in brief. *Eur J Immunol.* 37 Suppl 1: S46–52.

Wan YY, DeGregori J. 2003. The survival of antigen-stimulated T cells requires NFκB-mediated inhibition of p73 expression. *Immunity* 18: 331–342.

Wan YY, Chi H, Xie M, Schneider MD, Flavell RA. 2006. The kinase TAK1 integrates antigen and cytokine receptor signaling for T cell development, survival and function. *Nat Immunol* 7: 851–858.

Wang L, Wuerffel R, Kenter AL. 2006. NF-κ B binds to the immunoglobulin S γ 3 region in vivo during class switch recombination. *Eur J Immunol* 36: 3315–3323.

Wang J, Wang X, Hussain S, Zheng Y, Sanjabi S, Ouaaz F, Beg AA. 2007. Distinct roles of different NF-κ B subunits in regulating inflammatory and T cell stimulatory gene expression in dendritic cells. *J Immunol* 178: 6777–6788.

Waterfield MR, Zhang M, Norman LP, Sun SC. 2003. NF-κB1/p105 regulates lipopolysaccharide-stimulated MAP kinase signaling by governing the stability and function of the Tpl2 kinase. *Mol Cell* 11: 685–694.

Weih F, Caamaño J. 2003. Regulation of secondary lymphoid organ development by the nuclear factor-κB signal transduction pathway. *Immunol Rev* 195: 91–105.

Weih DS, Yilmaz ZB, Weih F. 2001. Essential role of RelB in germinal center and marginal zone formation and proper expression of homing chemokines. *J Immunol* 167: 1909–1919.

Weih F, Carrasco D, Durham SK, Barton DS, Rizzo CA, Ryseck RP, Lira SA, Bravo R. 1995. Multiorgan inflammation and hematopoietic abnormalities in mice with a targeted disruption of RelB, a member of the NF-κ B/Rel family. *Cell* 80: 331–340.

Weil R, Israël A. 2006. Deciphering the pathway from the TCR to NF-κB. *Cell Death Differ* 13: 826–833.

Wu L, Timmers C, Maiti B, Saavedra HI, Sang L, Chong GT, Nuckolls F, Giangrande P, Wright FA, Field SJ, et al. 2001. The E2F1-3 transcription factors are essential for cellular proliferation. *Nature* 414: 457–462.

Xiong N, Raulet DH. 2007. Development and selection of γδ T cells. *Immunol Rev* 215: 15–31.

Yamada T, Mitani T, Yorita K, Uchida D, Matsushima A, Iwamasa K, Fujita S, Matsumoto M. 2000. Abnormal immune function of hemopoietic cells from alymphoplasia (aly) mice, a natural strain with mutant NF-κ B-inducing kinase. *J Immunol* 165: 804–812.

Yamamoto M, Takeda K. 2008. Role of nuclear IκB proteins in the regulation of host immune responses. *J Infect Chemother* 14: 265–269.

Yano M, Kuroda N, Han H, Meguro-Horike M, Nishikawa Y, Kiyonari H, Maemura K, Yanagawa Y, Obata K, Takahashi S, et al. 2008. Aire controls the differentiation program of thymic epithelial cells in the medulla for the establishment of self-tolerance. *J Exp Med* 205: 2827–2838.

Yilmaz ZB, Weih DS, Sivakumar V, Weih F. 2003. RelB is required for Peyer's patch development: Differential regulation of p52-RelB by lymphotoxin and TNF. *EMBO J* 22: 121–130.

Yoshida H, Naito A, Inoue J, Satoh M, Santee-Cooper SM, Ware CF, Togawa A, Nishikawa S, Nishikawa S. 2002. Different cytokines induce surface lymphotoxin-αβ on IL-7 receptor-α cells that differentially engender lymph nodes and Peyer's patches. *Immunity* 17: 823–833.

Zazzeroni F, Papa S, Algeciras-Schimnich A, Alvarez K, Melis T, Bubici C, Majewski N, Hay N, De Smaele E, Peter ME, et al. 2003. Gadd45 β mediates the protective effects of CD40 costimulation against Fas-induced apoptosis. *Blood* 102: 3270–3279.

Zelazowski P, Shen Y, Snapper CM. 2000. NF-κB/p50 and NF-κB/c-Rel differentially regulate the activity of the 3'αE-hsl,2 enhancer in normal murine B cells in an activation-dependent manner. *Int Immunol* 12: 1167–1172.

Zhan Y, Gerondakis S, Coghill E, Bourges D, Xu Y, Brady JL, Lew AM. 2008. Glucocorticoid-induced TNF receptor expression by T cells is reciprocally regulated by NF-κB and NFAT. *J Immunol* 181: 5405–5413.

Zhang X, Wang H, Claudio E, Brown K, Siebenlist U. 2007. A role for the IκB family member Bcl-3 in the control of central immunologic tolerance. *Immunity* 27: 438–452.

Zheng Y, Vig M, Lyons J, Van Parijs L, Beg AA. 2003. Combined deficiency of p50 and cRel in CD4+ T cells reveals an essential requirement for nuclear factor κB in regulating mature T cell survival and in vivo function. *J Exp Med* 197: 861–874.

Zhu M, Chin RK, Christiansen PA, Lo JC, Liu X, Ware C, Siebenlist U, Fu YX. 2006. NF-κB2 is required for the establishment of central tolerance through an Aire-dependent pathway. *J Clin Invest* 116: 2964–2971.

NF-κB in the Immune Response of *Drosophila*

Charles Hetru and Jules A. Hoffmann

Centre National de la Recherche Scientifique, Institute of Molecular and Cellular Biology,
15 rue René Descartes, 67084 Strasbourg, France

Correspondence: j.hoffmann@ibmc.u-strasbg.fr

The nuclear factor κB (NF-κB) pathways play a major role in *Drosophila* host defense. Two recognition and signaling cascades control this immune response. The Toll pathway is activated by Gram-positive bacteria and by fungi, whereas the immune deficiency (Imd) pathway responds to Gram-negative bacterial infection. The basic mechanisms of recognition of these various types of microbial infections by the adult fly are now globally understood. Even though some elements are missing in the intracellular pathways, numerous proteins and interactions have been identified. In this article, we present a general picture of the immune functions of NF-κB in *Drosophila* with all the partners involved in recognition and in the signaling cascades.

The paramount roles of NF-κB family members in *Drosophila* development and host defense are now relatively well established and have been the subject of several in-depth reviews in recent years, including some from this laboratory (e.g., Hoffmann 2003; Minakhina and Steward 2006; Ferrandon et al. 2007; Lemaitre and Hoffmann 2007; Aggarwal and Silverman 2008). To avoid excessive duplication, we limit this text to the general picture that has evolved over nearly two decades—since the initial demonstration that the *dorsal* gene plays a role in dorsoventral patterning in embryogenesis of *Drosophila* and that it encodes a member of the NF-κB family of inducible transactivators (Nüsslein-Volhard et al. 1980; Steward 1987; Roth et al. 1989). In the early nineties, it became apparent that NF-κB also plays a role in the antimicrobial host defense of *Drosophila* (Engström et al. 1993; Ip et al. 1993; Kappler et al. 1993; Reichhart et al. 1993). We focus in this article on the immune functions of NF-κB and refer the reader to recent reviews for the roles of NF-κB in development (Roth 2003; Brennan and Anderson 2004; Moussian and Roth 2005; Minakhina and Steward 2006).

The *Drosophila* genome codes for three NF-κB family members (Fig. 1). Dorsal and DIF (for dorsal-related immunity factor) are 70 kDa proteins, with a typical Rel homology domain, which is 45% identical to that of the mammalian counterparts c-Rel, Rel A, and Rel B. Dorsal and DIF lie some 10 kbp apart on the second chromosome and probably arose from a recent duplication (Meng et al. 1999). Both proteins are retained in the cytoplasm by binding to the same 54-kDa inhibitor

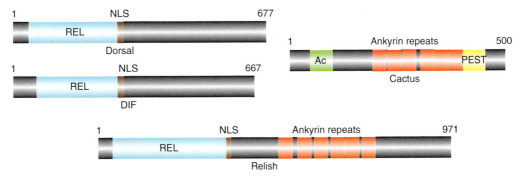

Figure 1. The NF-κB and IκB proteins in *Drosophila*. The length in amino acids is indicated by numbers. REL, Rel-homology domain; NLS, nuclear localization sequence; PEST, proline, glutamic acid, serine, and threonine-rich segment; Ac, acidic domain.

protein Cactus, which is homologous to mammalian IκBs (Schüpbach and Wieshaus 1989; Geisler et al. 1992). The single *Drosophila* Cactus gene is closest to mammalian IκBα (Huguet et al. 1997). The third member of the family in *Drosophila*, Relish, is a 100-kDa protein with an amino-terminal Rel domain and a carboxy-terminal extension with typical ankyrin repeats, as found in Cactus and mammalian IκBs. Relish is similar to mammalian p100 and p105 and its activation requires proteolytic cleavage as in the case for these mammalian counterparts (reviewed in Hultmark 2003).

Put in simple terms, NF-κB family members function in the host defense of *Drosophila* to control the expression of genes encoding immune-responsive peptides and proteins. Prominent among the induced genes are those encoding peptides with direct antimicrobial activity. To exert this function, Dorsal and DIF are translocated to the nucleus following stimulus-induced degradation of the inhibitor Cactus, whereas Relish requires stimulus-induced proteolytic cleavage for nuclear translocation of its amino-terminal Rel domain. This paradigm is similar to that observed in mammalian immunity. Again, for the sake of simplicity, we may say that the stimulus-induced degradation of Cactus, and the concomitant release of Dorsal or DIF, is primarily observed during Gram-positive bacterial and fungal infections and mediated by the Toll signaling pathway. In contrast, stimulus-induced

proteolytic cleavage of Relish, and concomitant nuclear translocation of its amino-terminal Rel domain, is the hallmark of the response to Gram-negative bacterial infection and mediated by the Imd signaling pathway. Whether these pathways are also involved in the multifaceted defense against viruses remains an open question (Zambon et al. 2005). The Toll pathway was further shown to be involved in hematopoiesis of flies (Qiu et al. 1998). Of note, the Cactus-NF-κB module also plays a central role in the elimination of *Plasmodium* parasites in infected mosquitoes (Frolet et al. 2006). In the following, we review our information of the two established signaling pathways, Toll and Imd, which lead to gene reprogramming through NF-κB in response to bacterial and fungal infections. We first consider the upstream mechanisms that mediate the recognition of infection and allow for a certain level of discrimination between invading microorganisms. Gene reprogramming in this context is best illustrated by the induction of the antimicrobial peptide genes, which serve as the most convenient readouts of the antimicrobial defense of *Drosophila* (see Samakovlis et al. 1990; Reichhart et al. 1992; Ferrandon et al. 1998). Flies produce at least seven families of mostly cationic, small-sized, membrane-active peptides, with spectra variously directed against Gram-positive (defensins) and Gram-negative (diptericins, attacins, and drosocin) bacteria, and against fungi (drosomycins and metchnikowins), or with overlapping spectra

Cite this article as *Cold Spring Harb Perspect Biol* 2009;1:a000232

(cecropins) (reviewed in Bulet et al. 1999; Hetru et al. 2003). The primary site of biosynthesis of these peptides is the fat body, a functional equivalent of the mammalian liver. Blood cells also participate in the production of antimicrobial peptides. As a rule, these molecules are secreted into the hemolymph where they reach remarkably high concentrations to oppose invading microorganisms (Hetru et al. 2003). This facet of the antimicrobial host defense is generally referred to as systemic immune response. Of note, the gut and the tracheae also produce antimicrobial peptides in response to microbes (see Tzou et al. 2000; Onfelt Tingvall et al. 2001; Liehl et al. 2006; Nehme et al. 2007).

During infection, the Toll and Imd pathways control the expression of hundreds of genes. In addition to the antimicrobial peptides, these genes encode proteases, putative cytokines, cytoskeletal proteins, and many peptides and proteins whose function in the host defense are still not understood (De Gregorio et al. 2001; Irving et al. 2001).

ACTIVATION OF NF-κB VIA THE TOLL SIGNALING CASCADE

The Toll signaling cascade is activated by Gram-positive bacterial peptidoglycan and by fungal β-(1,3)-glucan (Ochiai and Ashida 2000; Michel et al. 2001; Leulier et al. 2003; Gottar et al. 2006). In contrast to initial assumptions, these microbial inducers do not directly interact with the Toll transmembrane receptor, but with circulating proteins belonging to two distinct families: The peptidoglycan recognition proteins (PGRPs) (Kang et al. 1998; Werner et al. 2000; Steiner 2004) and the glucan-binding proteins (Lee et al. 1996; Gobert et al. 2003; Gottar et al. 2006; Wang et al. 2006) (GNBPs, formerly referred to as Gram-negative binding proteins—the acronym is preserved here for consistency in the literature and now noted as glucaN binding proteins). Binding of the microbial inducers to these proteins results in the activation of proteolytic cascades, which culminate in the cleavage of the cytokine Spaetzle

and binding of cleaved Spaetzle to the Toll receptor, thereby triggering the intracytoplasmic signaling cascade.

The *Drosophila* genome encodes 13 members of the peptidoglycan recognition protein family (Fig. 2A), and one of these can give rise to three splice isoforms (PGRP-LC, see later) (Werner et al. 2000; Kaneko et al. 2004; Piao et al. 2005). They all share an evolutionary conserved domain related to bacteriophage type II amidases. This amidase function is retained in the majority of PGRPs (Mellroth et al. 2003; Steiner 2004), whereas some have lost the amino acids required for amidase activity and function as recognition PGRPs (Kim et al. 2003). Recognition PGRPs can discriminate between the predominant PGN of Gram-positive bacteria, which is characterized by a lysine residue in position three of the stem peptide, and the PGN of Gram-negative bacteria (and of Gram-positive bacilli), which carries a diaminopimelic acid (DAP) in the same position of the stem peptide (Fig. 2B) (Leulier et al. 2003). It is remarkable that this small difference, in fact the addition in DAP-PGN of a carboxyl in α-position of an amine of Lysine, is sufficient to discriminate between two large groups of pathogens and to trigger distinct signaling cascades and gene expression programs. Prototypical PGRPs recognizing Gram-positive bacterial peptidoglycan are the circulating PGRP-SA (Michel et al. 2001) and PGRP-SD (Bischoff et al. 2004), whereas an essential recognition protein for Gram-negative peptidoglycan is the transmembrane protein PGRP-LC (Choe et al. 2002; Gottar et al. 2002; Ramet et al. 2002). Three GNBPs are present in the *Drosophila* genome (Fig. 3). They have in common an amino-terminal domain, which in GNBP-3 mediates recognition of polymeric glucans, and a carboxy-terminal glucanase-like domain, which has lost its catalytic activity (Wang et al. 2006; A. Roussel, pers. comm.).

Binding of Lys-PGN to PGRP-SA in circulation activates a cascade of CLIP-domain zymogens (Piao et al. 2005), which includes the serine proteases Grass and Spirit (Kambris et al. 2006; El Chamy et al. 2008), and leads

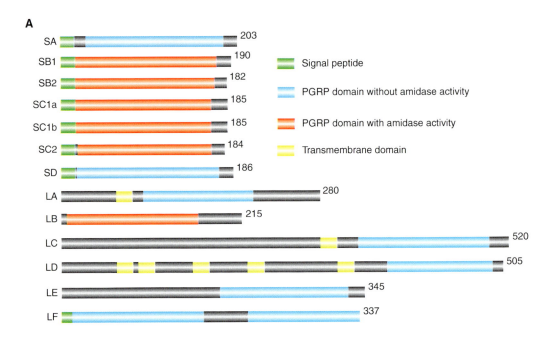

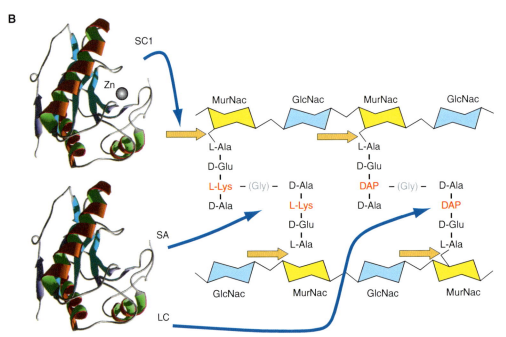

Figure 2. (*A*) The 13 PGRPs of *Drosophila melanogaster*. (*B*) Simplified representation of the various activities of PGRPs versus PGN. (*Top*) PGRP with an amidase activity cleaves PGN (SC1, arrow). (*Bottom*) PGRPs that have lost their amidase activities serve as recognition proteins for Lys-type PGN (PGRP-SA) or DAP-type PGN (PGRP-LC).

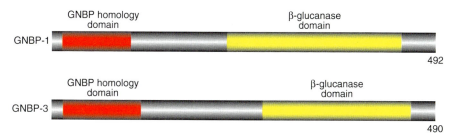

Figure 3. GNBP1 and GNBP3.

to the activation of the zymogen spaetzle-processing enzyme (SPE) (Jang et al. 2006). SPE in turn cleaves Spaetzle, a 37-kDa polypeptide with nine cysteines, engaged in four intramolecular disulfide bridges folded in a typical cystine-knot array (Fig. 4) (Mizuguchi et al. 1998). The additional cysteine residue allows for covalent dimerization of Spaetzle (DeLotto and DeLotto 1998). Cleavage of Spaetzle by SPE occurs amino-terminally outside the cystine-knot domain and the dimeric carboxy-terminal fragment binds to Toll, inducing the dimerization of the transmembrane receptor, thereby triggering activation of the downstream signaling cascade (Mizuguchi et al. 1998; Weber et al. 2003, 2005).

For reasons that are not fully understood, Lys-PGN activation of the PGRP-SA-dependent zymogen cascade requires the concomitant presence of one of the members of the GNBP family (GNBP-1) (Gobert et al. 2003; Pili-Floury et al. 2004; Wang et al. 2006). Further, PGRP-SD, another circulating member of the Lys-PGN recognition proteins, can substitute for PGRP-SA or potentiate its role (Bischoff et al. 2004). How recognition of the microbial ligands to the cognate binding proteins translates into activation of the downstream zymogen cascades remains to be established (Fig. 4).

Fungal or yeast glucans, which are also potent activators of the Toll pathway, interact with GNBP-3 and in turn trigger a zymogen cascade, which activates the serine protease SPE, and leads to the cleavage of Spaetzle and to Toll activation. This zymogen cascade shares some of the serine proteases identified downstream of PGRP-SA.

Of major potential interest are recent observations that fungal proteases can activate the circulating zymogen Persephone (Ligoxygakis et al. 2002) during the process of infection, mediating activation of SPE and of Toll. This pathogen-induced activation of Toll is not directly dependent on recognition of microbial cell wall components, but rather on microbial virulence factors (Gottar et al. 2006). It can be mimicked, again in a Persephone-dependent way, through injection of exogenous subtilisin, hence the proposal that Persephone is responsive to a danger signal (Fig. 4) (El Chamy et al. 2008).

The *Drosophila* transmembrane receptor Toll is a member of a family of nine receptors, which are involved in developmental processes during embryogenesis and probably later in development (see Eldon et al. 1994; Tauszig et al. 2000; Kambris et al. 2002; Gay and Gangloff 2007). Toll has an extracytoplasmic domain with numerous leucine-rich repeats (LRR) of which several contain cysteines. Of note, Toll9 has a single cystein-containing LRR in proximity of the plasma membrane, as is the case in mammalian Toll-like receptors (see Fig. 5). The intracytoplasmic domain of Toll is homologous to the intracytoplasmic signaling domain of the mammalian interleukin-1 receptor and of all mammalian TLRs (and is referred to as TIR domain [Hashimoto et al. 1988]). Interaction of dimeric, cleaved Spaetzle leads to dimerization of Toll and its TIR domains, which in turn interact with a platform of three distinct death domain containing proteins, dMyD88, Tube, and Pelle (Fig. 6) (Lemaitre et al. 1996; Tauszig-Delemasure et al. 2002). dMyD88 is homologous to mammalian MyD88, and interacts through

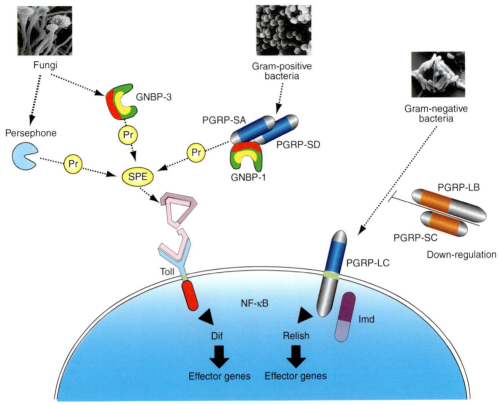

Figure 4. Induction of Toll and Imd pathway by pathogens. Gram-positive bacteria are recognized by PGRP-SA with the cooperation of GNBP1 and PGRP-SD. This recognition results in activation of a proteolytic cascade culminating in the cleavage of Spaetzle via the serine protease Spaetzle Processing Enzyme (SPE). Dimeric cleaved Spaetzle binds to the Toll receptor, activating the intracellular signaling cascade shown in Figure 6. Fungi are recognized by GNBP3 and induce a proteolytic cascade, which also activates SPE. Entomopathogenic fungi have been shown to also activate the Toll pathway through secreted proteases, which activate the circulating zymogen Persephone (virulence factor). Gram-negative bacteria are directly detected by the transmembrane receptor PGRP-LC, which recognizes the DAP-type peptidoglycan. The amidase PGRP-LB and PGRP-SC degrade PGN to nonimmunogenic moieties (Fig. 2).

its TIR domain with the TIR domain of Toll. Through its death domain, it associates with the death domain of Tube. Tube has a bifunctional death domain, which allows it to also interact with the death domain of Pelle (Sun et al. 2004), which is a member of the IL-1R associated kinase (IRAK) family of serine-threonine kinases. In the process of activation of the Toll signaling cascade, Cactus is phosphorylated by an as yet unidentified kinase; Pelle does apparently not fulfill this role. Phosphorylated Cactus undergoes K48 ubiquitination and is degraded by the proteasome (Belvin et al. 1995; Fernandez et al. 2001).

Dorsal and/or DIF are thus relieved from their inhibition and translocate to the nucleus where they bind to κB-response elements and trans-activate a specific set of genes (Bergmann et al. 1996; Reach et al. 1996), namely the drosomycin gene, which is often used as a readout for Toll pathway activation. Genetic evidence points to DIF as the major transactivator in Toll-dependent defenses in adults, whereas Dorsal can substitute for DIF in larvae (Rutschmann et al. 2000a).

NF-κB activation by Toll has relatively slow kinetics, with maximum levels at 24–48 h (if transcript levels of the drosomycin gene are

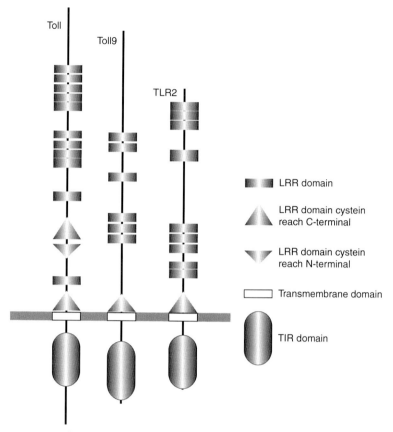

Figure 5. Representatives of the Toll and the TLR families.

used as readouts) (Lemaitre et al. 1997). Negative regulation of this response during infection has only been marginally investigated. The serpin necrotic, which blocks the serine protease Persephone upstream of Spaetzle, is strongly induced during Toll pathway activation, potentially contributing to down-regulation of the pathway (Irving et al. 2001).

This short overview of NF-κB activation by the Toll pathway in response to microbial infections calls for several comments: (1) the parallels between some developmental regulations and this immune response are remarkable. We now know that Spaetzle is a member of a family of several neurotrophins, some of which (including Spaetzle) have been shown to play neurotrophic functions (Parker et al. 2001; Zhu et al. 2008). Of the five presently established members of the *Drosophila* neurotrophin-Spaetzle family, and of the nine Tolls,

which are functional in flies, apparently only the Spaetzle-Toll couple was recruited to serve an immune function, in addition to its developmental roles; (2) neurotrophins (including Spaetzle) require, like many cytokines and growth factors, proteolytic cleavage to become active ligands. In the embryonic zymogen cascade (not discussed here, but see Moussian and Roth 2005, for instance), a small molecular weight compound initiates the cascade, whereas in the immune response, small bacterial or cell wall components similarly trigger a comparable cascade; (3) as regards the intracytoplasmic signaling cascade, the various players appear mostly identical, but it is obvious that our information here is still fragmentary. For one, the actual Cactus-kinase has not yet been identified and there are reasons to expect that K63 ubiquitination plays a major role in stabilization/scaffolding of the activating intracytoplasmic

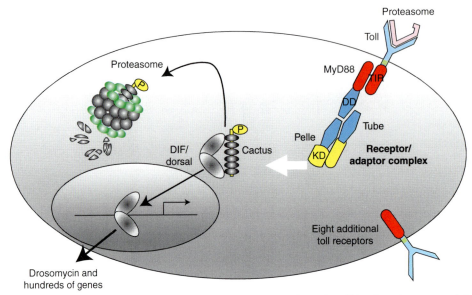

Figure 6. Toll pathway in adult *Drosophila melanogaster*. Signaling through Toll involves the receptor/adaptor complex, composed of three death-domain-containing proteins, MyD88, Tube, and Pelle. The trimeric complex assembles around the bipartite DD of Tube. It is unclear how the signal is transduced from the complex to Cactus, a homolog of IkB. Phosphorylated Cactus is degraded after polyubiquitination by the proteasome. The liberated DIF is translocated to the nucleus and binds to NF-κB response elements, inducing the expression of hundreds of genes, namely those encoding antimicrobial peptides, such as drosomycin.

complexes. Direct phosphorylation has also been proposed to confer transactivating potential to Dorsal, independently of the release from Cactus (Drier et al. 2000). In essence, however, our view remains still superficial and much additional work will be required to fully understand how this pathway functions, both in immunity and development.

ACTIVATION OF NF-κB BY THE IMD PATHWAY

Whereas activation of the Toll pathway is initiated by interaction of microbial ligands with circulating proteins, the Imd pathway is triggered on direct interaction of the transmembrane receptor PGRP-LC (Gottar et al. 2002; Choe et al. 2002; Ramet et al. 2002; Choe et al. 2005) with Gram-negative bacterial peptidoglycan (diaminopimelic peptidoglycan-DAP-PGN) (Fig. 3). The 55 kDa PGRP-LC consists of an extracytoplasmic part that harbors the conserved peptidoglycan recognition domain,

a short transmembrane domain of 22 amino acids, and an intracytoplasmic signaling domain. The PGRP-LC gene can generate three distinct splice isoforms, giving rise to three types of receptors with identical transmembrane and intracytoplasmic sequences and slightly different exodomains, and referred to as LCa, LCx, and LCy (Werner et al. 2000). It is now understood that PGRP-LCx homodimers sense polymeric DAP-PGN and that LCx-LCa heterodimers detect short PGN end fragments (Kaneko et al. 2004). Such short PGN fragments are released by bacteria during growth and division (for a detailed structural study of the interaction of PGRP-LC isoforms and DAP-PGN, see Chang et al. 2006; Lim et al. 2006). Binding of monomeric and multimeric DAP-PGN to the cognate receptors induces their dimerization or oligomerization and leads to signaling (Mellroth et al. 2005; Chang et al. 2006; Lim et al. 2006).

In addition to PGRP-LC, other members of the PGRP family appear to play an upstream role in the Imd pathway. One of these is PGRP-

LE (Takehana et al. 2002), which can act as an intracellular receptor for monomeric peptidoglycan (Kaneko et al. 2006). A naturally truncated form of LE, containing only the PGRP domain, functions extracellularly to enhance PGRP-LC-mediated recognition on the cell surface (Takehana et al. 2004). The membrane-associated PGRP-LF acts as a specific negative regulator of the Imd pathway: Reduction of PGRP-LF levels, in the absence of infection, is sufficient to trigger Imd pathway activation (Maillet et al. 2008). Furthermore, normal development is impaired in the absence of functional PGRP-LF, a phenotype which is mediated by the JNK pathway (Maillet et al. 2008) (see the following).

The intracytoplasmic cascade (Fig. 7) of the Imd pathway starts with the recruitment of the 25 kDa death domain protein Imd (of note, the sequence of this particular death domain is closest to that of mammalian RIP1 that is TNF-receptor interacting protein) (Lemaitre et al. 1995; Georgel et al. 2001). Both the intracytoplasmic domain of PGRP-LC and the adaptor protein Imd contain a so-called RHIM domain (for receptor interacting protein [RIP] homotypic interaction motif) required for signaling, but not for their respective interaction, which probably requires an additional as yet unidentified partner (Kaneko et al. 2006; Choe et al. 2005). Imd further associates with the mammalian homolog of FADD and with the caspase-8 homolog DREDD (Leulier et al. 2000, 2002). Through mechanisms not fully understood at present, but which are likely to involve K63 ubiquitination, this upstream receptor–adaptor complex activates the MAP3kinase TAK1 (Zhou et al. 2005). TAK1 is associated with the homolog of mammalian TAB2, which has a conserved sequence domain known to interact with K63 polyubiquitin chains (Boutros et al. 2002; Silverman et al. 2003; Gesellchen et al. 2005; Geuking et al. 2005; Kleino et al. 2005).

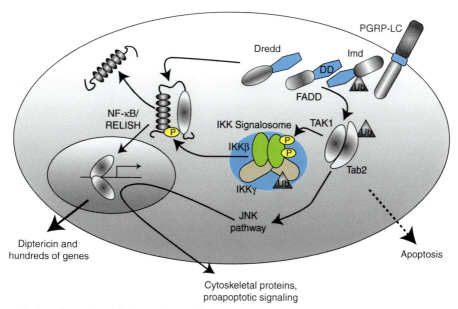

Figure 7. Imd pathway in adult *Drosophila melanogaster*. The transmembrane protein PGRP-LC senses the presence of DAP-type PGN and activates the Imd pathway Imd has an essential role in controlling the phosphorylation of Relish through the activation of TAK1 and the IKK signalosome. Cleavage of phosphorylated Relish is dependent on the caspase Dredd. The activation of TAK1 and the IKK signalosome requires several proteins, including FADD and Dredd. Cleaved Relish is translocated in the nucleus and binds to NF-κB response elements, inducing the expression of numerous genes encoding immune response proteins, namely antimicrobial peptides such as Diptericin. In addition to phosphorylating the IKK complex, TAK1 also acts as a JNK kinase and activates the JNK pathway.

Downstream of the TAK1/TAB2 protein complex, the Imd pathway branches into a signaling cascade, leading to Relish activation and, a second, to JNK activation (Vidal et al. 2001; Zhuang et al. 2006). In contrast to the Toll pathway, the Imd-Relish cascade relies on an IKK complex, which consists of homologs of both IKKβ (referred to as Ird5) (Wu and Anderson 1998; Silverman et al. 2000; Lu et al. 2001) and IKKγ/NEMO (referred to as kenny) (Rutschmann et al. 2000b). Once activated by TAK1, the IKK complex phosphorylates the NF-κB protein Relish on specific serine residues and phosphorylated Relish is cleaved into an amino-terminal transcriptional regulatory domain, which translocates to the nucleus where it binds to Relish response elements and directs expression of dedicated genes (Stoven et al. 2000; De Gregorio et al. 2002). Interestingly, a κB sequence code has been identified in the promoters of immune-responsive genes, which directs binding of the NF-κB family members of *Drosophila* and thereby induces specialized programs of gene expression in the Toll and Imd pathway (Busse et al. 2007). The carboxy-terminal fragment remains in the cytoplasm. It is probable that the caspase-8 homolog DREDD, after its initial association with the death domain protein dFADD, is responsible for cleavage of Relish (Leulier et al. 2002; Stöven et al. 2003). A recent study has identified a 32 kDa highly conserved protein, Akirin, which acts in conjunction with Relish to control Imd pathway dependent gene transcription (Goto et al. 2008).

TAK1 activates the JNK branch of the Imd pathway by signaling to Hemipterous, a *Drosophila* homolog of JNKK, which phosphorylates basket (dJNK), resulting in the activation of *Drosophila* AP-1 (Boutros et al. 2002; Silverman et al. 2003; Park et al. 2004). The exact role of the JNK branch in the host defense of *Drosophila* is not firmly established. The gene set up-regulated by this branch leads to production of cytoskeletal proteins (probably involved in wound repair) and in proapoptotic signaling (Delaney et al. 2006). These genes exhibit transient induction kinetics, with peak values at 1 h postinduction.

Negative regulation of JNK can be mediated by Relish-dependent degradation of TAK1 (Park et al. 2004).

The kinetics of the Imd-Relish pathway are rapid as compared to those of the Toll pathway, and activation, to judge from the transcription profiles of antimicrobial peptide genes, is relatively short-lived (peak values at 6 h postinduction) (Lemaitre et al. 1997). Recent investigations point to several levels of negative control of the Imd pathway. For one, down-regulation of the highly immunogenic DAP-PGN has been documented for several of the PGRPs that have retained their amidase activity and cleave PGN between the glucan chains and the stem peptides to nonimmunogenic fragments (Mellroth et al. 2003; Mellroth and Steiner 2006; Zaidman-Remy et al. 2006). This is in particular the case for PGRP-LB, SB1, and SC1. As noted above, PGRP-LF also serves as a negative regulator of this pathway. Other levels of negative regulation are the recently identified proteins Pirk (Aggarwal et al. 2008; Kleino et al. 2008; Lhocine et al. 2008) and Caspar (Kim et al. 2006). Pirk is a cytoplasmic protein that co-immunoprecipitates with Imd and the cytoplasmic tail of PGRP-LC. Down-regulation of Pirk results in hyperactivation of the Imd pathway on infection with Gram-negative bacteria, whereas overexpression of Pirk reduces the Imd pathway responses in vitro and in vivo. Further, Pirk-overexpressing flies are more susceptible to Gram-negative bacterial infection than wild-type flies. Caspar is a cytoplasmic homologous to human Fas-associated factor 1 (hFAF-1), which associates with various components of the TNF pathway, namely FAS, FADD, caspase-8, and NF-κB. Caspar mutants show constitutive activation of diptericin in the absence of infection (Kim et al. 2006). Overexpression of Caspar, on the other hand, inhibits the induction of antimicrobial peptides and it has been proposed that Caspar blocks Relish cleavage by interfering with the Caspase-8 homolog DREDD (Kim et al. 2006).

We would like to add several comments to this short overview of the activation of NF-κB by the Imd pathway: (1) the intracytoplasmic signaling cascade shows some remarkable similarities with that of the TNF-receptor pathway. This is in contrast to that of the Toll pathway previously discussed. Also in contrast to the Toll pathway, the Imd pathway has a transmembrane receptor with dedicated recognition domains for microbial cell wall components of Gram-negative bacterial origin. As a result, this pathway cannot rely on an amplification cascade similar to that of the zymogen cascade downstream of the circulating recognition PGRPs in Toll pathway activation. It has been proposed that the capacity of PGRP-LCx dimers to respond to DAP-PGN end chain fragments (monomeric PGN or tracheal cytotoxin) obviates the necessity of an amplification cascade, as these fragments are likely to be produced in high amounts during growth and division of infecting Gram-negative bacteria (Boneca 2005); (2) as in the TNF-receptor pathway, the Imd pathway branches downstream of the MAP3kinase TAK1 and also activates the JNKinase system, thereby leading to the remarkably rapid transcription of a gene set distinct from that induced by cleaved Relish. The detailed analysis of the recent literature, which is beyond the scope of the present review, points to a complex interplay between these two pathways, and namely to the possibility that gene products induced by one pathway tend to down-regulate the other pathway; (3) the Imd pathway was discovered through its role in the defense against Gram-negative bacterial infection (Lemaitre et al. 1995). However, several recent observations point to Imd pathway activation (as judged by induction of antimicrobial peptide genes) in the absence of infection. One striking example is the report by Mukae et al. (2002) that chromosomal DNA from apoptotic cells that escaped digestion by appropriate enzymes can activate the expression of diptericin and attacin (Imd pathway readouts). M. Lagueux (pers. comm.) had noted that during metamorphosis, significant transcription of these genes occurred in the absence of infection. In conclusion, we need to develop a better understanding of the Imd pathway, both regarding the signaling cascade itself and the roles and interactions of its various partners, and regarding the potential functions of this pathway in development and in inflammation.

CONCLUDING REMARKS

We have focused in this short overview on NF-κB in antimicrobial defenses of *Drosophila*. Obviously, in addition to this central role, NF-κB is centrally involved in embryonic development and differentiation in flies—as it is in other organisms (see Baldwin 1996). We now know that the Rel/NF-κB family of inducible transactivators originated at the dawn of the Metazoa and are present in Sponges (*Amphimedon queenslandica*) (Gauthier and Degnan 2008), Cnidaria (*Nematostella vectensis*) (Sullivan et al. 2007), and in all Bilateria. Genome sequencing data from early phyla indicate that many of the genes discussed in this article are present in these groups and we may anticipate that future studies will unravel NF-κB regulatory pathways that share essential characteristics with those described here.

Regarding the future of the studies related to NF-κB in flies, essential questions remain unresolved. For one, our information on the identities and roles of the players in these pathways is too fragmentary. To give but a few examples: We do not understand the mechanisms by which binding of microbial cell components to the recognition proteins, be it PGRPs or GNBPs, activate the downstream events; further, we suspect that K63 polyubiquitination plays a role in the signaling cascades but have very little reliable information in this regard. We do not fully understand how the signaling pathways discussed here interact with each other and with a variety of other cellular signaling pathways—even our knowledge on the interaction of the Imd-Relish pathway with the JNK pathway is sketchy. Whereas the studies on the Toll pathway in immune defenses have been crucially dependent on its role in embryonic development, it is not yet understood whether the Imd pathway similarly plays a role in development, apoptosis, and inflammation (as TNF does in mammals). Much additional work and new approaches are required if we are to fully understand the activation and roles of NF-κB in the flies. These are some of the frontiers in this field of research.

REFERENCES

Aggarwal K, Silverman N. 2008. Positive and negative regulation of the *Drosophila* immune response. *BMB Rep* **41:** 267–277.

Aggarwal K, Rus F, Vriesema-Magnuson C, Ertürk-Hasdemir D, Paquette N, Silverman N. 2008. Rudra interrupts receptor signaling complexes to negatively regulate the IMD pathway. *PLoS Pathog* **4:** e1000120.

Baldwin AS Jr. 1996. The NF-κ B and I κ B proteins: New discoveries and insights. *Annu Rev Immunol* **14:** 649–83.

Belvin MP, Jin Y, Anderson KV. 1995. Cactus protein degradation mediates *Drosophila* dorsal-ventral signaling. *Genes Dev* **9:** 783–793.

Bergmann A, Stein D, Geisler R, Hagenmaier S, Schmid B, Fernandez N, Schnell B, Nüsslein-Volhard C. 1996. A gradient of cytoplasmic Cactus degradation establishes the nuclear localization gradient of the dorsal morphogen in *Drosophila*. *Mech Dev* **60:** 109–123.

Bischoff V, Vignal C, Boneca IG, Michel T, Hoffmann JA, Royet J. 2004. Function of the *Drosophila* pattern-recognition receptor PGRP-SD in the detection of Gram-positive bacteria. *Nat Immunol* **5:** 1175–1180.

Boneca IG. 2005. The role of peptidoglycan in pathogenesis. *Curr Opin Microbiol* **8:** 46–53.

Boutros M, Agaisse H, Perrimon N. 2002. Sequential activation of signaling pathways during innate immune responses in *Drosophila*. *Dev Cell* **3:** 711–722.

Brennan CA, Anderson KV. 2004. *Drosophila*: the genetics of innate immune recognition and response. *Annu Rev Immunol* **22:** 457–483.

Bulet P, Hetru C, Dimarcq JL, Hoffmann D. 1999. Antimicrobial peptides in insects; structure and function. *Dev Comp Immunol* **23:** 329–344.

Busse MS, Arnold CP, Towb P, Katrivesis J, Wasserman SA. 2007. A κB sequence code for pathway-specific innate immune responses. *EMBO J* **26:** 3826–3835.

Chang CI, Chelliah Y, Borek D, Mengin-Lecreulx D, Deisenhofer J. 2006. Structure of tracheal cytotoxin in complex with a heterodimeric pattern-recognition receptor. *Science* **311:** 1761–1764.

Choe KM, Werner T, Stöven S, Hultmark D, Anderson KV. 2002. Requirement for a peptidoglycan recognition protein (PGRP) in Relish activation and antibacterial immune responses in *Drosophila*. *Science* **296:** 359–362.

Choe KM, Lee H, Anderson KV. 2005. *Drosophila* peptidoglycan recognition protein LC (PGRP-LC) acts as a signal-transducing innate immune receptor. *Proc Natl Acad Sci* **102:** 1122–1126.

De Gregorio E, Spellman PT, Rubin GM, Lemaitre B. 2001. Genome-wide analysis of the *Drosophila* immune response by using oligonucleotide microarrays. *Proc Natl Acad Sci* **98:** 12590–12595.

De Gregorio E, Spellman PT, Tzou P, Rubin GM, Lemaitre B. 2002. The Toll and Imd pathways are the major regulators of the immune response in *Drosophila EMBO J* **21:** 2568–2579.

Delaney JR, Stöven S, Uvell H, Anderson KV, Engström Y, Mlodzik M. 2006. Cooperative control of *Drosophila* immune responses by the JNK and NF-κB signaling pathways. *EMBO J* **25:** 3068–3077.

DeLotto Y, DeLotto R. 1998. Proteolytic processing of the *Drosophila* Spätzle protein by easter generates a dimeric NGF-like molecule with ventralising activity. *Mech Dev* **72:** 141–148.

Drier EA, Govind S, Steward R. 2000. Cactus-independent regulation of Dorsal nuclear import by the ventral signal. *Curr Biol* **10:** 23–26.

El Chamy L, Leclerc V, Caldelari I, Reichhart JM. 2008. Sensing of 'danger signals' and pathogen-associated molecular patterns defines binary signaling pathways 'upstream' of Toll. *Nat Immunol* **9:** 1165–1170.

Eldon E, Kooyer S, D'Evelyn D, Duman M, Lawinger P, Botas J, Bellen H. 1994. The *Drosophila* 18 wheeler is required for morphogenesis and has striking similarities to Toll. *Development* **120:** 885–899.

Engström Y, Kadalayil L, Sun SC, Samakovlis C, Hultmark D, Faye I. 1993. κ B-like motifs regulate the induction of immune genes in *Drosophila*. *J Mol Biol* **232:** 327–333.

Fernandez NQ, Grosshans J, Goltz JS, Stein D. 2001. Separable and redundant regulatory determinants in Cactus mediate its dorsal group dependent degradation. *Development* **128:** 2963–2974.

Ferrandon D, Imler JL, Hetru C, Hoffmann JA. 2007. The *Drosophila* systemic immune response: Sensing and signalling during bacterial and fungal infections. *Nat Rev Immunol* **7:** 862–874.

Ferrandon D, Jung AC, Criqui M, Lemaitre B, Uttenweiler-Joseph S, Michaut L, Reichhart J, Hoffmann JA. 1998. A drosomycin-GFP reporter transgene reveals a local immune response in *Drosophila* that is not dependent on the Toll pathway. *EMBO J* **17:** 1217–1227.

Frolet C, Thoma M, Blandin S, Hoffmann JA, Levashina EA. 2006. Boosting NF-κB-dependent basal immunity of *Anopheles gambiae* aborts development of *Plasmodium berghei*. *Immunity* **25:** 677–685.

Gauthier M, Degnan BM. 2008. The transcription factor NF-κB in the demosponge *Amphimedon queenslandica*: Insights into the evolutionary origin of the Rel homology domain. *Dev Genes Evol* **218:** 23–32.

Gay NJ, Gangloff M. 2007. Structure and function of Toll receptors and their ligands. *Annu Rev Biochem* 2007 **76:** 141–165.

Geisler R, Bergmann A, Hiromi Y, Nüsslein-Volhard C. 1992. cactus, a gene involved in dorsoventral pattern formation of *Drosophila*, is related to the I κ B gene family of vertebrates. *Cell* **71:** 613–621.

Georgel P, Naitza S, Kappler C, Ferrandon D, Zachary D, Swimmer C, Kopczynski C, Duyk G, Reichhart JM, Hoffmann JA. 2001. *Drosophila* immune deficiency (IMD) is a death domain protein that activates antibacterial defense and can promote apoptosis. *Dev Cell* **1:** 503–514.

Gesellchen V, Kuttenkeuler D, Steckel M, Pelte N, Boutros M. 2005. An RNA interference screen identifies Inhibitor of Apoptosis Protein 2 as a regulator of innate immune signalling in *Drosophila*. *EMBO Rep* **6**: 979–984.

Geuking P, Narasimamurthy R, Basler K. 2005. A genetic screen targeting the tumor necrosis factor/Eiger signaling pathway: Identification of *Drosophila* TAB2 as a functionally conserved component. *Genetics* **171**: 1683–1694.

Gobert V, Gottar M, Matskevich AA, Rutschmann S, Royet J, Belvin M, Hoffmann JA, Ferrandon D. 2003. Dual activation of the *Drosophila* toll pathway by two pattern recognition receptors. *Science* **302**: 2126–2130.

Goto A, Matsushita K, Gesellchen V, El Chamy L, Kuttenkeuler D, Takeuchi O, Hoffmann JA, Akira S, Boutros M, Reichhart JM. 2008. Akirins are highly conserved nuclear proteins required for NF-κB-dependent gene expression in *Drosophila* and mice. *Nat Immunol* **9**: 97–104.

Gottar M, Gobert V, Michel T, Belvin M, Duyk G, Hoffmann JA, Ferrandon D, Royet J. 2002. The *Drosophila* immune response against Gram-negative bacteria is mediated by a peptidoglycan recognition protein. *Nature* **416**: 640–644.

Gottar M, Gobert V, Matskevich AA, Reichhart JM, Wang C, Butt TM, Belvin M, Hoffmann JA, Ferrandon D. 2006. Dual detection of fungal infections in *Drosophila* via recognition of glucans and sensing of virulence factors. *Cell* **127**: 1425–1437.

Hashimoto C, Hudson KL, Anderson KV. 1988. The Toll gene of *Drosophila*, required for dorsal-ventral embryonic polarity, appears to encode a transmembrane protein. *Cell* **52**: 269–279.

Hetru C, Troxler L, Hoffmann JA. 2003. *Drosophila melanogaster* antimicrobial defense. *J Infect Dis* **187**: S327–334.

Hoffmann JA. 2003. The immune response of *Drosophila*. *Nature* **426**: 33–38.

Huguet C, Crepieux P, Laudet V. 1997. Rel/NF-κ B transcription factors and I κ B inhibitors: evolution from a unique common ancestor. *Oncogene* **15**: 2965–2974.

Hultmark D. 2003. *Drosophila* immunity: Paths and patterns. *Curr Opin Immunol* **15**: 12–19.

Ip YT, Reach M, Engstrom Y, Kadalayil L, Cai H, González-Crespo S, Tatei K, Levine M. 1993. Dif, a dorsal-related gene that mediates an immune response in *Drosophila*. *Cell* **75**: 753–763.

Irving P, Troxler L, Heuer TS, Belvin M, Kopczynski C, Reichhart JM, Hoffmann JA, Hetru C. 2001. A genome-wide analysis of immune responses in *Drosophila*. *Proc Natl Acad Sci* **98**: 15119–15124.

Jang IH, Chosa N, Kim SH, Nam HJ, Lemaitre B, Ochiai M, Kambris Z, Brun S, Hashimoto C, Ashida M, et al. 2006. A Spätzle-processing enzyme required for toll signaling activation in *Drosophila* innate immunity. *Dev Cell* **10**: 45–55.

Kambris Z, Brun S, Jang IH, Nam HJ, Romeo Y, Takahashi K, Lee WJ, Ueda R, Lemaitre B. 2006. *Drosophila* immunity: A large-scale *in vivo* RNAi screen identifies five serine proteases required for Toll activation. *Curr Biol* **16**: 808–813.

Kambris Z, Hoffmann JA, Imler JL, Capovilla M. 2002. Tissue and stage-specific expression of the Tolls in *Drosophila* embryos. *Gene Expr Patterns* **2**: 311–317.

Kaneko T, Goldman WE, Mellroth P, Steiner H, Fukase K, Kusumoto S, Harley W, Fox A, Golenbock D, Silverman N. 2004. Monomeric and polymeric gram-negative peptidoglycan but not purified LPS stimulate the *Drosophila* IMD pathway. *Immunity* **20**: 637–649.

Kaneko T, Yano T, Aggarwal K, Lim JH, Ueda K, Oshima Y, Peach C, Erturk-Hasdemir D, Goldman WE, Oh BH, et al. 2006. PGRP-LC and PGRP-LE have essential yet distinct functions in the *Drosophila* immune response to monomeric DAP-type peptidoglycan. *Nat Immunol* **7**: 715–723.

Kang D, Liu G, Lundström A, Gelius E, Steiner H. 1998. A peptidoglycan recognition protein in innate immunity conserved from insects to humans. *Proc Natl Acad Sci* **95**: 10078–10082.

Kappler C, Meister M, Lagueux M, Gateff E, Hoffmann A, Reichhart JM. 1993. Insect immunity. Two 17 bp repeats nesting a κ B-related sequence confer inducibility to the diptericin gene and bind a polypeptide in bacteria-challenged *Drosophila*. *EMBO J* **12**: 1561–1568.

Kim MS, Byun M, Oh BH. 2003. Crystal structure of peptidoglycan recognition protein LB from *Drosophila melanogaster*. *Nat Immunol* **4**: 787–793.

Kim M, Lee JH, Lee SY, Kim E, Chung J. 2006. Caspar, a suppressor of antibacterial immunity in *Drosophila*. *Proc Natl Acad Sci* **103**: 16358–16363.

Kleino A, Myllymäki H, Kallio J, Vanha-aho LM, Oksanen K, Ulvila J, Hultmark D, Valanne S, Rämet M. 2008. Pirk is a negative regulator of the *Drosophila* Imd pathway. *J Immunol* **180**: 5413–5422.

Kleino A, Valanne S, Ulvila J, Kallio J, Myllymäki H, Enwald H, Stöven S, Poidevin M, Ueda R, Hultmark D, et al. 2005. Inhibitor of apoptosis 2 and TAK1-binding protein are components of the *Drosophila* Imd pathway. *EMBO J* **24**: 3423–3434.

Lee WJ, Lee JD, Kravchenko VV, Ulevitch RJ, Brey PT. 1996. Purification and molecular cloning of an inducible gram-negative bacteria-binding protein from the silkworm, *Bombyx mori*. *Proc Natl Acad Sci* **93**: 7888–7893.

Lemaitre B, Hoffmann J. 2007. The host defense of *Drosophila melanogaster*. *Annu Rev Immunol* **25**: 697–743.

Lemaitre B, Reichhart JM, Hoffmann JA. 1997. *Drosophila* host defense: Differential induction of antimicrobial peptide genes after infection by various classes of microorganisms. *Proc Natl Acad Sci* **94**: 14614–14619.

Lemaitre B, Kromer-Metzger E, Michaut L, Nicolas E, Meister M, Georgel P, Reichhart JM, Hoffmann JA. 1995. A recessive mutation, immune deficiency (imd), defines two distinct control pathways in the *Drosophila* host defense. *Proc Natl Acad Sci* **92**: 9465–9469.

Lemaitre B, Nicolas E, Michaut L, Reichhartb JM, Hoffmann JA. 1996. The dorsoventral regulatory gene cassette spätzle/Toll/cactus controls the potent antifungal response in *Drosophila* adults. *Cell* **86**: 973–983.

Leulier F, Rodriguez A, Khush RS, Abrams JM, Lemaitre B. 2000. The *Drosophila* caspase Dredd is required to resist gram-negative bacterial infection. *EMBO Rep* **1**: 353–358.

Leulier F, Vidal S, Saigo K, Ueda R, Lemaitre B. 2002. Inducible expression of double-stranded RNA reveals a role for dFADD in the regulation of the antibacterial response in *Drosophila* adults. *Curr Biol* **12**: 996–1000.

Leulier F, Parquet C, Pili-Floury S, Ryu JH, Caroff M, Lee WJ, Mengin-Lecreulx D, Lemaitre B. 2003. The *Drosophila* immune system detects bacteria through specific peptidoglycan recognition. *Nat Immunol* **4**: 478–484.

Liehl P, Blight M, Vodovar N, Boccard F, Lemaitre B. 2006. Prevalence of local immune response against oral infection in a *Drosophila/Pseudomonas* infection model. *PLoS Pathog* **2**: e56.

Ligoxygakis P, Pelte N, Hoffmann JA, Reichhart JM. 2002. Activation of *Drosophila* Toll during fungal infection by a blood serine protease. *Science* **297**: 114–116.

Lim JH, Kim MS, Kim HE, Yano T, Oshima Y, Aggarwal K, Goldman WE, Silverman N, Kurata S, Oh BH. 2006. Structural basis for preferential recognition of diaminopimelic acid-type peptidoglycan by a subset of peptidoglycan recognition proteins. *J Biol Chem* **281**: 8286–8295.

Lhocine N, Ribeiro PS, Buchon N, Wepf A, Wilson R, Tenev T, Lemaitre B, Gstaiger M, Meier P, Leulier F. 2008. PIMS modulates immune tolerance by negatively regulating Drosophila innate immune signaling. *Cell Host Microbe* **4**: 147–158.

Lu Y, Wu LP, Anderson KV. 2001. The antibacterial arm of the *Drosophila* innate immune response requires an IκB kinase. *Genes Dev* **15**: 104–110.

Maillet F, Bischoff V, Vignal C, Hoffmann J, Royet J. 2008. The *Drosophila* peptidoglycan recognition protein PGRP-LF blocks PGRP-LC and IMD/JNK pathway activation. *Cell Host Microbe* **3**: 293–303.

Manfruelli P, Reichhart JM, Steward R, Hoffmann JA, Lemaitres B. 1999. A mosaic analysis in Drosophila fat body cells of the control of antimicrobial peptide genes by the Rel proteins Dorsal and DIF. *EMBO J* **18**: 3380–3391.

Mellroth P, Steiner H. 2006. PGRP-SB1: An N-acetylmuramoyl L-alanine amidase with antibacterial activity. *Biochem Biophys Res Commun* **350**: 994–9.

Mellroth P, Karlsson J, Steiner H. 2003. A scavenger function for a *Drosophila* peptidoglycan recognition protein. *J Biol Chem* **278**: 7059–7064.

Mellroth P, Karlsson J, Håkansson J, Schultz N, Goldman WE, Steiner H. 2005. Ligand-induced dimerization of *Drosophila* peptidoglycan recognition proteins *in vitro*. *Proc Natl Acad Sci* **102**: 6455–6460.

Meng X, Khanuja BS, Ip YT. 1999. Toll receptor-mediated *Drosophila* immune response requires Dif, an NF-κB factor. *Genes Dev* **13**: 792–797.

Michel T, Reichhart JM, Hoffmann JA, Royet J. 2001. *Drosophila* Toll is activated by Gram-positive bacteria through a circulating peptidoglycan recognition protein. *Nature* **414**: 756–759.

Minakhina S, Steward R. 2006. Nuclear factor-κ B pathways in *Drosophila*. *Oncogene* **25**: 6749–6757.

Mizuguchi K, Parker JS, Blundell TL, Gay NJ. 1998. Getting knotted: A model for the structure and activation of Spätzle. *Trends Biochem Sci* **23**: 239–242.

Moussian B, Roth S. 2005. Dorsoventral axis formation in the *Drosophila* embryo–shaping and transducing a morphogen gradient. *Curr Biol* **15**: R887–899.

Mukae N, Yokoyama H, Yokokura T, Sakoyama Y, Nagata S. 2002. Activation of the innate immunity in *Drosophila* by endogenous chromosomal DNA that escaped apoptotic degradation. *Genes Dev* **16**: 2662–2671.

Nehme NT, Liégeois S, Kele B, Giammarinaro P, Pradel E, Hoffmann JA, Ewbank JJ, Ferrandon D. 2007. A model of bacterial intestinal infections in *Drosophila melanogaster*. *PLoS Pathog* **3**: e173.

Nüsslein-Volhard C, Lohs-Schardin M, Sander K, Cremer C. 1980. A dorso-ventral shift of embryonic primordia in a new maternal-effect mutant of Drosophila. *Nature* **283**: 474–476.

Ochiai M, Ashida M. 2000. A pattern-recognition protein for β-1,3-glucan. The binding domain and the cDNA cloning of β-1,3-glucan recognition protein from the silkworm, *Bombyx mori J Biol Chem* **275**: 4995–5002.

Onfelt Tingvall T, Roos E, Engström Y. 2001. The imd gene is required for local Cecropin expression in *Drosophila* barrier epithelia. *EMBO Rep* **2**: 239–243.

Park JM, Brady H, Ruocco MG, Sun H, Williams D, Lee SJ, Kato T Jr, Richards N, Chan K, Mercurio F, et al. 2004. Targeting of TAK1 by the NF-κ B protein Relish regulates the JNK-mediated immune response in *Drosophila*. *Genes Dev* **18**: 584–594.

Parker JS, Mizuguchi K, Gay NJ. 2001. A family of proteins related to Spätzle, the toll receptor ligand, are encoded in the *Drosophila* genome. *Proteins* **45**: 71–80.

Piao S, Song YL, Kim JH, Park SY, Park JW, Lee BL, Oh BH, Ha NC. 2005. Crystal structure of a clip-domain serine protease and functional roles of the clip domains. *EMBO J* **24**: 4404–4414.

Pili-Floury S, Leulier F, Takahashi K, Saigo K, Samain E, Ueda R, Lemaitre B. 2004. *In vivo* RNA interference analysis reveals an unexpected role for GNBP1 in the defense against Gram-positive bacterial infection in Drosophila adults. *J Biol Chem* **279**: 12848–12853.

Qiu P, Pan PC, Govind S. 1998. A role for the *Drosophila* Toll/Cactus pathway in larval hematopoiesis. *Development* **125**: 1909–1920.

Rämet M, Manfruelli P, Pearson A, Mathey-Prevot B, Ezekowitz RA. 2002. Functional genomic analysis of phagocytosis and identification of a *Drosophila* receptor for *E. coli*. *Nature* **416**: 644–648.

Reach M, Galindo RL, Towb P, Allen JL, Karin M, Wasserman SA. 1996. A gradient of cactus protein degradation establishes dorsoventral polarity in the *Drosophila* embryo. *Dev Biol* **180**: 353–364.

Reichhart JM, Georgel P, Meister M, Lemaitre B, Kappler C, Hoffmann JA. 1993. Expression and nuclear translocation of the rel/NF-κ B-related morphogen dorsal during the immune response of *Drosophila*. *C R Acad Sci III* **316**: 1218–12124.

Reichhart JM, Meister M, Dimarcq JL, Zachary D, Hoffmann D, Ruiz C, Richards G, Hoffmann JA. 1992. Insect immunity: Developmental and inducible activity of the *Drosophila* dipteticin promoter. *EMBO J* **11**: 1469–1477.

Roth S. 2003. The origin of dorsoventral polarity in *Drosophila*. *Philos Trans R Soc Lond B Biol Sci* **358:** 1317–1329.

Roth S, Stein D, Nüsslein-Volhard C. 1989. A gradient of nuclear localization of the dorsal protein determines dorsoventral pattern in the *Drosophila* embryo. *Cell* **59:** 1189–1202.

Rutschmann S, Jung AC, Hetru C, Reichhart JM, Hoffmann JA, Ferrandon D. 2000a. The Rel protein DIF mediates the antifungal but not the antibacterial host defense in *Drosophila*. *Immunity* **12:** 569–580.

Rutschmann S, Jung AC, Zhou R, Silverman N, Hoffmann JA, Ferrandon D. 2000b. Role of *Drosophila* IKK γ in a toll-independent antibacterial immune response. *Nat Immunol* **1:** 342–347.

Samakovlis C, Kimbrell DA, Kylsten P, Engström A, Hultmark D. 1990. The immune response in *Drosophila*: Pattern of cecropin expression and biological activity. *EMBO J* **9:** 2969–2976.

Schüpbach T, Wieschaus E. 1989. Female sterile mutations on the second chromosome of *Drosophila melanogaster*. I. Maternal effect mutations. *Genetics* **121:** 101–117.

Silverman N, Zhou R, Erlich RL, Hunter M, Bernstein E, Schneider D, Maniatis T. 2003. Immune activation of NF-κB and JNK requires *Drosophila* TAK1. *J Biol Chem* **278:** 48928–48934.

Silverman N, Zhou R, Stöven S, Pandey N, Hultmark D, Maniatis T. 2000. A *Drosophila* IκB kinase complex required for Relish cleavage and antibacterial immunity. *Genes Dev* **14:** 2461–2471.

Steiner H. 2004. Peptidoglycan recognition proteins: On and off switches for innate immunity. *Immunol Rev* **198:** 83–96.

Steward R. 1987. Dorsal, an embryonic polarity gene in *Drosophila*, is homologous to the vertebrate proto-oncogene, c-rel. *Science* **238:** 692–694.

Stöven S, Ando I, Kadalayil L, Engström Y, Hultmark D. 2000. Activation of the *Drosophila* NF-κB factor Relish by rapid endoproteolytic cleavage. *EMBO Rep* **1:** 347–352.

Stöven S, Silverman N, Junell A, Hedengren-Olcott M, Erturk D, Engstrom Y, Maniatis T, Hultmark D. 2003. Caspase-mediated processing of the *Drosophila* NF-κB factor Relish. *Proc Natl Acad Sci U S A* **100:** 5991–5996.

Sullivan JC, Kalaitzidis D, Gilmore TD, Finnerty JR. 2007. Rel homology domain-containing transcription factors in the cnidarian *Nematostella vectensis*. *Dev Genes Evol* **217:** 63–72.

Sun H, Towb P, Chiem DN, Foster BA, Wasserman SA. 2004. Regulated assembly of the Toll signaling complex drives *Drosophila* dorsoventral patterning. *EMBO J* **23:** 100–110.

Takehana A, Katsuyama T, Yano T, Oshima Y, Takada H, Aigaki T, Kurata S. 2002. Overexpression of a pattern-recognition receptor, peptidoglycan-recognition protein-LE, activates imd/relish-mediated antibacterial defense and the prophenoloxidase cascade in *Drosophila* larvae. *Proc Natl Acad Sci* **99:** 13705–13710.

Takehana A, Yano T, Mita S, Kotani A, Oshima Y, Kurata S. 2004. Peptidoglycan recognition protein (PGRP)-LE and PGRP-LC act synergistically in *Drosophila* immunity. *EMBO J* **23:** 4690–4700.

Tauszig-Delamasure S, Bilak H, Capovilla M, Hoffmann JA, Imler JL. 2002. *Drosophila* MyD88 is required for the response to fungal and Gram-positive bacterial infections. *Nat Immunol* **3:** 91–97.

Tauszig S, Jouanguy E, Hoffmann JA, Imler JL. 2000. Toll-related receptors and the control of antimicrobial peptide expression in *Drosophila*. *Proc Natl Acad Sci* **97:** 10520–10525.

Tzou P, Ohresser S, Ferrandon D, Capovilla M, Reichhart JM, Lemaitre B, Hoffmann JA, Imler JL. 2000. Tissue-specific inducible expression of antimicrobial peptide genes in Drosophila surface epithelia. *Immunity* **13:** 737–748.

Vidal S, Khush RS, Leulier F, Tzou P, Nakamura M, Lemaitre B. 2001. Mutations in the *Drosophila* dTAK1 gene reveal a conserved function for MAPKKKs in the control of rel/NF-κB-dependent innate immune responses. *Genes Dev* **15:** 1900–1912.

Wang L, Weber AN, Atilano ML, Filipe SR, Gay NJ, Ligoxygakis P. 2006. Sensing of Gram-positive bacteria in *Drosophila*: GNBP1 is needed to process and present peptidoglycan to PGRP-SA. *EMBO J* **25:** 5005–5014.

Weber AN, Moncrieffe MC, Gangloff M, Imler JL, Gay NJ. 2005. Ligand-receptor and receptor-receptor interactions act in concert to activate signaling in the *Drosophila* toll pathway. *J Biol Chem* **280:** 22793–22799.

Weber AN, Tauszig-Delamasure S, Hoffmann JA, Lelièvre E, Gascan H, Ray KP, Morse MA, Imler JL, Gay NJ. 2003. Binding of the *Drosophila* cytokine Spätzle to Toll is direct and establishes signaling. *Nat Immunol* **4:** 794–800.

Werner T, Liu G, Kang D, Ekengren S, Steiner H, Hultmark D. 2000. A family of peptidoglycan recognition proteins in the fruit fly *Drosophila* melanogaster. *Proc Natl Acad Sci* **97:** 13772–13777.

Wu LP, Anderson KV. 1998. Regulated nuclear import of Rel proteins in the *Drosophila* immune response. *Nature* **392:** 93–97.

Zaidman-Rémy A, Hervé M, Poidevin M, Pili-Floury S, Kim MS, Blanot D, Oh BH, Ueda R, Mengin-Lecreulx D, Lemaitre B. 2006. The *Drosophila* amidase PGRP-LB modulates the immune response to bacterial infection. *Immunity* **24:** 463–473.

Zambon RA, Nandakumar M, Vakharia VN, Wu LP. 2005. The Toll pathway is important for an antiviral response in *Drosophila*. *Proc Natl Acad Sci* **102:** 7257–7262.

Zhou R, Silverman N, Hong M, Liao DS, Chung Y, Chen ZJ, Maniatis T. 2005. The role of ubiquitination in *Drosophila* innate immunity. *J Biol Chem* **280:** 34048–34055.

Zhu B, Pennack JA, McQuilton P, Forero MG, Mizuguchi K, Sutcliffe B, Gu CJ, Fenton JC, Hidalgo A. 2008. *Drosophila* neurotrophins reveal a common mechanism for nervous system formation. *PLoS Biol* **6:** e284.

Zhuang ZH, Sun L, Kong L, Hu JH, Yu MC, Reinach P, Zang JW, Ge BX. 2006. *Drosophila* TAB2 is required for the immune activation of JNK and NF-κB. *Cell Signal* **18:** 964–970.

NF-κB in the Nervous System

Barbara Kaltschmidt[1] and Christian Kaltschmidt[2]

[1]Molecular Neurobiology, University of Bielefeld, Universitätsstr. 25, D-33501 Bielefeld

[2]Cell Biology, University of Bielefeld, Universitätsstr. 25, D-33501 Bielefeld

Correspondence: barbara.kaltschmidt@uni-bielefeld.de

The transcription factor NF-κB has diverse functions in the nervous system, depending on the cellular context. NF-κB is constitutively activated in glutamatergic neurons. Knockout of p65 or inhibition of neuronal NF-κB by super-repressor IκB resulted in the loss of neuroprotection and defects in learning and memory. Similarly, p50−/− mice have a lower learning ability and are sensitive to neurotoxins. Activated NF-κB can be transported retrogradely from activated synapses to the nucleus to translate short-term processes to long-term changes such as axon growth, which is important for long-term memory. In glia, NF-κB is inducible and regulates inflammatory processes that exacerbate diseases such as autoimmune encephalomyelitis, ischemia, and Alzheimer's disease. In summary, inhibition of NF-κB in glia might ameliorate disease, whereas activation in neurons might enhance memory. This review focuses on results produced by the analysis of genetic models.

In vertebrates, the nervous system is composed of the central nervous system (CNS) and the peripheral nervous system (PNS). The CNS comprises the brain and spinal cord, whereas peripheral nerves are part of the PNS. The CNS coordinates different tasks: integration of all stimuli that are presented from outside or inside the organism, coordination of all motor processes, regulation of hormone systems, and organ control. The most fascinating function of the brain is the coordination of learning and memory. This review will focus on the function of NF-κB in the nervous system, especially in learning and memory in rodent genetic models. The most abundant cell types in the vertebrate nervous system are neurons (about 100 billion) and glia (10–50 times more).

Typical glial cells are astrocytes, microglia, and the nerve fiber ensheathing cells, such as Schwann cells, which insulate nerves in the PNS and oligodendrocytes in the CNS. Neurons are highly polarized cells. Dendrites and/or soma receive electrochemical signals that are transmitted by axons to other neurons. Synapses are major sites of information input or output. These are formed by the presynaptic boutons derived from axons and the postsynaptic sites mainly localized on dendrites.

In a short introduction to the neurobiology of learning, we present a reductionist view (Dudai 1989). Although it is not easy to formally define learning (see Dudai 1989), here learning is defined simply as the capability of an animal to acquire novel skills. Learning can lead to a lasting modification of the internal representations of the outer world. Memory is thus the retention of the outer world experience-dependent internal representations. Memory retrieval is the use of memory in behavioral tasks. Comparison of learning capabilities in different species has suggested that learning has developed to provide organisms with an improved coping mechanism against adverse environments. Thus the ability to learn is encoded within the genome.

Over the years, a biochemical/cellular concept of memory has emerged that can be summarized as a dialogue between genes and synapses (Kandel 2001). Central to this concept is synaptic transmission, which is the release of, for example, excitatory neurotransmitters (axon potential-inducing) from a presynaptic release site. Transformation of an axon potential to neurotransmitter release takes place at presynaptic sites and is a unidirectional process. Two major transmitter systems in the CNS are predominant: glutamate, released from excitatory glutamatergic neurons, and γ-aminobutyric acid (GABA), released from inhibitory GABAergic neurons. Glutamate can direct the opening of Na^+ ion channels at the postsynaptic site, inducing an excitatory neuronal response. In contrast, GABA directs the opening of Cl^- channels, thus inducing an inhibitory response. Activity-dependent release of glutamate from presynaptic sites leads to the activation of AMPA receptors and to the depolarization of the postsynaptic neuron. Depolarization can occur by action potentials in the millisecond range locally at the synapse. Depolarization of postsynaptic neurons leads to the removal of NMDA receptor inhibition by Mg^{++} and then to Ca^{2+} influx through the receptor. This activates voltage-gated calcium channels (VGCCs), another source of synaptic Ca^{2+} ions. This short-lived membrane depolarization process can be transformed to changes in gene expression and to long-term morphological changes such as synaptic plasticity, leading to additional or more efficient synapses (Lamprecht and LeDoux 2004). It is thought that by the modification of synapses the internal

representations are modified and thus memory is enhanced, retained, or lost.

The role of NF-κB in the nervous system has gained interest because of its involvement in synaptic processes, neurotransmission, and neuroprotection. Furthermore, inducible NF-κB plays a crucial role in brain inflammation and neural stem cell proliferation. The role of NF-κB in the cellular context of the nervous system is depicted in Figure 1.

All DNA-binding subunits of NF-κB have been detected within the CNS, and in the adult rodent brain the major DNA-binding complexes are p50/p65 (Kaltschmidt et al. 1993; Bakalkin et al. 1993; Schmidt-Ullrich et al. 1996; Meffert et al. 2003). In contrast, in the developing nervous system complexes consisting of cRel/p65, p50/p65, and p50 homodimers were reported (Bakalkin et al. 1993).

NEURONS

Initially, constitutive (high basal) NF-κB activity was found in glutamatergic neurons of the CNS, such as the hippocampus (granule cells and pyramidal neurons of CA1 and CA3) and cerebral cortex (layers 2, 4, and 5), by antibody staining (Kaltschmidt et al. 1993; Kaltschmidt et al. 1994; Kaltschmidt et al. 1995), gelshift assays, and Western blotting. Transgenic reporter mouse models verified these data and also showed constitutive NF-κB activity in several rodent brain regions such as the cerebral cortex, hippocampus, amygdala, olfactory lobes, cerebellum, and hypothalamus (Schmidt-Ullrich et al. 1996; Bhakar et al. 2002). Here, constitutive NF-κB activity shall be defined as activity that makes cells blue in transgenic reporter mice (mainly in neurons, but also endothelial cells). Constitutive NF-κB activity can be suppressed by pharmacological inhibitors such as glutamate antagonists and L-type Ca^{2+} channel blockers (Lilienbaum and Israel 2003; Meffert et al. 2003). This suggests that constitutive NF-κB results from physiological basal synaptic transmission. Furthermore, constitutive NF-κB activity in endothelial cells, the roof plate (a dorsal signaling center of the developing nervous

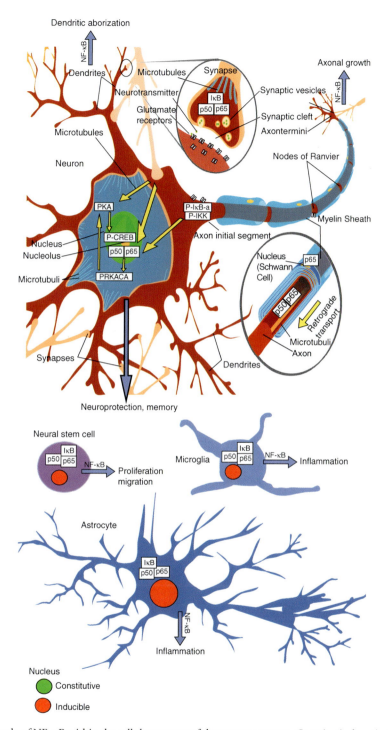

Figure 1. The role of NF-κB within the cellular context of the nervous system. Constitutively activated NF-κB is detected mostly in glutamatergic neurons (green nuclei), whereas NF-κB in glia has a lower basal activity and is heavily inducible (red nuclei). For details, see text.

system), and floor plate (part of the neural tube) is dependent on TRAF6, as shown by crossing lacZ reporter mice with TRAF6 $-/-$ mice (Dickson et al. 2004). Inducible NF-κB was detected in biochemically purified synapses (synaptosomes) (Kaltschmidt et al. 1993; Meberg et al. 1996; Meffert et al. 2003). Interestingly, analysis of TNFRI $-/-$, p65 $-/-$ double knockout mice showed no presence of p65, p50, and IκB α and β in synapses. These data suggest that p65 is the driving subunit for synaptic localization and transport processes. It cannot be replaced by alternative subunits such as c-Rel or RelB.

A p65/GFP fusion protein can be retrogradely transported from active synaptic sites back to the nucleus (Wellmann et al. 2001; Meffert et al. 2003) after glutamatergic stimulation. This retrograde transport is dependent on the p65 nuclear localization signal (NLS) and involves a dynein–dynactin motor protein complex gliding on microtubules (Mikenberg et al. 2007; Shrum et al. 2009). p65 NLS is also important for the contact of p65 and the motor protein complexes. In summary, these observations indicate that NF-κB has a central role in translating short-term synaptic events into changes in gene expression.

LEARNING AND MEMORY

Genetic evidence[*] for the involvement of NF-κB in learning and memory was provided for the first time in TNFRI $-/-$, p65 $-/-$ mice by Meffert et al. 2003. In a radial arm maze, TNFRI $-/-$, p65 $-/-$ mice made significantly more trial errors than control mice. A second, conditional mouse model uses a neuronal promoter-specific ablation of NF-κB in the basal forebrain (CamKII-tTA/ tetOtnIκB-α). In this model, all NF-κB subunits were repressed in vivo as measured by triple

---

*This review focuses on evidence from genetic mouse models. Readers interested in pharmacological approaches to NF-κB function in the nervous system should consult earlier reviews by Kaltschmidt et al. (2005), Meffert and Baltimore (2005), and Romano et al. (2006) for further information.

transgenic lacZ reporter mice in hippocampal neurons (Fridmacher et al. 2003). These mice showed impairments in spatial learning. The mice had to learn the position of a platform useful for resting that was submersed within a water basin (Morris water maze). Mice expressing the super-repressor (tnIκB-α) showed reduced long-term potentiation (LTP) and long term depression (LTD) (Kaltschmidt et al. 2006). Furthermore, learning during training sessions in a Morris water maze strongly induced NF-κB binding activity in hippocampal extracts of wild-type mice (O'Mahony et al. 2006). To obtain clues about NF-κB-regulated changes in gene expression, transcriptome profiling was performed. The catalytic subunit α of the protein kinase A gene (PRKACA) was identified as a novel NF-κB target gene (Kaltschmidt et al. 2006). Recent research has shown that protein kinase A (PKA)–CREB signaling is an essential pathway for learning and memory (Kandel 2001). Connected with this, repression of NF-κB in transgenic mice resulted in reduced PKA activity, leading to a strong reduction in forskolin-induced CREB phosphorylation (forskolin activates adenylcylase and thus raises intracellular cAMP levels). A functional NF-κB binding element was identified within intron 2 of mouse and human PKA genes (Kaltschmidt et al. 2006). These findings identified a novel transcriptional signaling cascade in neurons in which NF-κB regulates the PKA/CREB pathway function in learning and memory. Synapse density is an important parameter for learning processes. After NF-κB ablation in the hippocampus, the density of synapses was strongly reduced within a hippocampal subfield, the stratum lucidum (Kaltschmidt et al., unpubl.). In addition, in mice with neuronal NF-κB ablation the formation of axonal mossy fiber projections was impaired (Kaltschmidt et al. unpublished). These fibers normally connect granule cells with CA3 pyramidal cells. To our knowledge, this is the first in vivo model showing NF-κB dependent structural plasticity.

Furthermore, dendritic arborization is regulated by NF-κB in peripheral and cortical neurons (Gutierrez et al. 2005). Thus both the receiving structure (dendrite) and the sending structure (axon) seem to be regulated by NF-κB. Current literature suggests that dendritic and axonal outgrowth can be regulated by NF-κB in different ways in different neurons. In cortical neurons, dendritic arborization is repressed by IκB (Gutierrez et al. 2005), whereas in hippocampal granule cells the axonal outgrowth depends on NF-κB activity (Kaltschmidt et al. unpublished). An important region of neuronal information processing is the axon initial segment (AIS). This region is extremely rich in membrane-embedded voltage-gated sodium channels and is the axon potential generating region. Interestingly, the phosphorylated forms of IKK-1, IKK-2, and IκB-α are concentrated within the AIS (Schultz et al. 2006). The node of Ranvier is also enriched in action potential generating sodium channels and also in phosphorylated forms of IKK 1, 2, and IκB-α (Politi et al. 2008). The localization of phosphorylated NF-κB regulators might indicate the major sites of intracellular NF-κB activation. Pharmacological blockade of IKK function in cultured hippocampal neurons interfered with the localization of phosphorylated IκB-α and IKK within the AIS (Sanchez-Ponce et al. 2008). In summary, these data show that NF-κB is an important regulator of neuronal morphology and shapes brain structures that are important for learning and memory.

A critical function of NF-κB in inhibitory GABAergic interneurons was observed in a recent transgenic mouse model (O'Mahony et al. 2006). Cell-type specific expression of super-repressor IκB-α (Prion promoter-driven tTA/tetO super-repressor IκB-α) in both inhibitory GABAergic interneurons (robust expression) and hippocampal excitatory glutamatergic neurons (low level expression) resulted in a phenotype opposite to that resulting from the inhibition of NF-κB in glutamatergic neurons only (Kaltschmidt et al. 2006). Expression of glutamate decarboxylase (GAD65), a rate-limiting enzyme required for the synthesis of the inhibitory neurotransmitter, GABA, was down-regulated in these mice. Super-repressor IκB mice completed a radial maze in less time

and made fewer errors than control mice, indicating enhanced spatial learning and memory. Furthermore, LTP was enhanced. These findings might be explained by a model in which excitatory neurons are the motor of learning and GABAergic neurons provide the brake. A blockade of the brake (by expression of the super-repressor) would result in overactivation of the motor (excitatory neurons) and would explain the enhanced learning. Consistent with this is the observation that Baclofen, a pharmacological agonist of GABA, negatively affected learning (McNamara and Skelton 1996). A major advantage of super-repressor expression is the inhibition of all NF-κB subunits. On the other hand, overexpression might have unexpected gain of function effects. The p50−/− mice had defects in novel task acquisition (Kassed et al. 2002), decreased anxiety (Kassed and Herkenham 2004), and reduced short-term memory (Denis-Donini et al. 2008). Homodimers of p50, which lack a transactivation domain, might act as a repressor of gene transcription on a set of specific promoters. Thus, some of the effects observed in p50−/− animals might be in addition to the loss of p50 function, a result of the derepression of NF-κB target genes.

The function of NF-κB was analyzed in fear-conditioning paradigms, a form of learning in which an unconditioned stimulus like a tone is coupled to noxious stimulation such as electric shock. Fear-conditioning seems to depend on the amygdala, a region in which constitutive NF-κB activity has already been reported. Moreover, a c-Rel knockout was analyzed in a cued fear-conditioning paradigm and analyzed 24 h later. No amygdala-dependent changes in long-term memory were reported in c-Rel knockout animals (Levenson et al. 2004). The presentation of landmarks in the fear-conditioning paradigm gave quite different results, with deficits in freezing being observed. In addition, c-Rel−/− mice showed lower activity in an open field test. Bioinformatic analysis suggested an overrepresentation of NF-κB binding sites in the promoters of genes potentially involved in memory consolidation (Levenson et al. 2004), but this was not tested experimentally with, for example, ChIP. Likewise, a recent bioinformatic study has suggested an enrichment of NF-κB and E2F binding sites in genes potentially important for the development of neurons from neural precursors, and thus additional experiments are necessary (Greco et al. 2008).

Recently, a pharmacological study using DDTC (Diethyldithiocarbamate) and SN50 (a cyclic peptide, spanning the NLS of p50) showed impaired memory reconsolidation (Lubin and Sweatt 2007).

Overall, the results of behavioral analyses of mice with reduced NF-κB activity in brain (see Table 1) can be summarized as follows: Deletion of DNA-binding subunits in all cell types, including neurons and glia, resulted in lower performance in different behavioral tests. This phenotype was also seen following repression

of NF-κB in glutamatergic neurons (CamKII tTA / tetO super-repressor), suggesting that NF-κB function in glutamatergic neurons is responsible for learning and memory. On the other hand, changing the balance of glutamatergic and GABAergic neurons by the higher expression of super-repressor in GABAergic neurons enhances the learning of spatial clues.

HYPOTHALAMUS

Previous reports analyzing transgenic NF-κB lacZ reporter mice have described constitutive NF-κB activity within the hypothalamus (Schmidt-Ullrich et al. 1996; Bhakar et al. 2002). The hypothalamus contains the dominant neuroendocrine center for the control of food intake and energy expenditure. Recently,

Table 1. Genetic mouse models interfering with NF-κB activity in the nervous system

Genotype	Cell type afflicted	Cognitive defect	Additional phenotype	References
p50−/−	All	Defect in novel task acquisition; decreased anxiety; reduced short-term memory	Reduced neuroprotection; hearing loss; reduced neurogenesis; reduced ischemic damage; impaired acute and inflammatory nociception	Yu et al. 1999; Yu et al. 2000; Kassed et al. 2002; Kassed and Herkenham 2004; Duckworth et al. 2006; Denis-Donini et al. 2008; Schneider et al. 1999, Niederberger et al. 2007
p65−/−	Isolated sensory neurons	na[*]	Reduced neuroprotection	Middleton et al. 2000
p65−/−	Isolated Schwann cells	na[*]	Reduced myelination of peripheral nerves;	Nickols et al. 2003
p65−/− /TnfrI−/−	All	Delayed spatial learning in radial maze	No synaptic NF-κB	Meffert et al. 2003
CamKII tTA/tetO super-repressor IκB-α	Glutamatergic forebrain neurons	Impairments in spatial memory; reduced LTP and LTD	Reduced neuroprotection; decreased PKA expression and P-CREB	Fridmacher et al. 2003; Kaltschmidt et al. 2006
Prion-tTA/tetO super-repressor IκB-α	Glutamatergic and inhibitory neurons	Enhanced spatial learning; enhanced LTD	Reduced GAD65 expression	O'Mahony et al. 2006

(Continued)

Table 1. *Continued*

Genotype	Cell type afflicted	Cognitive defect	Additional phenotype	References
GFAP-super-repressor IκB-α	Glia: astrocytes	Deficits in learning only in females: delayed spatial learning, impaired cued fear memory	LTP reduced in females; LTP enhanced in males; reduction of mGluR5 in females; better recovery after spinal cord injury; reduced pain sensitivity	Bracchi-Ricard et al. 2008; Brambilla et al. 2005
c-Rel−/−	All	Impaired late phase LTD; impaired long-term memory; impaired cued fear memory	Reduced neuroprotection	Pizzi et al. 2002; Levenson et al. 2004; Ahn et al. 2008
LysM-Cre/IKK-2$^{FL/FL}$	Glia: microglia; macrophages	na*	30% reduction of neuronal death: 10-fold reduced infarct size after MCAO	Cho et al. 2008
Nestin-Cre/IKK-2$^{FL/FL}$	Neural (glia and neuron)	na*	25% reduction of infarct size after MCAO; amelioration of EAE	Herrmann et al. 2005; van Loo et al. 2006
Nestin-Cre/Ikk-1$^{FL/FL}$	Neural (glia and neuron)	na*	No effect on EAE	van Loo et al. 2006
Nestin-Cre/Nemo$^{FL/FL}$	Neural (glia and neuron)	na*	Amelioration of EAE	van Loo et al. 2006
NSE-SR-IκB-α	Neuronal	na*	Improved LPS-induced hypothermia and survival	Juttler et al. 2007

na*: not analyzed.

it was reported that a high fat chow increased the already activated NF-κB in the hypothalamus neurons two- to fourfold (De Souza et al. 2005). Zhang and coworkers reported high expression of IKK-2 and IκB-α in neurons of the mediobasal hypothalamus, a brain region involved in nutrition sensing. High fat chow, acute administration of glucose, or intraventricular injection of oleic acid hyper-activated NF-κB two- to fourfold (Zhang et al. 2008). Tissue-specific knockout of IKK-2 with Nestin CRE mice or virally transmitted CRE (lentivirus or adenovirus) injected into the hypothalamus reduced food consumption and weight gain in transgenic mice fed with high fat chow. In contrast, the neural expression of a constitutive form of IKK-2 enhanced weight gain after a high-fat diet and impaired hypothalamic sensitivity to insulin and leptin. Ablation of IKK-2 in specific AGRP hypothalamic neurons protected against induction of central insulin and leptin resistance after a high-fat diet. Zhang and coworkers identified SOCS3 (a suppressor of cytokine signaling) as a novel NF-κB target gene and a crucial regulator of diet-induced obesity. Hypothalamic neurons are now also considered to form a key center involved in sleep regulation (Mignot et al. 2002). Sleep-inducing substances include proinflammatory cytokines such as TNF and IL-1. TNFR1 knockout animals cannot sleep in response to TNF-α application (Fang et al. 1997).

A unique population of neurons immunoreactive for the p65 subunit of NF-κB was previously localized within the caudal dorsolateral hypothalamus of rats.

In relation to this, analysis of NF-κB-driven lacZ reporter mice has shown that sleep deprivation increases the number of cells expressing NF-κB-dependent β-galactosidase in the magnocellular lateral hypothalamus and zona incerta dorsal, as well as the adjacent subthalamus in transgenic mice (Brandt et al. 2004). Intracerebral injection of cell permeable NF-κB-inhibiting peptide spanning the NLS of p50 (SN50) inhibited IL-1-induced sleep in rats and rabbits (Kubota et al. 2000). LPS-induced hypothermia activated κB lacZ expression in brain, and neuronal expression of the IkB[α] super-repressor suppressed hypothermia and increased survival (Juttler et al. 2007).

NEUROPROTECTION

Initially observed in cerebellar granule neurons, a subtoxic dose of a neurotoxin (Aβ or Fe^{++}) led to long-lasting NF-κB activation. This was protective against a higher dose of neurotoxins (Kaltschmidt et al. 1999; Kaltschmidt et al. 2002). This process was described by the 16th-century physician Theophrastus Bombastus von Hohenheim (Paracelsus): "All Ding sein Gift allein die Dosis machts" (all things are poisons only the dose is important). Preconditioning or hormesis (Mattson 2008) was dependent on activated NF-κB and might be a subcellular vaccination strategy, as suggested by D. Baltimore (Baltimore 1988). Further in vitro studies provided evidence that activation of NF-κB can protect neurons against amyloid β peptide toxicity (Barger et al. 1995) and excitotoxic or oxidative stress (Goodman and Mattson 1996; Mattson et al. 1997). Adenoviruses expressing super-repressor or dominant-negative NIK reduced survival of cortical neurons, whereas overexpression of p65 protected cortical neurons against apoptotic cell death induced by etopside (Bhakar et al.

2002). In vivo experiments in models of brain preconditioning with kainate or ischemia or linolenic acid showed NF-κB dependent protection against neuronal death (Blondeau et al. 2001). Later, transgenic inhibition of NF-κB by neuronal overexpression of the IkB[α] super-repressor decreased neuroprotection after kainic acid or Fe^{++} application (Fridmacher et al. 2003). In hippocampal acute slices derived from cRel$-/-$ mice treatment with NMDA and IL-1β increased neurodegeneration (Pizzi et al. 2002). Surprisingly, under wild-type conditions no c-Rel containing DNA binding complexes were detected in adult brain (Kaltschmidt et al. 1993; Schmidt-Ullrich et al. 1996; Bakalkin et al. 1993; Meffert et al. 2003). Only activation of metabotropic glutamate receptors with, for example, 1 mM (R,S)-2-chloro-5-hydroxyphenylglycine (CHPG) uncovered the protective action of c-Rel, whereas RNA against the c-Rel or c-Rel$-/-$ genotype had no cell death enhancing effect in neurons without pharmacological activation of the metabotropic glutamate receptor (Pizzi et al. 2005). Injection of neurotoxins in p50$-/-$ mice such as trimethyltin (Kassed et al. 2004), excitotoxins (Yu et al. 1999), or mitochondrial toxin 3-nitropropionic acid (Yu et al. 2000) increased neuronal damage. Furthermore, p50$-/-$ animals suffered from hearing loss because of degeneration of spiral ganglion neurons (Lang et al. 2006). Ischemia induced in p50$-/-$ mice showed a clear neuroprotective role of NF-κB in the hippocampus and striatum, in which degenerating neurons were detected 4 d after 1 h of ischemia (MCAO paradigm) (Duckworth et al. 2006). Degenerating neurons did not show NF-κB-dependent reporter gene expression, and in addition p50$-/-$ mice experienced a 2.4-fold increase in postoperative death in comparison to controls (Duckworth et al. 2006). At a first glance, these data appear to conflict with reports of reduced infarct volume in p50$-/-$ mice and a 25% reduction in models with neuronal- or brain-specific IKK-2 ablation (Schneider et al. 1999). However, a recent analysis showed a 10-fold reduced infarct size in mice with microglial-specific IKK-2 ablation (Cho et al.

2008). Thus NF-κB-dependent microglia activation might be a crucial contributor to ischemia. Unfortunately, neuroprotective strategies developed in animal models do not work in human stroke (Rother 2008). There could be many reasons for this, such as translation arrest and/or increased vulnerability in human white matter. A summary of the neuroprotective effects observed in genetic models is presented in Table 1.

GLIA

No constitutive NF-κB activity was detected in glial cells of untreated animals (Schmidt-Ullrich et al. 1996; Bhakar et al. 2002), suggesting that inducible NF-κB activity might be linked to pathological events. Initially, it was reported that NF-κB activated in astrocytes via amyloid β peptide (Aβ) led to the production of nitric oxide (Akama et al. 1998). This astroglial neurodegenerative role of Aβ contrasts with the neuroprotective activation of NF-κB in neurons by nanomolar concentrations of Aβ (see previous discussion) (Kaltschmidt et al. 1997; Kaltschmidt et al. 1999). GFAP promoter-driven super-repressor IκB expression was present in astrocytes of brain, spinal cord, and peripheral nerves (Brambilla et al. 2005). Spinal cord contusion injury increased GFAP expression in activated astrocytes at the lesion site. Activated astrocytes release inflammatory mediators such as chemokines and cytokines. Unexpectedly, expression of the NF-κB target gene IL-6 was significantly up-regulated in GFAP super-repressor mice, whereas TNF expression remained unchanged (Brambilla et al. 2005). However, the expression of the chemokines CXCL10 and CCL2 was significantly down-regulated in comparison to wild-type mice after spinal-cord injury in GFAP SR mice. Outcome after spinal-cord injury, as measured by locomoter activity, was significantly ameliorated, suggesting that inducible astrocytic NF-κB does not function in cell survival but exacerbates chemokine-driven defects in spinal-cord recovery. Likewise, NF-κB inhibition resulted in reduced glial scarring,

which perhaps allows better axon regeneration (Brambilla et al. 2005). Axotomy of the entorhinal perforant path projection resulted in astrocytic STAT2 up-regulation and phosphorylation and concomitant expression of CCL2. Using the same mouse model (GFAP-SR), it was shown that the lesion-induced expression of CCL2 and STAT2 was significantly blunted (Khorooshi et al. 2008). These results suggest that NF-κB signaling in astrocytes might regulate chemokine-induced infiltration of immune cells into the lesioned brain. Similarly, Nemo and IKK-2 deletion within the CNS resulted in a significant reduction of infiltration of inflammatory cells (van Loo et al. 2006). Experimental autoimmune encephalomyelitis (EAE), a model of human multiple sclerosis, was ameliorated after Nemo or IKK-2 deletion in astrocytes. A 10-fold reduction of nuclear p65 in astrocytes after IL-1 or TNF treatment resulted in a more than 10% reduction in chemokine production (CXCL10). These data underscore the pivotal role of astrocytes in EAE and the recruitment of infiltrating inflammatory cells. The role of microglia in neuroprotection or neurodegeneration is heavily debated. Recently, analysis of LysM-Cre/IKK$^{FL/}$^{FL} mice showed 36% deletion of the IKK β gene in cultured neonatal microglia (Cho et al. 2008), but only 4% deletion in adult microglia. Kainic acid injection resulted in the deletion of IKK-2 in 73% of microglial cells. Kainic acid-induced cell death was reduced, presumably by decreased activation of NF-κB-regulated microglial proinflammatory genes, such as TNF-α and IL-1β. A remarkable result was the 10-fold reduction of the infarct area after MCAO (Cho et al. 2008). In summary, IKK-2-mediated microglia activation potentiated neuronal excitotoxicity.

PAIN

Recently, a role of glial NF-κB in pain has emerged. Pain can arise from the activation of specific high-threshold PNS neurons (nociceptors) and could serve as a sensing mechanism to prevent further damage. However, clinical pain can arise from damage to the nervous system (neuropathic pain) or chronic inflammation (inflammatory pain). Analysis of p50$-$/$-$ mice revealed an impairment of acute and inflammatory nociception (Niederberger et al. 2007). Recent data suggest an important role of astroglial NF-κB in pain perception: Inhibition of the expression of pain mediators (TNF-α, IL-6, and iNOS) by lentiviral delivery of IκB-α super-repressor injected into the dorsal horn reduced pain (Meunier et al. 2007). Similarly, GFAP super-repressor mice have decreased formalin pain sensation (Fu et al. 2007).

NEURAL STEM CELLS

During adulthood, neural stem cells continue to proliferate in the subventricular zone (SVZ) and within the hilus of the hippocampus. In adult neural stem cells isolated from the SVZ, p65 regulates proliferation via NF-κB target genes c-myc and cylin D1 (Widera et al. 2006). Furthermore, proliferation was strongly inhibited in neural stem cells prepared from the ganglionic eminence of p50$-$/$-$ p65$-$/$-$ embryos (Young et al. 2006). Expression of p65 and p50 persists into adulthood, particularly in subventricular zone astrocyte-like cells and in migrating neuronal precursors, respectively.

In particular, p65 and p50 are expressed in radial glial cells, in migrating neuronal precursors, and in a population belonging to the astrocytic lineage (Shingo et al. 2001). RelB, on the other hand, is only expressed in migrating neuronal precursors, whereas c-Rel is present in a few cells located at the edges of the rostral migratory stream (Denis-Donini et al. 2005).

In p50$-$/$-$ animals, no defects in proliferation of hippocampal stem cells was detected (Denis-Donini et al. 2008). However, survival of neural progenitors after 21 days post BrdU injection was significantly reduced in p50$-$/$-$ animals.

Taken together, current literature suggests that NF-κB fulfils very different functions in different cell types. A lot of controversial results were obtained in the last decade by neglecting the cell type-specific effect of NF-κB and the interplay between different neuronal cell types with glia.

In this line, the generation and analysis of sophisticated cell type-specific knockout models might be necessary to unravel all of the mysteries of NF-κB in the nervous system.

ABBREVIATIONS

PNS: peripheral nervous system
CNS: central nervous system
MCAO: middle cerebral artery occlusion
EAE: experimental autoimmune
        encephalomyelitis
VGCC: voltage-gated calcium channels
GABA: γ-aminobutyric acid
AIS: axon initial segment
ChIP: chromatin immunoprecipitation
LTD: long term depression
LTP: long term potentiation

REFERENCES

Ahn HJ, Hernandez CM, Levenson JM, Lubin FD, Liou HC, Sweatt JD. 2008. c-Rel, an NF-κB family transcription factor, is required for hippocampal long-term synaptic plasticity and memory formation. *Learn Mem* **15:** 539–549.

Akama KT, Albanese C, Pestell RG, Van Eldik LJ. 1998. Amyloid β-peptide stimulates nitric oxide production in astrocytes through an NFκB-dependent mechanism. *Proc Natl Acad Sci* **95:** 5795–5800.

Bakalkin G, Yakovleva T, Terenius L. 1993. NF-κ B-like factors in the murine brain. Developmentally-regulated and tissue-specific expression. *Brain Res Mol Brain Res* **20:** 137–146.

Baltimore D. 1988. Gene therapy. Intracellular immunization. *Nature* **335:** 395–396.

Barger SW, Horster D, Furukawa K, Goodman Y, Krieglstein J, Mattson MP. 1995. Tumor necrosis factors α and β protect neurons against amyloid β-peptide toxicity: Evidence for involvement of a κ B-binding factor and attenuation of peroxide and Ca^{2+} accumulation. *Proc Natl Acad Sci* **92:** 9328–9332.

Bhakar AL, Tannis LL, Zeindler C, Russo MP, Jobin C, Park DS, MacPherson S, Barker PA. 2002. Constitutive nuclear factor-κ B activity is required for central neuron survival. *J Neurosci* **22:** 8466–8475.

Blondeau N, Widmann C, Lazdunski M, Heurteaux C. 2001. Activation of the nuclear factor-κB is a key event in brain tolerance. *J Neurosci* **21:** 4668–4677.

Bracchi-Ricard V, Brambilla R, Levenson J, Hu WH, Bramwell A, Sweatt JD, Green EJ, Bethea JR. 2008. Astroglial nuclear factor-κB regulates learning and memory and synaptic plasticity in female mice. *J Neurochem* **104:** 611–623.

Brambilla R, Bracchi-Ricard V, Hu WH, Frydel B, Bramwell A, Karmally S, Green EJ, Bethea JR. 2005. Inhibition of astroglial nuclear factor κB reduces inflammation and improves functional recovery after spinal cord injury. *J Exp Med* **202:** 145–156.

Brandt JA, Churchill L, Rehman A, Ellis G, Memet S, Israel A, Krueger JM. 2004. Sleep deprivation increases the activation of nuclear factor κ B in lateral hypothalamic cells. *Brain Res* **1004:** 91–97.

Cho IH, Hong J, Suh EC, Kim JH, Lee H, Lee JE, Lee S, Kim CH, Kim DW, Jo EK, et al. 2008. Role of microglial IKKβ in kainic acid-induced hippocampal neuronal cell death. *Brain* **131:** 3019–3033.

De Souza CT, Araujo EP, Bordin S, Ashimine R, Zollner RL, Boschero AC, Saad MJ, Velloso LA. 2005. Consumption of a fat-rich diet activates a proinflammatory response and induces insulin resistance in the hypothalamus. *Endocrinology* **146:** 4192–4199.

Denis-Donini S, Caprini A, Frassoni C, Grilli M. 2005. Members of the NF-κB family expressed in zones of active neurogenesis in the postnatal and adult mouse brain. *Brain Res Dev Brain Res* **154:** 81–89.

Denis-Donini S, Dellarole A, Crociara P, Francese MT, Bortolotto V, Quadrato G, Canonico PL, Orsetti M, Ghi P, Memo M, et al. 2008. Impaired adult neurogenesis associated with short-term memory defects in NF-κB p50-deficient mice. *J Neurosci* **28:** 3911–3919.

Dickson KM, Bhakar AL, Barker PA. 2004. TRAF6-dependent NF-kB transcriptional activity during mouse development. *Dev Dyn* **231:** 122–127.

Duckworth EA, Butler T, Collier L, Collier S, Pennypacker KR. 2006. NF-κB protects neurons from ischemic injury after middle cerebral artery occlusion in mice. *Brain Res* **1088:** 167–175.

Dudai Y. 1989. *The Neurobiology of Memory.* Oxford University Press, New York.

Fang J, Wang Y, Krueger JM. 1997. Mice lacking the TNF 55 kDa receptor fail to sleep more after TNFα treatment. *J Neurosci* **17:** 5949–5955.

Fridmacher V, Kaltschmidt B, Goudeau B, Ndiaye D, Rossi FM, Pfeiffer J, Kaltschmidt C, Israel A, Memet S. 2003. Forebrain-specific neuronal inhibition of nuclear factor-κB activity leads to loss of neuroprotection. *J Neurosci* **23:** 9403–9408.

Fu ES, Zhang YP, Sagen J, Yang ZQ, Bethea JR. 2007. Transgenic glial nuclear factor-κ B inhibition decreases formalin pain in mice. *Neuroreport* **18:** 713–717.

Goodman Y, Mattson MP. 1996. Ceramide protects hippocampal neurons against excitotoxic and oxidative insults, and amyloid β-peptide toxicity. *J Neurochem* **66:** 869–872.

Greco D, Somervuo P, Di Lieto A, Raitila T, Nitsch L, Castren E, Auvinen P. 2008. Physiology, pathology and relatedness of human tissues from gene expression meta-analysis. *PLoS ONE* **3:** e1880.

Gutierrez H, Hale VA, Dolcet X, Davies A. 2005. NF-κB signalling regulates the growth of neural processes in the developing PNS and CNS. *Development* **132:** 1713–1726.

Herrmann O, Baumann B, de Lorenzi R, Muhammad S, Zhang W, Kleesiek J, Malfertheiner M, Kohrmann M, Potrovita I, Maegele I, et al. 2005. IKK mediates ischemia-induced neuronal death. *Nat Med* **11:** 1322–1329.

Juttler E, Inta I, Eigler V, Herrmann O, Maegele I, Maser-Gluth C, Schwaninger M. 2007. Neuronal NF-κB influences thermoregulation and survival in a sepsis model. *J Neuroimmunol* **189:** 41–49.

Kaltschmidt C, Kaltschmidt B, Baeuerle PA. 1993. Brain synapses contain inducible forms of the transcription factor NF-κ B. *Mech Dev* **43:** 135–147.

Kaltschmidt C, Kaltschmidt B, Neumann H, Wekerle H, Baeuerle PA. 1994. Constitutive NF-κ B activity in neurons. *Mol Cell Biol* **14:** 3981–3992.

Kaltschmidt C, Kaltschmidt B, Baeuerle PA. 1995. Stimulation of ionotropic glutamate receptors activates transcription factor NF-κ B in primary neurons. *Proc Natl Acad Sci* **92:** 9618–9622.

Kaltschmidt B, Uherek M, Volk B, Baeuerle PA, Kaltschmidt C. 1997. Transcription factor NF-κB is activated in primary neurons by amyloid β peptides and in neurons surrounding early plaques from patients with Alzheimer disease. *Proc Natl Acad Sci* **94:** 2642–2647.

Kaltschmidt B, Uherek M, Wellmann H, Volk B, Kaltschmidt C. 1999. Inhibition of NF-κB potentiates amyloid β-mediated neuronal apoptosis. *Proc Natl Acad Sci* **96:** 9409–9414.

Kaltschmidt B, Heinrich M, Kaltschmidt C. 2002. Stimulus-dependent activation of NF-κB specifies apoptosis or neuroprotection in cerebellar granule cells. *Neuromolecular Med* **2:** 299–309.

Kaltschmidt B, Widera D, Kaltschmidt C. 2005. Signaling via NF-κB in the nervous system. *Biochim Biopyhs Acta* **1745:** 287–299.

Kaltschmidt B, Ndiaye D, Korte M, Pothion S, Arbibe L, Prullage M, Pfeiffer J, Lindecke A, Staiger V, Israel A, et al. 2006. NF-κB regulates spatial memory formation and synaptic plasticity through protein kinase A/CREB signaling. *Mol Cell Biol* **26:** 2936–2946.

Kandel ER. 2001. The molecular biology of memory storage: A dialogue between genes and synapses. *Science* **294:** 1030–1038.

Kassed CA, Herkenham M. 2004. NF-κB p50-deficient mice show reduced anxiety-like behaviors in tests of exploratory drive and anxiety. *Behav Brain Res* **154:** 577–584.

Kassed CA, Willing AE, Garbuzova-Davis S, Sanberg PR, Pennypacker KR. 2002. Lack of NF-κB p50 exacerbates degeneration of hippocampal neurons after chemical exposure and impairs learning. *Exp Neurol* **176:** 277–288.

Kassed CA, Butler TL, Patton GW, Demesquita DD, Navidomskis MT, Memet S, Israel A, Pennypacker KR. 2004. Injury-induced NF-κB activation in the hippocampus: Implications for neuronal survival. *FASEB J* **18:** 723–724.

Khorooshi R, Babcock AA, Owens T. 2008. NF-κB-driven STAT2 and CCL2 expression in astrocytes in response to brain injury. *J Immunol* **181:** 7284–7291.

Kubota T, Kushikata T, Fang J, Krueger JM. 2000. Nuclear factor-κB inhibitor peptide inhibits spontaneous and interleukin-1β-induced sleep. *Am J Physiol Regul Integr Comp Physiol* **279:** R404–413.

Lamprecht R, LeDoux J. 2004. Structural plasticity and memory. *Nat Rev Neurosci* **5:** 45–54.

Lang H, Schulte BA, Zhou D, Smythe N, Spicer SS, Schmiedt RA. 2006. Nuclear factor κB deficiency is associated with auditory nerve degeneration and increased noise-induced hearing loss. *J Neurosci* **26:** 3541–3550.

Levenson JM, Choi S, Lee SY, Cao YA, Ahn HJ, Worley KC, Pizzi M, Liou HC, Sweatt JD. 2004. A bioinformatics analysis of memory consolidation reveals involvement of the transcription factor c-rel. *J Neurosci* **24:** 3933–3943.

Lilienbaum A, Israel A. 2003. From calcium to NF-κ B signaling pathways in neurons. *Mol Cell Biol* **23:** 2680–2698.

Lubin FD, Sweatt JD. 2007. The IκB kinase regulates chromatin structure during reconsolidation of conditioned fear memories. *Neuron* **55:** 942–957.

Mattson MP. 2008. Hormesis defined. *Ageing Res Rev* **7:** 1–7.

Mattson MP, Goodman Y, Luo H, Fu W, Furukawa K. 1997. Activation of NF-κB protects hippocampal neurons against oxidative stress-induced apoptosis: Evidence for induction of manganese superoxide dismutase and suppression of peroxynitrite production and protein tyrosine nitration. *J Neurosci Res* **49:** 681–697.

McNamara RK, Skelton RW. 1996. Baclofen, a selective GABAB receptor agonist, dose-dependently impairs spatial learning in rats. *Pharmacol Biochem Behav* **53:** 303–308.

Meberg PJ, Kinney WR, Valcourt EG, Routtenberg A. 1996. Gene expression of the transcription factor NF-κB in hippocampus: Regulation by synaptic activity. *Brain Res Mol Brain Res* **38:** 179–190.

Meffert MK, Chang JM, Wiltgen BJ, Fanselow MS, Baltimore D. 2003. NF-κ B functions in synaptic signaling and behavior. *Nat Neurosci* **6:** 1072–1078.

Meffert MK, Baltimore D. 2005. Physiological functions for brain NF-κB. *Trends Neurosci* **28:** 37–43.

Meunier A, Latremoliere A, Dominguez E, Mauborgne A, Philippe S, Hamon M, Mallet J, Benoliel JJ, Pohl M. 2007. Lentiviral-mediated targeted NF-κB blockade in dorsal spinal cord glia attenuates sciatic nerve injury-induced neuropathic pain in the rat. *Mol Ther* **15:** 687–697.

Middleton G, Hamanoue M, Enokido Y, Wyatt S, Pennica D, Jaffray E, Hay RT, Davies AM. 2000. Cytokine-induced nuclear factor κ B activation promotes the survival of developing neurons. *J Cell Biol* **148:** 325–332.

Mignot E, Taheri S, Nishino S. 2002. Sleeping with the hypothalamus: Emerging therapeutic targets for sleep disorders. *Nat Neurosci* **5 Suppl:** 1071–1075.

Mikenberg I, Widera D, Kaus A, Kaltschmidt B, Kaltschmidt C. 2007. Transcription factor NF-κB is transported to the nucleus via cytoplasmic dynein/dynactin motor complex in hippocampal neurons. *PLoS ONE* **2:** e589.

Nickols JC, Valentine W, Kanwal S, Carter BD. 2003. Activation of the transcription factor NF-κB in Schwann cells is required for peripheral myelin formation. *Nat Neurosci* **6:** 161–167.

Niederberger E, Schmidtko A, Gao W, Kuhlein H, Ehnert C, Geisslinger G. 2007. Impaired acute and inflammatory nociception in mice lacking the p50 subunit of NF-κB. *Eur J Pharmacol* **559:** 55–60.

O'Mahony A, Raber J, Montano M, Foehr E, Han V, Lu SM, Kwon H, LeFevour A, Chakraborty-Sett S, Greene WC. 2006. NF-κB/Rel regulates inhibitory and excitatory neuronal function and synaptic plasticity. *Mol Cell Biol* **26:** 7283–7298.

Pizzi M, Goffi F, Boroni F, Benarese M, Perkins SE, Liou HC, Spano P. 2002. Opposing roles for NF-κ B/Rel factors p65 and c-Rel in the modulation of neuron survival elicited by glutamate and interleukin-1β. *J Biol Chem* **277:** 20717–20723.

Pizzi M, Sarnico I, Boroni F, Benarese M, Steimberg N, Mazzoleni G, Dietz GP, Bahr M, Liou HC, Spano PF. 2005. NF-κB factor c-Rel mediates neuroprotection elicited by mGlu5 receptor agonists against amyloid β-peptide toxicity. *Cell Death Differ* **12:** 761–772.

Politi C, Del Turco D, Sie JM, Golinski PA, Tegeder I, Deller T, Schultz C. 2008. Accumulation of phosphorylated I κB α and activated IKK in nodes of Ranvier. *Neuropathol Appl Neurobiol* **34:** 357–365.

Romano A, Freudenthal R, Merlo E, Routtenberg A. 2006. Evolutionarily-conserved role of the NF-kappaB transcription factor in neural plasticity and memory. *Eur J Neurosci* **24:** 1507–1516.

Rother J. 2008. Neuroprotection does not work! *Stroke* **39:** 523–524.

Sanchez-Ponce D, Tapia M, Munoz A, Garrido JJ. 2008. New role of IKK α/β phosphorylated I κ B α in axon outgrowth and axon initial segment development. *Mol Cell Neurosci* **37:** 832–844.

Schmidt-Ullrich R, Memet S, Lilienbaum A, Feuillard J, Raphael M, Israel A. 1996. NF-κB activity in transgenic mice: Developmental regulation and tissue specificity. *Development* **122:** 2117–2128.

Schneider A, Martin-Villalba A, Weih F, Vogel J, Wirth T, Schwaninger M. 1999. NF-κB is activated and promotes cell death in focal cerebral ischemia. *Nat Med* **5:** 554–559.

Schultz C, Konig HG, Del Turco D, Politi C, Eckert GP, Ghebremedhin E, Prehn JH, Kogel D, Deller T. 2006. Coincident enrichment of phosphorylated IκBα, activated IKK, and phosphorylated p65 in the axon initial segment of neurons. *Mol Cell Neurosci* **33:** 68–80.

Shingo T, Sorokan ST, Shimazaki T, Weiss S. 2001. Erythropoietin regulates the in vitro and in vivo production of neuronal progenitors by mammalian forebrain neural stem cells. *J Neurosci* **21:** 9733–9743.

Shrum CK, Defrancisco D, Meffert MK. 2009. Stimulated nuclear translocation of NF-κB and shuttling differentially depend on the dynactin complex. *PNAS* **106:** 2647–2652.

van Loo G, De Lorenzi R, Schmidt H, Huth M, Mildner A, Schmidt-Supprian M, Lassmann H, Prinz MR, Pasparakis M. 2006. Inhibition of transcription factor NF-κB in the central nervous system ameliorates autoimmune encephalomyelitis in mice. *Nat Immunol* **7:** 954–961.

Wellmann H, Kaltschmidt B, Kaltschmidt C. 2001. Retrograde transport of transcription factor NF-κ B in living neurons. *J Biol Chem* **276:** 11821–11829.

Widera D, Mikenberg I, Elvers M, Kaltschmidt C, Kaltschmidt B. 2006. Tumor necrosis factor α triggers proliferation of adult neural stem cells via IKK/NF-κB signaling. *BMC Neurosci* **7:** 64.

Young KM, Bartlett PF, Coulson EJ. 2006. Neural progenitor number is regulated by nuclear factor-κB p65 and p50 subunit-dependent proliferation rather than cell survival. *J Neurosci Res* **83:** 39–49.

Yu Z, Zhou D, Bruce-Keller AJ, Kindy MS, Mattson MP. 1999. Lack of the p50 subunit of nuclear factor-κB increases the vulnerability of hippocampal neurons to excitotoxic injury. *J Neurosci* **19:** 8856–8865.

Yu Z, Zhou D, Cheng G, Mattson MP. 2000. Neuroprotective role for the p50 subunit of NF-κB in an experimental model of Huntington's disease. *J Mol Neurosci* **15:** 31–44.

Zhang X, Zhang G, Zhang H, Karin M, Bai H, Cai D. 2008. Hypothalamic IKKβ/NF-κB and ER stress link overnutrition to energy imbalance and obesity. *Cell* **135:** 61–73.

Use of Forward Genetics to Discover Novel Regulators of NF-κB

Tao Lu and George R. Stark

Department of Molecular Genetics, Lerner Research Institute, Cleveland Clinic Foundation, Cleveland, Ohio 44195

Correspondence: starkg@ccf.org

Forward and reverse genetic experiments have both played important roles in revealing critical aspects of mammalian signal transduction pathways in cell culture experiments. Only recently have we begun to comprehend the depth, breadth, and complexity of these pathways and of their interrelationships. Here, we summarize successful examples in which different forward genetic approaches have led to novel discoveries in NF-κB signaling. We believe that forward genetics will continue to play an irreplaceable role in advancing our understanding of the complexities of the pathways that regulate the functions of this key transcription factor.

FORWARD GENETICS IN MAMMALIAN CELLS IN TISSUE CULTURE

Genetic approaches to investigate signaling mechanisms in cell culture fall into two broad categories: In forward genetics, one creates random mutations in a population of cells, whereas in reverse genetics, one manipulates a known gene in a single cell clone or in a population of similar clones. Both approaches have played critical roles in revealing the depth and complexity of mammalian signal transduction pathways. It is fair to say that only in recent years have we begun to comprehend the depth, breadth, and complexity of these pathways and of their interrelationships. Many of the intricate networks that provide sophisticated regulation of signaling pathways would have been very difficult to recognize or understand in the absence of powerful genetic

techniques. Forward genetics seeks to associate a specific protein with a biological phenotype in a pathway of interest without necessarily relying on any previous knowledge. Typical steps are: (1) Create cell libraries containing millions of random mutations; (2) apply a selective pressure or sorting technique to isolate rare cells in which the targeted phenotype has been altered; (3) identify the mutated gene or gene product; and (4) characterize the function in the pathway of the altered, overexpressed, or missing protein.

A LETHAL SELECTION SYSTEM TO IDENTIFY MUTANTS IN WHICH NF-κB-DEPENDENT SIGNALING IS ALTERED

Starting with HEK293 cells, which express a high level of transfected IL-1 receptor subunits

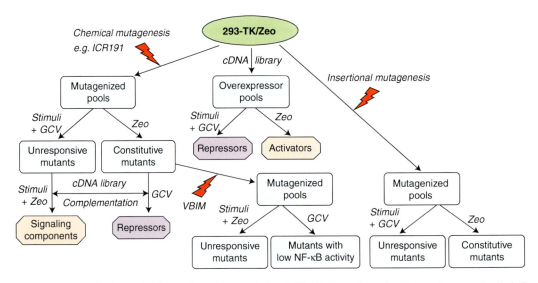

Figure 1. General scheme for forward genetics analysis of NF-κB-dependent signaling pathway, using lethal selection to identify regulative proteins. 293-TK/Zeo cells, carrying both TK and Zeo selectable markers, can be used in conjunction with chemical mutagenesis (*left*) or insertional mutagenesis (*right*) to obtain mutants unresponsive to a stimulus or constitutive mutants, and then to identify either stimulus-specific signaling components or general activators or repressors. A cDNA library (*middle*) can also be overexpressed to screen directly for activators or repressors. Additionally, constitutive mutants can be further mutated by insertional mutagenesis using VBIM (*bottom*) to screen for mutants with low NF-κB to identify activators or repressors.

(Cao et al. 1996), we introduced two separate selectable markers, both driven by the NF-κB-dependent E-selectin promoter (Fig. 1). One is a gene whose protein product confers resistance to zeocin (Zeo) and the other is the herpes simplex thymidine kinase (TK) gene (Li et al. 1999). The E-selectin promoter has low basal activity and can be induced strongly by activators of NF-κB, such as IL-1. Clone 293-TK/Zeo survives in ganciclovir (GCV, converted to a toxic metabolite by TK) and dies in GCV plus IL-1; it dies in Zeo and survives in Zeo plus IL-1. This clone has been used extensively, for many different experiments.

An important feature of the GCV-TK selection is that one can manipulate the concentration of GCV so that cells with a low basal TK expression survive but are killed after induction. There is a difference of 10-fold or more between basal and induced levels of TK in 293-TK/Zeo cells (Li et al. 1999). Moreover, because GCV is a poor substrate for mammalian TK, the GCV selection does not require the use of a TK-null cell line. Therefore, 293-TK/Zeo cells, with their low basal level of NF-κB activity,

survive in GCV and die in Zeo. When these cells are treated with a ligand that activates NF-κB or are mutated so that they have high constitutive activation, the expression of both TK and Zeo are induced, so the cells die in GCV and survive in Zeo (Fig. 1). Using this dual selection system, we have used several different forward genetic approaches to randomly alter the expression of several different proteins, thus identifying both positive and negative novel regulators of NF-κB (Fig. 1).

EXAMPLES OF DIFFERENT FORWARD GENETIC APPROACHES TO STUDY NF-κB

The approaches that we have used to dissect NF-κB-dependent signaling pathways, using the 293-TK/Zeo cell system, include: (1) Chemical mutagenesis, followed by identification of the affected gene (Li et al. 1999; Sathe et al. 2004); (2) expression of cDNA libraries (Li et al. 2000); (3) retroviral mutagenesis by promoter insertion (Kandel et al. 2005); (4) transposon-based insertional mutagenesis (Dasgupta et al. 2008); and (5) lentiviral-based insertional mutagenesis (Lu et al. 2009a). Methods three to five include strategies to remove or silence the inserted promoter after a mutant clone has been isolated, to validate the dependence of the mutant phenotype on the function of the promoter.

Use of Chemical Mutagenesis to Help Elucidate the Role of IRAK1 in IL-1-dependent Signaling to NF-κB

The chemical mutagen ICR191, an intercalating agent that causes frame-shift mutations and deletions, has been extremely useful in generating recessive mutants by randomly inactivating both alleles of a specific target gene. We first used this approach to select eight different mutant cell lines defective in responding to interferons, contributing in this way to the discovery of the JAK/STAT pathway and to the elucidation of many details of how cells respond to interferons (Pellegrini et al. 1989; Velazquez et al. 1992; Darnell et al. 1994; Borden et al. 2007). To extend this method to an analysis of NF-κB, 293-TK/Zeo cells were subjected to five rounds of chemical mutagenesis with ICR191, followed by selection in GCV plus IL-1. Four independent mutant cell lines were isolated that failed to respond to IL-1. Attempts to complement these mutant cell lines with cDNA libraries were unsuccessful. Therefore, each mutant was tested with antibodies against known IL-1 signaling components. Mutant I1A lacks the expression of IRAK1 (Li et al. 1999). Two limitations became apparent during the course of this work, for reasons that we still understand incompletely. First, despite repeated attempts, we were unable to obtain mutant cell lines lacking additional proteins already known to be required for IL-1-dependent signaling. Second, we could not functionally complement three mutant cell lines that still express all proteins known to be required for response to IL-1, despite extensive attempts using different cDNA libraries.

Use of the human null cell line I1A helped to position IRAK1 within the IL-1 signaling pathway. IRAK1 is required for the activation of both NF-κB and Jun kinase in response to IL-1, functioning between MyD88 and TRAF6

(Li et al. 1999). Putting the kinase-dead IRAK1 mutant K239A back into I1A cells allowed us to make the surprising discovery that the kinase activity of IRAK is not required for it to function in IL-1-dependent signaling (Li et al. 1999). Later, by expressing a series of IRAK1 deletion constructs in I1A cells, the relationship between the structure of IRAK1 and its function was analyzed in great detail (Li et al. 2001; Jiang et al. 2002). These findings suggested that both the amino-terminal death domain and the carboxy-terminal region of IRAK are required for IL-1-induced NF-κB and JNK activation, whereas the amino-proximal undetermined domain is required for the activation of NF-κB but not JNK. The phosphorylation and ubiquitination of IRAK deletion mutants correlate tightly with their ability to activate NF-κB in response to IL-1, but IRAK can mediate IL-1-induced JNK activation without being phosphorylated. These studies reveal that the IL-1-induced signaling pathways leading to NF-κB and JNK activation diverge either at IRAK or at a point nearer to the receptor.

Chemical Mutagenesis Has Led to the Generation of a Group of Mutant Cells with Constitutive Activation of NF-κB

As shown in Figure 1, a single chemically mutated pool can be selected either in GCV plus activator, for mutants that do not respond to a specific stimulus, or in Zeo in the absence of a stimulus, for mutants with abnormal constitutive activation of NF-κB. Using the latter strategy, we isolated eight mutant cell lines in which NF-κB is constitutively activated (Sathe et al. 2004). These cells have different properties and belong to eight different complementation groups, showing that there must be many more than eight different mutations that can lead to this phenotype. All but one of the mutants is recessive, indicating that a negative regulatory function had been lost because of mutagenesis in these seven cases. The eight mutants represent at least five different biochemical phenotypes, differing in the sets of upstream kinases that were abnormally activated, including IKK, JNK, Akt, p90[rsk1], and ERK (Sathe et al. 2004).

The constitutive activation of NF-κB in these mutant cells, and also in many different cancer cell lines, is almost always caused by the constitutive secretion of one or more factors that activate NF-κB in autocrine fashion from outside the cell (Lu et al. 2004a). Interestingly, in the mutant cell line Z12, TGF-β2 was highly expressed and secreted. This observation led to the discovery that TGF-β activates NF-κB in several different tumors and mutant cell lines, and that the basis of increased TGF-β secretion is an increased steady-state level of the corresponding mRNA (Lu et al. 2004a; Lu et al. 2004b; Lu and Stark 2004). In addition, we found that relatively high concentrations of TGF-β (about ten times higher than the concentrations required to activate SMADs) activate NF-κB by recruiting and activating the IL-1 receptor and, conversely, that IL-1 activates SMADs similarly, by recruiting the TGF-β receptor. We propose that this unusual cross talk is especially important in the immediate vicinity of tumors or at sites of inflammation, where the concentrations of TGF-β or IL-1 are likely to be high (Lu et al. 2007).

Use of a cDNA Library to Discover the NF-κB Regulator Act1

A popular approach in forward genetics is to randomly overexpress in a target cell's population proteins, protein fragments, or anti-sense RNAs from libraries constructed in expression vectors, followed by recovery of the causative constructs from mutant clones. Several types of libraries have been used: full-length cDNAs (Miki and Aaronson 1995), anti-sense cDNAs (Deiss and Kimchi 1991), truncated cDNA fragments (genetic suppressor elements or GSEs) (Gudkov et al. 1994), small interfering RNAs (Berns et al. 2004), and hammerhead ribozymes (Wadhwa et al. 2004). These approaches, still actively used in many laboratories, continue to contribute new information to our understanding of complex signaling pathways (Neznanov et al. 2003; Wang et al. 2002). In all of these approaches, the effect of the cloned element on the phenotype of interest can be confirmed by expressing it in naïve cells, not necessarily

the same as the ones used for the initial isolation. Significant disadvantages of these approaches include the difficulty of constructing, maintaining, and delivering a comprehensive library of the required complexity.

In our laboratory, a human keratinocyte cDNA library was delivered into 293-TK/Zeo cells, followed by selection in Zeo (Fig. 1). The 2.6 kb cDNA recovered from one of the resistant clones corresponds to a gene of previously unknown function (Li et al. 2000). The predicted 60-kDa polypeptide was named Act1 (NF-κB activator 1). At about the same time, the same protein was discovered by another group and named CIKS (connection to IKK and SAPK/JNK) or TRAF3IP2 (TRAF3 interacting protein 2) (Leonardi et al. 2000).

Since its discovery, Act1 has been found to play major roles in NF-κB-dependent signaling, with about 300 publications describing its functions. Act1 turns out to be an important regulator of CD40-dependent signaling (Qian et al. 2004). Genetic deficiency in Act1 results in a dramatic increase in peripheral B cells, which culminates in lymphadenopathy and splenomegaly, hyper-γ-globulinemia, and auto-antibodies. Although the B-cell specific Act1 knockout mice displayed a similar phenotype with less severity, the pathology of the Act1-deficient mice was mostly blocked in CD40-Act1 and BAFF-Act1 double knockout mice. CD40- and BAFF-mediated survival is significantly increased in Act1-deficent B cells, with stronger IκB phosphorylation, processing of NF-κB2 (p100/p52), and activation of JNK, ERK, and p38 pathways, indicating that Act1 negatively regulates CD40- and BAFF-mediated signaling events. These findings demonstrate that Act1 plays an important role in the homeostasis of B cells by attenuating CD40 and BAFFR signaling.

Recently, Act1 has been linked to IL-17-dependent signaling, where it acts as an essential adaptor in this newly discovered pathway that is associated with autoimmune and inflammatory diseases when misregulated (Hunter 2007; Qian et al. 2007). T-helper cells that produce IL-17 are associated with inflammation and the control of certain bacteria. After stimulation with IL-17, recruitment of Act1 to IL-17R required the IL-17R conserved cytoplasmic "SEFIR" domain, followed by recruitment of the kinase TAK1 and the E3 ubiquitin ligase TRAF6, which mediate downstream activation of NF-κB. IL-17-induced expression of inflammation-related genes was abolished in Act1-deficient primary astroglial and gut epithelial cells. This reduction was associated with much less inflammatory disease in vivo in both autoimmune encephalomyelitis and dextran sodium-sulfate-induced colitis (Qian et al. 2007).

Thus, the adapter molecule Act1 regulates autoimmunity through its impact on both T- and B-cell mediated immune responses. Whereas Act1 is an important negative regulator for B-cell-mediated humoral immune responses through its function in CD40L and BAFF signaling, it is also a key positive signaling component for IL-17 signaling pathway, critical for T(H)17-mediated autoimmune and inflammatory responses. The dual, seemingly opposite, functions of Act1 in CD40- BAFFR- and IL-17R-dependent signaling are orchestrated by different domains. Whereas Act1 interacts with the IL-17R through the carboxy-terminal SEFIR domain, it is recruited to CD40 and BAFFR indirectly, mediated by TRAF3 through the TRAF binding site in Act1. Such delicate regulatory mechanisms may provide a common means to promote balance between host defense to pathogens and tolerance to self (Qian et al. 2004, 2007 and 2008).

Insertional Mutagenesis

In this group of forward genetic methods, a defined DNA fragment is inserted approximately randomly throughout the genome. Thus, these methods have the potential to target any gene and can be applied in many different experimental systems. If the inserted DNA includes a strong promoter, dominant mutants can be obtained by driving transcription into an adjacent gene, leading to the overexpression of an mRNA encoding a full-length or truncated protein, or an antisense RNA, depending on the position and orientation of the inserted

promoter. We have developed three types of insertional mutagenesis vectors. All use physical removal or silencing of the inserted promoter to prove that the mutant phenotype depends on its function ("validation").

Retroviral Insertional Mutagenesis Has Led to New Information about p65 and TAB3

We used this technique (Fig. 2A) to search for proteins that regulate NF-κB-dependent signaling in 293-TK/Zeo cells. Two reversible mutants were characterized and the affected genes were identified as *relA*, encoding the NF-κB p65 subunit, and *act1* (Kandel et al. 2005). Conditioned medium from the mutant overexpressing p65 activates NF-κB, and the ability to secrete factors that activate NF-κB was reduced sharply in medium from cells in which the phenotype was reversed by introducing Cre, revealing that the overexpression of p65 could activate NF-κB and that the secretion of NF-κB-activating factors occurs in a p65-dependent manner (Kandel et al. 2005).

Jin et al. (2004) used a similar stratagem to identify TAB3 as an NF-κB-activating protein when overexpressed in human cells. The activation of NF-κB by TAB3 could be blocked by the NF-κB inhibitor SN50 and by the expression of dominant–negative forms of TRAF6 (TNFα-associated factor 6) and TAK1 (TGF-β-activated kinase 1), suggesting that TAB3 is a component of an NF-κB-dependent signaling pathway functioning upstream of TRAF6/TAK1. Furthermore, overexpression of TAB3 in NIH3T3 cells resulted in transformation; TAB3 overexpression was also found in some cancer tissues, suggesting a causative link between elevated TAB3 expression, constitutive NF-κB activation, and oncogenesis (Jin et al. 2004).

Use of Transposon-mediated Insertional Mutagenesis to Discover that Short RIP is an Activator of NF-κB

The benefits of transposons include a DNA-only life cycle, avoiding some constraints imposed by retroviral insertional vectors, and lack of known endogenous homolog in mammalian cells. Because the integration biases

A

B

C

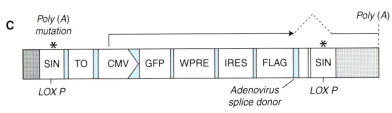

Figure 2. Structures of different promoter insertion vectors. (*A*) Retroviral vectors. Retroviral vectors optimized for insertional mutagenesis were generated by using the MMLV (Moloney murine leukemia virus) backbone of pBabe vectors. The recognition sequence for Cre recombinase was placed inside the 3′ LTR so that, on reverse transcription, this motif would be copied into the 5′-LTR (Kandel et al. 2005). Abbreviations: TORE, tetracycline-regulated promoter; SV40, promoter and enhancer from the SV40 virus; LTR, long terminal repeat. Symbols: ⊠, the LTR modification, including a LoxP site for cre-mediated recombination, leading to deletion of the promoter; grid lines, host genomic DNA. (*B*) Sleeping Beauty transposon-based vectors. This series of mutagenesis vectors contain a coding region for GFP controlled by TORE, which is placed in the vicinity of a divergent SV40 promoter, followed by a mini-exon, which ends with an unpaired adenoviral splice site (Kandel et al. 2005). These vectors can be cotransfected with the Sleeping Beauty transposase to randomly generate mutations. Because the promoter is tet-regulated, mutant phenotypes caused by the inserted promoter can be reverted on promoter shutdown on treatment with the tetracycline analog doxycycline (Dasgupta et al. 2008). Abbreviations: IR/DR, terminal repeat from Sleeping Beauty; TORE, modified tetracycline-regulated promoter; GFP, coding region of enhanced GFP. Symbol: grid lines, host genomic DNA. (*C*) VBIM vectors. The VBIM vectors use a lentiviral backbone with polyadenylation mutations in both 5′ and 3′ LTRs and a lox P site in the 3′ LTR. This design allows excision of all but 238 bp of inert proviral DNA, lacking both promoter activity and polyadenylation signals, following cleavage by Cre, a critical feature for complete and consistent phenotypic reversion. The polyadenylation mutations also permit the mutagenic promoter to be placed in the same direction as transcription from the 5′ LTR during virus packaging, resulting in high virus titers that are comparable to those obtained with standard lentiviral vectors, eliminating promoter conflicts that occur with alternative designs. Furthermore, eliminating promoter interference also permits the use of a strong full-length CMV promoter rather than a minimal tetracycline-regulated promoter, which requires the tet activator protein tTA to be present in the target cells (Kandel et al. 2005). Thus, primary and even differentiated or senescent cells can be mutated without prior manipulation to express tTA or an ecotropic receptor, as previously required (Kandel et al. 2005). Besides these, a tetracycline-binding element upstream of the full-length CMV promoter is introduced to allow tetracycline-regulated control of the mutagenic promoter after the mutant has been created, by using a TR-KRAB fusion protein. Detailed description of this series of vectors has been reported recently (Lu et al. 2009a). Abbreviations: CMV, cytomegalovirus promoter; GFP, green fluorescent protein; IRES, internal ribosome entry sequence; LoxP, site for cre-mediated recombination; SIN, self-inactivating LTR; TO, tetracycline operon; WPRE, woodchuck hepatitis virus post-transcriptional regulatory element. Symbol: grid lines, host genomic DNA.

of transposons and various retroviruses differ (Yant et al. 2005), the two approaches may be complementary in attempts to achieve full-genome coverage. The cell-autonomous nature of transposition is reflected in in vivo mutagenesis. Transposons have been successfully used to search for oncogenes in mice using insertion of transposon constitutive promoters (Collier et al. 2005; Dupuy et al. 2005).

We used a vector based on the Sleeping Beauty transposon to search for constitutive activators of NF-κB in cultured cells. Dominant mutations were produced by random insertion of the tetracycline-regulated promoter TORE (Fig. 2B), which provided robust and exceptionally well-regulated expression of downstream genes. The ability to regulate the mutant phenotype was used to attribute it to the insertional event. In one such mutant, the promoter was inserted in the middle of the gene encoding receptor-interacting protein kinase 1 (RIP1). The protein encoded by the hybrid transcript lacks the putative kinase domain of RIP1, but potently stimulates NF-κB activity (Dasgupta et al. 2008). Similar to TNFα treatment, the expression of short RIP1 is toxic to cells that fail to up-regulate NF-κB. The effect of short RIP1 did not require endogenous RIP1 or cytokine treatment. More importantly, a similar short RIP1 is produced naturally from the *ripk1* locus. Elevated expression of short RIP1 resulted in a loss of full length RIP1 from cells, indicating a novel mechanism through which the abundance of RIP1 and the related signals could be regulated (Dasgupta et al. 2008).

The previous results demonstrate the successful use of promoter insertion to isolate mutants in which constitutive activators are overexpressed. However, because many different activators of NF-κB are already known, the probability that novel activators will be found against such a high background is not great.

Use of the VBIM Technique to Discover FBXL11 as a Negative Regulator of NF-κB

To improve the features of reversible promoter insertional technique to facilitate the creation of dominant mutants in which a strong

promoter, inserted into the genome approximately randomly, drives high-level expression of downstream genes, we designed a set of lentiviral validation-based insertional mutagenesis (VBIM) vectors (Fig. 2C) that extend the application of reversible promoter insertion previously described by our laboratory (Fig. 2A) to nearly any type of mammalian cell, even cells that are not dividing (Lu et al. 2009a). The VBIM lentiviruses are designed to increase the expression of downstream genomic sequences that encode full-length proteins, truncated proteins, or antisense RNAs, and potentially even microRNAs. These dominant mutations can thus identify either positive or negative regulators from the same genetic screen. The VBIM vectors also allow the mutant phenotype to be reversed, either by removing the inserted promoter with Cre recombinase or by silencing it with the transcriptional repressor Kruppel-associated box (KRAB) domain of the human Kox1 zinc finger protein. One can then readily associate a mutant phenotype with a specific target gene by cloning sequences flanking the insertion site. In several different screens, we obtained validated mutant clones with frequencies of the order of 10^{-6} to 10^{-5}, a high yield that allows the selection of multiple mutants in experiments of very reasonable scale. Therefore, the VBIM technique is a powerful tool for gene discovery that has broad applications in many different systems.

To use VBIM to screen for negative regulators of NF-κB, we started with mutant Z3 cells, in which constitutive NF-κB activation had been generated by chemical mutagenesis (Sathe et al. 2004). These cells survive in Zeo and die in GCV (Fig. 1). Constitutive overexpression of a protein in Z3 cells that causes the constitutive NF-κB activity to shut down can be selected for by requiring the cells to survive in GCV and die in Zeo (Fig. 1).

Z3 cells infected with VBIM viruses were selected in GCV for suppression of NF-κB activity (Fig. 1). Three reversible mutants were obtained, at a frequency of ca. 10^{-6}. The mutant SD1-11 was characterized in detail, and characterization of the other two mutants is still in progress. SD1-11 cells were infected with a vector encoding Cre recombinase to verify that the phenotype is reversible. A variety of approaches (including drug selections, electrophoretic mobility gel shift assays, luciferase assays, and Western and Northern assays) were used to confirm that the phenotype of SD1-11 is reversed by Cre. A Southern experiment showed that there is only one insertion in this mutant, and inverse PCR was used to clone the flanking genomic sequences. The gene up-regulated by the inserted promoter is F-box leucine repeat rich protein 11 (*FBXL11*), encoding a known histone H3 lysine 36 (H3K36) demethylase (Tsukada et al. 2006). The insertion in the second intron of FBXL11 produces a Flag-tagged fusion protein lacking only 14 of 1162 amino acid residues at its amino terminus, leaving all the functional domains intact (Lu et al. 2009a).

Mechanistic studies show that a point mutation that knocks out the demethylase activity of full-length FBXL11 also abolishes its ability to inhibit NF-κB activation. Knocking down the expression of FBXL11 activates NF-κB, as does overexpressing the histone H3K36 methylase NSD1 (the nuclear receptor-binding SET domain-containing protein 1). In cells with constitutively active NF-κB, or in cytokine-treated cells, the p65 subunit binds to both FBXL11 and NSD1 and significant mono-methylation of K218 and di-methylation of K221 of p65 occurs. Single K-A mutations of K218 or K221 reveal that these two residues are targets of FBXL11, and the K-A double mutants are much less active than wild-type p65. Importantly, the *FBXL11* gene is transcribed in response to NF-κB activation and thus, similarly to the well known inhibitor IκB, FBXL11 participates in an autoregulatory negative-feedback loop (Fig. 3) (Lu et al. 2009b). In summary, we have uncovered a novel regulatory pathway for NF-κB that is driven by cycles of lysine methylation and demethylation of its p65 subunit. The discovery of reversible lysine methylation of NF-κB represents an important new contribution to our understanding of how fine control of this key transcription factor is achieved, and the discovery of FBXL11 as a negative regulator of NF-κB by utilizing

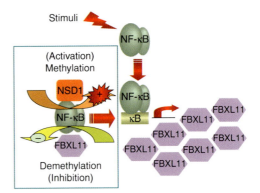

Figure 3. Regulation of NF-κB by lysine methylation. NF-κB is regulated through reversible methylation of p65, catalyzed by the NSD1–FBXL11 enzyme pair. The FBXL11 gene is activated by NF-κB, forming a negative-feedback loop that down-regulates NF-κB.

the VBIM technique vividly demonstrates its power.

Recently, Huang et al. (2007) showed that p53 is demethylated on K370 by LSD1, a histone H3K4 or H3K9 demethylase. The dual function of LSD1 in regulating both histone H3 and p53 is strikingly similar to our finding that FBXL11 regulates both histone H3 and NF-κB. These two examples suggest that the regulation of transcription factors by histone methylases and demethylases may be a general phenomenon in mammalian cells, and probably in other biological systems as well.

PERSPECTIVES

A crucial question follows from the mechanism represented in Figure 3: What controls the activating methylation of NF-κB in response to a ligand such as IL-1? A strong clue is given by the observations that NF-κB (our study) and p53 (Huang and Berger 2008; Huang et al. 2007) are methylated by enzymes that also modify histones in the context of chromatin. A likely possibility is that NF-κB and NSD1 are in different compartments of untreated cells and that activation by the H3K36 methylase NSD1 occurs only after NF-κB enters the nucleus and binds to target promoters (Fig. 4). In support of this possibility, we have recent evidence that the signal transducer and activator of transcription 3 (STAT3) is methylated only when it is bound to DNA (unpubl. data). In response to IL-6, lysine 140 of STAT3 is reversibly di-methylated, reducing its ability to bind to the promoters of a subset of IL-6-induced genes. Methylation of K140 does not occur on variant STAT3 proteins that are tyrosine-phosphorylated normally but that are prevented by point mutations from entering the nucleus or binding to DNA. Consistently, wild-type di-methyl STAT3 remains in the nucleus and is not observed in the cytoplasm, even following long exposure of cells to IL-6. Furthermore, modifications other than methylation may also occur only following the binding of transcription factors to promoters. The well known IL-6-induced phosphorylation of STAT3 on S727 occurs only when tyrosine-phosphorylated STAT3 can enter the nucleus and bind to DNA (unpubl. data), similar to an earlier observation concerning the serine phosphorylation of STAT1 (Sadzak et al. 2008).

It is well known that activated inducible transcription factors such as NF-κB drive chromatin remodeling (Natoli 2009). Genes induced in response to NF-κB without new protein synthesis (primary response genes) fall into "fast" and "slow" subclasses. NF-κB binds to the promoters of fast genes rapidly, because the chromatin landscape is typical of genes "poised for immediate activation, including high levels of histone H3/H4 acetylation and trimethylation of histone H3K4, a histone modification specifically enriched at active or poised transcription start sites." "Conversely, slow genes in unstimulated cells are associated with hypo-acetylated histones and are negative for H3K4me3. In response to activation, both acetylation and H3K4 trimethylation progressively increase, with a kinetics that apparently precedes NF-κB recruitment." (Natoli 2009).

Taken together, the previous observations lead to the following working hypotheses: (1) Reversible methylation and serine phosphorylation (and probably other modifications) of NF-κB and STAT3 occur, in concert with histone modifications, *only* when these transcription factor targets bind to specific

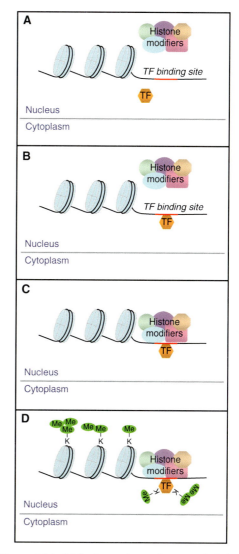

Figure 4. Model for transcription factor methylation by histone-modifying enzymes. (*A*) The activated transcription factor (TF) enters the nucleus and (*B*) binds to a promoter. (*C*) The histone-modifying complex is recruited. (*D*) This complex then methylates *both* the histones and the promoter-bound TF. Multiple sites are mono-, di-, and tri-methylated, profoundly affecting the functions of both the histones and the TF.

promoters and thus become available as substrates for the local chromatin remodeling machinery (Fig. 4); (2) these modifications profoundly affect NF-κB and STAT3 functions at these promoters, thus affecting the strength and duration of inducible expression; (3)

these effects may be *gene-specific*, leading to differential effects on gene expression that give plasticity to the dependent biological responses. For example, (1) interpretation of a single extracellular signal might be different for different genes in the same cell; (2) the downstream effects of different mechanisms of activating NF-κB (for example, in response to IL-1 *vs.* TNF, or *vs.* toll-like receptor [TLR] ligands) might be interpreted differently in the same cell; (3) a given gene might be regulated differentially in response to a given signal in different cell types.

We have discussed several different applications of forward genetics to some of the pathways of NF-κB activation. In the future, these methods could also be applied to uncover novel positive and negative regulators of additional pathways in which NF-κB plays a role, for example the TLRs, major mediators of innate immunity whose regulation and function are under intense investigation in many laboratories. Upon stimulation by double-stranded RNA (dsRNA), NF-κB and the interferon regulatory factor 3 (IRF3) are activated, both through the membrane-bound receptor TLR3 and through the internal cytoplasmic retinoic acid-inducible gene I (RIG-I) pathway. The expression of TLR3 in 293-TK/Zeo cells should allow the use of insertional mutagenesis to reveal additional details of the pathway through which dsRNA activates NF-κB.

The VBIM technique generates dominant mutants in which RNAs encoded downstream of the inserted promoter are overexpressed. Most of the "positive" phenotypes described here and in additional unpublished studies result from the overexpression of mRNAs encoding either full-length proteins or truncated proteins that function similarly to their full-length counterparts. Overexpressed truncated proteins might also have dominant–negative functions or unregulated positive functions. Additionally, in one case (unpubl. data), we observed the overexpression of a dominant–negative RNA, resulting from the intragenic insertion of the promoter in the antisense direction. It would be extremely valuable to have a strategy in which *recessive* mutants, in which

the expression of a protein required for signaling has been ablated, are obtained from a validation-based insertional mutagenesis approach. To do this, one might use a gene-trap vector that inactivates the expression of one allele of a gene when inserted into any intron or exon, and promote the loss of the remaining unmutated allele, for example by stimulating chromosome nondisjunction so that the entire homologous chromosome is lost, by stimulating mitotic recombination, or by performing experiments in hypo-diploid cells (Guo et al. 2004; Nobukuni et al. 2005; Banks and Bradley 2007; Kim et al. 2007). It is certain that more powerful and user-friendly methods of using forward genetics in mammalian cells will continue to be developed, leading to many additional exciting discoveries.

REFERENCES

Banks DJ, Bradley KA. 2007. SILENCE: A new forward genetic technology. *Nat Methods* **4:** 51–53.

Berns K, Hijmans EM, Mullenders J, Brummelkamp TR, Velds A, Heimerikx M, Kerkhoven RM, Madiredjo M, Nijkamp W, Weigelt B, et al. 2004. A large-scale RNAi screen in human cells identifies new components of the p53 pathway. *Nature* **428:** 431–437.

Borden EC, Sen GC, Uze G, Silverman RH, Ransohoff RM, Foster GR, Stark GR. 2007. Interferons at age 50: Past, current and future impact on biomedicine. *Nat Rev Drug Discov* **6:** 975–990.

Cao Z, Henzel WJ, Gao X. 1996. IRAK: A kinase associated with the interleukin-1 receptor. *Science* **271:** 1128–1131.

Collier LS, Carlson CM, Ravimohan S, Dupuy AJ, Largaespada DA. 2005. Cancer gene discovery in solid tumours using transposon-based somatic mutagenesis in the mouse. *Nature* **436:** 272–276.

Darnell JE Jr, Kerr IM, Stark GR. 1994. Jak-STAT pathways and transcriptional activation in response to IFNs and other extracellular signaling proteins. *Science* **264:** 1415–1421.

Dasgupta M, Agarwal MK, Varley P, Lu T, Stark GR, Kandel ES. 2008. Transposon-based mutagenesis identifies short RIP1 as an activator of NF-κB. *Cell Cycle* **7:** 2249–2256.

Deiss LP, Kimchi A. 1991. A genetic tool used to identify thioredoxin as a mediator of a growth inhibitory signal. *Science* **252:** 117–120.

Dupuy AJ, Akagi K, Largaespada DA, Copeland NG, Jenkins NA. 2005. Mammalian mutagenesis using a highly mobile somatic Sleeping Beauty transposon system. *Nature* **436:** 221–226.

Gudkov AV, Kazarov AR, Thimmapaya R, Axenovich SA, Mazo IA, Roninson IB. 1994. Cloning mammalian genes by expression selection of genetic suppressor elements: Association of kinesin with drug resistance and cell immortalization. *Proc Natl Acad Sci* **91**: 3744–3748.

Guo G, Wang W, Bradley A. 2004. Mismatch repair genes identified using genetic screens in Blm-deficient embryonic stem cells. *Nature* **429**: 891–895.

Huang J, Berger SL. 2008. The emerging field of dynamic lysine methylation of non-histone proteins. *Curr Opin Genet Dev* **18**: 152–158.

Huang J, Sengupta R, Espejo AB, Lee MG, Dorsey JA, Richter M, Opravil S, Shiekhattar R, Bedford MT, Jenuwein T, et al. 2007. p53 is regulated by the lysine demethylase LSD1. *Nature* **449**: 105–108.

Hunter CA. 2007. Act1-ivating IL-17 inflammation. *Nat Immunol* **8**: 232–234.

Jiang Z, Ninomiya-Tsuji J, Qian Y, Matsumoto K, Li X. 2002. Interleukin-1 (IL-1) receptor-associated kinase-dependent IL-1-induced signaling complexes phosphorylate TAK1 and TAB2 at the plasma membrane and activate TAK1 in the cytosol. *Mol Cell Biol* **22**: 7158–7167.

Jin G, Klika A, Callahan M, Faga B, Danzig J, Jiang Z, Li X, Stark GR, Harrington J, Sherf B. 2004. Identification of a human NF-κB-activating protein, TAB3. *Proc Natl Acad Sci* **101**: 2028–2033.

Kandel ES, Lu T, Wan Y, Agarwal MK, Jackson MW, Stark GR. 2005. Mutagenesis by reversible promoter insertion to study the activation of NF-κB. *Proc Natl Acad Sci* **102**: 6425–6430.

Kim SO, Ha SD, Lee S, Stanton S, Beutler B, Han J. 2007. Mutagenesis by retroviral insertion in chemical mutagen-generated quasi-haploid mammalian cells. *Biotechniques* **42**: 493–501.

Leonardi A, Chariot A, Claudio E, Cunningham K, Siebenlist U. 2000. CIKS, a connection to IκB kinase and stress-activated protein kinase. *Proc Natl Acad Sci* **97**: 10494–10499.

Li X, Commane M, Burns C, Vithalani K, Cao Z, Stark GR. 1999. Mutant cells that do not respond to interleukin-1 (IL-1) reveal a novel role for IL-1 receptor-associated kinase. *Mol Cell Biol* **19**: 4643–4652.

Li X, Commane M, Nie H, Hua X, Chatterjee-Kishore M, Wald D, Haag M, Stark GR. 2000. Act1, an NFκB-activating protein. *Proc Natl Acad Sci* **97**: 10489–10493.

Li X, Commane M, Jiang Z, Stark GR. 2001. IL-1-induced NF-κB and c-Jun N-terminal kinase (JNK) activation diverge at IL-1 receptor-associated kinase (IRAK). *Proc Natl Acad Sci* **98**: 4461–4465.

Liu D, Yang X, Yang D, Zhou S. 2000. Genetic screens in mammalian cells by enhanced retroviral mutagens. *Oncogene* **19**: 5964–5972.

Lu T, Stark GR. 2004. Cytokine overexpression and constitutive NF-κB in cancer. *Cell Cycle* **3**: 1114–1117.

Lu T, Burdelya LG, Swiatkowski SM, Boiko AD, Howe PH, Stark GR, Gudkov AV. 2004a. Secreted transforming growth factor β2 activates NF-κB, blocks apoptosis, and is essential for the survival of some tumor cells. *Proc Natl Acad Sci* **101**: 7112–7117.

Lu T, Sathe SS, Swiatkowski SM, Hampole CV, Stark GR. 2004b. Secretion of cytokines and growth factors as a general cause of constitutive NF-κB activation in cancer. *Oncogene* **23**: 2138–2145.

Lu T, Tian L, Han Y, Vogelbaum M, Stark GR. 2007. Dose-dependent cross-talk between the transforming growth factor β and interleukin-1 signaling pathways. *Proc Natl Acad Sci* **104**: 4365–4370.

Lu T, Jackson MW, Singhi AD, Kandel ES, Yang M, Zhang Y, Gudkov AV, Stark GR. 2009a. Validation-based insertional mutagenesis identifies lysine demethylase FBXL11 as a negative regulator of NF-κB. *Proc Natl Acad Sci* **106**: 16339–16344.

Lu T, Jackson MW, Wang B, Yang M, Chance MR, Miyagi M, Gudkov AV, Stark GR. 2009b. Regulation of NF-B by NSD1/FBXL11-dependent reversible lysine methylation of p65. *Proc Natl Acad Sci* (in press).

Miki T, Aaronson SA. 1995. Isolation of oncogenes by expression cDNA cloning. *Methods Enzymol* **254**: 196–206.

Natoli G. 2009. Control of NF-κB-dependent transcriptional responses by chromatin organization. *Cold Spring Harb Perspect Biol* **1**: a000224.

Neznanov N, Neznanova L, Kondratov RV, Burdelya L, Kandel ES, O'Rourke DM, Ullrich A, Gudkov AV. 2003. Dominant negative form of signal-regulatory protein-α (SIRPα /SHPS-1) inhibits tumor necrosis factor-mediated apoptosis by activation of NF-κB. *J Biol Chem* **278**: 3809–3815.

Nobukuni Y, Kohno K, Miyagawa K. 2005. Gene trap mutagenesis-based forward genetic approach reveals that the tumor suppressor OVCA1 is a component of the biosynthetic pathway of diphthamide on elongation factor 2. *J Biol Chem* **280**: 10572–10577.

Pellegrini S, John J, Shearer M, Kerr IM, Stark GR. 1989. Use of a selectable marker regulated by α interferon to obtain mutations in the signaling pathway. *Mol Cell Biol* **9**: 4605–4612.

Qian Y, Qin J, Cui G, Naramura M, Snow EC, Ware CF, Fairchild RL, Omori SA, Rickert RC, Scott M, et al. 2004. Act1, a negative regulator in CD40- and BAFF-mediated B cell survival. *Immunity* **21**: 575–587.

Qian Y, Liu C, Hartupee J, Altuntas CZ, Gulen MF, Jane-Wit D, Xiao J, Lu Y, Giltiay N, Liu J, et al. 2007. The adaptor Act1 is required for interleukin 17-dependent signaling associated with autoimmune and inflammatory disease. *Nat Immunol* **8**: 247–256.

Qian Y, Giltiay N, Xiao J, Wang Y, Tian J, Han S, Scott M, Carter R, Jorgensen TN, Li X. 2008. Deficiency of Act1, a critical modulator of B cell function, leads to development of Sjögren's syndrome. *Eur J Immunol* **38**: 2219–2228.

Sadzak I, Schiff M, Gattermeier I, Glinitzer R, Sauer I, Saalmüller A, Yang E, Schaljo B, Kovarik P. 2008. Recruitment of Stat1 to chromatin is required for interferon-induced serine phosphorylation of Stat1 transactivation domain. *Proc Natl Acad Sci* **105**: 8944–8999.

Sathe SS, Sizemore N, Li X, Vithalani K, Commane M, Swiatkowski SM, Stark GR. 2004. Mutant human cells with constitutive activation of NFκB. *Proc Natl Acad Sci U S A* **101**: 192–197.

Tsukada Y, Fang J, Erdjument-Bromage H, Warren ME, Borchers CH, Tempst P, Zhang Y. 2006. Histone demethylation by a family of JmjC domain-containing proteins. *Nature* **439**: 811–816.

Velazquez L, Fellous M, Stark GR, Pellegrini S. 1992. A protein tyrosine kinase in the interferon α/β signaling pathway. *Cell* **70:** 313–322.

Wadhwa R, Yaguchi T, Kaur K, Suyama E, Kawasaki H, Taira K, Kaul SC. 2004. Use of a randomized hybrid ribozyme library for identification of genes involved in muscle differentiation. *J Biol Chem* **279:** 51622–51629.

Wang D, You Y, Case SM, McAllister-Lucas LM, Wang L, DiStefano PS, Nuñez G, Bertin J, Lin X. 2002. A requirement for CARMA1 in TCR-induced NF-κB activation. *Nat Immun* **3:** 830–835.

Yant SR, Ehrhardt A, Mikkelsen JG, Meuse L, Pham T, Kay MA. 2002. Transposition from a gutless adeno-transposon vector stabilizes transgene expression *in vivo. Nat Biotechnol* **20:** 999–1005.

Yant SR, Wu X, Huang Y, Garrison B, Burgess SM, Kay MA. 2005. High-resolution genome-wide mapping of transposon integration in mammals. *Mol Cell Biol* **25:** 2085–2094.

Index